Queueing Theory and its Applications

Liber Amicorum for J. W. Cohen

CWI Monographs

Centrum voor Wiskunde en Informatica
Centre for Mathematics and Computer Science
P.O.Box 4079, 1009 AB Amsterdam, The Netherlands

The CWI is a research institute of the Stichting Mathematisch Centrum, which was founded on February 11, 1946, as a nonprofit institution aiming at the promotion of mathematics, computer science, and their applications. It is sponsored by the Dutch Government through the Netherlands Organization for the Advancement of Pure Research (Z.W.O.)

CWI Monograph 7

Queueing Theory and its Applications

Liber Amicorum for J. W. Cohen

Edited by
O. J. Boxma
R. Syski

1988

North-Holland
Amsterdam • New York • Oxford • Tokyo

ISBN: 0 444 70497 3

Publishers:

Elsevier Science Publishers B.V.
P.O. Box 1991
1000 BZ Amsterdam
The Netherlands

Sole distributors for the U.S.A. and Canada:

Elsevier Science Publishing Company, Inc.
52 Vanderbilt Avenue
New York, N.Y. 10017
U.S.A.

LIBRARY OF CONGRESS

Library of Congress Cataloging-in-Publication Data

Queueing theory and its applications / edited by O.J. Boxma, R. Syski.
p. cm. -- (CWI monograph ; 7)
Festschrift for Jacob Willem Cohen.
Bibliography: p.
ISBN 0-444-70497-3
1. Queueing theory. I. Boxma, O.J., 1952- . II. Syski, R., 1924- .
III. Cohen, Jacob Willem. IV. Series.
T57.9.Q48 1988
519. 8'2--dc19

88-22613
CIP

Printed in the Netherlands

Preface

Books containing a collection of papers are usually edited with a preface which explains why the collection was compiled. In this case, we are pleased to state that this book was produced to honor a man - a friend and colleague - whose ideas and results have to a large extent shaped queueing theory in its present form.

Wim Cohen celebrates his sixty-fifth birthday on August 27, 1988. According to present Dutch law, he will have to retire as Professor of Operations Research from the University of Utrecht on September 1, 1988. In view of his outstanding contributions to the theory of stochastic processes in general, and to queueing theory, teletraffic and performance evaluation in particular, over a time span of more than thirty years, it seems appropriate to let his birthday and retirement not pass unnoticed.

A Liber Amicorum - a book of friends - dedicated to J.W. Cohen has appeared to us the most suitable form to mark the occasion. In August 1986, we issued invitations to a number of long-time friends, colleagues and former students of Wim Cohen to contribute a paper in the area of queueing theory and its applications. The positive response and enthusiastic cooperation of these distinguished scientists have resulted in the present collection of twenty papers. On September 1, 1988, at the end of a one-day symposium in honor of J.W. Cohen, the book will be presented to him.

The papers in this Liber Amicorum have been grouped into five parts, papers being ordered alphabetically by author within each part. Our classification stresses mathematical techniques used, so there is some overlap between parts as far as applications are concerned.

Part I consists of survey papers which present a broad picture of the developments in several areas of queueing theory and performance evaluation. Parts II-V contain research papers dealing with problems of current interest, in four areas in which Wim Cohen has made a significant contribution. Part II carries the appropriate title 'The Single Server Queue', and Part III is devoted to 'Analytic Methods', a subject that has always been very dear to Cohen. Part IV treats queueing networks and their applications to communication and computer systems, and Part V discusses various topics in probability and statistics with their implications for queueing theory. Several authors have added personal comments expressing their high esteem and respect for Wim Cohen as a researcher, colleague and friend. These sentiments are well described by Prof. Arne Jensen in the Epilogue of this book.

A large number of individuals have helped to make this Liber Amicorum a worthy scientific monument for J.W. Cohen. First among these are the authors. As an extra insurance for high quality, all papers were subjected to refereeing, revision and editing. We are indebted to both authors and referees.

The book has been published in the North-Holland Series of CWI Monographs. We should like to express our gratitude to the Centre for Mathematics and Computer Science (CWI) for its continuous support of the project, and in particular to Wim Aspers (CWI desk editor) and Josi Kustina and Mini Middelberg from the CWI typesetting department for their effort in producing a manuscript of high typographical quality, with as much uniformity as is possible with twenty different authors. Finally, we thank North-Holland for the production of this book.

The editors

J.W. Cohen: his Scientific Career

This book celebrates the 65th birthday of Jacob Willem Cohen. The following brief account of his life and work may help the reader to put him and his work in perspective. For further details on his life and his scientific views we refer to the autobiographical paper [A87], which has also been beneficial to us in writing this account.

Wim Cohen was born in Leeuwarden, The Netherlands, on August 27, 1923. After completing highschool he spent a large part of World War II in hiding, with relatives and friends. During these years he acquired a deep knowledge of large parts of physics and mathematics by self-study. After the war, he decided to study mechanical engineering at Delft University of Technology. In 1949 he completed his Master Thesis (cum laude), and in 1955 he graduated cum laude, again at Delft University of Technology, with a Ph.D. Thesis [B1] entitled *On Stress Calculations in Helicoidal Shells and Propeller Blades.* Just as Cohen's later work, this thesis research had a fundamental character, but was also directly related to practical applications.

In the period 1950-1957 Cohen worked as a 'teletraffic engineer' in the Philips Telecommunication Group. Around 1950 Philips entered the field of telephone engineering. Teletraffic, or congestion theory as it was then called, had already built up a tradition, and several theoretical and experimental techniques had been developed. The available theoretical studies were nearly all of a pioneering type, using seemingly *ad hoc* techniques. The book of Th.C. Fry, *Probability and its Engineering Uses,* of 1928 was the only available text which tried to present a coherent account of the available models and to explain what are currently being referred to as birth-and-death processes. The publication of Feller's famous *An Introduction to Probability Theory and its Applications* in 1950 was a landmark in the development of stochastic modeling. It removed several hazy notions in the basic definitions of probabilities and stochastic variables. At the same time, the possibilities for Operations Research became clear. In the late fifties probabilistic modeling became a tool in industrial engineering and management, and randomness was accepted as a phenomenon that could be modeled.

During his Philips years, Wim Cohen published several papers on telephone systems. His 1957 papers [A10-15], in particular, make fundamental contributions to the existing congestion theory, and some of them are still cited quite often. E.g., in [A14] Cohen investigated the loss system with a finite number N of servers and a finite number $M>N$ of sources (subscribers), every source having its individual characteristics with respect to calling rate and service time distributions. Using Kosten's technique of the supplementary variable, Cohen obtained the stationary distribution of specified idle and busy sources. This distribution appeared to be *insensitive* in the sense that only the first moments of the service time distributions were involved. Moreover, the distribution exhibited a *product form.* Some twenty years later, when the concepts of insensitivity and product forms had been firmly established as basic elements of queueing network theory, Cohen was to make further fundamental contributions to these concepts [A68,B3].

In the late fifties research workers in congestion or queueing theory had a fairly good set of basic models and appropriate mathematical techniques at their disposal. At that stage, professional mathematicians became interested in the subject. In line with this development, in 1957 Wim Cohen was appointed to a chair of Applied Mathematics in the Mathematics Department of the Delft University of Technology. He did not completely leave Philips, though; until 1987 he has served as a consultant for the Philips Telecommunication Group.

Among the papers he wrote in his early Delft period were several studies on derived Markov chains [A20-23]. The randomization or uniformization technique, which has recently become a popular tool in the analysis of the transient behavior of complex systems, is strongly related to these derived Markov chains.

In 1964 Wim Cohen married Annette Waterman, the daughter of a distinguished Dutch professor in chemistry. She created an ideal atmosphere for research and scholarship, which enabled Cohen to undertake the immense task of writing *The Single Server Queue* [B2] (the author index contains the name A.B. Waterman, as a tribute). The book contains much original research in the theory of stochastic processes; one of its many virtues is also that it beautifully demonstrates the power of the method of Pollaczek in queueing theory. This method basically describes a physical queueing situation in a natural way by analytic expressions (in terms of simple contour integrals), and thus transforms the probabilistic problem into a problem in complex analysis. In [B2] Cohen converts this approach into a powerful practical tool. The book is still considered to be the standard work in the area of mathematical queueing theory; a second revised edition appeared in 1982.

Soon after the completion of the book, in 1968, Wim Cohen became head of the Mathematics Department. As this occasion coincided with the student revolt (in The Netherlands as well as elsewhere), which led to a drastic change in the hierarchical structure of the University of Delft, Cohen did not have an easy time. The atmosphere in the department became such that he found it increasingly hard to devote his energy to research, and he took the difficult decision to leave his Alma Mater; on September 1, 1973, he took up the chair of Operational Analysis at the Mathematics Department of the University of Utrecht. He has stayed in Utrecht ever since, enjoying its special atmosphere which enables one to concentrate almost fully on research and teaching.

Cohen's scientific curiosity and creativity seemed to increase rather than decline over the years, and he grabbed the opportunity offered by the Utrecht environment to start novel ambitious research projects. Characteristically, these were mainly solo operations; in the period 1973-1978 Boxma was his only local graduate student and, later, cooperator.

Cohen's first research project in Utrecht concerned level crossings for regenerative processes. In queueing theory many of the involved stochastic processes are regenerative, and quite often this regenerative character provides the key to the analysis of the queueing model to be studied. In the monograph [B3], J.W. Cohen derives a number of *stochastic mean value theorems.* Such theorems express the stationary distribution of a regenerative process, if it exists, as a time average of the process over a regenerative cycle. The formulation of these stochastic mean value theorems makes it possible to apply immediately the sample function relations which exist between the involved stochastic processes. By this approach many well-known basic relations in queueing theory, like Little's formula, can be derived in an extremely simple and elegant way. A large part of [B3] is devoted to the application of stochastic mean value theorems to the M/G/K queueing and loss systems. In particular, a lengthy but essentially simple proof is presented of the insensitivity property of Erlang's formula for the distribution of the number of busy servers in the M/G/K loss system. The last part of [B3] starts to explore the possibility of obtaining relations between queueing quantities by using a *level crossing analysis.* In a series of papers [A61,A63-66,A71] this technique is further developed. The level crossing method finds its applications in such fields as inventory and production theory, data communications and reliability.

At the end of the sixties queueing theory got a new stimulus from the field of computer engineering. At that time, computer technology had reached a level of development which required a good insight into the data flow inside a computer, as well as in computer networks. For the latter in particular, the classical single service facilities did not suffice; networks of service facilities had to be analyzed. The pioneering work of J.R. Jackson turned out to be extremely useful in this respect. Jackson had studied open queueing networks with Poisson arrival streams (and later on, basically, closed networks) in which the service time distributions at all service centres are exponential. He showed that the stationary joint queue length distribution at the centres has a simple *product form,* which in principle allows a straightforward determination of all quantities of interest. However, the usefulness of Jackson's results for computer performance analysis was still severely limited by the restrictions that all customers follow the same rules of behavior, and that only exponential service time distributions occur. Furthermore, computer technology created a new type of service discipline, viz. the time sharing discipline, in which each job receives a quantum Δq of service and, when the total service request is not yet fulfilled, returns to the end of the queue. Kleinrock made the important observation that in the limit $\Delta q \to 0$ a queueing model results which is more amenable to a detailed mathematical analysis; the resulting service discipline is coined *processor sharing.* The stationary queue length distribution of an M/G/1 queue with processor sharing is independent of the service time distribution, apart from its first moment: the intriguing insensitivity phenomenon reappears! Baskett, Chandy, Muntz and Palacios, in their celebrated 1975 paper, proved that the product form property is not destroyed if, in a Jackson network, the service discipline at one or more of the centres is processor sharing - with not necessarily negative exponentially distributed service times. In fact they also allowed multiple classes of customers, with different characteristics. Cohen's interest in the concepts of product form and insensitivity was strongly revived by this BCMP paper, and by the beautiful studies of the young Cambridge mathematician Frank Kelly. A short but vigorous research effort led to the publication of 'The multiple phase service network with generalized processor sharing' [A68] and two other papers [A70,72]. We consider [A68] to be one of Cohen's best papers. It develops a general theory for a broad class of processor sharing models. The generalization of the ordinary processor sharing concept leads to a pleasing unification and extension of many of the known product form and insensitivity results.

Over the many years that J.W. Cohen had been involved in the analysis of queueing models, he had often come across apparently unsolvable, deceptively simply formulated 'two-dimensional' queueing models: the M/G/2 queue; the model of two servers and two queues, arriving customers choosing the shortest queue; the model of two queues, alternatively attended by one server; two servers with related speeds ('two coupled processors'); etc. In a fundamental publication in 1977, Fayolle and Iasnogorodski showed that the coupled processor problem can be analyzed by applying a boundary value technique from mathematical physics. The idea is to transform the problem of determining a two-dimensional Laplace transform or generating function into the problem of solving a (Dirichlet, Riemann or Riemann-Hilbert) boundary value problem. In its simplest form this requires the determination of a function which is analytic inside (and in some cases also outside) a simple closed contour, with prescribed discontinuities on that contour.

Cohen became absolutely fascinated by this technique. He started to study the books of Russian authors like Gakhov and Mushkelishvili on boundary value problems and (related) singular integral equations. Applying and extending the theory developed in those books, he was finally able to develop a method for the analysis of a large class of 'two-dimensional' random walk and queueing models. One of us (OJB) participated in the project, but Cohen was the driving force; countless mathematical difficulties were encountered, and they would not have been solved if Cohen had not devoted all his ingenuity and analytical skills to them. As a result the monograph [B4] was published (a Russian translation appeared in 1987). Cohen also supervised the Ph.D. theses of Hans Blanc (1982), Jan de Klein (1988) and Henk Nauta (1988) in the same area. The boundary value technique is still in its infancy, but it has disclosed an important research area and already several long-standing open problems have been successfully tackled; moreover, the approach allows for a fairly straightforward numerical evaluation. A major drawback is that there is so far no real clue as to an extension to 'higher-dimensional' problems. In the last few years Cohen has spent a tremendous amount of time and energy trying to accomplish such an extension, but he was not able to obtain a really manageable solution for, say, 'three-dimensional' problems (cf. [A81,A85]). Characteristically, on several occasions in private he optimistically expressed his hope of 'almost being there', only to say the next day that something crucial was still missing. As is not unusual in fundamental research, there were unexpected spin-offs; in this case, some very elegant identities for entrance times in random walks [A88-90,A92].

On September 1, 1988, following Dutch regulations, J.W. Cohen will retire from the University of Utrecht. In view of the fact that his interest in queueing theory remains undiminished and that, anyway, he has done most of his research in his study at home, it seems likely that his 65th birthday and retirement will not mark the end of his research. Shortly before this retirement, in June 1988, Cohen received a honorary doctorate from the Technion Israel Institute of Technology in recognition of his research achievements. Two years earlier, the Dutch Society of Sciences had honoured Cohen by awarding him the AKZO prize. This prestigious prize is awarded once a year for scientific research done over the preceding ten-year period. There are no particular restrictions on the area of research.

This short account of Cohen's scientific career would not be complete without a few words about his organizational activities in the international scientific community. He has been a member of the International Advisory Council for Teletraffic Congresses from the very beginning, in 1955; the ITC has always been particularly dear to him. He has also been one of the founding editors of *Stochastic Processes and their Applications,* and he has served as an editor for a number of other leading journals. In these ways, and also as an organizer of workshops and conferences, and as teacher of a few generations of Dutch probabilists, he has made a strong contribution to the scientific community.

However, there are two even more lasting contributions, we believe. First of all, in his scientific writings he has set a standard of mathematical rigour which has strongly enhanced the maturing of queueing theory as a mathematical discipline. And secondly, due to the combination of an engineering background and a strong mathematical ability he was able to understand all those people interested in queueing theory: probabilists, teletraffic engineers, operations researchers and computer performance analysts - and to bridge the gaps between them. In this way he has helped to make queueing theory an indispensable tool in the design and performance analysis of computer and communication systems. For these contributions, the queueing community will always owe a debt to J.W. Cohen. Wim Cohen feels at home in this diverse community, counting many colleagues among his closest friends - and this is reflected in the contributions in this book.

The editors

Scientific Publications of J.W. Cohen

A. Papers

1952 1. With H. Grosser. Calculation of the leakage flux of the pendant-armature relay. *Communication News 12,* 125-131.

1953 2. A graphical method of calculating the strains in stacked contact banks of selectors used in automatic telephony. *Communication News 14,* 1-11.

1954 3. With W.Th. Bahler and M.M. Jung. Calculation of the number of first group selectors for telephone systems provided with first and second concentration stages, taking into account the internal blocking. *Communication News 14,* 51-58.

4. With P. Harkema. A study of the delay encountered in the telegraph tape-relay switching. *Communication News 15,* 47-58.

1955 5. Some examples of the use of implication in switching algebra. *Communication News 16,* 2-10.

6. Berechnung der Verkehrsgröszen im Wartezeitsystem aus den Verkehrsgröszen eines Verlustsystems. *FTZ 8,* 139.

7. Das Warteproblem für das vollkommene Bündel mit einer endlichen Quellenzahl. *NTZ 8,* 641-646.

1956 8. Certain delay problems for a full availability trunk group loaded by two traffic sources. *Communication News 16,* 105-113.

1957 9. *Wachten,* Inaugurele rede, Waltman, Delft (in Dutch).

10. The full availability group of trunks with an arbitrary distribution of the inter-arrival times and a negative exponential holding time distribution. *Simon Stevin 26,* 169-181.

11. With M.M. JUNG. Calculations on a method for signal setting. *Philips Telecommunication Review 17,* 81-89.

12. With B.J. BEUKELMAN. Call congestion of transposed multiples. *Philips Telecommunication Review 17,* 145-154.

13. Basic problems of telephone traffic theory and the influence of repeated calls. *Philips Telecommunication Review 18,* 49-100.

14. The generalized Engset formulae. *Philips Telecommunication Review 18,* 158-170.

15. A survey of queueing problems occurring in telephone and telegraph traffic theory. *Proc. 2nd Int. Conference on Operational Research,* Oxford, 138-146.

1958 16. Some recent results on traffic theory. *P.T.T. Bedrijf 9,* 175-177. (Proc. 2nd ITC)

17. With R. SYSKI. List of references to publications on the application of probability theory to telephone trunking and related problems. *ITC Bibliography Committee.*

1960 18. The inadequacy of the classical stress-strain relations for the right helicoidal shell. W.T. KOITER (ed.). *Proc. I.U.T.A.M. Symposium on the Theory of Thin Elastic Shells,* North-Holland, Amsterdam, 415-433.

1961 19. Laplace and Fourier transformations. *Simon Stevin 35,* 14-30.

1962 20. Sur les chaines de Markov derivées et leur application à la théorie des queues. *Ann. des Telecommunication 17,* 164.

21. On derived and nonstationary Markov chains. *Theory of Probab. and Applications 7,* 402-423.

22. Derived Markov chains I, II, III. *Indag. Math. 24,* 55-92.

23. A note on skeleton chains. *Nieuw Arch. v. Wiskunde 10,* 180-186.

24. Enige beschouwingen over stochastische processen in het wegverkeer. *De Ingenieur 74,* 39-47 (in Dutch).

1963 25. A note on the delay problem in crossing a traffic stream. *Statistica Neerlandica 17,* 3-11.

26. Applications of derived Markov chains in queueing theory. *Appl. Sci. Res. B 10,* 269-303.

27. Laplace-transformaties. L. KUIPERS, R. TIMMAN (eds.). *Bijdrage 'Handboek der Wiskunde' I,* Scheltema & Holkema N.V., Amsterdam, 682-748 (in Dutch).

1964 28. Waarschijnlijkheidsrekening in techniek en bedrijfsvoering. *Water 48,* 185-191 (in Dutch).

29. A note on queueing processes with group arrival. *Post Office Telecommunications,* 24. (Proc. 4th ITC)

30. *Fundamental Methods in Queueing Theory,* Report 2. Ser. II, Center of Appl. Stochastics, Purdue Univ.

1965 31. Some comments on: 'Another note on the delay problem in crossing a traffic stream'. *Statistica Neerlandica 19,* 1-2.

32. Mathematical theory of traffic flow, a review. *Proc. NATO Seminar on Decision Problems in Connection with Traffic Planning,* Techn. Univ. Denmark, 48-62.

1967 33. With S.J. DE LANGE. Numerical results for queueing for gaps in a traffic stream. *Proc. Third Int. Symp. on the Theory of Traffic Flow,* Am. Elsevier Publ. Co., New York, 306-307.

34. Distribution of the maximum number of customers present simultaneously during a busy period for the system M/G/1 and for the system G/M/1. *J. Appl. Probab. 4,* 162-179.

35. On two integral equations of queueing theory. *J. Appl. Probab. 4,* 343-355.

36. Review of recently obtained results in queueing theory. *Proc. 5th ITC,* 59-65.

1968 37. With I. GREENBERG. Distribution of crossings of level K in a busy cycle of the M/G/1 queue. *Ann. Inst. H. Poincaré B IV,* 75-81.

38. Extreme value distribution for the M/G/1 and the G/M/1 queueing systems. *Ann. Inst. H. Poincaré B IV,* 83-98.

39. Single server queue with uniformly bounded virtual waiting time. *J. Appl. Probab. 5,* 93-122.

1969 40. Discussie bijdrage. *Statistica Neerlandica 23,* 209-212 (in Dutch).

41. Single server queues with restricted accessibility. *Journ. Engin. Math. 3,* 265-284.

1970 42. Aspecten en principes van de verkeerstheorie in telecommunicatie en informatie-verwerkende systemen. *De Ingenieur 82,* E159-E165 (in Dutch).

1971 43. On the busy periods for the M/G/1 queue with finite and with infinite waiting room. *J. Appl. Probab. 8,* 821-827.

44. On a formula dual to Erlang's loss formula. *Oper. Res. 19,* 1759-1760.

45. On memory units with finite capacity. *Proc. 6th ITC,* 321/1-321/7.

1972 46. Stochastische Processen. L. KUIPERS, R. TIMMAN (eds.). *Bijdrage 'Handboek der Wiskunde' III,* Scheltema & Holkema N.V., Amsterdam, 171-238 (in Dutch).

47. The suprema of the actual and virtual waiting times during a busy cycle of the $K_m/K_n/1$ queueing system. *Adv. Appl. Probab. 4,* 339-356.

48. With H. CALLAERT. A lemma on regular variation of a transient renewal function. *Z. Wahrsch. Verw. Gebiete 24,* 275-278.

49. On the tail of the stationary waiting time distribution and limit theorems for the M/G/1 queue. *Ann. Inst. H. Poincaré B VIII,* 255-263.

1973 50. Some results on regular variation for distributions in queueing and fluctuation theory. *J. Appl. Probab. 10,* 343-353.

51. Asymptotic relations in queueing theory. *Stoch. Proc. Appl. 1,* 107-124.

1974 52. Some ideas and models in reliability theory (Invit. Lect. NATO Conf. The Hague, 1972). *Statistica Neerlandica 28,* 1-10.

53. Superimposed renewal processes and storage with gradual input. *Stoch. Proc. Appl. 2,* 31-57.

54. Some aspects of queueing theory (Invit. Lect. 7th ITC). *Statistica Neerlandica 28,* 55-67.

55. De betekenis van de wachttijdtheorie voor het ontwerp en het gebruik van computers en computerbestuurde systemen. *Informatie 16,* 652-664 (in Dutch).

1975 56. Plastic-elastic torsion, optimal stopping and free boundaries. *Journ. Engin. Math. 9,* 219-226.

57. Contribution panel discussion Conf. Stoch. Proc. Appl. Toronto, 1974. *Stoch. Proc. Appl. 3,* 237-239.

1976 58. The Wiener-Hopf technique in applied probability. J. GANI (ed.). *Perspectives in Probability & Statistics. Papers in honour of M.S. Bartlett,* Academic Press, London, 145-156.

59. On a single-server queue with group arrivals. *J. Appl. Probab. 13,* 619-622.

60. With E.W.B. VAN MARION. Analysis of storage hierarchy. *Journ. Engin. Math. 10,* 263-283.

61. On the optimal switching level for an M/G/1 queueing system. *Stoch. Proc. Appl. 4,* 297-316.

62. On Erlang's formula for the loss system M/G/K. *Proc. 8th ITC,* 124/1-124/6.

1977 63. On up- and downcrossings. *J. Appl. Probab. 14,* 405-410.

64. With M. RUBINOVITCH. On level crossings and cycles in dam processes. *Math. Oper. Res. 2,* 297-310.

1978 65. On the maximal content of a dam and logarithmic concave renewal functions. *Stoch. Proc. Appl. 6,* 291-304.

66. Properties of the process of level crossings during a busy cycle of the M/G/1 queueing system. *Math. Oper. Res. 3,* 133-144.

67. Commentary on section 9 of L. Kosten's doctoral thesis. E.M.J. BERTIN, H.J.M. BOS, A.W. GROOTENDORST (eds.). *Two Decades of Mathematics in the Netherlands 1920-1940,* Mathematical Centre, Amsterdam, 372-379.

1979 68. The multiple phase service network with generalized processor sharing. *Acta Informatica 12,* 245-284.

69. With O.J. BOXMA and N. HUFFELS. Approximations of the mean waiting time in an M/G/s queueing system. *Oper. Res. 27,* 1115-1127.

70. On networks with generalized processor sharing and a new property of the Erlang B-formula. *Proc. 9th ITC.*

1980 71. With M. RUBINOVITCH. Level crossings and stationary distributions for general dams. *J. Appl. Probab. 17,* 218-226.

72. Sensitivity and insensitivity (Lecture Symp. Professor Kosten retirement). *Delft Progress Report 5,* 159-173.

1981 73. With G. HOOGHIEMSTRA. Brownian excursion, the M/M/1 queue and their occupation times. *Math. Oper. Res. 6,* 608-629.

74. With O.J. BOXMA. The M/G/1 queue with alternating service formulated as a Riemann-Hilbert problem. F.J. KYLSTRA (ed.). *Performance '81,* North-Holland, Amsterdam, 181-199. (Extracted from the more detailed report of the same title, Report 177, Univ. Utrecht, 1980).

75. Obituary: Félix Pollaczek. *J. Appl. Probab. 18,* 959-963.

76. *Analysis of the Left-Continuous Random Walk on the Lattice in the First Quadrant,* Report 214, Math. Institute, Univ. Utrecht.

1982 77. On the M/G/2 queueing model. *Stoch. Proc. Appl. 12,* 231-248.

1983 78. Queueing problems with a two-dimensional state space. *Proc. 10th ITC,* 1-6.

79. On queueing theory and two-dimensional queueing models. P.J. KUEHN, K.M. SCHULZ (eds.). *Messung, Modellierung und Bewertung von Rechen Systeme,* Springer Verlag, Berlin, 107-121.

80. On the analysis of two-dimensional queueing problems. G. IAZEOLLA, P.J. COURTOIS, A. HORDIJK (eds.). *Mathematical Computer Performance and Reliability,* North-Holland, Amsterdam, 17-32.

1984 81. *On a Functional Relation in Three Complex Variables; Three Coupled Processors,* Report 359, Math. Institute, Univ. Utrecht.

82. On processor sharing and random service. *J. Appl. Probab. 21,* 937.

1985 83. With O.J. BOXMA. A survey of the evolution of queueing theory. *Statistica Neerlandica 39,* 143-158.

84. Who introduced the birth- and death technique? M. AKIYAMA (ed.). *Proc. 11th ITC,* North-Holland, Amsterdam, 36-39.

85. *On the Analysis of Parallel, Independent Processors,* Report 374, Math. Institute, Univ. Utrecht.

86. *On Renewal Processes with Discontinuous Sample Functions,* Report 401, Math. Institute, Univ. Utrecht.

1986 87. Some samples of modelling. J. GANI (ed.). *The Craft of Probabilistic Modelling,* Springer Verlag, Berlin, 88-108.

88. *On Entrance Times of Homogeneous Semi-Bounded N-Dimensional Random Walks: an Identity,* Report 440, Math. Institute, Univ. Utrecht.

89. *On Entrance Times of a Homogeneous N-Dimensional Random Walk; an Identity,* Report 442, Math. Institute, Univ. Utrecht.

1987 90. On entrance time distributions for two-dimensional random walks. G. IAZEOLLA, P.J. COURTOIS, O.J. BOXMA (eds.). *Computer Performance and Reliability,* North-Holland, Amsterdam, 25-41.

91. A two-queue, one-server model with priority for the longer queue. *Queueing Systems 2,* 261-283.

92. Continuation of: *On Entrance Times of Homogeneous Semi-Bounded N-Dimensional Random Walks: an Identity,* Report 460, Math. Institute, Univ. Utrecht.

93. *Boundary Value Problems and Random Walk Analysis in Queueing System Analysis,* Report 479, Math. Institute, Univ. Utrecht (to appear in *Queueing Systems*).

94. *On a Two-Dimensional Workload Process,* Report 490, Math. Institute, Univ. Utrecht.

1988 95. A two-queue model with semi-exhaustive alternating service. P.J. COURTOIS, G. LATOUCHE (eds.). *Performance '87,* North-Holland, Amsterdam, 19-37.

B. BOOKS

1955 1. *On Stress Calculations in Helicoidal Shells and Propeller Blades,* Thesis, Waltman, Delft.

1969 2. *The Single Server Queue,* North-Holland, Amsterdam, 2nd revised edition 1982.

1976 3. *On Regenerative Processes in Queueing Theory,* Lect. Notes in Econ. and Math. Systems, Vol. 121, Springer Verlag, Berlin.

1983 4. With O.J. BOXMA. *Boundary Value Problems in Queueing System Analysis,* North-Holland, Amsterdam. Russian transl. Mir Publishers, Moscow, 1987.

CONTENTS

PART I

SURVEYS

PART II

THE SINGLE SERVER QUEUE

PART III

ANALYTIC METHODS IN QUEUEING THEORY

PART IV

NETWORKS AND COMPUTERS

PART V

QUEUEING MODELS AND RELATED TOPICS

EPILOGUE

PART I
SURVEYS

Performance Evaluation of

Distributed Computer-Communication Systems

Leonard Kleinrock[1]
Computer Science Department, University of California
Los Angeles, California 90024-1596, USA

The computer-communication field is rich with extremely challenging problems for the queueing theorist. In this paper, we describe a number of computer-communication applications in wide-area networks, packet radio networks and local area networks. All of these involve sophisticated queueing theoretic models which have led to advanced applications of existing theory and, in some cases, to new methods in queueing theory A significant component in many of these applications is that of multi-access to a common resource (typically, a communication channel); we devote a significant portion of this paper to multi-access systems. In addition to identifying problems of interest to the queueing theorist, this paper summarizes the latest results for the mean response time of many current computer-communication applications.

1. Introduction

The advent of the digital computer brought about a major rebirth in applied queueing theory. Indeed, since the early 1960's we have seen a sequence of applications in the information processing field which have challenged the capabilities of that theory and which have forced the theory to be extended in new directions. There is no reason to believe that that challenge will subside for some time to come.

The response to these challenges has produced some important extensions in queueing theory. For example, there has been a major (and largely successful) assault on queueing networks of various kinds. Some of the results are quite beautiful, and exhibit robustness in various directions (see [81]). We have also seen the development of some very important approximation techniques. Multi-access and broadcast communication problems have forced us to develop further analytical techniques. Many of the applications lead to coupled queues which involve two-dimensional queueing problems in the simplest cases, and higher dimensions in the usual cases; here too, new approaches in queueing theory have had to be developed (e.g., the solution of Riemann-Hilbert problems). Due to the difficulty of many of these problems, we find

1. This work was supported by the Defense Applied Research Projects Agency under Contract No. MDA 903-82-C-0064.

that the theory has had to tread a very fine line between approximation and attention to detail.

We organize this paper as follows. In Section 2, we discuss the problems inherent in sharing resources in a distributed environment. In Section 3, we discuss response time and 'power' as measures of system performance. Computer network performance analysis, design and buffer sharing are discussed in Section 4. A large number of multi-access schemes are described and evaluated in Section 5 (the largest section of this paper). Finally, in Section 6, a few ideas from the distributed processing field are presented.

2. Distributed resource sharing as a problem in queueing theory

The earliest problems from information processing which we addressed using queueing theory were simple isolated problems (e.g., memory access). Shortly thereafter, the first major class of problems arose in an attempt to analyze and design computer networks [55]. It was recognized immediately that the exact analysis of networks was hopelessly intractable and so some key (and sweeping) assumptions had to be introduced to allow for a solution (see Section 4.1). But this analysis was, in some sense, premature; in the mid-60's, the computer industry was not yet aware of the need for networks, and so very little work continued in this direction at that time. Instead, the advent of time sharing caught the fancy of the data processing industry. It was found that these systems presented another major opportunity for the use of queueing theory to provide the tools for a proper analytic treatment (see Lavenberg [81] for an excellent summary and history of this area). It was only in the late 60's that networking became fashionable and began to attract large numbers of researchers. Then, in the mid-70's, wireless communication systems (e.g., broadcast satellite communications, packet radio, etc.) presented another set of challenges, and once again models, analyses and approximations were developed. The study of local area networks (LANs) followed in the early 80's and continues to occupy our attention up to the present time.

A major driving force in all these developments was micro-electronics, i.e., integrated chip technology. Of course, the major impact of VLSI was to radically drop the cost of computing and this led to advances such as supercomputers and personal computers; these, in turn, have led to parallel and distributed information processing systems. Indeed, distributed systems have emerged during this period in a variety of forms: distributed communications (e.g., packet switching); distributed processing (e.g., multiprocessing); distributed data base; and distributed control (e.g., multi-access communications).

There is a common theme shared by all of these systems, namely, the fact that they are all distributed. It is this that gives rise to some fascinating new problems for the queueing analyst. A distributed system represents a collection of resources (servers) which are provided to serve a set of users (customers). What makes these problems new is that users may not have convenient access to (nor knowledge of) all of these resources.

Let us elaborate on this issue. In a classic single (or multiple) server system, all users have access to all the resources (i.e., any server can serve any customer) and, in addition, all users can 'see' the queue (i.e., the users themselves can observe which position they currently occupy in the queue). This is an ideal situation in that the server acts as a perfect dynamically shared resource. In a distributed system, we do not ordinarily have the luxury of such a perfectly organized queue. Let us list the ways in which a user may not have proper access to a resource:

1. *The resource may be* **forbidden** *(completely inaccessible) to a user*
For example, if we had two separate single server systems, the second server is inaccessible to the first server's customer stream. A simple study to quantify the cost of this inaccessibility was published in [82]. The model was that of M identical and independent M/M/1 queues, each operating at a utilization factor, ρ. The quantity $Q = P$[at least one customer is waiting in some queue and at least one server is idle] was found to be equal to

$$Q = 1 - \rho^M + (1 - \rho)^M[\rho^M - (1+\rho)^M]. \tag{2.1}$$

Fig. 1 shows this function and clearly demonstrates that this distributed system invokes a large cost (as measured by Q) when $M \gg 1$ for all values of ρ inside the unit interval.

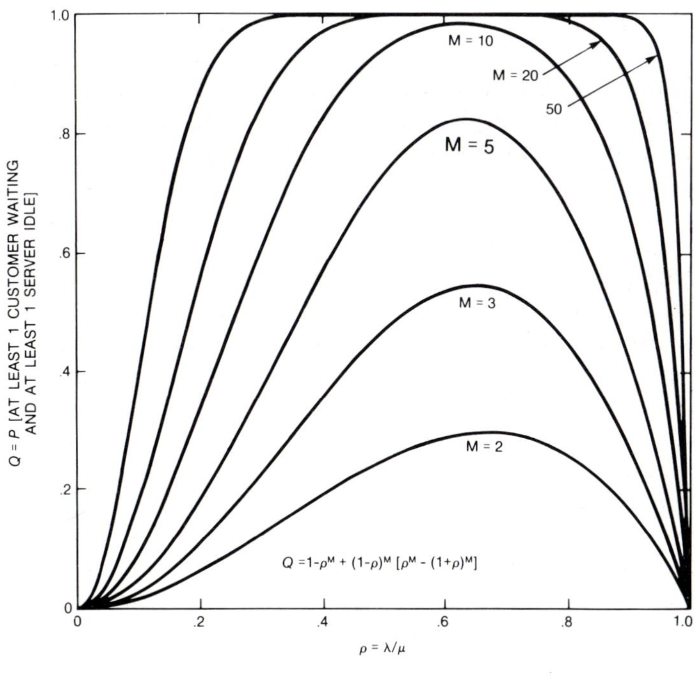

FIGURE 1. *Performance of M M/M/1 queueing systems*

A second (simple) way to quantify this cost of inaccessibility is to compare the mean response time, T_M, for these M systems to the mean response time, T_1, of a single M/M/1 system serving the total input from the M streams and with a large server who is M times faster than each of those above. That is, if each of the M separate systems has an input rate λ and a mean service time $\bar{x}$, then the single large server system has an input rate $M\lambda$ and a mean service time $\bar{x}/M$. In both cases, the server utilization is $\rho = \lambda\bar{x}$. From M/M/1 theory [59], we find this ratio to be

$$\frac{T_M}{T_1} = M \tag{2.2}$$

for all $0 \leqslant \rho < 1$. This is another measure of the cost of forbidden resources, and is referred to as the 'economy of scale' principle in [64].

2. *The user may lack* **precise** *information about the state (i.e., the congestion) of a resource*
For example, in a computer network, a path must be chosen for a message to travel. The current congestion situation for channels which are many hops away from the current location of the message may only be known approximately (due, say, to the method by which this information is collected and dispersed).

3. *The user may not have* **current** *information about the state of a resource*
The same example as for (2) applies here. State information may be precise, but may be delayed before it is available to the message.

4. *Immediate and precise information provided to a user about a resource may become* **outdated** *by the time the user has an opportunity to use the resource*
The same example as for (2) applies here as well. The difficulty here is that it takes some time for the message to make its way along the journey to distant channels.

In order to provide users improved access to resources in distributed systems, a number of procedures have been developed. These procedures are referred to as 'multi-access' procedures and fall into three canonical classes:
1. Static assignment algorithms;
2. Dynamic assignment algorithms;
3. Random access algorithms.

(See for example, [64].) Each access method is an attempt to overcome the inaccessibility problem created by the distributed nature of the system. Each is trying, in some sense, to make all users aware of the distributed queue in which they reside so as to allow the users to coordinate their access to the resources. Each method incurs a price for this coordination attempt as follows:

ALGORITHM	EXAMPLE	PRICE
Static	FDMA	Idle resources
Dynamic	Token Passing	Control overhead
Random	Ethernet	Collisions

TABLE 1. *The price for distributed access*

By idle, we mean there is a resource available which some customer could have used, but was disallowed since it was assigned to another user (or group of users) who did not, at that time, need the resource; for example, a collection of separate M/M/1 queues. Control overhead refers to the cost of the dynamic allocation of resources; for example, the time it takes to move the token around in token passing schemes. By collisions, we mean that two or more users attempt to use the same resource simultaneously, in which case, neither gets served; for example, this happens with the Ethernet access scheme when two or more users transmit on the common bus.

With multi-access communications, another key quantity must be taken into account. This quantity involves three important system variables, namely,

C = capacity of the communication channel (bits/sec);
b = the (average) number of bits in the packet to be transmitted;
L = 'length' of the channel (measured in seconds and representing the time it takes from when the first bit is pumped into the channel until that bit appears at the output of the channel, i.e., L is the channel propagation delay).

The key quantity referred to above is denoted by a, and is defined as

$$a = LC/b \tag{2.3}$$

which is readily seen to be the (dimensionless) ratio of the channel propagation delay (L) to the time (b/C) it takes to pump the packet into the channel. It is also equal to the number of packets which can be 'pipelined' in the channel. The value of a turns out to be extremely important in determining system performance and therefore in system design. For example, let us calculate the value of a for four important computer communication cases: a local area network; a long-haul packet-switched network; a broadcast satellite system; and a long-haul fiber optic link. The assumptions and values for a corresponding to these cases are shown in Table 2:

SYSTEM	C (Mbps)	b (bits)	L (microsec)	a
Local Net	10	1000	5	.05
Packet Net	.05	1000	20,000	1
Satellite	.05	1000	250,000	12.5
Fiber Link	1000	1000	20,000	20,000

TABLE 2. *Some typical values for* $a = LC/b$

Thus we see that the value of a varies by a factor of 400,000 over this set of interesting problems. As we said, this value is a key determinant in system design. As an example, whereas the Ethernet local area network access scheme (carrier sense multiple access with collision detection - CSMA/CD - see Section 5.5.4) works well for $a \ll 1$, it would be a disaster in all the other systems.

Considerations such as those listed in this section have motivated a number of new queueing theoretic problems and solutions. We discuss some of these below. In the discussion of these applications, however, we focus on those aspects of distributed resource sharing for which queueing theory has a role to play; we will not address a number of other topics which may be of significance to an application area but which do not have a queueing theoretic component.

3. Performance measures

3.1. Response time

We define the response time (sojourn time) of a system to be the interval from when a user generates a request for service until that service is complete. The mean response time is $T = E$ [response time] where $E[X]$ denotes the expected value of the random variable X. This is the major performance measure used in most queueing systems studies, particularly in computer communications. T is usually given as a function of the system load which measures the efficiency of the server. The load in queueing theory is usually denoted by ρ. In computer communications, we use the same symbol (ρ), but sometimes we use S, particularly when we discuss multi-access systems of certain types. Indeed, when the mean service time, $(\overline{x})$, is not a system design parameter, then we often consider T to be a function only of λ, the system throughput (measured in units of customers per second).

In many of the systems we discuss below, the evaluation of $T = T(\rho)$ is very difficult, and an explicit expression for this function is not available. In [65], a simple approximation for this function is offered which provides three degrees of freedom which allow the analyst to adjust the shape of $T(\rho)$ to match the problem at hand. This approximation is

$$T(\rho) = A\frac{Z - \rho}{P - \rho}, \tag{3.1}$$

and is referred to as the 'ZAP' approximation. The three parameters are used as follows:

1. The value of P (the pole) is equal to the maximum load the system can support (in classical queueing theory, this is $P=1$; in the ALOHA system discussed below, it is $P=1/e$).
2. The value of Z (the zero) is selected to adjust the relative shape of the growing delay function. If $Z = P + \epsilon$ (for small, positive ϵ), then we will have a mean response time function which has a very flat behavior over $0 \leqslant \rho < P$ and a very sharp rise to infinity as $\rho \to P$. On the other hand, if $Z \gg P$ or if $Z \ll 0$, then the mean response time will vary more smoothly over its range.
3. A is selected to match the value of $T(0)=AZ/P$. We note that A and Z must both have the same sign.

See Fig. 2 for typical cases. Often, this expression can be made to exactly match $T(\rho)$ as, for example: M/D/1 ($Z=2$, $A=\bar{x}/2$, $P=1$); FDMA with M users ($Z=2$, $A=M\bar{x}/2$, $P=1$); TDMA with M users ($Z=1+M/2$, $A=\bar{x}$, $P=1$); and others.

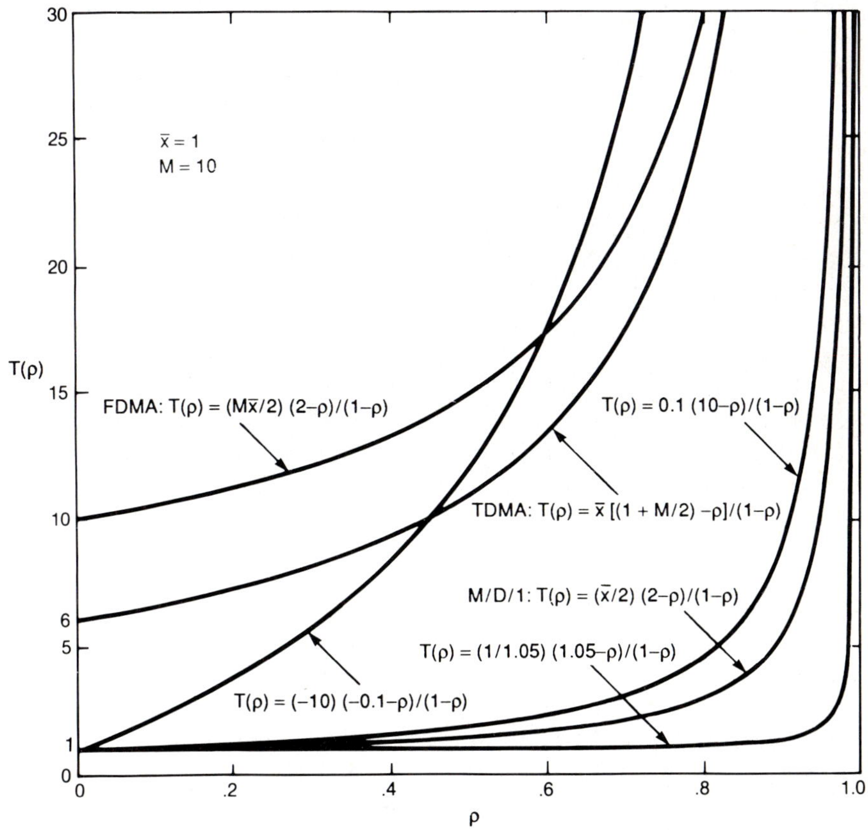

FIGURE 2. *The ZAP approximation*

3.2. Power

Ideally, one would love to operate the system such that it would exhibit minimum delay and maximum throughput at the same time. In fact, this is possible for D/D/1. However, it is far from the case for most systems of interest. Therefore, one wonders where the 'appropriate' operating point should be. If you hate delay much more than you value efficiency, then you would be happy to operate in the vicinity of ρ near zero (giving minimum delay, but poor efficiency). On the other hand, if you consider efficiency much more important than delay, you would work much closer to the maximum allowable ρ (and suffer a large delay). Where, indeed should you operate?

A quantity known as *power*, and denoted by P (not to be confused with the pole, P, from Section 3.1 nor with the number of processors, P, from Section 6.2), was introduced in [36] and [140] which combined both mean response time and throughput into a single measure. Their definition is $P = \lambda/T =$ throughput/mean response time. Clearly, if this measures your relative preferences, then you would like to maximize P. In [66], this optimization problem was addressed and it was shown that maximum power occurs whenever

$$dT(\lambda)/d\lambda = T(\lambda)/\lambda. \tag{3.2}$$

It is trivial to see that in the $T-\lambda$ plane, this maximum power point occurs at that value of λ where a ray out of the origin is tangent to the response time curve (and where this ray makes the smallest angle with the horizontal axis). See Fig. 3 for an example. In fact, this can be at a point where the mean response time has no derivative.

It turns out for all M/G/1 systems, that maximum power occurs at that λ which produces $\overline{N} = E$[number of customers in system]$=1$ exactly! In [68], this observation is discussed in terms of deterministic reasoning. It is found that $\overline{N} = 1$ is a common invariant in these problems; this fact is exploited in a detailed study of power in [33] and in [47].

4. Computer networks

An enormous literature on computer networks has been created since the first comprehensive work was published by Kleinrock in 1964 [55]. In that work, he solved many of the basic network analysis and design problems.

4.1. Analysis

The two most significant quantities to solve for in a computer network are γ, the network throughput (the average number of messages per second delivered to destinations) and T, the mean time a message spends in the network (the mean sojourn time in the network). Messages originate at their source node, follow some path through the network, and arrive at their destination nodes. When they reach an intermediate node along their path, they join a queue while awaiting transmission (over, say, channel i) to the next node in their journey. The mean time they spend waiting for, plus transmitting over, channel i is denoted by T_i. Let λ_i denote the average number of messages per second passing over channel i, and let M denote the total number of communication

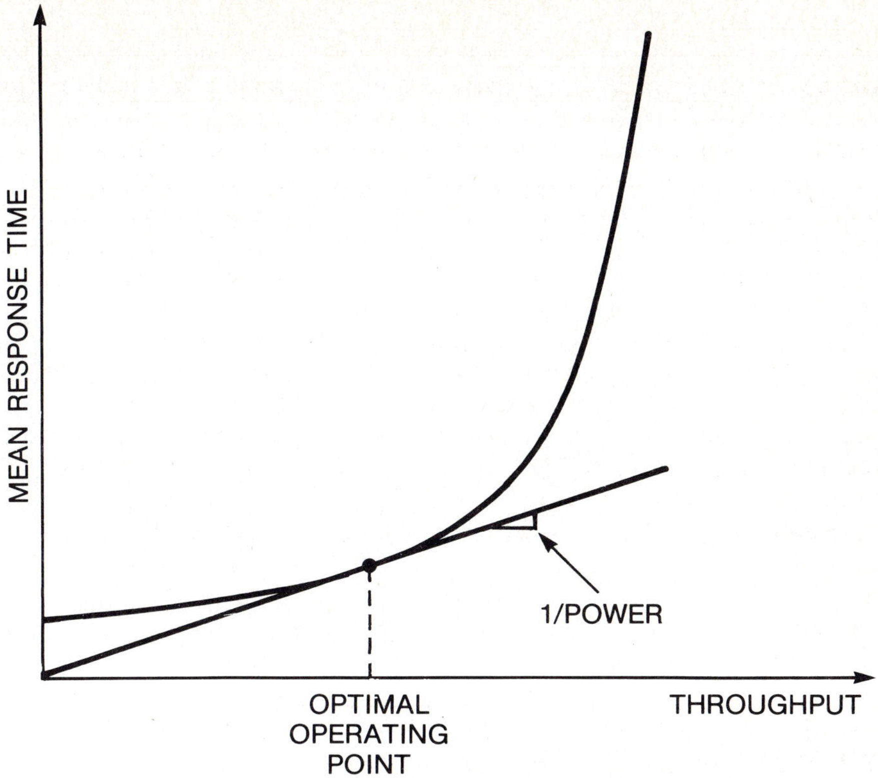

FIGURE 3. *The key system profile and power*

channels in the network. The following expression gives the exact relationship among these quantities [55]:

$$T = \sum_{i=1}^{M} \frac{\lambda_i}{\gamma} T_i. \tag{4.1}$$

This is easily proven with two applications of Little's result [59] as follows. Let $\overline{N} = E$[number of messages in the network]. The first application of Little's result at the network level gives

$$\gamma T = \overline{N}. \tag{4.2}$$

Let $\overline{N}_i = E$[number of messages waiting for or transmitting over channel i]. By assumption, if a message is in the network, it is either waiting for or transmitting over some network channel. Thus

$$\overline{N} = \sum_{i=1}^{M} \overline{N}_i. \tag{4.3}$$

The second application of Little's result at each network channel gives

$$\overline{N} = \sum_{i=1}^{M} \lambda_i T_i. \tag{4.4}$$

From equations (4.2) and (4.4) we arrive at the result of equation (4.1)! Equation (4.1) is the key equation for the *exact* analytical behavior of extremely general computer networks.

In (4.1), the network throughput γ can be found by summing all elements in the (given) traffic matrix; that is, we assume we are given the entries, γ_{jk}, in the traffic matrix (average number of messages entering the network per second with origin j and destination k). The channel traffic λ_i can be calculated from the traffic matrix and the routing procedure. The difficult part of the evaluation is to calculate the mean response time, T_i, for each channel queueing system. The source of the difficulty has to do with the correlation of a message's (i.e., a customer's) service time in successive network nodes.

For example, consider the simple two-node series network shown in Fig. 4. Here we assume that messages arrive to Node 1 from an external Poisson source at a rate of γ messages per second, each with an exponentially chosen length (with an average of $1/\mu$ bits). The messages queue (FCFS) while awaiting transmission (service) from Node 1 to Node 2 over a noiseless communication channel whose capacity is C bits/sec. Thus, the average transmission time on the channel is $1/\mu C$ seconds. When a message is finished being transmitted from Node 1, it then joins a queue while awaiting transmission out of Node 2; we assume that the channel supporting this transmission also has a capacity of C bits per sec. After this second transmission, the message leaves the system. Of interest is the response time (sojourn time) of the message in the network. Clearly, the first node acts as a simple M/M/1 queue. However, the second node presents some difficult problems which come about since the service time of a message on the second channel is *exactly the same* as its service time on the first channel. This not only increases the dimensionality of the state space, but it also introduces a dependence between the sequence of interarrival times and service times in the second node. This two-node problem was posed in [55], and was not solved until 1979 when BOXMA [11] published a solution as the content of his doctorate work under J.W. Cohen. This solution applied only to the case where both channels had the same capacity! His solution was so involved that it provided further evidence that the general network (with arbitrary capacities, paths and topology) would likely never yield to an exact solution.

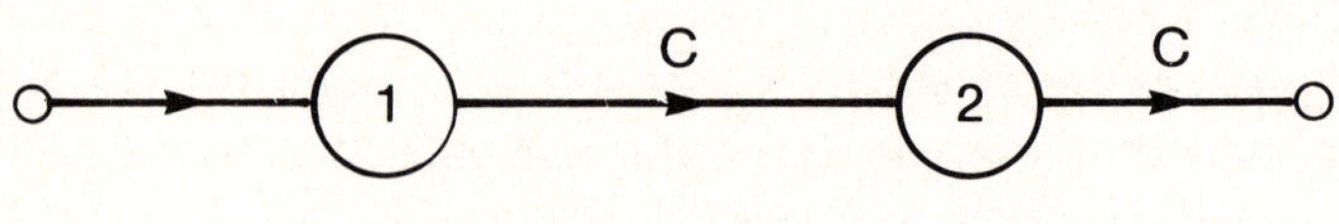

FIGURE 4. *Two-node tandem net*

Recognizing this essential difficulty, Kleinrock introduced a bold 'Independence Assumption' which cracked the problem wide open and admitted a straightforward solution. This (admittedly false) assumption states that a message's length is *independently* selected from the same exponential distribution each time it enters a node. Using this assumption, Kleinrock could then model the network as an open queueing network (a Jackson network [49]) whose solution for the joint distribution of customer occupancy in each node has a product form. See [81] for a discussion of queueing networks. Considerable experimental evidence supports the use of this model insofar as it has been used to calculate the mean sojourn time of messages in computer networks. Indeed, using this assumption, each channel may be modeled as an independent M/M/1 queueing system whose mean response time is simply

$$T_i = \frac{1}{\mu C_i - \lambda_i}, \quad i = 1,2, \ldots, M. \tag{4.5}$$

Equations (4.1) and (4.5) form the simplest model for the mean response time in computer networks. The Jackson network solution also gives the joint distribution for customer occupancy, but does not give the distribution of sojourn time. WONG [137] found the sojourn distribution for a limited class of networks (networks with no 'overtaking'); see also [54,89,136].

Equation (4.5) may be refined in a number of ways to account for a number of practical factors which we have neglected. These include, for example, the nodal processing time, the propagation delay, the transmission time from the exit node to the attached Host computer, the interference due to control traffic, etc. These refinements may be found in [64].

RUBIN [108] also studied the tandem queue (see Fig. 4). However, he allowed arbitrary channel capacities, an arbitrary number of nodes in the tandem, and a *constant* length for each arriving message. He was able to show that this tandem system could be modeled by an M/G/1 queue and so the Pollaczek-Khinchin results for M/G/1 solve this system completely. Unfortunately, his results do not extend to more complex topologies nor to randomly chosen message lengths.

4.2. Design

There are many aspects to the design of a computer network. Among these are: the choice of node locations; hardware structure of the node; software and protocol design; layered communication structure; topological structure; routing procedure; and capacity of communication circuits. In addition, one must make a number of model assumptions such as: channel cost function (tariff); traffic matrix; message arrival process; and message length distribution. Depending on the formulation of the design problem, the objective function and design constraints can exchange roles. In any case, one must consider: T, the mean system response time; D, the total dollar budget available to spend on the design; and some measure of network reliability. For a complete discussion of these issues see, for example, [64].

We may select two (equivalent) versions of the (optimal) design problem. The first is:

PROBLEM 1

Minimize: $$T = \sum_{i=1}^{M} \frac{\lambda_i}{\gamma} T_i = \sum_{i=1}^{M} \frac{\lambda_i/\gamma}{\mu C_i - \lambda_i}$$

With respect to: Topological design
Routing procedure
Channel capacity assignment

Subject to: $$D = \sum_{i=1}^{M} d_i(C_i)$$
and deliver the traffic specified in the traffic matrix

Where: C_i = Capacity of i-th channel
$d_i(C_i)$ = cost (dollars) to supply C_i units of capacity to the i-th channel
M = Number of channels in net

An alternate (dual) to this problem statement is:

PROBLEM 2

Minimize: $$D = \sum_{i=1}^{M} d_i(C_i)$$

With respect to: Topological design
Routing procedure
Channel capacity assignment

Subject to: $T \leq T_{MAX}$
and deliver the traffic specified in the traffic matrix

Where: T_{MAX} is a given allowable maximum mean response time

Clearly, the first problem asks us to minimize network delay at a fixed maximum cost while the dual version asks us to minimize network cost while meeting a maximum network delay criterion. The reliability constraint can be stated in a variety of ways; we do not discuss reliability in this paper.

For design purposes, we usually make the simplest possible model assumptions. Typically, these are:

Exponentially distributed message lengths (mean $1/\mu$ bits)
Independence assumption
Poisson message arrivals (at a rate of γ_{jk} messages/sec originating at node j and destined for node k)

Nodal switch locations preselected
Nodal switches supplied at no cost
Nodal processing times are negligible (infinite nodal processing speed)
Propagation delay is zero (speed of light is infinite)
No control traffic

These assumptions lead us to a Jackson open network model as discussed above. Nevertheless, the optimal design problems stated above are still far too difficult to permit an exact solution. Therefore, the problem is usually decomposed into simpler sub-problems. The 'cleanest' and most commonly discussed sub-problem is the 'Capacity Assignment Problem' which is Problem 1 above, in which the topology and the routing procedure (i.e., the λ_i's) are assumed to be given, and the only issue is to optimally spend D dollars to supply each of the M channels in the topology with a capacity so that the overall mean response time is minimized. The solution depends upon the form of the cost function $d_i(C_i)$. The simplest case considered is the case of a linear cost function, i.e.,

$$d(C_i) = d_i C_i. \tag{4.6}$$

For this optimal assignment, the (minimum) message delay is

$$C_i = \frac{\lambda_i}{\mu} + \left(\frac{D_e}{d_i}\right) \frac{\sqrt{\lambda_i d_i}}{\sum_{j=1}^{M} \sqrt{\lambda_j d_j}}, \quad i = 1,2, \ldots ,M. \tag{4.7}$$

D_e is defined below. For this case, the known solution [55] for the minimum T is:

$$T = \frac{\bar{n}}{\mu D_e} \left[\sum_{i=1}^{M} \sqrt{\lambda_i \, d_i / \lambda} \right]^2, \tag{4.8}$$

with $\bar{n} = \Sigma_{i=1}^{M} \lambda_i / \gamma$ (is average path length). MEISTER, MULLER, and RUDIN [88] generalized this linear cost-capacity case by changing the objective function from that given in PROBLEM 1 to the following

$$T^{(k)} = \left[\sum_{i=1}^{M} \frac{\lambda_i}{\gamma} (T_i)^k \right]^{1/k} \tag{4.9}$$

where T_i is once again given in (4.5). The form for the optimum capacity and delay may be found in [64,88].

We can generalize the capacity-cost function beyond that of linear to obtain other assignments and other performance equations. However, in all cases, we observe two invariants. First, all optimal capacity assignments require that the i-th channel be provided a minimum capacity of λ_i/μ bits per sec. (otherwise the utilization factor for this channel would exceed unity). Second, a strict lower bound on the number of dollars necessary to provide a feasible solution to this problem for any cost function $d_i(C_i)$ is

$$D_{\min} = \sum_{i=1}^{M} d_i(\frac{\lambda_i}{\mu}). \tag{4.10}$$

In (4.7) and (4.8), D_e is defined as $D_e = D - D_{\min}$. Without this minimum dollar expenditure, some channel will be overloaded, and then, from (4.1), we see that the mean response time will blow up. Thus, for all capacity-cost functions, the system feasibility condition is $D > D_{\min}$.

A second subproblem assumes that the topology as well as the capacity assignment are given and one wishes to optimally route the traffic through the network to minimize delay; this is the 'Flow Assignment Problem'. That is, we wish to find the optimal λ_i's subject to the constraint that we satisfy the requirements of the traffic matrix. This problem turns out to be that of minimizing a convex function over a convex set; consequently, almost any valley-finding or steepest descent type of algorithm will work. A particularly effective method, known as the Flow Deviation Method, may be found in [31]; this method is based on queueing-theoretic principles.

In order to 'solve' the more general problem, e.g., PROBLEM 2 above, one must resort to sub-optimal methods. We will not discuss these here, except to point out that a number of different heuristic methods have been developed, e.g., [30,38,39,40,64] and these all tend to give the same profile in the throughput versus network-cost plane.

4.3. *Buffer sharing*

An issue of importance in computer network implementation is that of buffer sharing. As messages (or packets) arrive at a node, they must be buffered while awaiting transmission out of that node. These nodes have a finite number of message buffers and so one would like to allocate these buffers in some appropriate fashion.

The comprehensive model developed in [52] is as follows. Consider a single node with a storage capacity of B buffers. This node is subject to M independent Poisson input streams, the m-th of which has a rate λ_m messages per second, and which must be transmitted out on the m-th channel (of capacity C_m) leaving this node ($m = 1,2, \ldots ,M$). Message lengths are exponentially distributed with a mean of $1/\mu$ bits per message. The average message transmission (service) time is, therefore, simply $1/\mu C_m$ seconds on the m-th channel.

Thus, we have M M/M/1 queueing systems which are coupled only through the fact that they share a common finite storage capacity of B buffers. If no space is available for an incoming message, this message is lost; all accepted messages are served in a first-come-first-serve fashion on their respective outgoing channels. Five buffer sharing schemes are considered in [52]. The first is Complete Partitioning (CP) in which the buffer space is permanently partitioned into M separate regions; no sharing at all is allowed in CP. At the other extreme is the second scheme, Complete Sharing (CS) in which all of the storage space is available to any incoming message, independent of which M/M/1 system it belongs to. Neither of these is satisfactory for all loading conditions. In order to provide some, but not excessive, sharing, a third scheme is introduced known as Sharing with Maximum Queue Length (SMXQ). In this scheme, a maximum is placed on how many buffers may be utilized by each M/M/1 system. SMXQ may not provide for full storage utilization under heavy traffic, and so a fourth scheme, known as Sharing with Minimum Allocation (SMA) is also included. This last provides a guaranteed minimum number of buffers for each of the traffic streams in addition to a common shared pool of buffers. However, the shared pool in SMA may be monopolized by a single stream unfairly, and so the fifth scheme considered is a combination of the last two, namely, Sharing with a Maximum Queue and Minimum Allocation (SMQMA).

The sharing of the finite storage capacity introduces a dependency among the M queueing systems. However, the entire system is a birth-death queueing process [59] whose state vector is $\mathbf{n} = (n_1, n_2, \ldots, n_M)$ where n_m is an integer random variable equal to the number of customers in the node from the m-th queueing system. The equilibrium probability associated with this vector is given by the well-known product form solution for closed queueing networks [4] as follows:

$$P(\mathbf{n}) = P(n_1, n_2, \ldots, n_M) = \begin{cases} C_x \rho_1^{n_1} \rho_2^{n_2} \ldots \rho_M^{n_M}, & \text{for } \mathbf{n} \in F_x \\ 0, & \text{otherwise} \end{cases} \tag{4.11}$$

where $\rho_m = \lambda_m / \mu C_m$. The subscript x refers to one of the five particular buffer allocation schemes described above, and F_x represents the set of allowed states for scheme x. The quantity C_x is simply the probability of an empty system for scheme x and is given by

$$C_x^{-1} = \sum_{\mathbf{n} \in F_x} \rho_1^{n_1} \rho_2^{n_2} \ldots \rho_M^{n_M}. \tag{4.12}$$

Having obtained the joint probability distribution for each scheme, one can then obtain the probability of blocking, the throughput and the average delay for the messages which successfully pass through this shared node.

In [52], the five schemes are compared relative to blocking, throughput and mean response time. It is shown that for most loading situations, some restriction on the shared space is desirable.

LAM and WONG [80] used closed queueing networks to model finite buffer behavior as well. They also used these tools to model window-controlled virtual channels in computer networks.

5. MULTI-ACCESS COMMUNICATIONS

Whenever we have a resource (a communications channel, a computer, a storage facility, a server of any sort) which may be requested by more than one user at a time, then we refer to this as a multi-access situation; clearly, one then requires a method for resolving this conflict, as discussed in Section 2.

In Section 4, we discussed computer communication systems in which the connectivity was provided by point-to-point links. However, there are a number of communication systems which provide a broadcast link in which all users hear every transmission. Examples of these systems are satellite broadcast systems, packet radio systems, and local area networks. Some authors reserve the term *multi-access* for such broadcast links, and we adopt that distinction in this section. The rules used for resolving conflicts in this environment are referred to as 'multi-access protocols'.

The algorithms in this section (Section 5) represent practical implementations in this environment and each must pay its price to nature (as described in Section 2). (See also [65]).

5.1. Definitions and model assumptions

Here we summarize the usual definitions and assumptions for models of multi-access systems. A number of these are relaxed or altered in specialized papers.

1. The time to transmit a fixed length packet (of say, b bits) is referred to as a 'slot'. This duration is equal to b/C seconds.
2. A 'minislot' is the time required for all users in the broadcast environment to hear any user's transmission; that is, it is the system propagation delay, L seconds.
3. The key parameter 'a' is the ratio of a minislot duration to a slot duration. See (2.3).
4. New arrivals to the total system come from a Poisson stream in the continuous time case. In discrete time we allow a more general arrival process. In both cases, the arrival rate is λ messages per second, or, in its normalized form, S packets per slot.
5. Packets which are not successful must be retransmitted. The retransmissions plus the new transmissions form a Poisson process at a rate G packets per slot. This is known as the 'bold' Poisson assumption and is a gross approximation to the actual state of affairs.
6. A user (terminal) with a packet which must be retransmitted is said to be 'backlogged'.
7. The channel is a broadcast channel, i.e., all users are in range and line-of-sight of all other users (this is said to be a 'single-hop' environment).
8. If more than one user transmits at the same time (in a single-hop environment), a collision occurs. In this case, no transmissions are successful, and the identity of the transmitting users is not available to any users that hear the collision. If exactly one user transmits, it is assumed to be perfectly received by all.
9. When the user population is of infinite size we assume that the users are unbuffered. In this case, at most one packet may be held for transmission by a user.

The notation we use is as follows.
M = number of users
C = channel capacity (bits/sec)
b = (fixed) packet length (bits)

b/C = slot duration = packet transmission time (sec)
$\overline{x}$ = E[message transmission time using the full capacity C] (sec)
β = E[message transmission time] (slots) = $\overline{x}C/b$ (slots)
$\overline{x^2}$ = second moment of message transmission time (sec^2)
λ = arrival rate to all M users (λ/M to each user) (messages/sec)
$\rho = \lambda\overline{x}$ = utilization factor
$S = \rho$ = throughput (packets/slot)
σ^2 = variance of number of messages (bulk) arriving in a slot
L = propagation delay (sec)
a = LC/b (slots)
T = E[message response time] (sec)
W = E[message waiting time] $= T - \overline{x}$ (sec)

5.2. Taxonomy

One useful way to distinguish the many multi-access protocols is to divide them into the three groups described in Section 2, namely, static, dynamic, and random. A static assignment algorithm assigns portions of the resource to the users in a predetermined (static) fashion. A dynamic algorithm makes the allocation 'on the fly' according to the current demand of the users. A random algorithm allows users to simultaneously access the resource (e.g., the channel) to a greater or lesser extent (depending on the algorithm); whenever more than one user attempts to transmit in the same bandwidth at the same time in the same range, then a 'collision' occurs, and none of these attempts are successful. Below, we discuss some examples from each of these methods, focusing on the queueing aspects of the problems.

Static assignment schemes perform very poorly at light load (e.g., at $S = 0$, FDMA has a mean response time equal to $M\overline{x}$) but do perform well at high load (indeed, FDMA can achieve a throughput approaching $S = 1$, the maximum throughput for our channel). On the other hand, random access schemes (such as ALOHA) perform beautifully at light load, but very rapidly degrade as the load increases (e.g., ALOHA only supports a throughput of $S < 1/2e$).

We now examine a number of these schemes broken down into the three groups discussed. Our major objective is to define the algorithm, provide some references, and (when possible) give the mean waiting time W.

5.3. Static assignment algorithms

In these schemes, the input traffic is partitioned into substreams and each substream is assigned a portion of the communication resource on a permanent basis. This assignment may be done in a number of ways, the two most well-known being

FDMA - Frequency Division Multiple Access
TDMA - Time Division Multiple Access

In FDMA, the communication bandwidth is split into, say, M subchannels, each of which is assigned to one of M substreams into which the input traffic has been partitioned. The m-th substream (say with an arrival rate of λ_m messages per second, each with message lengths drawn from some general distribution with a mean message length of $1/\mu$ bits) can only be served by the m-th subchannel (say of capacity C_m bits per second). The load on the m-th queue is equal to

$$\rho_m = \frac{\lambda_m}{\mu C_m} = \lambda_m \bar{x}_m. \qquad (5.1)$$

For example, if each input traffic stream is Poisson at the rate $\lambda_m = \lambda/M$, if the messages are all of a fixed size ($1/\mu$ bits), and $C_m = C/M$, then each subsystem is an M/D/1 system where the average time to transmit (service) a message is $M/\mu C$ seconds. Note from the definition of W in Section 5.1 that it equals $T - \bar{x}$ where $\bar{x} = 1/\mu C$, the time to service the message when the full capacity, C, of the channel is used. Thus, using [59], we have:

$$W = T - 1/\mu C = \frac{(2-\rho)M/\mu C}{2(1-\rho)} - 1/\mu C,$$

and so

$$W = \left[\frac{(M-2)(1-\rho)+M}{2(1-\rho)}\right]\bar{x}, \qquad (5.2)$$

where $\rho = \lambda/\mu C$ and $\bar{x} = 1/\mu C$.

In TDMA, the time axis is partitioned into an infinite sequence of adjacent intervals. M subsequences are defined, with the m-th such subsequence preassigned for the exclusive use of the m-th substream. Whenever an interval is assigned to the m-th substream, that implies that the entire channel bandwidth is to be used by that substream during that interval. Usually, the intervals are each of a fixed length (i.e., a slot), and the subsequences are derived in a 'round-robin' fashion; that is, every m-th slot is assigned to the m-th substream. This is usually referred to as 'synchronous' TDMA. If the resultant capacity assigned to the m-th substream is, again, C_m, then (5.1) applies to this substream. Again, if the input traffic is Poisson ($\lambda_m = \lambda/M$), the message transmission time is constant (and equal to a slot duration), and $C_m = C/M$, then we see that an arrival must wait, on average, M/2 (slots) until his substream's first assigned slot becomes available. If he finds, on average, $\bar{N}$ messages from his substream upon his arrival, then he must wait an additional $\bar{N}M$ slots, on average, before he finally enters service; this is because each round-robin cycle takes exactly M slots. Thus his mean wait W is

$$W = (\frac{M}{2} + \bar{N}M)\,\bar{x},$$

where $\bar{x}$ = slot duration. But, by Little's result, $\bar{N} = (\lambda/M)W$. Thus we find:

$$W = \frac{M\bar{x}}{2(1-\rho)}, \qquad (5.3)$$

where $\rho = \lambda\bar{x}$. See (5.12) for a generalization of this result.

In both cases, more general inputs produce a G/G/1 queue for the subsystem. As stated in Section 2, the penalty for static assignment is that some of the system capacity may lie idle when the *m*-th subsystem goes empty while another subsystem may have a queue of waiting messages.

5.4. Dynamic assignment algorithms

5.4.1. Asynchronous TDMA. This is a TDMA scheme in which the time axis is not slotted and is not preassigned, but rather the entire input stream forms a single queue which is served by the channel in some order (typically first-come-first-serve). Nowadays this is referred to as a statistical multiplexer. It is nothing more than a single-server queueing system and represents a perfect resource sharing device.

Of course, we have assumed that all the users can form a single queue with no help from the system, which is an unrealistic assumption in most multi-access environments. In that sense, then, it represents an ideal model for multi-access and gives the minimum wait. In the case of M/D/1, we have the following [59]

$$W = \frac{\rho\,\bar{x}}{2\,(1-\rho)}. \tag{5.4}$$

5.4.2. Polling, Mini-Slotted Alternating Priority (MSAP) and token passing. Polling is a well-known technique in telephony and many of these polling methods have been adopted in data communications. The classical method, *roll-call polling,* is such that a central station coordinates traffic from many users by broadcasting unique user addresses one at a time; when a user hears his address, he may transmit any data he has waiting [44,112,119]. A second technique, *hub-go-ahead polling,* reduces the polling overhead by allowing control to transfer between users directly [64,119]. This last technique was the predecessor to the token ring. Indeed, the major application of the mathematical analysis of polling systems has been to local area network performance.

An extensive literature has been developed on polling systems in the last few years. An excellent survey of many of the results from these studies is given in [119] and we refer the reader to this work for a complete list of references. Some of the early basic work on applying these techniques to analyzing computer-communication systems was done by KAYE [53] and also by KONHEIM and MEISTER [73] in which they exploited the theory of random walks.

Let us now present some of the known results for polling systems, after which we apply these results to certain computer-communication systems.

The typical polling model is that of a 'patrolling repairman' in which a server serves a finite number, *M*, of users by 'walking' from user to user looking for work to be done (messages to transmit). Usually, it is assumed that he patrols the users in a cyclic fashion (this is the assumption we use for most of this paper). The users may be modeled as unbuffered terminals or buffered terminals. The walking time between users may be fixed or random, with possibly different parameters for user pairs; the walking time in the model corresponds to the polling overhead in computer-communication systems. The amount of work that the server accepts upon finding a busy user may be: *exhaustive* (the server serves a given user until that user's work is emptied); *gated* (the server serves all that work he finds upon arriving to the user, but accepts none of the work that arrives while he is serving that user - this work will be done on his next visit); or *limited* (the server will serve up to a fixed maximum amount of work (or number of customer arrivals at the user) each time he visits a user). New arrivals to a user arrive at a rate λ/M messages per second, the rate summed over all M users being λ messages per second.

Of interest to the analysis is the *polling cycle* (the time from when a given user is polled until that user is polled again) and the *message response time* (the interval from when a message arrives until it is delivered).

Below, we list some of the known results for users that have infinite buffers and which are statistically identical (i.e., all users have the same distributions for the walking time between users, for the arrival process and for the message service time). The additional notation is as follows:

$r = E$[walking time between two terminals] (sec)
$R = Mr = E$[walking (polling) time in a polling cycle] (sec)
$D = E$[polling cycle] (sec)
$\delta^2 =$ variance of polling time between two terminals (sec^2)
$T_m = E$[time spent serving user m in a polling cycle] (sec)

We begin by deriving an expression for the mean polling cycle duration of infinite buffer polling systems with an exhaustive service policy. For more generality in this derivation, we temporarily let the walking time between terminals m and $(m + 1)$ modulo M be equal to r_m and we let λ_m be the arrival rate to terminal m. Since the cycle time is the sum of the polling (walking) time between users and the time spent transmitting messages at each stop in the cycle, we have

$$D = \sum_{m=1}^{M} r_m + \sum_{m=1}^{M} T_m.$$

But T_m is simply the time needed to transmit all messages that arrived in the previous polling cycle (whose mean duration is D). In D seconds, we will have, on average, $\lambda_m D$ arrivals, each of which will take an average of $\bar{x}$ seconds to transmit. Thus,

$$T_m = \lambda_m D \bar{x}.$$

Defining the utilization factor for the m-th user as $\rho_m = \lambda_m \bar{x}$, we have

$$D = \frac{\sum_{m=1}^{M} r_m}{1 - \sum_{m=1}^{M} \rho_m}.$$

In the identical user case, $r_m = r$, $R = rM$, $\lambda_m = \lambda / M$ and $\rho_m = \rho / M = \lambda \bar{x} / M$ and so

$$D = \frac{R}{1 - \rho}. \tag{5.5}$$

This result also applies to the gated service policy [74].

A useful conservation law was published in [56] which expressed a linear constraint on the set of mean waiting times (W_m) for M (work-conserving) priority groups in an M/G/1 environment ($m = 1,2, \ldots ,M$). That law states that

$$\sum_{m=1}^{M} \frac{\rho_m}{\rho} W_m = \frac{\sum_{m=1}^{M} \lambda_m \overline{x^2}/2}{1 - \rho} \tag{5.6}$$

where the subscripts refer to the individual priority groups in the obvious way. For the three service policies we consider in this section (and more generally for mixtures of them), there have been developed pseudo-conservation laws as reported in [12,13]; these laws resemble that in (5.6).

Next, we consider a discrete time system (with the time unit equal to one slot) using *exhaustive* polling in an environment with an arbitrary arrival process (with mean $\lambda b / C$, and variance σ^2, for the number of messages (i.e., the bulk size) arriving in a slot), an arbitrary discrete message service time distribution (with a mean $\bar{x}$ seconds and second moment $\overline{x^2}$), an arbitrary discrete walking time distribution (with mean r seconds and variance δ^2) and users with an infinite buffering capacity. The mean waiting time is given by [119]

$$W = \frac{\delta^2}{2r} + \frac{\lambda \overline{x^2} + R(1 - \frac{\rho}{M}) + (\frac{M \sigma^2 \beta}{\lambda} - 1)\, \bar{x}}{2(1 - \rho)} - \frac{b}{2C}. \tag{5.7}$$

We comment that the study of the gambler's ruin problem is the key to analyzing this system [73].

For the continuous time version of the above discrete time problem, but with (non-bulk) Poisson message arrivals at rate λ / M per second to each user, we have [119]

$$W = \frac{\delta^2}{2r} + \frac{\lambda \overline{x^2} + R(1 - \frac{\rho}{M})}{2(1 - \rho)}. \tag{5.8}$$

We now consider the same discrete time system discussed above, but here we adopt the *gated* service policy (as opposed to the exhaustive policy). The mean message waiting time is given by [119]

$$W = \frac{\delta^2}{2r} + \frac{\lambda \overline{x^2} + R(1 + \frac{\rho}{M}) + (\frac{M \sigma^2 \beta}{\lambda} - 1)\, \bar{x}}{2(1 - \rho)} - \frac{b}{2C}. \tag{5.9}$$

Note that the mean wait for the gated system exceeds that of the exhaustive system by the amount $r\rho/(1-\rho)$.

For the continuous time version of this gated policy, we use the same assumptions as in the continuous time exhaustive problem discussed above and obtain [119]

$$W = \frac{\delta^2}{2r} + \frac{\lambda\overline{x^2} + R(1+\frac{\rho}{M})}{2(1-\rho)}. \tag{5.10}$$

As in the discrete time case, we note that the mean wait for the gated system exceeds that of the exhaustive system by the amount $r\rho/(1-\rho)$.

Let us now consider a discrete time system as above, but with a *limited* service policy. In particular, each time the m-th user is polled, at most one (the limit) message will be transmitted. We find the mean wait to be [119]

$$W = \frac{\delta^2}{2r} + \frac{\lambda\overline{x^2} + R(1+\frac{\rho}{M}) + (\frac{b}{C}+R)(\frac{M\sigma^2\beta}{\lambda}-1) + \lambda\delta^2}{2(1-\rho-r\lambda)} - \frac{b}{2C}. \tag{5.11}$$

The continuous time version of this limited service policy, with (non-bulk) Poisson arrivals, gives the following mean wait [119]:

$$W = \frac{\delta^2}{2r} + \frac{\lambda\overline{x^2} + R(1+\frac{\rho}{M}) + \lambda\delta^2}{2(1-\rho-r\lambda)}. \tag{5.12}$$

Let us now compare the mean wait for the three discrete time systems considered above. The exhaustive policy is given in (5.7), the gated policy is given in (5.9), and the limited policy is given in (5.11). We observe that if $\sigma^2 \geqslant \lambda/M\beta$ (e.g. the case $\sigma^2 = \lambda/M\beta$ corresponds to Poisson arrivals), then

$$W_{\text{exhaustive}} \leqslant W_{\text{gated}} \leqslant W_{\text{limited}}. \tag{5.13}$$

Moreover, if we compare the mean wait for the three continuous time policies in the non-bulk cases, we compare (5.8), (5.10) and (5.12) and find that the inequalities given in (5.13) hold here as well.

For the three continuous cases, we observe that in the limit as the walking time shrinks to zero ($r\to 0$) and $M\to\infty$, such that R and ρ remain constant, we have

$$W = \frac{\lambda\overline{x^2} + R}{2(1-\rho)}. \tag{5.14}$$

Note that this is simply the classic M/G/1 mean wait plus half the mean polling cycle duration.

The polling methods considered above all adhered to a *cyclic* polling order. In [72], a set of *random* polling systems is considered. The assumptions are the same as above (namely, M terminals with infinite buffers, discrete time, constant service time per packet equal to the slot size, general bulk arrival process, arbitrarily distributed walking time, and three service policies - exhaustive, gated and limited). However, the messages are all assumed to be exactly one packet long (and therefore $\bar{x} = b/C$ seconds, $\overline{x^2} = \bar{x}^2 = (b/C)^2$ and $\beta=1$ slot). The essentially new component is the polling order. We now introduce $P_m = P$[terminal m is polled next]. This is clearly a memoryless random polling policy. In the symmetrical case, $P_m = p$ and the following results can be obtained [72]. First, we find that the mean duration of the polling cycle, D, for the exhaustive and gated service policies once again is given as in (5.5). Next, we give expressions for the mean wait for the three service policies; we write these equations to allow an easy term-by-term comparison with the cyclic case. For the exhaustive case, we have

$$W = \frac{\delta^2}{2r} + \frac{\lambda(\bar{x})^2 + R(1-\frac{\rho}{M}) + (\frac{M\sigma^2}{\lambda} - 1)\bar{x} + (R-r)}{2\,(1-\rho)} - \frac{b}{2C}. \tag{5.15}$$

For the gated case, we have

$$W = \frac{\delta^2}{2r} + \frac{\lambda(\bar{x})^2 + R(1+\frac{\rho}{M}) + (\frac{M\sigma^2}{\lambda} - 1)\bar{x} + (R-r)}{2\,(1-\rho)} - \frac{b}{2C}. \tag{5.16}$$

Lastly, for the limited case, we have

$$W = \frac{\delta^2}{2r} + \frac{\lambda(\bar{x})^2 + R(1+\frac{\rho}{M}) + (\frac{b}{C} + R)(\frac{M\sigma^2}{\lambda} - 1) + \lambda\delta^2 + (R-r)}{2(1-\rho-r\lambda)} - \frac{b}{2C}. \tag{5.17}$$

Note that the result comparing these three expressions with those for the cyclic case (see (5.13)), also holds for this random case. Moreover, the differences among the three mean waits in the cyclic case and in the random case are the same.

As mentioned earlier, these polling system results have been applied to a number of computer-communication problems. In the remainder of this section, we discuss a few such applications.

In [125] roll-call polling is used to model discrete-time message flow from a set of M packet radio terminals to a central receiving station. The central station sends a fixed length poll to each terminal inquiring about traffic. The (one-way) propagation delay is L seconds. The constant walking (polling) time is $r = (2L + T_p)$ where T_p is the time to transmit the polling request; clearly, also, $\delta^2 = 0$. The message packets have a constant length of b bits (and take $\bar{x} = b/C$ seconds to transmit over the channel of C bits/sec) and arrive according to a Poisson process at a rate of λ/M messages per second at each terminal. As in (2.3), $a = LC/b$. The system throughput is simply $S = \lambda\bar{x} = \rho$. For this application, the mean wait is given by (5.7) which yields

$$W = \frac{S\bar{x} + (1-\frac{S}{M})R}{2(1-S)} + \frac{L}{2}. \tag{5.18}$$

If we convert this application to continuous time, then (5.8) gives

$$W = \frac{S\bar{x} + (1 - \frac{S}{M})R}{2(1-S)}. \tag{5.19}$$

Hub-go-ahead polling has been adapted in [69] to create the Mini Slotted Alternating Priority access scheme (MSAP) which was the predecessor of the token ring access method. (The Broadcast Recognizing Access Method (BRAM) [20] is similar to MSAP.) It is implemented by preassigning a cyclic polling sequence to the terminals and synchronizing their clocks. When the m-th terminal finishes transmitting its messages, all other terminals will hear a silent period of L seconds (a minislot). At that point, terminal $(m + 1)$ modulo M will recognize that it is his turn to transmit. If he has data, he transmits it and the process of 'token-passing' continues. If he has no data to send, then he remains silent and after another minislot of silence, terminal $(m + 2)$ modulo M receives his turn, etc. In this case, $r = L$ and again, $\delta^2 = 0$. We then find from (5.7) that the normalized mean wait is

$$W = \frac{S\bar{x} + (1 - \frac{S}{M})ML + (1-S)L}{2(1-S)}. \tag{5.20}$$

The token ring has been treated by Bux [14] by applying (5.8). In the token ring, as in MSAP, a control token is circulated around a ring. Upon receiving a free token, a user has the right to transmit his message (using the exhaustive service policy); if he has any messages, he marks the token as busy and transmits his data on the ring local area network, addressing this data to his intended receiver. When the data returns to him after circulating around the ring, this user removes the data he transmitted and then marks the token as free. Applying our result, we have

$$W = \frac{\lambda\overline{x^2} + (1 - \frac{S}{M})L}{2(1 - S)}. \tag{5.21}$$

In the case of a gated service policy for the token ring, we may use (5.10) to obtain

$$W = \frac{\lambda\overline{x^2} + (1 + \frac{S}{M})L}{2(1 - S)}. \tag{5.22}$$

In the case of a limited service policy, we have from (5.12)

$$W = \frac{\lambda\overline{x^2} + (1 + \frac{S}{M})L}{2(1 - S - r\lambda)}. \tag{5.23}$$

These last two results were also given in [119] and [26]. In [26], De Moraes and Rubin also give the mean wait for the token bus, in which a token passes along a bidirectional local area network bus among terminals which may not necessarily be adjacent on that bus. These expressions differ from those above only by the propagation delay which now increases to the entire length of the bus rather than the inter-terminal propagation delay used in the ring.

We close this section by listing a few of the many other applications of these polling results to computer-communication problems. Polling with finite storage capacity is treated in [133]. Multiple servers are treated in [96] and [104]. The Fasnet system is evaluated in [45] and also in [130] which also evaluates Expressnet. Automatic repeat request (ARQ) retransmission systems may be found in [75]. Newhall loops (one of the first loops proposed) are discussed in [18] and [19]. The register-insertion ring is analyzed in [46]. Nonsymmetric token rings are treated in [29].

5.4.3. Reservation protocols. Polling schemes (and Tree Access Schemes - see Section 5.5.3) are basically passive schemes in which users wait to be asked if they have any data to send. Reservation schemes allow the user to actively request access to the channel. The idea is that a portion of the channel time or bandwidth is set aside as a reservation subchannel. Users place reservations on the reservation subchannel for space on the data subchannel. Once the reservation is accepted, then the user is guaranteed space on the data subchannel, and everyone in the system knows that he has that space reserved.

There are a large number of different reservation schemes described in the literature. Some of them appear as hybrid schemes which incorporate certain features of static and random assignment algorithms. In this section, we briefly describe a few of these.

One of the first reservation schemes described was that of ROBERTS [107]. Here, the channel is slotted into data slots, with the occasional introduction of a reservation slot. A reservation slot is subdivided into a number of smaller slots (which are used for reserving data slots), and these slots are accessed by the users via slotted ALOHA (see Section 5.5.1). When a reservation makes it past the ALOHA contention, all users hear the request and schedule that user for the number of data slots he requested using the first available unassigned data slots; thus a 'queue in the sky' is maintained and does not require a central controller to do the scheduling.

The Split channel Reservation Multiple Access scheme (SRMA) divides the channel by frequency division into the data and reservation channels [125]. It operates in one of two modes, depending on whether the reservation channel is used as a single channel (which carries both requests for space by the user and replies to these requests which give the user permission to begin transmitting) or as two control channels (one for requests and one for replies which schedule the request in the future). Control signals may compete for space on the request channel by any access scheme (e.g., some random access scheme).

Global Scheduling Multiple Access (GSMA) creates a reservation channel (for communicating to a central controller) and a data channel, both with no collisions using TDMA [83]. This is accomplished by using two subframes in each major frame. Each subframe consists of a number of slots. The control subframe has one slot assigned to each user for making his reservations. The central controller assigns slots in the data subframe according to the requests.

In a satellite environment, Reservation-ALOHA [24] uses a frame consisting of a fixed number of equal sized slots; the frame is longer than the propagation delay. A user who successfully transmitted in a given slot in the previous frame has the exclusive use of that slot in the next frame. If a slot was either idle or contained a collision in the previous frame, then it is 'up-for-grabs' by any user; such a slot is accessed using ALOHA.

A similar scheme, the Round-Robin Reservation Scheme [7], uses the same frame structure, but guarantees that the number of slots in a frame exceeds the number of users. One slot in each frame is permanently assigned to each user. In addition to sending his data packets, a user appends to each packet a count of how many more packets he has queued. When a slot contains a count of zero, then it is assigned by some collision-free algorithm (in this case, by round robin) to another user. When the original 'owner' of a slot wishes to transmit and another user has been assigned his slot, then the original owner transmits, causing a collision, thereby announcing that he is taking his slot back.

In [10] the Split Reservation Upon Collision (SRUC) access scheme is described. It uses fixed length slots, each of which is partitioned into a data subslot and a control subslot. Ordinarily, the data subslots are in the contention mode, in which case, any user may access them. Whenever a collision occurs, however, all data subslots switch to the reserved mode which means that they will all be used to resolve the conflict only among those users who created the collision; no new users may enter the fray until the conflict is fully resolved. The control subchannel is used by a subset of the conflicting users to assist in the conflict resolution. The Tree access schemes of Section 5.5.3 operate in a similar fashion.

The Priority-Oriented Demand Assignment access method (PODA) uses implicit as well as explicit reservations, priorities, and centralized as well as distributed control [50]. The channel is divided into two subframes - a data subframe and a control subframe. The data subframe contains data and piggybacked control information. The control subframe is used to hasten the delivery of urgent control information. Control frame access can be any of a number of suitable types. The fraction of control versus data subframes can be dynamically adapted to loading conditions.

Group Random Access (GRA) uses a TDMA access method, where groups of users (rather than individual users) share fixed TDMA slots using (possibly) different access schemes for each group [109,110].

5.5. Random assignment algorithms

In these schemes, more than one user may attempt to use the channel at the same time, resulting in a collision; in this case no successful transmission occurs. On the other hand, one of the users may be lucky and find no one else using the channel, in which case he will be successful. Such schemes are discussed in this section.

5.5.1. ALOHA. ALOHA access schemes are, in some sense, the most random. Essentially no constraints are placed on the users; they are allowed to transmit any time they wish. Packets are assumed to be of fixed length (exactly one slot long). The ALOHA access algorithm was introduced by ABRAMSON [1] in an unslotted environment (i.e., users were completely unsynchronized). The relationship between the input S and the channel traffic G (recall we are using the bold Poisson assumption) is

$$S = Ge^{-2G}. \tag{5.24}$$

This leads to a maximum throughput of $S = 1/2e = 0.184...$ (packets per slot) which occurs at $G = 0.5$.

An improvement was recommended by ROBERTS [106] by going to a slotted system in which a minimal constraint was placed on users whereby they could begin transmission only at a slot boundary. That is, time was discretized into intervals exactly one slot in duration. With this modification, the S, G relationship becomes

$$S = Ge^{-G}. \tag{5.25}$$

The maximum throughput is now equal to $S = 1/e = 0.368...$ and this occurs when $G = 1$. At $G=1$, the expected number of transmissions is exactly one (satisfying one's intuition regarding optimality). We saw this optimality point occur in our discussion of power in Section 3.2; it occurs in many other situations as well (see, for example, [138]). We note further that the expected number of transmissions required by a packet is simply G/S; the expected number of unsuccessful transmissions is $(G/S) - 1$. In 1973, ABRAMSON [2] analyzed the throughput for slotted ALOHA systems with a *finite* number of users. He found, among other things, that the maximum throughput profile occurred when $G = 1$, exactly as with the infinite population model.

In 1973, KLEINROCK and LAM [57,76] established the equations for the mean response time of the slotted ALOHA system. (It was no surprise to find that the behavior at light load is excellent, but that it badly degrades as the load increases toward $S = 1/e$).

This analysis for an infinite population slotted ALOHA model used the bold Poisson assumption and assumed that retransmissions were uniformly selected over the next K slots once a user learned that his packet was unsuccessful; among the results obtained there was the following for the mean wait:

$$W = \frac{1-q}{q_t}\left[L + \frac{b}{C}\left(\frac{K+1}{2}\right)\right] + L, \tag{5.26}$$

where

$$q = \left[e^{\frac{-G}{K}} + \frac{G}{K}e^{-G}\right]^K e^{-S}, \tag{5.27}$$

and

$$q_t = \left[\frac{e^{\frac{-G}{K}} - e^{-G}}{1 - e^{-G}} \right] \left[e^{\frac{-G}{K}} + \frac{G}{K} e^{-G} \right]^{K-1} e^{-S}, \tag{5.28}$$

and

$$S = G \frac{q_t}{1 - q + q_t}. \tag{5.29}$$

The detailed behavior of slotted ALOHA with a finite population of M unbuffered terminals may be described by a discrete-time Markov chain given by the number of backlogged terminals at the slot boundaries [60,76]. Only when a terminal's buffer is empty may a new packet be generated (by the terminal's external source) and this will occur with probability σ in a slot. The combined input rate into all terminals at time (slot) t is denoted by $S(t)$. For this Markov chain analysis, we define p to be the probability that a backlogged packet retransmits in the next slot (of duration $\bar{x}$ seconds) and often is taken as a given parameter of the problem. Thus, the retransmission delay is geometrically distributed with a mean of $\bar{x}/p$ seconds. We let $N(t)$ be a random variable representing the total number of backlogged terminals at slot t (referred to as the *channel backlog* at t). The vector $[N(t), S(t)]$ is the channel load at time t. For $N(t) = n$ we have

$$[\, N(t), S(t) \,] = [\, n, (M - n)\,\sigma \,], \tag{5.30}$$

giving rise to what we call the *linear feedback model.* Let

$$p_{ij} = P[N(t+1) = j | N(t) = i].$$

Then, these one-step transition probabilities are simply [60,76]

$$p_{ij} = \begin{cases} 0, & j \leq i-2, \\ ip(1-p)^{i-1}(1-\sigma)^{M-i}, & j = i-1, \\ [1-ip(1-p)^{i-1}](1-\sigma)^{M-i} + \\ (M-i)\sigma(1-\sigma)^{M-i-1}(1-p)^i, & j = i, \\ (M-i)\sigma(1-\sigma)^{M-i-1}[1-(1-p)^i], & j = i+1, \\ \binom{M-i}{j-i} \sigma^{j-i}(1-\sigma)^{M-j}, & j \geq i+2. \end{cases} \tag{5.31}$$

One then does the usual Markov chain analysis to solve for the equilibrium distribution for $N(t)$, namely,

$$p_n = \lim_{t \to \infty} P[N(t) = n],$$

by solving the system of equations,

$$p_n = \sum_{i=0}^{M} p_i p_{in},$$

and, of course, requiring the p_n's to sum to unity.

Given $N(t) = n$, there is no guarantee that the channel input rate, $S(t)$ and the channel output rate, which we denote by $S_{\text{out}}(t)$, will be the same. In fact, given $N(t) = n$, we may solve for $S_{\text{out}}(t)$ (which we denote by $S_{out}(n)$ when $N(t) = n$) to obtain

$$S_{\text{out}}(n) = (1-p)^n(M-n)\sigma(1-\sigma)^{M-n-1} + np(1-p)^{n-1}(1-\sigma)^{M-n}. \tag{5.32}$$

In the limit when $M \to \infty$ and $\sigma \to 0$ (such that $M\sigma = S$), we have the infinite population model in which new packets are generated for transmission over the channel at the constant Poisson rate S (packets per slot); in this case, (5.32) gives

$$S_{\text{out}}(n) = (1 - p)^n Se^{-S} + np(1 - p)^{n-1}e^{-S}. \tag{5.33}$$

Given that the input and output rates may differ, it was natural [58] to investigate the difference $S(t) - S_{out}(t)$ which has since come to be known as the *drift*; the drift is simply the expected growth of the channel backlog in a slot. If we assume a steady state (more on this later), then, following [128] the expected output rate, $\bar{S}$, is given by

$$\bar{S} = \sum_{n=0}^{M} S_{out}(n)p_n, \tag{5.34}$$

and $\bar{N}_q$ the average number of backlogged terminals may be found from

$$\bar{N}_q = \sum_{n=0}^{M} np_n. \tag{5.35}$$

If we use Little's result, we find the mean wait in this slotted ALOHA system to be

$$W = \frac{\bar{N}_q \bar{x}}{\bar{S}}. \tag{5.36}$$

Since the output and input rate, on average, must be equal in equilibrium, we have

$$\bar{S} = (M - \bar{N}_q)\sigma \tag{5.37}$$

and so

$$W = [(M/\bar{S}) - (1/\sigma)]\,\bar{x}. \tag{5.38}$$

Let us examine the dynamic behavior somewhat further. If we assume the channel backlog is $N(t) = n$, then we may calculate $S_{out}(n)$ as a function of n and $S(n)$. If we further identify the *equilibrium contour* to be the projection of the curve $S_{\text{out}}(n) = S(n)$ onto the $(n, S(n))$ plane (where we denote $S(t)$ by $S(n)$ when $N(t) = n$), then we obtain the curve as shown in Fig. 5.

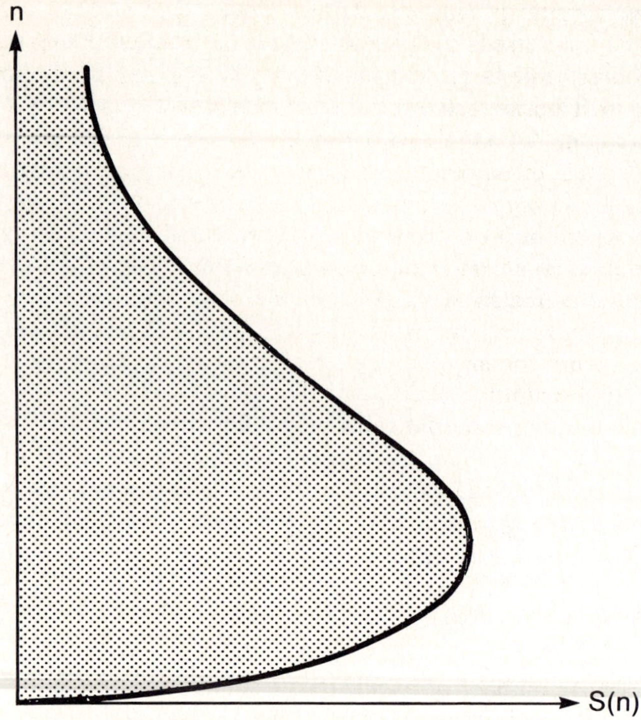

FIGURE 5. *The equilibrium contour*

In this figure, the shaded area corresponds to the region where the drift is negative (i.e., $S_{out}(n) > S(n)$) and $N(t)$ tends to reduce; outside the shaded area, the drift is positive ($S_{out}(n) < S(n)$) and $N(t)$ tends to increase. In Fig. 6, we show four such curves along with the channel *load line*, $n = M - S(n)/\sigma$ (see (5.30)); these curves are obtained by selecting different values for the retransmission probability, p. The system is constrained to lie on the load line. If the system state is in the shaded region, then the backlog decreases down the load line; if not, it rises up the load line. Since we expect the system to lie in the vicinity of the equilibrium contour, and since we require it to lie on the channel load line, then any intersection of the load line with the equilibrium contour is a point to be studied. In Fig. 6a, we observe that there is only one such intersection which we refer to as the *channel operating point.* We also observe that the drift will drive us back to this point if the system takes any excursion along the load line away from this point. Such a point (where the drift pulls us back) is a stable equilibrium point. If the system has only one such stable point, the channel is said to be *stable.* See Fig. 7a for the distribution of backlog for a stable channel. On the other hand, in Fig. 6b, we see a case with three intersections, two of which are stable. The third (middle) point is such that the drift will drive us away from the intersection if we take a small excursion along the load line; such a point is called an unstable equilibrium point. A channel with two stable points and one unstable point is referred to as a *bistable* or an *unstable channel;* such a channel will spend its time largely in the vicinity of the two stable points and will occasionally drift between the two (as can be seen from the backlog distribution shown in Fig. 7b). Of the two stable points in Fig. 6b, the one with large backlog and small throughput is clearly undesirable and is referred to as the *channel saturation point;* the one with large throughput and small backlog is the (channel operating) point at which we would like to operate. In Fig. 6c, we show the case of an infinite population channel; clearly all infinite population channels are unstable unless some dynamic control policy is used (see below). An *overloaded* or *saturated channel* is one with a single stable operating point which occurs for large backlog and small throughput, as shown in Fig. 6d; its backlog distribution may be seen in Fig. 7c.

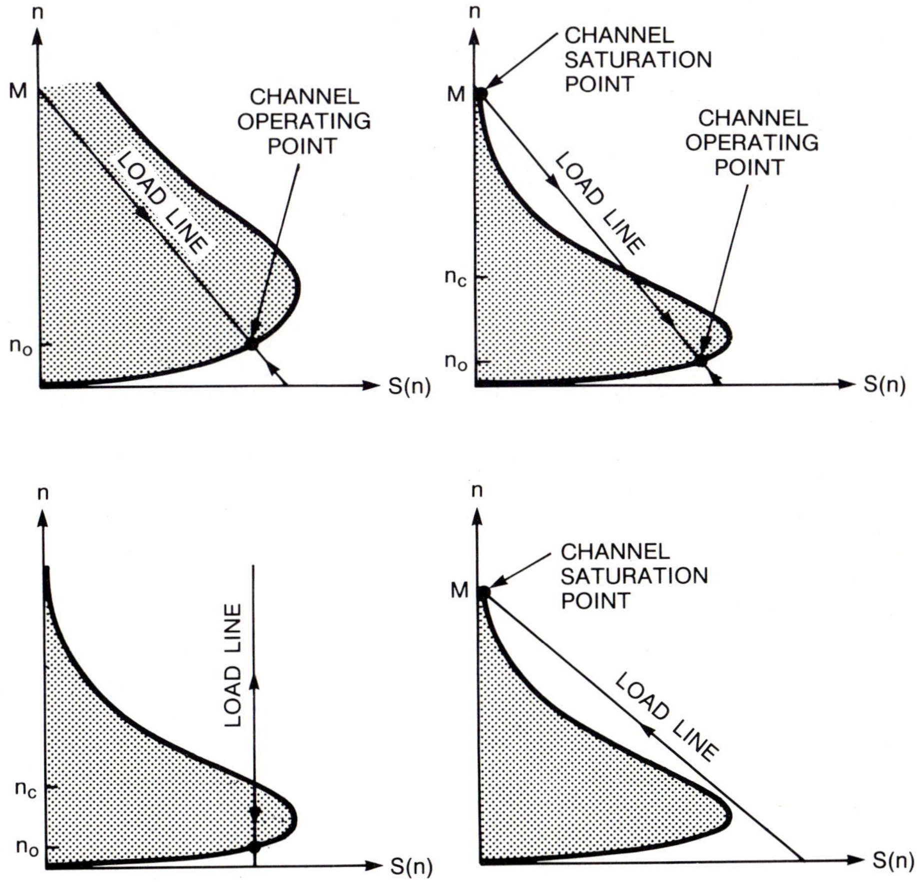

FIGURE 6. *Stable and unstable channels*

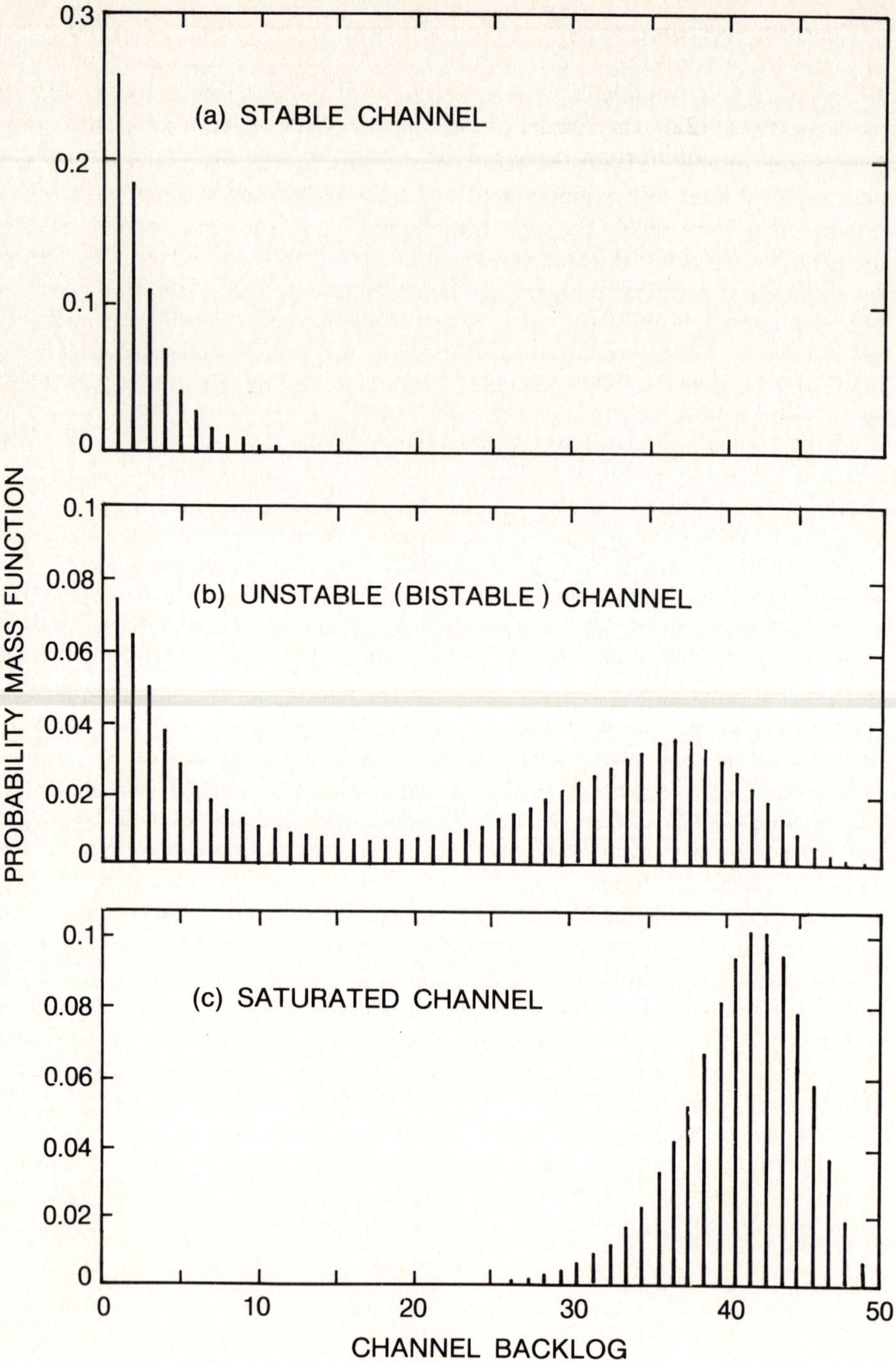

FIGURE 7. *Distribution of backlog*

The ideas regarding stability and equilibrium points described in [60,76] led FUKUDA [32] to develop a method known as *Equilibrium Point Analysis* (EPA); this method was extended by Tasaka who wrote a comprehensive monograph on the subject [122]. The formal idea behind EPA is to assume that the system is always at an equilibrium point (i.e. at a point of zero drift). Furthermore, in the EPA approach, the expectation of any random variable taken with respect to the channel backlog, $N(t)$, is approximated by the value of that random variable at the equilibrium point. In [122], this method is applied to evaluate the throughput and mean delay for a number of multi-access schemes (Slotted ALOHA, Reservation ALOHA, TDMA, SRUC, buffered CSMA/CD - see below, and BRAM).

For an analysis of slotted ALOHA with a finite number of *buffered* terminals, see [111,122].

The fundamental instability of ALOHA systems we have identified above was studied in [27,60,76]. In order to stabilize ALOHA, a number of dynamic control algorithms have been introduced and studied (e.g., [7,41,42,60,64,76,77,78,79,90,105,111]). These stabilization methods typically control either the input rate to the system or the retransmission parameters as a function of the backlog (or an estimate of the backlog). As an example of retransmission control, let us observe the behavior of the otherwise unstable channel shown in Fig. 6b. We see that as we move along the load line (due to statistical fluctuations), the drift is negative as long as $n < n_c$. Suppose we track (perhaps only an estimate of) the backlog and find that $n = n_c$; we are now in danger of entering the (unstable) region of positive drift for $n > n_c$. However, if we now reduce p, the retransmission probability, then the equilibrium contour will change to that shown in Fig. 6a and we will once again find ourselves in a (stable) negative drift situation! Of course, this reduced value of p implies a longer delay between retransmissions which is not a desirable situation. In order to remedy this, we switch back to the original value for p when n falls below n_c giving us negative drift and smaller retransmission delay once again. Thus, by tracking the backlog, n, we dynamically control the channel in a fashion which guarantees stability.

Since ALOHA is fundamentally unstable, the performance of stabilized ALOHA is of interest. The distribution of delay is found numerically in [21] for a stabilized slotted ALOHA system. In [35], an approximate expression for the mean wait of a stabilized slotted ALOHA is given as

$$W = \frac{b}{C}\left[\frac{e - \frac{1}{2}}{1 - eS} - \frac{(e^S - 1)(e - 1)}{S\left[1 - (e^S - 1)(e - 1)\right]}\right] + L. \quad (5.39)$$

5.5.2. The urn model. The URN scheme [67] was devised to behave like ALOHA at light load and TDMA at heavy load; thus, in some sense, it is an adaptive access scheme. Its principle of operation is to divide the population of conflicting users into subgroups each of which is likely to contain only one busy user and then to select one of the subgroups to transmit. In this scheme, we assume that we have a slotted channel supporting M user terminals. At the beginning of each slot, we assume that every user knows N, the exact *number* of users that have packets ready to send; they simply do not know the *identity* of these busy users. Based on this knowledge, what is the optimum way for this information to be used?

If we require all users to behave in a symmetric fashion, i.e., each busy user will transmit into the next slot with an equal probability, say P, then the optimum policy is $P = 1/N$. However, symmetric policies are clearly not optimum as can be seen in Fig. 8. This figure shows the throughput for $M = 2$ and $N = 2$ as a function of the probability of transmission for each of the two users. Indeed, the optimum (asymmetric) policy for this case is for one of the busy users to surely transmit ($P_i = 1$) and for the other to keep quiet ($P_j = 0$) where $i,j = 1,2$, and $i \neq j$. However in a distributed environment it is difficult to select a single busy user when one only knows how many in the total population are busy.

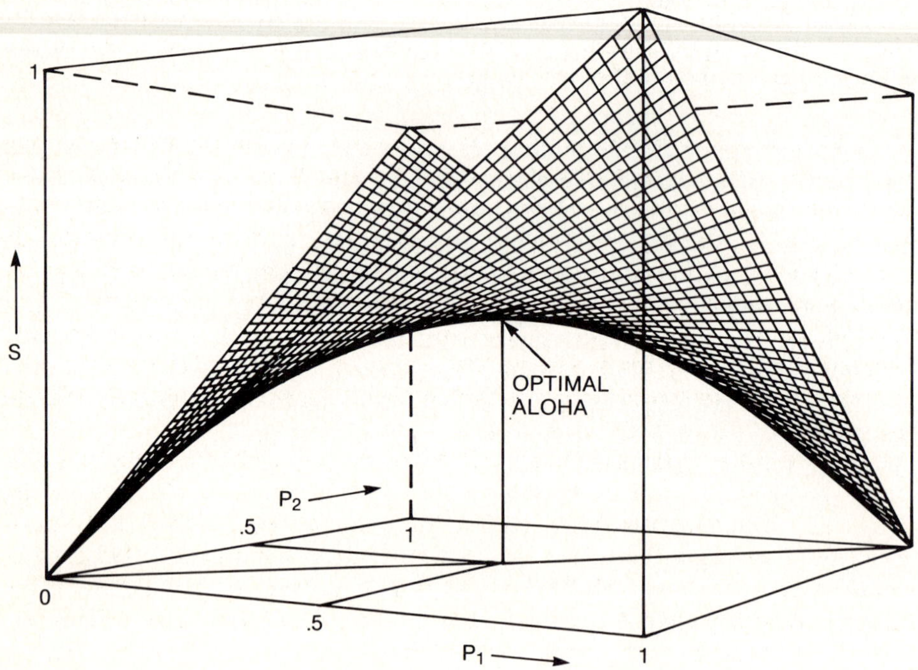

FIGURE 8. *Throughput surface for two busy terminals*

As shown in [67], the optimum policy, then, is to select a critical number of users, k, and to give each of these permission to transmit; the only users who will transmit are those that are both busy and selected. The value of k is chosen to maximize the throughput. This is equivalent to maximizing the probability of selecting exactly one busy user; in this case, it will also be true that the expected number of busy users selected will be equal to unity. The value of k is found from the following expression

$$k = \left\lfloor \frac{M}{N} \right\rfloor, \tag{5.40}$$

where $\lfloor X \rfloor$ denotes the integer part of X. In order to guarantee that all users select the same set of k users, one can require that each user use the same pseudorandom number generator to make the selection; in fact, it is only required that each user decide whether he, himself, is to be selected (again, using the same pseudorandom generator). The throughput is given by

$$S = \left\lfloor \frac{M}{N} \right\rfloor \frac{\text{bin}\left(M - \left\lfloor \frac{M}{N} \right\rfloor, N - 1\right)}{\text{bin}(M,N)}, \tag{5.41}$$

where we have denoted the binomial coefficient as $\text{bin}(a,b) = \dfrac{a!}{b!(a-b)!}$. The algorithm is quite robust, not even requiring that all users have exactly the same estimate of the number of backlogged users.

5.5.3. Tree access schemes. Tree access algorithms (also known as splitting algorithms) represent a variation of the key idea in the URN model. Once again, we are interested in dividing up the potentially conflicting users into smaller groups, thereby reducing the difficulty of resolving the conflict. (Of course, polling and token passing represent a mechanism for reducing the number of competing users down to a single user.) The idea here is to find some property of the competing users which allows them to form subgroups; each subgroup will be resolved separately in some sequential fashion. The process may be applied recursively to further divide each subgroup into subsubgroups, etc.

In 1978, a few months after the URN model was published, Hayes introduced a method known as 'probing' [43]. Independently, and at about the same time, CAPETANAKIS [15,16] and TSYBAKOV and MIKHAILOV [134] introduced the Tree Algorithm which has since had many refinements and improvements. Sometime later, Molle and Kleinrock published the Virtual Time CSMA protocol [93,94]. All of these fall into the class of Tree Access Schemes.

In the probing algorithm, subgroups of users are polled rather than polling individual users. The idea is to attempt to eliminate many idle users simultaneously. Each user is assumed to have a unique identifier (usually a binary address) which is used to form the subgroups. For example, suppose we had a population of 16 users. By means of a centralized poller (or some equivalent distributed poller), the entire population is simultaneously asked if anyone has a message to send. Any users with packets to transmit would respond in the affirmative by transmitting energy on the common channel. If a response is heard on the common channel, then the poller splits the population into two equal sized (sub)groups (say users 1 to 8 and users 9 to 16). The first group is then polled. If no answer returns (indicating that users 1 to 8 are idle), then the second group (9 to 16) is polled. Suppose, in fact, only the second group responds; it is then split further (into 9 to 12 and into 13 to 16). The process continues until a responding subgroup contains only one user, in which case the (single) user transmits his message. Then, the most recent responding subgroup is polled until it, too, is reduced to one responding user, etc. With such a procedure, large groups of idle users are eliminated very easily. It is also possible to adjust the size of the groups polled according to the traffic load on the system. Indeed, the optimum size group to poll is such that the probability of finding that group idle is 0.5 based on information theoretic arguments. Hayes provides an analytic treatment of probing in his text [44]; there he solves for the cycle time behavior and for the mean message delay.

The original Tree algorithm [16,134] operates as follows. When a collision occurs, say in the k-th slot, then all users not involved in that collision suspend their activity until those involved in the collision fully resolve their conflict. The conflicting users randomly split into two subgroups (say using the same pseudorandom coin flips or based on the user's address or based on the arrival time of the user's packet). The first of these subgroups transmits in slot $k + 1$. If this results in an idle or a success, then the second subgroup transmits in slot $k + 2$; otherwise, the second subgroup suspends (until the first subgroup resolves their conflict) and the first subgroup splits again into two further subsubgroups, etc. When all packets from this first subgroup are successfully transmitted, the second subgroup proceeds. When the second subgroup finishes, then all (or possibly, a subset whose size is selected optimally as a function of the traffic) of the originally suspended busy users, plus any new ones which turned busy during the previous conflict resolution period, will transmit and the process repeats. As shown in [16], the maximum throughput (for the case where number of users that come out of suspension is chosen optimally) is 0.43... packets per slot. A Tree access algorithm using *ternary* feedback (terminals can distinguish idle, success and collision slots), immediate transmission by newly generated packets and m-ary, as opposed to binary, splitting using minislots is presented in [99]; assuming $\overline{x} = b/C$ and $L = 0$, they obtain

$$W = \frac{b}{C}\left[\frac{\alpha\, e^{2\alpha}\,(1 - \frac{1+\alpha}{2}\, e^{-\alpha})}{(1 - \alpha\, e^{\alpha})\,(1 - \alpha)}\right] + \frac{b}{C}\left[\frac{e^{\alpha} - \alpha}{1 - \alpha} - \frac{1}{2}\right], \tag{5.42}$$

where $\alpha = \lambda b/C$ (packets per slot) and where we see the maximum throughput is 0.567... packets per slot.

Improvements and variations to the basic Tree algorithm followed quickly after its introduction. For example, it can be seen that if a collision is followed by an idle slot from the first of two subgroups, then the second subgroup will surely generate a collision on the next slot (all the colliding users from the first collision were clearly assigned to the second subgroup); in such a case, the second subgroup should be split in two immediately [84]. This increases the maximum system throughput to 0.46... packets per slot. Another improvement is to recognize that if one collision is followed by a second collision, then the subgroup that did not transmit in the second collision can properly be considered as untested users and placed in the suspended group as if they had never transmitted. This leads to a first-come-first-serve Tree algorithm whose maximum throughput is 0.4871... ; see [34,35,92,102,135] for analyses of this first-come-first-serve algorithm and improvements of it which show that the maximum stable throughput achievable by any tree algorithm lies between 0.4878 and 0.587. An approach to bounding the mean delay for a class of tree algorithms is given in [37]. In [103] a Tree algorithm using a packet admission algorithm based on windows in time is analyzed; there, they solve for the mean response time for a finite non-homogeneous Poisson user population. A study of adaptive polling which allows the polled terminals to respond using a richer alphabet (than binary) is given in [132].

We observe that the Tree algorithms described above take some advantage of the fact that users may be distinguished by their unique time of arrival (i.e., the time when they generate a new packet for transmission). This observation leads us to the Virtual Time CSMA (VTCSMA) protocol [93,94] which takes exquisite advantage of this distinction. In this protocol, which is actually a variation of the CSMA protocols described below in Section 5.5.4, each user keeps two clocks running. The first clock keeps track of real time. The second keeps track of virtual time (which, ideally, should be the same for all users). The virtual time clock stops running whenever the channel is sensed busy; whenever the channel is idle it runs either at a rate $\eta > 1$ sec/sec (if virtual time is less than real time) or at a rate equal to unity (if virtual time and real time are the same). Thus the virtual time clock is never ahead of real time (and is often behind it). Whenever the arrival time of user's packet is passed by the virtual time clock, then that user will transmit his packet. This protocol greatly increases the probability that a transmission will be successful on its first attempt. It is also quite robust, even allowing the individual virtual time clocks to drift relative to each other. The details as to what should be done when a collision occurs can be resolved by any of a number of existing protocols; VTCSMA is designed to assist the first transmission. MOLLE [95] presents a delay model for VTCSMA (and other 'Moving Server' random access protocols); he finds that the mean wait until the *first* transmission is given by

$$W_F = \frac{\lambda \overline{x^2} \left[\dfrac{\eta}{\eta - 1}\right]^2}{1 - \left[\dfrac{\eta}{\eta - 1}\right] S}. \qquad (5.43)$$

To approximate the overall mean wait for VTCSMA, one may use the expression

$$W \approx W_F + (\frac{G}{S} - 1)\, T_R + L\,, \tag{5.44}$$

where we recall that $(G/S - 1)$ is the average number of retransmissions and T_R is the mean delay caused by the retransmission protocol. In [86], a performance analysis of VTCSMA is given which studies the system stability and throughput; in [87], this analysis is extended and also an explicit expression for the mean wait is given.

5.5.4. Carrier Sense Multiple Access - CSMA. When the key parameter a from Section 2 is small ($a \ll 1$), then we may add some further intelligence to the simple ALOHA protocol to greatly improve its performance. The addition is to require that each user listen to ('sense' the carrier of) the channel before attempting any transmission. If the channel is sensed busy (either due to an ongoing successful transmission or a collision), then the user surely will not transmit, but will follow one of several protocols described below. If the channel is sensed idle, then the user will follow a transmission policy which also depends upon the particular protocol. In all cases, we succeed in preventing a sure collision (as would have occurred in ALOHA) when a user begins transmitting while another user can be sensed to already be in the process of transmitting. These systems were first studied (in the environment of packet radio systems) by KLEINROCK and TOBAGI [59,123].

In all of the CSMA protocols, it is possible for two (or more) users to collide if a user senses the channel idle when, in fact, another user has just begun a transmission which has not yet 'reached' the listening user (i.e., the listening user senses the channel at a time that is less than a propagation delay from when the transmitting user began his transmission). Whenever a user detects that his transmission has been ruined by a collision, that user schedules the retransmission (i.e., he will once again sense the channel) at some later time according to a retransmission delay distribution.

The first CSMA protocol is known as *non-persistent CSMA.* In this case, if the channel is sensed idle, the user will transmit his packet immediately. If the channel is sensed busy, the user acts as though there was a collision (see above) and at the rescheduled time, repeats the algorithm.

The next CSMA protocol is known as *1-persistent CSMA.* Here, again, the user will transmit if he senses the channel idle. If he senses the channel busy, then he continues to sense the channel until it goes idle and then, with probability one, he will transmit. We observe that if more than one user senses the channel busy, then, when the channel goes idle, there will be a guaranteed collision. One way to help prevent the latter from occurring is given in the following protocol.

The third protocol is referred to as *p-persistent CSMA.* It is identical to 1-persistent CSMA except that when the channel goes from busy to idle, a waiting user will transmit with probability p. With probability $1-p$, he will reschedule his transmission by L seconds (see Section 2), that is, by one propagation delay (a minislot). If the channel is idle at this next attempt (i.e., after one minislot), then he will transmit with probability p, etc. If it is sensed busy after this minislot, the user will act as if his packet met with a collision and will reschedule his transmission according to the retransmission delay distribution.

The $S:G$ relationships for these three protocols (and variations) are as follows (see [61,62,123]):

Non-persistent CSMA

$$S = \frac{Ge^{-aG}}{G(1+2a)+e^{-aG}}; \tag{5.45}$$

Slotted non-persistent CSMA

$$S = \frac{aGe^{-aG}}{1-e^{-aG}+a}; \tag{5.46}$$

1-persistent CSMA

$$S = \frac{G[1+G+aG(1+G+aG/2)]e^{-G(1+2a)}}{G(1+2a)-(1-e^{-aG})+(1+aG)e^{-G(1+a)}}; \tag{5.47}$$

Slotted 1-persistent CSMA

$$S = \frac{Ge^{-G(1+a)}[1+a-e^{-aG}]}{(1+a)(1-e^{-aG})+ae^{-G(1+a)}}; \tag{5.48}$$

p-persistent CSMA

$$S(G,p,a) = \frac{(1-e^{-aG})[P_s'\pi_0+P_s(1-\pi_0)]}{(1-e^{-aG})[a\bar{t}'\pi_0+a\bar{t}(1-\pi_0)+1+a]+a\pi_0}; \tag{5.49}$$

where P_s', P_s, $\bar{t}'$, $\bar{t}$ and π_0 are defined in [62,123].

A delay analysis for CSMA is given in [127,128]. Stability considerations for CSMA protocols may be found in [25,35,85,126]. The analysis in [127,128] for non-persistent CSMA leads to a system of equations which must be evaluated numerically. However, as was the case with ALOHA, we find that CSMA is fundamentally unstable, and some form of control procedure must be used to render the system stable. In [35] a stabilization procedure is discussed which adjusts the probability of retransmitting a backlogged packet as a function of the estimated number of backlogged packets in the system; this procedure is based on the procedure given in [77]. Upon optimizing this retransmission probability, the mean wait is approximated by

$$W \approx \frac{(S+2\sqrt{2a})\,\bar{x}}{2\,[1-S(1+\sqrt{2a})]}. \tag{5.50}$$

An approximate analysis for p-persistent slotted CSMA with a finite number of buffered terminals is given in [115] which requires the numerical evaluation of a set of equations.

The CSMA protocols were first introduced for application in packet radio networks [51]. In such radio networks, it is relatively easy to sense the channel before you transmit over it; however, once a user begins transmission, it is imperative that his receiver be turned off. Thus, the user cannot monitor the channel while he is transmitting. However, in a wire-based local area network, it is perfectly feasible to 'listen while transmitting' as well as to 'listen before transmitting'. If one does sense the channel during transmission, then a user can detect if his transmission is destroyed by a collision that hits him after he starts. This is referred to as 'collision detection' (CD). This is often used with CSMA to produce yet another protocol, Carrier Sense Multiple Access with Collision Detection - CSMA/CD. In fact, this is the media access protocol used with Ethernet [91]. The form of carrier sense used with Ethernet is, in fact, 1-persistent; the guaranteed collisions which can occur with this CSMA protocol are quickly aborted using the CD capability. The throughput analysis for CSMA/CD is nicely treated in a unified fashion in [117,120]; an earlier treatment was given in [127]. For example, the $S:G$ relationship for unslotted 1-persistent CSMA/CD with an infinite population of users, is

$$\begin{aligned} S = {} & G(1+bG)e^{-(a+b)G} / \\ & \Big[G(a+b)\left[1-(1+G)e^{-(1+a)G}\right] - (1+bG)\left[1-G(1-a-b)\right]e^{-(a+b)G} \\ & + e^{-bG}\left[2+\frac{G}{4}(a+5b)+\frac{1}{2}b(a+b)G^2\right] \\ & + \frac{G}{4}(2bG+3+2aG)(a+b)e^{-(2a+b)G} \\ & - G(1-b)e^{-(1+a+b)G} - \frac{1}{4}e^{-2bG}\left[1-(1+2aG)e^{-2aG}\right]\Big], \end{aligned} \tag{5.51}$$

where b = collision detection time (normalized to an average packet transmission time).

The first delay analyses for CSMA/CD were carried out under slightly different assumptions and approximations in [79,127]. The analysis in [127] for a finite population slotted non-persistent CSMA/CD system requires the numerical evaluation of a series of equations (as for CSMA). The analysis in [79] for an infinite population slotted p-persistent CSMA/CD system leads to the following explicit equation for the mean wait:

$$\begin{aligned} W = {} & 2L(1+\frac{1}{p}) - \left[\frac{1-e^{-2\lambda L}}{pC^*(\frac{S}{\bar{x}})-(1-e^{-2\lambda L})}\right]\left[\frac{1}{p}+L(p-3)\right] \\ & + \frac{S\frac{\overline{x^2}}{2\bar{x}}+SL+\lambda\frac{L^2}{2}+2(\bar{x}+L)\frac{2L}{p}+4L^2(1+2\frac{1-p}{p^2})}{1-S-\lambda L(1+\frac{2}{p})}. \end{aligned} \tag{5.52}$$

Here, $C^*(s) = B^*(s)e^{-sL}$ and $B^*(s)$ is the Laplace transform of the service time probability density function. Analysis of a finite population of buffered terminals using CSMA/CD is provided in [23,115,116,121,122]. We point out that CSMA/CD is also unstable and requires a control procedure to stabilize it. In [35], the mean wait for an infinite population stabilized and optimized slotted p-persistent CSMA/CD system is given as

$$W \approx \frac{\dfrac{S\overline{x^2}}{2\overline{x}} + L(2.31+S)}{1 - S(1+3.31a)}. \tag{5.53}$$

An approach for the analysis of a number of random access schemes with a finite number of buffered terminals is given in [98].

5.5.5. Output processes. Let X be the time between successful packet transmissions. If we can find its mean, $\overline{X}$, then the inverse of this will be the system throughput; that is, $S = 1/\overline{X}$. This approach was taken by Ferguson for ALOHA [28] and by Tobagi for ALOHA and CSMA [128]. An extensive treatment using this 'output process' approach is given by TAKAGI and KLEINROCK [117] for a number of random access protocols. In that treatment, memoryless protocols are defined to be those for which a user acts independently of the previous history of the system whenever he senses the channel idle. For these protocols, we view the channel as alternating between 'transmission periods' (defined as a continuous interval during which at least one user is transmitting, or any transmission is being sensed) and 'idle periods' (defined as a continuous interval during which no transmissions are taking place or being sensed). An unsuccessful transmission period has a duration F, and a successful transmission period has duration T; an idle period has duration I. These random variables have means $\overline{F}$, $\overline{T}$ and $\overline{I}$ and pdf's whose Laplace transforms are $F^*(s)$, $T^*(s)$ and $I^*(s)$, respectively. Further, let $X^*(s)$ be the Laplace transform of the pdf for X, the interdeparture time of successful packets. Lastly, let γ be the probability of a successful transmission once it has been started by breaking an idle period (γ depends upon the protocol and other system parameters). As shown in [117], $X^*(s)$ is given by

$$X^*(s) = \frac{\gamma\, T^*(s)}{\dfrac{1}{I^*(s)} - (1-\gamma)\, F^*(s)}, \tag{5.54}$$

and the system throughput is given by

$$S = \left[\frac{1-\gamma}{\gamma}(\overline{I} + \overline{F}) + \overline{I} + \overline{T} \right]^{-1}. \tag{5.55}$$

These last two fundamental results may be applied to a number of random access protocols as shown in [117].

5.5.6. Multi-hop systems. Essentially all of the material so far in this section has been devoted to the multi-access broadcast channel. An underlying assumption in these models is that all users can hear all other users. In a number of environments, particularly, in packet radio systems, this broadcast assumption is false. For example, users may not be in line-of-sight of each other or, they may not be in range of each other. In this case the connectivity is less than complete and the users are connected via a graph which describes who can hear whom. Thus, in order to send a packet from one user to another, it may be necessary to relay the packet through intermediate users (acting as repeaters for such packets). Whereas this routing problem creates many analytic (and operational) problems, it also permits more than one successful transmission to take place simultaneously as long as these transmissions do not overlap at any of the receivers; this is referred to as 'spatial reuse' of the channel. This environment is known as a 'multi-hop system'.

The analytical problems associated with multi-hop systems are ferocious! A number of studies have been published which solve some very specialized aspects of certain models of the system, but to date, no satisfactory complete solution is available.

One of the earliest studies of this problem was given in [63]. Another early study [124] investigated the 'hidden terminal problem' in which CSMA was shown to degrade badly if even a small fraction of the users could not be heard; in that same study, an access protocol known as 'Busy Tone Multiple Access' [124] was devised to solve this problem. Another study of BTMA is given in [113].

One must also study the effect of different transmission ranges on the system performance. Among the published works here, we include [3,4,8,97,114,118]. A general rule for optimizing the system throughput was given in [138]; this rule states that a user should transmit at a rate and range such that the additional throughput he achieves is exactly equal to the rate at which he destroys successful transmissions by other users.

A nice approach to solving for the throughput and blocking probability of multi-hop packet radio systems was first presented in 1980 [8] and refined in [9]. The key idea in this approach is to identify a set of busy (i.e., transmitting) terminals; under the appropriate assumptions regarding the channel traffic and the access method (e.g., CSMA), the authors show that each such set of busy nodes represents a state in a Markov process. They show that the stationary probabilities for sets of busy terminals have a simple product form solution which depends on the message transmission times only through their mean (and not on the distribution); this is similar to the usual queueing network situation [4]. In [129], the underlying reversible Markov process was studied and, using the conditions necessary for an access protocol to have a product-form solution, they analyzed the throughput for a number of multi-hop channel access schemes.

5.5.7. ***A macroscopic approach.*** Coupled queueing systems usually present enormous analytic complexity. Multi-hop packet radio networks represent one such example. (One notable exception is the class of product-form queueing networks.) It does seem hopeless to attempt to track the exact behavior of each terminal in a multi-hop environment as the number of terminals grows. Yet queueing theory has traditionally required a fine-grained analysis of many interacting elements in order to evaluate coarse-grained equilibrium performance measures. However, a recent and novel approach to dealing with large distributed queueing systems is given in [100,101,139].

A sketch of the approach via an example follows. Consider a multi-hop packet radio system serving M terminals which use CSMA as their access method. Further assume that each terminal schedules new or retransmitted messages for transmission (i.e., listens to the channel for transmission of a message) according to a Poisson process at a rate λ. Message lengths are assumed to be exponentially distributed with a mean of b bits and the channel speed is assumed to be C bits per second; thus the average message transmission time (if successful) is $\bar{x} = b/C$ seconds. Further, assume $L = 0$ (i.e., zero propagation delay). The equilibrium probability, P_k of having k simultaneous transmissions in the system, is given by

$$P_k = \frac{\rho^k}{Z_M}, \tag{5.56}$$

where $\rho = \lambda\bar{x}$ and Z_M is the *partition function* of the system defined as

$$Z_M = \sum_{k=0}^{M} \alpha_M^{(k)} \rho^k. \tag{5.57}$$

Here, $\alpha_M^{(k)}$ is the number of ways in which k terminals out of M can be transmitting at the same time. We recognize that the partition function is the generating function for the number of simultaneous transmissions in the system. Once the partition function is computed for a system, then one can obtain the system throughput, among other measures. Note the similarity of this approach to that of [8].

In [139], an analogy between this approach and that of statistical mechanics is made. Drawing upon an observation of Beneš [6], it is observed that statistical mechanics need not track the dynamics of the system to obtain the equilibrium behavior. In a series of papers ([100,101,139]), we are treated by the application of these ideas to a number of computer and communication examples and to a correspondence between quantities from statistical mechanics and multi-access systems: microstate energy <=> size of set of independent transmissions; global energy <=> system throughput; volume occupied by a terminal <=> E[number of transmissions blocked by that terminal when it transmits]; pressure of a multi-access system <=> average rate of blocked transmissions; critical phase transition <=> discontinuous throughput behavior. One wonders if this intriguing approach will lead to a kinetic theory of packet motion in networks of many types.

6. Distributed processing

No discussion of performance evaluation would be complete without discussing distributed processing systems; these play an important role in today's distributed environment and create a greater need for suitable computer-communication networks. Moreover, they present a number of fascinating queueing theoretic problems of great interest.

6.1. A simple model for distributed processing

In [70], the question is posed, 'Is distributed processing any good?'. Surprisingly, the analysis there shows that it never improves performance. Let us review that analysis.

The model is a rather simple one. We assume that a Poisson stream of arrivals (at rate λ jobs per second) requires work from a processing system whose total processing power is C operations per second. Each arriving job requires an exponentially distributed number of operations, with an average of $1/\mu$ operations. However, rather than assuming a single processor of capacity C, we allow ourselves to consider a general series-parallel configuration as shown in Fig. 9.

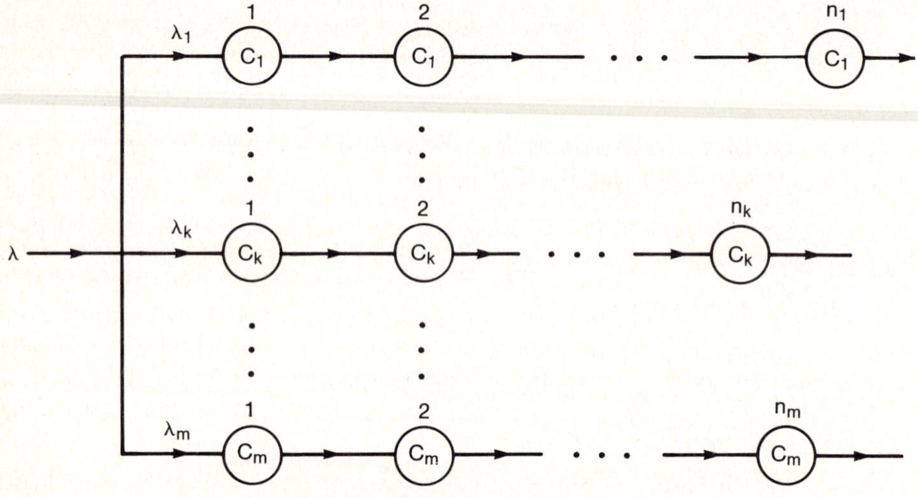

Figure 9. *The general series-parallel topology*

There are m parallel chains, the k-th of which contains n_k processors in series. We have allocated the total system capacity among smaller processors such that each processor in the k-th chain has the same capacity, C_k. Clearly, the *total* system capacity (in operations/sec) is

$$C = \sum_{k=1}^{m} n_k C_k. \tag{6.1}$$

We assume that an entering job will randomly select the k-th chain in which to be processed; the result of this selection produces a net Poisson flow of λ_k jobs per second on the k-th chain. We further assume that each of the n_k processors in this chain will do $1/n_k$ of the work necessary for this job and that each such processor takes an exponentially distributed time to complete the work (with mean $1/\mu n_k C_k$ seconds). This assumption of individual exponentiality contradicts the earlier assumption of an exponentially distributed total job length, but we use it here for simplicity; an analysis which accounts for the Erlangian nature of these series chains may be found in [131]. Each processor in the k-th chain has a utilization factor given by

$$\rho_k = \lambda_k / \mu n_k C_k. \tag{6.2}$$

Define

$T(\rho) =$ E [response time for a job in the series-parallel network],
$T_0(\rho) =$ E [response time for a job served in a central processor of capacity C].

We wish to calculate $T(\rho)/T_0(\rho)$. Clearly, if distributed processing is to reduce the mean response time for jobs, then this ratio should be less than unity. Unfortunately, this is never true! As shown in [70],

$$\frac{T(\rho)}{T_0(\rho)} = \sum_{k=1}^{m} n_k \frac{\rho_k/(1 - \rho_k)}{\rho/(1 - \rho)}, \tag{6.3}$$

where $\rho = \lambda/\mu C$.

Some examples will illustrate the performance. For the case $m = 1$ and $n_1 = n$ (the pure series chain), we have

$$T(\rho)/T_0(\rho) = n. \tag{6.4}$$

For the case $n_k = 1$, $C_k = C/m$ and $\lambda_k = \lambda/m$ for $k = 1,2, \ldots, m$ (the pure parallel case), we have

$$T(\rho)/T_0(\rho) = m. \tag{6.5}$$

For the case $n_k = n$, $C_k = C/mn$ and $\lambda_k = \lambda/m$ for $k = 1, 2, \ldots, m$ (the symmetrical series-parallel case), we have

$$T(\rho)/T_0(\rho) = mn. \tag{6.6}$$

Things are just getting worse! It appears as though the mean response time is directly proportional to the number of processors over which we distribute the work. Indeed, for the general series-parallel case where all processors have the same utilization factor $\rho = \lambda/\mu C$, we have

$$T(\rho)/T_0(\rho) = \sum_{k=1}^{m} n_k. \tag{6.7}$$

As discussed in [70], the practical justification for the great current interest in distributed processing (in spite of the pessimistic results just described) is due to the enormously reduced cost of microprocessors as opposed to mainframes.

6.2. A new class of queueing problems

A computer job is often modelled as a collection of tasks which depend upon each other in some sequential fashion. These dependencies are often drawn as a partially ordered tree known as a computation graph for jobs. The level of resolution for tasks varies, as does the architecture for processing the job (see [71]). An example of a computation graph is given in Fig. 10.

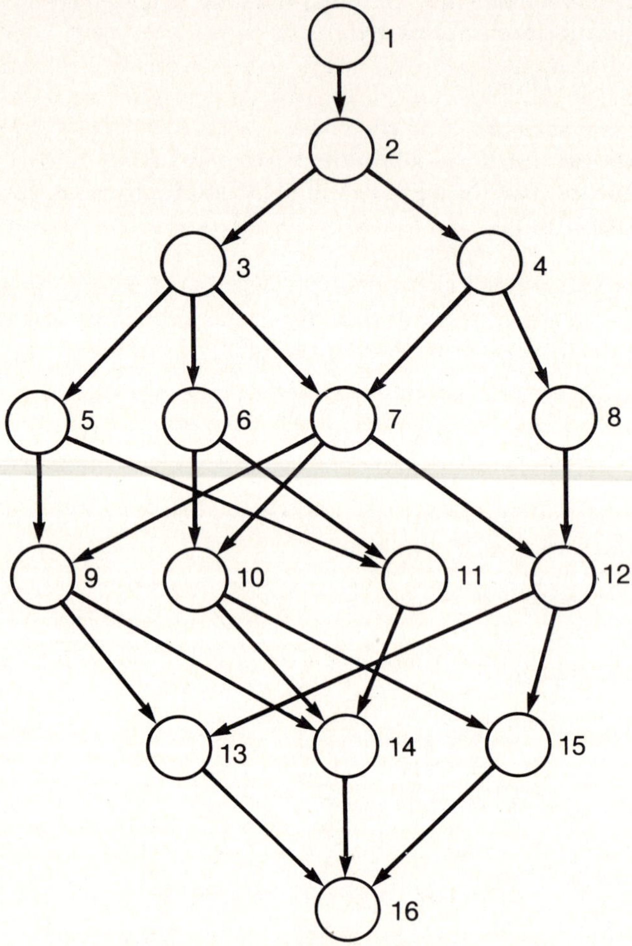

FIGURE 10. *The graph model of computation*

The nodes represent tasks and the arrows represent dependencies in the sense that a task at the head end of an arrow cannot begin execution until the task at the tail end of the arrow is completed. Let us assume that each task requires a random amount of time to be processed (for simplicity, assume that each is exponentially distributed with rate μ). Assume we have P processors which are available to work on the job and that not more than one processor may work on a given task. When node (task) 1 is being processed, the job proceeds at a rate μ. When it completes, node 2 begins and again the job proceeds at a rate μ. When it finishes, nodes 3 and 4 both become activated and, assuming at least two processors are available, the job then proceeds at a rate 2μ, etc. What we see is that the rate at which a job can absorb work depends upon where it is in its processing cycle (i.e., which tasks are currently being processed) and also depends upon how many processors are available to work on the job (other jobs may be competing for a finite number of processors). Of course, we would like to handle the case of an arbitrary distribution of task times. The classic queueing model, shown in Fig. 11a assumes that the rate at which a job can absorb work is constant, and that the total amount of work is a random variable, $\tilde{x}$. Specifically, we plot $r(x)$, the desired rate of service (seconds per second) as a function of the elapsed time in service (assuming no competition for service). In Fig. 11b, we show a general picture for our new class of problems where the rate varies with the amount of service received.

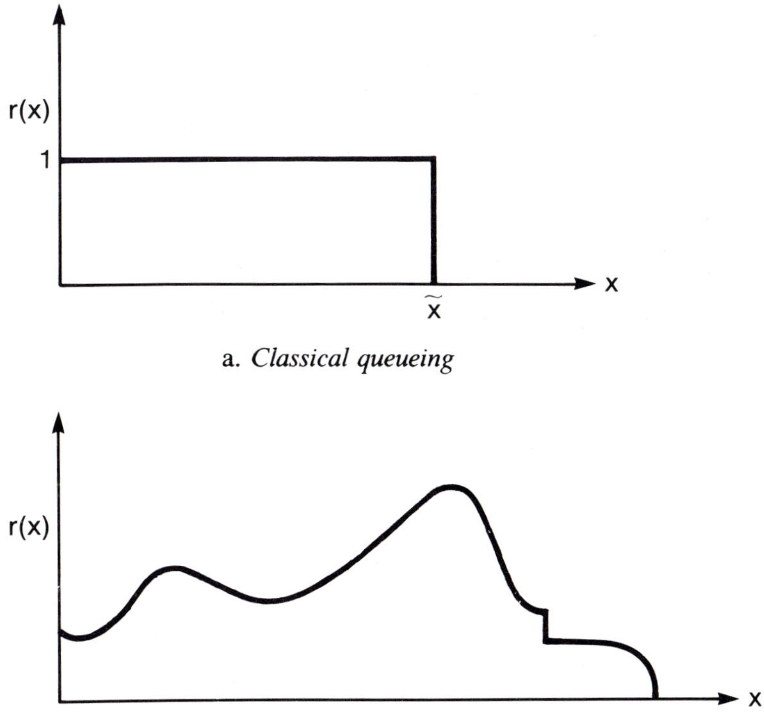

a. *Classical queueing*

b. *New queueing class*

FIGURE 11. *Desired rate of service as a function of the elapsed time in service*

To date, the class of queueing systems shown in Fig. 11b has not been solved. We point out that the work of COHEN [22] offers an approach to this problem. Further, it has been shown in [5] for any computation graph (possibly randomly selected among the jobs), any service time distribution, any arrival process and any work-conserving processor scheduling algorithm, that $\overline{B}$, the average number of busy processors, is given by

$$\overline{B} = \begin{cases} \lambda\overline{X} & \text{for } \lambda\overline{X} < P, \\ P & \text{for } \lambda\overline{X} \geqslant P, \end{cases} \tag{6.8}$$

where P is the number of processors available, λ is the rate at which jobs arrive, and $\overline{X}$ is the average number of seconds of processing required by a job.

7. Conclusions

An enormous literature is devoted to the modeling, analysis and design of computer-communication systems. The thrust for this activity comes from the practical application areas of data communications and data processing. This same practical environment has forced the theoretician/engineer to find working approximate (if not exact) solutions to some very difficult problems.

Nevertheless, from a theoretical point of view, many of the interesting problems are still unsolved. This has caused the queueing theorist to develop new techniques for many of these problems. The next few years will only increase the pressure and challenge to the theorist to keep up with the breakneck pace at which new problems arise.

References

1. N. Abramson (1970). The ALOHA system - another alternative for computer communications. *1970 Fall Joint Computer Conference, AFIPS Conference Proceedings, Vol. 37,* Montvale, NJ:AFIPS Press, 281-285.
2. N. Abramson (1973). Packet switching with satellites. *1973 National Computer Conference, AFIPS Conference Proceedings, Vol. 42,* 695-702.
3. G. Akavia, L. Kleinrock (1984). On a self-adjusting capability of ALOHA networks. *IEEE Trans. Commun. COM-32,* 40-47.
4. F. Baskett, K.M. Chandy, R.R. Muntz, F. Palacios-Gomez (1975). Open, closed, and mixed networks of queues with different classes of customers. *J. Assoc. Comput. Mach. 22,* 248-260.
5. A. Belghith, L. Kleinrock (1984). *Analysis of the Number of Occupied Processors in a Multi-Processing System,* Technical Report No. 850027, Computer Science Dept., School of Engineering and Applied Science, University of California, Los Angeles.
6. V.E. Beneš (1963). A 'thermodynamic' theory of traffic in connecting networks. *The Bell System Tech. J. 42,* 567-607.
7. R. Binder (1975). A dynamic packet switching system for satellite broadcast channels. *International Conference on Communications '75, Vol. 3,* San Francisco, California, 41-1 - 41-5.
8. R. Boorstyn, A. Kershenbaum (1980). Throughput analysis of

multihop packet radio. *International Conference on Communications '80, Vol. 1,* Seattle, WA, 13.6.1 - 13.6.6.

9. R. BOORSTYN, A. KERSHENBAUM, B. MAGLARIS, V. SAHIN (1987). Throughput analysis in multihop CSMA packet radio networks. *IEEE Trans. Commun. COM-35,* 267-274.
10. F. BORGONOVO, L. FRATTA (1978). SRUC: A technique for packet transmission on multiple access channels. *Proceedings of the Fourth International Conference on Computer Communications,* Kyoto, 601-607.
11. O.J. BOXMA (1979). On a tandem queueing model with identical service times at both counters, I, II. *Adv. in Appl. Probab. 11,* 616-643 and 644-659.
12. O.J. BOXMA, W.P. GROENENDIJK (1987). Pseudo-conservation laws in cyclic-service systems. *J. Appl. Probab. 24,* 949-964.
13. O.J. BOXMA, W.P. GROENENDIJK (1988). Waiting times in discrete-time cyclic-service systems. *IEEE Trans. Commun. COM-36,* 164-170.
14. W. BUX (1981). Local-area subnetworks: a performance comparison. *IEEE Trans. Commun. COM-29,* 1465-1473.
15. J.I. CAPETANAKIS (1977). *The Multiple Access Broadcast Channel: Protocol and Capacity Considerations,* Ph.D. Dissertation, Dept. of Electrical Engineering and Computer Science, MIT.
16. J.I. CAPETANAKIS (1979). Generalized TDMA: the multi-accessing tree protocol. *IEEE Trans. Commun. COM-27,* 1476-1484.
17. A.B. CARLEIAL, M.E. HELLMAN (1975). Bistable behavior of slotted ALOHA-type systems. *IEEE Trans. Commun. COM-23,* 401-410.
18. R.T. CARSTEN, E.E. NEWHALL, M.J.M. POSNER (1977). A simplified analysis of scan times in an asymmetrical Newhall loop with exhaustive service. *IEEE Trans. Commun. COM-25,* 951-957.
19. R.T. CARSTEN, M.J.M. POSNER (1978). Simplified statistical models of single and multiple Newhall loops. *Conference Record, 1978 National Telecommunications Conference, IEEE78CH1354-0,* Birmingham, Alabama, 44.5.1 - 44.5.7.
20. I. CHLAMTAC, W.R. FRANTA, K.D. LEVIN (1979). BRAM: the broadcast recognizing access method. *IEEE Trans. Commun. COM-27,* 1183-1190.
21. L.P. CLARE (1986). Delay analysis of stable slotted ALOHA systems. *Proceedings, IEEE INFOCOM '86,* Miami, Florida, 10-19.
22. J.W. COHEN (1979). The multiple phase service network with generalized processor sharing. *Acta Informatica 12,* 245-284.
23. E.J. COYLE, B. LIU (1982). Calculation of the stability characteristics and buffer requirements of asynchronous CSMA/CD networks. *International Conference on Communications '82,* Philadelphia, Pennsylvania, 7F.1.1 - 7F.1.5.
24. W.R. CROWTHER, R. RETTBERG, D. WALDEN, S. ORNSTEIN, F. HEART (1973). A system for broadcast communication: reservation-ALOHA. *Proceedings, 6th Hawaii International Conference on System Science.*
25. G. CUNNINGHAM, J. MEDITCH (1987). Distributed retransmission

controls for slotted nonpersistent and virtual time CSMA. *Proceedings, IEEE INFOCOM '87,* San Francisco, California, 754-762.

26. L.F.M. DE MORAES, I. RUBIN (1984). Analysis and comparison of message queuing delays in token-rings and token-buses local area networks. *International Conference on Communications, IEEE 84CH2028-9,* 130-135.
27. G. FAYOLLE, E. GELENBE, J. LABETOULLE (1977). Stability and optimal control of the packet switching broadcast channel. *J. Assoc. Comput. Mach. 24,* 375-386.
28. M.J. FERGUSON (1977). A bound and approximation of delay distribution for fixed-length packets in an unslotted ALOHA channel and a comparison with Time Division Multiplexing (TDM). *IEEE Trans. Commun. COM-25,* 136-139.
29. M.J. FERGUSON, Y.J. AMINETZAH (1985). Exact results for nonsymmetric token ring systems. *IEEE Trans. Commun. COM-33,* 223-231.
30. H. FRANK, H.W. CHOU (1972). Topological optimization of computer networks. *Proceedings IEEE 60,* 1385-1397.
31. L. FRATTA, M. GERLA, L. KLEINROCK (1973). The flow deviation method: An approach to store-and-forward communication network design. *Networks 3,* 97-133.
32. A. FUKUDA, K. MUKUMOTO, T. HASEGAWA (1978). Adaptive retransmission randomization schemes for a packet switched random-access broadcast channel. *Proceedings of the Fourth International Conference on Computer Communications,* Kyoto, 543-548.
33. H.R. GAIL (1983). *On the Optimization of Computer Network Power,* Ph.D. Dissertation, UCLA-CSD-830922, Computer Science Dept., School of Engineering and Applied Science, University of California, Los Angeles.
34. R. GALLAGER (1978). Conflict resolution in random access broadcast networks. *Proceedings AFOSR Workshop on Communication Theory Applications,* Provincetown, MA, 74-76.
35. R. GALLAGER, D. BERTSEKAS (1987). *Data Networks,* Prentice Hall, Inc., Englewood Cliffs, NJ.
36. A. GIESSLER, J. HANLE, A. KONIG, E. PADE (1978). Free buffer allocation - an investigation by simulation. *Computer Networks 1,* 191-204.
37. L. GEORGIADES, L.F. MERAKOS, P. PAPANTONI-KAZAKOS (1987). A method for the delay analysis of random multiple-access algorithms whose delay process is regenerative. *IEEE Journal on Selected Areas in Communications SAC-5,* 1051-1062.
38. M. GERLA (1973). *The Design of Store and Forward (S/F) Networks for Computer Communications,* Technical Report UCLA-ENG-7319, Computer Science Dept., School of Engineering and Applied Science, University of California, Los Angeles.
39. M. GERLA, H. FRANK, W. CHOU, J. ECKL (1974). A cut saturation algorithm for topological design of packet-switched communication networks. *Conference Record, 1974 National Telecommunications*

Conference, 1074-1085.

40. M. GERLA, L. KLEINROCK (1977). On the topological design of distributed computer networks. *IEEE Trans. Commun. COM-25,* 48-60.
41. M. GERLA, L. KLEINROCK (1977). Closed loop stability controls for S-ALOHA satellite communications. *Proceedings of the Fifth Data Communications Symposium,* Snowbird, Utah, 2-10 - 2-19.
42. B. HAJEK, T. VAN LOON (1982). Decentralized dynamic control of a multi-access broadcast channel. *IEEE Trans. Aut. Control AC-27,* 559-569.
43. J.F. HAYES (1978). An adaptive technique for local distribution. *IEEE Trans. Commun. COM-26,* 1178-1186.
44. J.F. HAYES (1984). *Modelling and Analysis of Computer Communications Networks,* Plenum Press, New York.
45. D.P. HEYMAN (1983). Data-transport performance analysis of Fasnet. *The Bell System Tech. J. 62,* 2547-2560.
46. W. HILAL, J.F. CHIU (1987). A study of register-insertion rings. *Proceedings, IEEE INFOCOM '87,* San Francisco, California, 479-491.
47. J.-H. HUANG (1988). *On the Behavior of Algorithms in a Multi-Processing Environment,* Ph.D. Dissertation, Computer Science Dept., School of Engineering and Applied Science, University of California, Los Angeles.
48. T.C. HOU, V.O.K. LI (1986). Transmission range control in multihop packet radio networks. *IEEE Trans. Commun. COM-34,* 38-44.
49. J.R. JACKSON (1957). Networks of waiting lines. *Oper. Res. 5,* 518-521.
50. I.M. JACOBS, R. BINDER, E.V. HOVERSTEN (1978). General purpose packet satellite networks. *Proceedings IEEE 66,* 1448-1467.
51. R.E. KAHN, S.A. GRONEMEYER, J.BURCHFIEL, R.C. KUNGELMAN (1978). Advances in packet radio technology. *Proceedings IEEE 66,* 1468-1496.
52. F. KAMOUN, L. KLEINROCK (1980). Analysis of shared finite storage in a computer network node environment under general traffic conditions. *IEEE Trans. Commun. COM-28,* 992-1003.
53. A.R. KAYE (1972). Analysis of a distributed control loop for data transmission. *Proceedings of the Symposium on Computer-Communications Networks and Teletraffic,* Polytechnic Institute of Brooklyn, 47-58.
54. F.P. KELLY, P.K. POLLETT (1983). Sojourn times in closed queueing networks. *Adv. in Appl. Probab. 15,* 638-658.
55. L. KLEINROCK (1972). *Communication Nets; Stochastic Message Flow and Delay,* McGraw-Hill Book Company, New York, 1964 (out of print), reprinted by Dover Publications, New York.
56. L. KLEINROCK (1965). A conservation law for a wide class of queuing disciplines. *Naval Research Logistics Quarterly 12,* 181-192.
57. L. KLEINROCK, S. LAM (1973). Packet switching in a slotted satellite channel. *1973 AFIPS Conference Proceedings, National Computer Conference and Exposition, Vol. 42,* New York, NJ:AFIPS Press, 703-710.

58. L. KLEINROCK, S. LAM (1974). On stability of packet switching in a random multi-access broadcast channel. *Computer Nets, a Supplement to the Proceedings of the Seventh Hawaii International Conference on System Sciences,* Honolulu, Hawaii, 74-77.

59. L. KLEINROCK (1975). *Queueing Systems, Volume I: Theory,* Wiley Interscience, New York.

60. L. KLEINROCK, S. LAM (1975). Packet switching in multi-access broadcast channel: Performance evaluation. *IEEE Trans. Commun. COM-23,* 410-423.

61. L. KLEINROCK, F.A. TOBAGI (1975). Packet switching in radio channels: part I - carrier sense multiple-access modes and their throughput-delay characteristics. *IEEE Trans. Commun. COM-23,* 1400-1416.

62. L. KLEINROCK, F.A. TOBAGI (1975). Random access techniques for data transmission over packet switched radio channels. *1975 National Computer Conference, AFIPS Conference Proceedings, Vol. 44,* Anaheim, NJ:AFIPS Press, 187-201.

63. L. KLEINROCK (1975). On giant stepping in packet radio networks. *UCLA, Packet Radio Temporary Note #5, PRT 136.*

64. L. KLEINROCK (1976). *Queueing Systems, Volume II: Computer Applications,* Wiley Interscience, New York.

65. L. KLEINROCK (1977). Performance of distributed multi-access computer-communication systems. *Information Processing '77, Proceedings of IFIP Congress '77,* North-Holland, Amsterdam, 547-552.

66. L. KLEINROCK (1978). On the flow control in computer networks. *International Conference on Communications, Vol. II,* Toronto, Ontario, 27.2.1 - 27.2.5.

67. L. KLEINROCK, Y. YEMINI (1978). An optimal adaptive scheme for multiple access broadcast communication. *International Conference on Communications, Vol. I,* Toronto, Ontario, 7.2.1 - 7.2.5.

68. L. KLEINROCK (1979). Power and deterministic rules of thumb for probabilistic problems in computer communications. *International Conference on Communications,* Boston, Massachusetts, 43.1.1 - 43.1.10.

69. L. KLEINROCK, M. SCHOLL (1980). Packet switching in radio channels: New conflict-free multiple access schemes. *IEEE Trans. Commun. COM-28,* 1015-1029.

70. L. KLEINROCK (1984). On the theory of distributed processing. *Proceedings of the Twenty-Second Annual Allerton Conference on Communication, Control and Computing,* Urbana-Champaign, 60-70.

71. L. KLEINROCK (1985). Distributed systems. *ACM/IEEE-CS Joint Special Issue, Commun. ACM 28,* 1200-1213, and *Computer 18,* 90-103.

72. L. KLEINROCK, H. LEVY (1988). The analysis of random polling systems. To be published in *Operations Research.*

73. A.G. KONHEIM, B. MEISTER (1974). Waiting lines and times in a system with polling. *J. Assoc. Comput. Mach. 21,* 470-490.

74. P.J. KUEHN (1979). Multiqueue systems with non-exhaustive cyclic service. *The Bell System Tech. J. 58,* 671-698.

75. P.J. KUEHN (1981). Performance of ARQ-protocols for HDX transmission in hierarchical polling systems. *Performance Evaluation 1,* 19-30.
76. S.S. LAM (1974). *Packet Switching in a Multi-Access Broadcast Channel with Applications to Satellite Communication in a Computer Network,* Computer Science Department, School of Engineering and Applied Science, Engineering Report UCLA-ENG-7429, University of California, Los Angeles.
77. S.S. LAM, L. KLEINROCK (1975). Dynamic control schemes for a packet switched multi-access broadcast channel. *1975 National Computer Conference, AFIPS Conference Proceedings, Vol. 44,* Anaheim, NJ:AFIPS Press, 143-153.
78. S.S. LAM (1979). On protocols for satellite packet switching. *International Conference on Communications,* Boston, 58.6.1 - 58.6.6.
79. S.S. LAM (1980). A carrier sense multiple access protocol for local networks. *Computer Networks 4,* 21-32.
80. S.S. LAM, J.W. WONG (1982). Queuing network models of packet-switching networks - part 2: networks with population size constraints. *Performance Evaluation 2,* 161-180.
81. S.S. LAVENBERG (1988). A perspective on queueing models of computer performance. O.J. BOXMA, R. SYSKI (eds.). *Queueing Theory and its Applications - Liber Amicorum for J.W. Cohen,* North-Holland, Amsterdam.
82. M. LIVNY, M. MELMAN (1982). Load balancing in homogeneous broadcast distributed systems. *Proceedings ACM Computer Network Performance Symposium,* 47-55.
83. J.W. MARK (1978). Global scheduling approach to conflict-free multiaccess via a data bus. *IEEE Trans. Commun. COM-26,* 1342-1352.
84. J.L. MASSEY (1980). *Collision-Resolution Algorithms and Random Access Communications,* Technical Report No. UCLA-ENG-8016, School of Engineering and Applied Science, University of California, Los Angeles.
85. J.S. MEDITCH, C.T.A. LEA (1983). Stability and optimization of the CSMA and CSMA/CD channels. *IEEE Trans. Commun. COM-31,* 763-774.
86. J.S. MEDITCH, C.T.A. LEA (1985). Stability and throughput in virtual time CSMA. *Computer Networks and ISDN Systems 10,* 19-26.
87. J.S. MEDITCH, D.H.L. YIN (1986). Performance analysis of virtual time CSMA. *Proceedings, IEEE INFOCOM '86,* Miami, Florida, 242-251.
88. B. MEISTER, H. MULLER, H. RUDIN (1971). New optimization criteria for message-switching networks. *IEEE Trans. Commun. COM-19,* 256-260.
89. B. MELAMED (1982). Sojourn times in queueing networks. *Math. of Oper. Res. 7,* 223-244.
90. R.M. METCALFE (1973). Steady-state analysis of a slotted and controlled ALOHA system with blocking. *Proceedings 6th Hawaii International Conference on Systems Science,* Honolulu, 375-378.
91. R.M. METCALFE, D.R. BOGGS (1976). ETHERNET: Distributed packet

switching for local computer networks. *Commun. ACM 19,* 395-403.

92. V.A. Mikhailov, B.S. Tsybakov (1981). Upper bound for the capacity of a random multiple access system. *Prob. Peredachi Infor. (USSR) 17,* 90-95.

93. M.L. Molle (1981). *Unifications and Extensions of the Multiple Access Communications Problem,* Ph.D. Dissertation, Report No. 810730, (UCLA-ENG-8118), Computer Science Dept., School of Engineering and Applied Science, University of California, Los Angeles.

94. M.L. Molle, L. Kleinrock (1985). Virtual time CSMA: Why two clocks are better than one. *IEEE Trans. Commun. COM-33,* 919-933.

95. M.L. Molle (1987). Modelling the delay in 'moving server' random access protocols. Y. Yemini (ed.). *Current Advances in Distributed Computing and Communications,* Computer Science Press, Inc., 17-32.

96. R.J.T. Morris, Y.T. Wang (1984). Some results for multi-queue systems with multiple cyclic servers. H. Rudin, W. Bux (eds.). *Performance of Computer Communication Systems,* North-Holland, Amsterdam, 245-258.

97. R. Nelson, L. Kleinrock (1985). Spatial TDMA: A collision free multihop channel access protocol. *IEEE Trans. Commun. COM-33,* 934-944.

98. M. Niktash, J.S. Riordon, S.A. Mahmoud (1987). A model for analysis of channel access protocols. *Proceedings, IEEE INFOCOM '87,* San Francisco, California, 381-389.

99. Y. Oie, T. Suda, H. Miyahara, T. Hasegawa (1987). Throughput and delay analysis of free access tree algorithms with mini-slots. *Proceedings, IEEE INFOCOM '87,* San Francisco, California, 258-267.

100. E. Pinsky (1985). *Canonical Approximation in the Performance Analysis of Distributed Systems,* Ph.D. Dissertation, Computer Science Department, Columbia University.

101. E. Pinsky, Y. Yemini, M. Sidi (1987). The canonical approximation in the performance analysis of packet radio networks. Y. Yemini (ed.). *Current Advances in Distributed Computing and Communications,* Computer Science Press, Inc., 140-160.

102. N. Pippenger (1981). Bounds on the performance of protocols for a multiple access broadcast channel. *IEEE Trans. Information Theory IT-27,* 145-151.

103. G. Polyzos, M. Molle, A. Venetsanopoulos (1985). Delay analysis of three conflict resolution algorithms: The non-homogeneous case. *Proceedings of IEEE Global Telecommunications Conference '85,* New Orleans, Louisiana, 1504-1509.

104. T. Raith (1985). Performance analysis of multibus interconnection networks in distributed systems. M. Akiyama (ed.). *Eleventh International Teletraffic Congress,* North-Holland, Amsterdam, 4.2A-5-1 - 4.2A-5-7 .

105. R.L. Rivest (1985). *Network Control by Bayessian Broadcast,* Report MIT/LCS/TM-285, Cambridge, MA: MIT, Laboratory for Computer Science.

106. L.G. Roberts (1972). ALOHA packet system with and without slots and capture. *ARPANET Satellite System Note 8 (NIC 11290);* reprinted in *Computer Communication Review 5 (1975),* 28-42.
107. L.G. Roberts (1973). Dynamic allocation of satellite capacity through packet reservation. *1973 National Computer Conference and Exposition, AFIPS Conference Proceedings, Vol. 42,* New York, NJ:AFIPS Press, 711-716.
108. I. Rubin (1974). Communication networks: Message path delays. *IEEE Trans. Information Theory IT-20,* 738-745.
109. I. Rubin (1977). A group random-access procedure for multi-access communication channels. *Conference Record, 1977 National Telecommunication Conference,* Los Angeles, 12:5-1 - 12:5-7.
110. I. Rubin (1977). *Integrated Random-Access Reservation Schemes for Multi-Access Communication Channels,* Technical Report UCLA-ENG-7752, School of Engineering and Applied Science, University of California, Los Angeles.
111. T.N. Saadawi, A. Ephremides (1981). Analysis, stability, and optimization of slotted ALOHA with a finite number of buffer users. *IEEE Trans. Aut. Control AC-26,* 680-689.
112. M. Schwartz (1977). *Computer-Communication Network Design and Analysis,* Prentice-Hall, Englewood Cliffs, NJ.
113. M. Sidi, A. Segall (1981). A busy-tone-multiple-access-type scheme for packet-radio networks. G. Pujolle (ed.). *Performance of Data Communication Systems and Their Applications,* North-Holland, Amsterdam, 1-10.
114. J. Silvester, L. Kleinrock (1978). Optimum transmission radii for packet radio networks or why six is a magic number. *Conference Record, 1978 National Telecommunications Conference,* Birmingham, Alabama, 4.3.2 - 4.3.5.
115. J. Silvester, I. Lee (1982). Performance modeling of buffer CSMA - an iterative approach. *Globecom '82, IEEE Global Telecommunications Conference, Vol. 3,* Miami, Florida, F1.6.1 - F1.6.5.
116. H. Takagi, L. Kleinrock (1984). Diffusion process approximation for the queuing delay in contention packet broadcasting system. H. Rudin, W. Bux (eds.). *Performance of Computer Communication Systems,* North-Holland, Amsterdam, 111-124.
117. H. Takagi, L. Kleinrock (1985). Output processes in contention packet broadcasting systems. *IEEE Trans. Commun. COM-33,* 1191-1199.
118. H. Takagi, L. Kleinrock (1985). Throughput-delay characteristics of some slotted-ALOHA multihop packet radio networks. *IEEE Trans. Commun. COM-33,* 1200-1207.
119. H. Takagi (1986). *Analysis of Polling Systems,* The MIT Press, Cambridge, Massachusetts. Also H. Takagi and L. Kleinrock (1985). *Analysis of Random Polling Systems,* JSI Research Report, TR87-0002, IBM Japan Science Institute.

120. H. TAKAGI, L. KLEINROCK (1987). Throughput analysis for persistent CSMA systems. *IEEE Trans. Commun. COM-35,* 243-245.
121. T. TAKINE, Y. TAKAHASHI, T. HASEGAWA (1987). Performance analysis of a buffered CSMA/CD. L.F.M. DE MORAES, E. DE SOUZA E SILVA, L.F.G. SOARES (eds.). *Proceedings of the Third International Conference on Data Communication Systems and Their Performance,* Editora Campus, Ltda. (Rio de Janeiro), 241-255.
122. S. TASAKA (1986). *Performance Analysis of Multiple Access Protocols,* The MIT Press, Cambridge, Massachusetts.
123. F.A. TOBAGI, L. KLEINROCK (1974). Carrier sense multiple-access for packet switched radio channels. *International Conference on Communications,* Minneapolis, Minnesota, 191-215.
124. F.A. TOBAGI, L. KLEINROCK (1975). Packet switching in radio channels: part II - the hidden terminal problem in carrier sense multiple-access and the busy-tone solution. *IEEE Trans. Commun. COM-23,* 1417-1433.
125. F.A. TOBAGI, L. KLEINROCK (1976). Packet switching in radio channels: part III - polling and dynamic split-channel reservation multiple channel access. *IEEE Trans. Commun. COM-24,* 832-845.
126. F.A. TOBAGI, L. KLEINROCK (1977). Packet switching in radio channels: part IV - stability considerations and dynamic control in carrier sense multiple access. *IEEE Trans. Commun. COM-25,* 1103-1119.
127. F.A. TOBAGI, V.B. HUNT (1980). Performance analysis of carrier sense multiple access with collision detection. *Computer Networks 4,* 245-259.
128. F.A. TOBAGI (1982). Distribution of packet delay and interdeparture time in slotted ALOHA and carrier sense multiple access. *J. Assoc. Comput. Mach. 29,* 907-927.
129. F.A. TOBAGI, J.M. BRAZIO (1983). Throughput analysis of multihop packet radio networks under various channel access schemes. *Proceedings of the IEEE INFOCOM '83,* San Diego, California, 381-389.
130. F.A. TOBAGI, M. FINE (1983). Performance of undirectional broadcast local area networks: Expressnet and Fasnet. *IEEE Journal on Selected Areas in Communications SAC-1,* 913-926.
131. F.A. TOBAGI, H. KANAKIA (1986). A comparison of distributed and centralized processing systems. T. HASEGAWA, H. TAKAGI, Y. TAKAHASHI (eds.). *Computer Networking and Performance Evaluation,* North-Holland, Amsterdam, 483-493.
132. D. TOWSLEY, J.K. WOLF (1984). On adaptive tree polling algorithms. *IEEE Trans. Commun. COM-32,* 1294-1298.
133. P. TRAN-GIA, T. RAITH (1986). Multi queue systems with finite capacity nonexhaustive cyclic service. T. HASEGAWA, H. TAKAGI, Y. TAKAHASHI (eds.). *Computer Networking and Performance Evaluation,* North-Holland, Amsterdam, 213-225.
134. B.S. TSYBAKOV, V.A. MIKHAILOV (1978). Free synchronous packet access in a broadcast channel with feedback. *Prob. Peredachi Infor. (USSR) 14, 32-59.*

135. B.S. Tsybakov, V.A. Mikhailov (1980). Random multiple access of packets: part and try algorithm. *Prob. Peredachi Infor. (USSR) 16, 65-79.*

136. J. Walrand, P. Varaiya (1980). Sojourn times and the overtaking condition in Jacksonian networks. *Adv. in Appl. Probab. 12,* 1000-1018.

137. J.W. Wong (1978). Distribution of end-to-end delay in message-switched networks. *Computer Networks 2,* 44-49.

138. Y. Yemini, L. Kleinrock (1979). On a general rule for access control or, silence is golden... . *Proceedings of the International Symposium on Flow Control in Computer Networks,* Versailles, France, 335-347.

139. Y. Yemini (1983). A statistical mechanics of distributed resource sharing mechanisms. *Proceedings of the IEEE INFOCOM '83,* San Diego, California, 531-540.

140. Y. Yoshioka, T. Nakamura, R. Sato (1977). An optimum solution of the queuing system. *Electronics and Communications in Japan 60-B,* 590-591.

A Perspective on Queueing Models of Computer Performance

S.S. Lavenberg
IBM Research Division, T.J. Watson Research Center
P.O. Box 704, Yorktown Heights, New York 10598, USA

We will survey analytical computer performance modeling with an emphasis on developments of practical importance. In doing so the influence that computer performance modeling has had on the field of queueing theory, in particular on queueing networks, will be made evident. The survey will proceed historically, starting with developments in the mid-1960s, when the first published results on queueing models of multiprogramming and time sharing systems appeared. The computer systems we will consider include conventional single processor systems as well as highly parallel multiprocessor systems. Both subsystem models and overall system models will be discussed. One purpose of this paper is to expose queueing theorists and applied queueing analysts to the practical queueing analysis problems that arise in computer performance modeling. However, this survey should also be useful for computer performance modeling practitioners and researchers.

1. Introduction

Performance is one of the key factors that influences the design, development, configuration and tuning of a computer system. Thus, computer performance evaluation is required throughout the life cycle of a system. (The evaluation of a computer system also involves such factors as cost, ease of use, function, availability, reliability and security.) Computer performance evaluation methods can be divided into three areas, namely measurement, simulation modeling and analytical modeling. Performance measurement is possible once a system is built, instrumented and running. Otherwise performance modeling is required to predict performance. We will not discuss the relative merits of simulation and analytical modeling here. Needless to say simulation has played an extremely important practical role in computer performance modeling, as partly attested to by the many simulation packages that have been developed primarily for this purpose, e.g. Sauer et al. [86,87], Merle et al. [66], Neuse et al. [75] and Veran and Potier [106]. A recent survey of measurement, simulation modeling and analytical modeling methods for computer performance evaluation can be found in Heidelberger and Lavenberg [37]. We will only consider analytical modeling in this paper.

Analytical computer performance modeling, as we know it today, came into being in the 1960s with the emergence of general purpose multiuser computer systems, i.e. multiprogramming and/or time sharing systems. Except under very light loads the performance of such systems is strongly affected by the queueing delays encountered by multiple users attempting to access and receive service from the resources that comprise the system. Thus, from the beginning, the performance models used have been queueing models. We will survey analytical computer performance modeling with an emphasis on developments of practical importance. In doing so the influence that computer performance modeling has had on the field of queueing theory, in particular on queueing networks, will be made evident. One purpose of this paper is to expose queueing theorists and applied queueing analysts to the practical queueing analysis problems that arise in computer performance modeling. The computer systems we will consider include conventional single processor systems as well as more recent highly parallel multiprocessor systems. Both subsystem models and overall system models will be discussed. Queueing models have also played an important role in modeling communication networks and manufacturing systems, the latter of which are becoming more and more computer controlled. These system modeling areas are, however, outside the scope of this paper. A survey of queueing models of communication networks can be found elsewhere in this book, KLEINROCK [50].

In the rest of this paper we will proceed more or less historically. In Section 2 we will review pre-1975 developments, highlighting developments that became important in subsequent years. During this period the first published results on queueing models of multiprogramming and time sharing systems appeared. Also the first successful uses of queueing models to predict computer system performance occurred for some early time sharing systems. Between 1975 and 1980 several fundamental results appeared in the published literature and computer performance modeling needs led to significant advancements in queueing network analysis. Also during this period the practical applicability of analytical queueing network models was established. We will review this period in Section 3. In Section 4 we will review post-1980 developments. This period has been characterized by the widespread development and application of a variety of techniques for the approximate solution of queueing network models of computer performance. These approximations are in large part based on the fundamental results that were obtained during the 1975-1980 period. We will present our conclusions in Section 5.

The performance measures we will consider are all stationary measures. This is not surprising since transient solutions are extremely difficult to obtain for queueing models. The main performance measures that are usually obtained are resource utilizations, throughputs, mean queue lengths (number in service plus number waiting in the queue) and mean response times (service time plus time waiting in the queue). Sometimes queue length distributions and response time distributions are also obtained.

2. PRE-1975 DEVELOPMENTS

2.1. Single server models of time sharing systems

What may be the earliest published analysis of a queueing model of a time sharing system appeared in KLEINROCK [47] and numerous papers on such models followed. These earliest models were single server queues with Poisson arrivals in which the only computer system resource that was modeled was the processor. These models were used to represent a variety of time sharing processor scheduling algorithms in which a job receives continuous service on the processor for at most an interval of time called a quantum. A job that does not complete its service during a quantum queues again to receive the smaller of its remaining service time and another quantum of service. There may be several levels of queues for the processor, with a job starting in the highest level queue and moving to the next highest level queue when its quantum expires. A job in a queue can only receive service when all higher level queues are empty and it is at the head of its queue. In addition the quantum size may be larger the lower the level of the queue. An extensive bibliography and an excellent review for models of this type are given in Chapter 4 of KLEINROCK [49]. Analysis of models of this type provided insight into the design of processor scheduling algorithms for time sharing systems, although most real schedulers are more complex than those that have been analyzed. In the simplest of these scheduling algorithms, called round robin, all quantum sizes are fixed and equal and a job that fails to complete its service during a quantum rejoins the tail of the queue. A mathematical idealization of round robin scheduling called processor sharing, in which the quantum size shrinks to zero, was introduced in KLEINROCK [48]. With the processor sharing (PS) queueing discipline all jobs receive simultaneous service from the processor at a rate that is inversely proportional to the number of jobs at the processor. The mean response time for the PS queue has the property that it depends on the service time distribution only through its mean. The PS queueing discipline came to play an important role in the subsequent development of queueing networks as we shall see. (In KLEINROCK [49] the less practically interesting last come first served preemptive resume (LCFSPR) queueing discipline was shown to yield the same mean response time as the PS discipline.)

2.2. The machine repair model of a time sharing system

In SCHERR [88] the queueing model, shown in Fig. 1, was used as a finite source model of a time sharing system. This model was borrowed from the queueing literature where it had been used to analyze machine breakdown and repair. Hence, it is commonly called the machine repair model. In applying the model to a time sharing system there is one job in the model for each interactive user of the system. A user alternates between thinking, entering input, queueing for and receiving service at the system and receiving output. The times to receive output, think and enter input are added together and called the think time which is the service time of a job at a terminal. The think times and the service times at the single server that represents the system were assumed to be exponentially distributed and the queueing discipline used by the system was assumed to be first come first served (FCFS). Scherr used this model to analyze the Compatible Time Sharing System (CTSS) that was developed at MIT by Project MAC, a government funded research group. This model was a reasonable representation of the system since CTSS allows only one job to use the system at a time. However, the exponential and FCFS assumptions were known to be unrealistic for this system. Despite this fact, the model predicted mean response time quite accurately. This model was also successfully used in the late 1960s to predict performance during the development of IBM's OS/360 operating system running its Time Sharing Option. This use is reported on in detail in LASSETTRE and SCHERR [58]. A shorter account in a more modern context can be found in Chapter 3 of LAVENBERG [59]. It was later shown why the mean response time predictions of this simple model proved to be reasonably accurate even though some of the basic assumptions of the model were violated. In particular it was shown in BASKETT [7] that the same expression for the mean response time is obtained when the queueing discipline at the single server is PS (the real systems used a form of round robin scheduling) and the think time and service time distributions are nonexponential with rational Laplace transforms. This important insensitivity result is a precursor to similar results that were subsequently obtained for much more general queueing networks as will be discussed in Section 3.

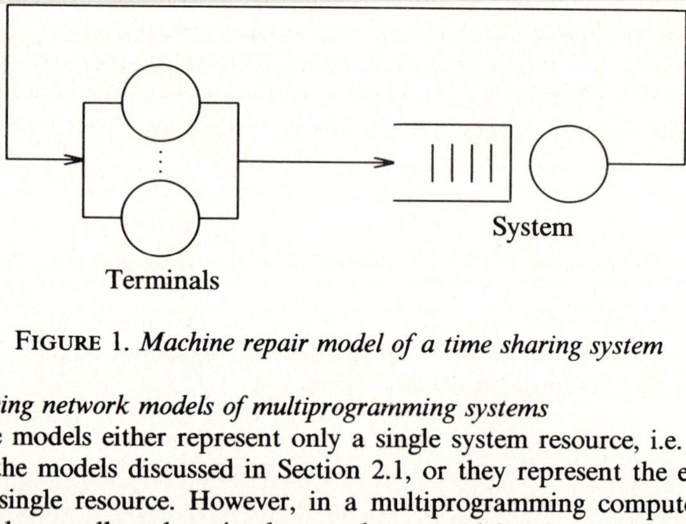

FIGURE 1. *Machine repair model of a time sharing system*

2.3. Queueing network models of multiprogramming systems

The above models either represent only a single system resource, i.e. the processor in the models discussed in Section 2.1, or they represent the entire system as a single resource. However, in a multiprogramming computer system multiple jobs are allowed to simultaneously contend for the multiple resources, e.g. processor and I/O devices, that comprise the system. An early queueing network model of a multiprogramming system can be found in GAVER [34]. The system was modeled as a closed network of two queues, a single server queue representing the processor and a multiserver queue representing multiple identical I/O devices. (A closed network is one in which jobs neither enter nor leave but circulate indefinitely. In an open network jobs arrive, spend time in the network and eventually depart.) The fixed number of jobs in the network corresponds to a fixed level of multiprogramming. The processor service times were allowed to have an arbitrary distribution while the I/O service times were assumed to be exponentially distributed. Semi-Markov process analysis techniques were used to solve for the processor utilization. Unfortunately the semi-Markov process approach is quite limited for analyzing queueing networks. It typically allows only one service time distribution to be arbitrary, but more importantly it becomes intractable for all practical purposes for networks with more than two queues. (Another early semi-Markov process analysis of a two queue closed multiprogramming model can be found in LEWIS and SHEDLER [63].)

Multiprogramming systems maintain a separate queue for each I/O device, while the model in GAVER [34] assumed a single queue for the I/O devices. A more appropriate representation is the closed queueing network shown in Fig. 2, called the central server model, which appeared in a 1971 Ph.D. thesis, BUZEN [14]. All service times were assumed to be exponentially distributed and the routing of a job completing service at the processor was determined by a fixed set of probabilities. The feedback path at the processor represents the completion of a job and its instantaneous replacement by a probabilistically identical job. The central server model is a special case of the more general closed queueing networks with exponential service times and probabilistic routing analyzed earlier in JACKSON [43] and GORDON and NEWELL [35]. These networks consist of single server queues, multiserver queues and, in JACKSON [43], queues with queue length dependent service rates. All jobs are probabilistically identical and the sequence of queues visited by each job is a discrete time irreducible Markov chain. (Both papers also consider other networks.)

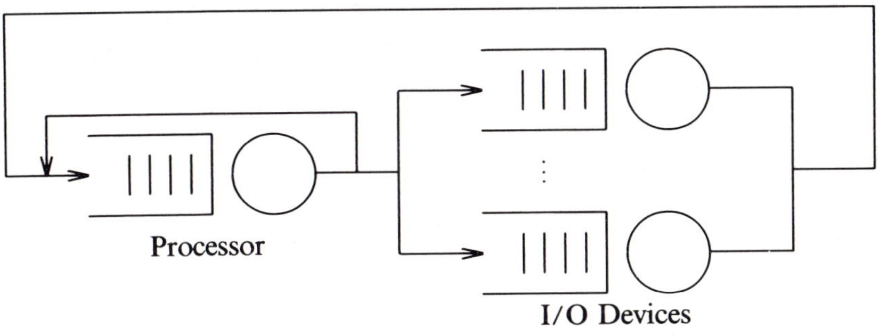

FIGURE 2. *Central server model of a multiprogramming system*

The stationary state distribution for these networks is a product of simple terms, one for each queue, divided by a normalizing constant. Hence they have come to be called product form networks. The central server model was used in BUZEN [14] to study static load balancing among I/O devices and in a Ph.D. thesis, MOORE [70], was found useful in predicting the performance of a real system. In JACKSON [43] and GORDON and NEWELL [35] computational issues were not considered. In BUZEN [14,15] an algorithm was presented to compute the normalizing constant for closed product form queueing networks in a recursive manner that is more efficient than summing the product form terms over the state space. (A generalization of this algorithm to product form networks with multiple types of jobs will be given in Section 3.1.2.) The performance measures of interest were shown to be directly computable from terms computed during the recursive computation of the normalizing constant. These major results for product form networks laid the groundwork for much future computer performance modeling work. In Sections 3.1.1 and 3.1.2 we will give the product form solution and computational algorithms for a more general class of product form networks.

The closed queueing networks discussed above have a fixed number of jobs competing for the processor and I/O devices, corresponding to a fixed level of multiprogramming. This may be a reasonable assumption for a multiprogramming system operating with a fixed number of memory partitions, as some early systems did, and under heavy load so that all memory partitions are occupied. Usually however the level of multiprogramming will vary. For example in a system with a fixed number M of memory partitions it will vary from zero up to M. If a job arrives when there are M jobs in the system, i.e. all M memory partitions are occupied, then it must wait in a queue in order to be allocated a memory partition. Once it receives a memory partition it can then contend for the processor and I/O devices. After the job completes its required number of processor and I/O device services it releases its memory partition and leaves the system. Each memory partition is an example of what came to be called a passive resource in the computer performance modeling literature, e.g. see SAUER and CHANDY [85]. A passive resource does not have a service time directly associated with it, but it must be acquired before service can be obtained from other resources and held during these services. An open queueing network model that represents the system that we have been discussing is shown in Fig. 3. The rectangle contains the fixed number of memory partitions. The upright triangle indicates a node where a memory partition, if one is available, is instantaneously allocated to a job. The inverted triangle indicates a node where a memory partition is instantaneously released. This pictorial representation is commonly used for a passive resource queue, e.g. SAUER and CHANDY [85]. The effect of contention for memory partitions is to limit to M the number of jobs that can simultaneously be in the subnetwork that represents the processor and I/O devices. Even under the assumptions of Poisson arrivals and exponential service times this network is not analytically tractable. In particular it is not a product form network due to the queue for passive resources. In AVI-ITZHAK and HEYMAN [1] this model, and similar ones, were solved approximately using the following two-step decomposition method. In the first step the subnetwork was treated in isolation as a closed queueing network with a fixed number of jobs n as shown in Fig. 4a. For each $n, n = 1, \ldots, M$, the closed network was solved for the throughput of jobs at the processor $\lambda(n)$. Under the assumption of exponential service times and probabilistic routing this closed network has a product form solution so that the throughputs are easy to obtain. Let C denote the mean number of times a job cycles through the subnetwork in Fig. 3 before departing. Then the throughput of the subnetwork is $1/C$ times the processor throughput. In the second step the subnetwork in the original network was replaced by a single server with exponential service times and queue length dependent service rates $\mu(n)$, where $\mu(n)=\lambda(n)/C$, $n=1,\ldots,M$ and $\mu(n)=\lambda(M)/C$, $n=M+1, M+2,\ldots,$ as shown in Fig. 4b. The queue length of the resulting single server queue is a birth and death process and hence it is trivial to solve for its stationary queue length distribution and mean response time. It was found in special cases that the approximation is reasonably accurate when C is large or when M is large. Subsequent work on decomposition provides an explanation for why this approximate analysis should be accurate for large C or for large M. We will discuss decomposition further in Section 3.2.

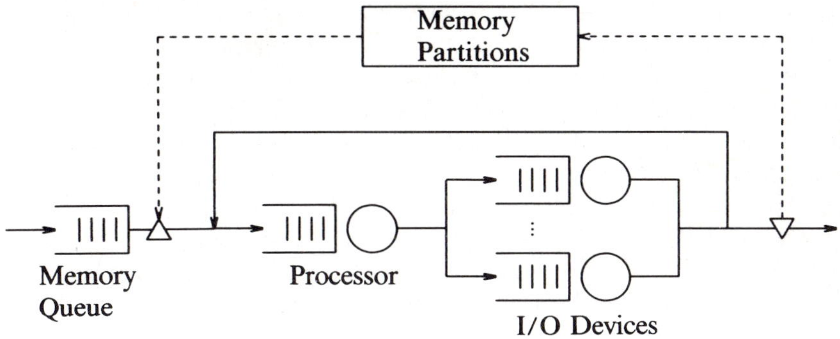

FIGURE 3. *Open queueing network model of a multiprogramming system*

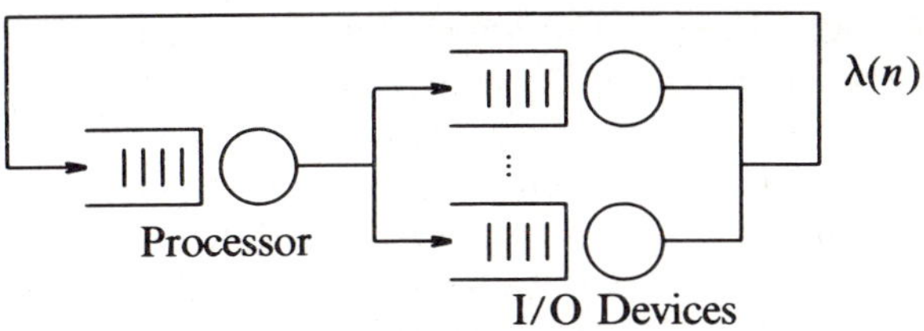

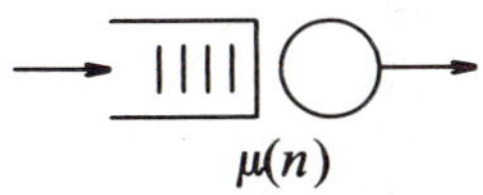

FIGURE 4a.
Subnetwork with n jobs

FIGURE 4b.
Single server with queue length dependent service rates

3. 1975-1980 DEVELOPMENTS

3.1. Product form queueing networks

3.1.1. Product form solution. Computer systems typically serve different types of jobs that can have quite different resource requirements, e.g. batch jobs with long processing times and interactive jobs with short processing times. It is therefore desirable to explicitly represent these different job types in a model and to predict performance measures for each job type. For the product form networks in JACKSON [43] and GORDON and NEWELL [35] all jobs are assumed to be probabilistically identical. In BASKETT ET AL. [8], motivated by computer system modeling applications, the class of product form networks was greatly extended to allow, among other features, different types of jobs. In addition, queues with PS and LCFSPR queueing disciplines and queues with an infinite number of servers, called infinite server (IS) queues, were allowed as well as FCFS queues. Different types of jobs can have different general service time distributions at PS, LCFSPR and IS queues provided the service time distributions have rational Laplace transforms. However, the service time distribution must be exponential at an FCFS queue and have the same mean for each job type. The sequence of queues visited by a job is a Markov chain, often called a routing chain, which can be different for each job type. In a closed network all routing chains are irreducible and jobs never leave the network. Closed networks are commonly used as computer system models with an IS queue used to represent users at terminals as in the machine repair model. In an open network all routing chains have a single absorbing state corresponding to a departure from the network. In a mixed network both types of routing chains exist. We will restrict our consideration to closed networks, although open and mixed networks were also considered in BASKETT ET AL. [8]. As is commonly done we will use the term chain synonymously with job type. We next introduce the notation that appears in the stationary distribution of the number of jobs of each chain at each queue for closed multiple chain product form networks.

Let J denote the number of queues, let K denote the number of chains (equals the number of job types) and let $\mathbf{N}=(N_1, \ldots, N_K)$ denote the network population vector where N_k denotes the number of jobs in chain k. Let ρ_{jk} denote the product of the mean service time and the relative visit ratio for chain k at queue j. (The relative visit ratio for chain k at queue j equals the stationary probability at queue j of Markov routing chain k, possibly times a constant that can depend on k but not on j.) Let $\mu_j(n)$ denote the service rate at queue j when there are n jobs present relative to the service rate when there is one job present. If a queue has a single server and $\mu_j(n)=1$ for all $n \geqslant 1$ then it is said to be a single server fixed rate (SSFR) queue. For an IS queue $\mu_j(n)=n$ for all $n \geqslant 1$. Any other queue is said to be a queue length dependent (QLD) queue. Let n_{jk} denote the number of chain k jobs at queue j, let $\mathbf{n}_j=(n_{j1}, \ldots, n_{jK})$ and let $|\mathbf{n}_j|=n_{j1}+\cdots+n_{jK}$. Then the stationary distribution has a product form given by

$$p(\mathbf{n}_1, \ldots, \mathbf{n}_J) = X_1(\mathbf{n}_1) \ldots X_J(\mathbf{n}_J)/G(\mathbf{N}) \tag{3.1a}$$

where

$$X_j(\mathbf{n}_j) = \prod_{n=1}^{|\mathbf{n}_j|} (n/\mu_j(n)) \prod_{k=1}^{K} (\rho_{jk}^{n_{jk}}/n_{jk}!) \tag{3.1b}$$

and $G(\mathbf{N})$ is a normalizing constant. Note that the stationary distribution depends on the general service time distributions at PS, IS and LCFSPR queues only through their means. This is a generalization of the insensitivity result in BASKETT [7] that was mentioned in Section 2.2.

The probabilistic routing can be made more general without changing the product form solution given above by the use of job classes as in BASKETT ET AL. [8]. There can be multiple job classes at a queue and the sequence of classes visited by a job is a Markov chain. Further a job's service time distribution at a queue can depend on the class it visits provided the queue is not FCFS. The relative visit ratio for a chain at a queue is then obtained by summing the relative visit ratios for the chain over all classes the chain visits at the queue. The mean service time for a chain at a queue is obtained as a weighted average of the mean service times for all classes the job visits at the queue, where the weights sum to one and are proportional to the class relative visit ratios at the queue for the chain.

The product form networks in BASKETT ET AL. [8] are often called BCMP networks after the initials of the authors' last names. They have been widely used in practice and are a landmark development in computer performance modeling. It should be mentioned that a somewhat different, but closely related, formulation of product form queueing networks appeared in KELLY [45,46]. Also the technical restriction that the general service time distributions have rational Laplace transforms was later shown not to be necessary, e.g. BARBOUR [2], CHANDY ET AL. [18] and COHEN [21]. The class of product form networks has been extended somewhat beyond BCMP networks. Several of these extensions are discussed in Chapter 3 of LAVENBERG [59] in which product form networks are discussed in detail. One interesting extension in TOWSLEY [101] allows certain special forms of state dependent routing. However, this extension and others have been rarely used in practice.

3.1.2. Computational algorithms. Computational issues were not discussed in BASKETT ET AL. [8]. The normalizing constant that appears in (3.1a) equals the summation of the numerator in (3.1a) over the set of JK-dimensional vectors with non-negative integral components given by

$$S(\mathbf{N}) = \{(\mathbf{n}_1, \ldots, \mathbf{n}_J): \mathbf{n}_1 + \cdots + \mathbf{n}_J = \mathbf{N}\}.$$

Computing this sum directly requires on the order of

$$\prod_{k=1}^{K} \binom{N_k + J - 1}{J - 1}$$

arithmetic operations. A more efficient method for computing $G(\mathbf{N})$, called the Convolution algorithm, was developed in REISER and KOBAYASHI [80], and is a generalization to BCMP networks of the algorithm developed in BUZEN [15] for single chain networks. Let $G_j(\mathbf{n})$ denote the normalizing constant for a network consisting of queues $1, \ldots, j$ of the original network and having population vector $\mathbf{n} \leqslant \mathbf{N}$ (component by component inequality). Note that $G_J(\mathbf{N})$ equals $G(\mathbf{N})$, the normalizing constant that we wish to compute. The Convolution algorithm is based on the following recursion. For $j = 2, \ldots, J$

$$G_j(\mathbf{n}) = \sum_{\mathbf{i}=\mathbf{0}}^{\mathbf{n}} G_{j-1}(\mathbf{i}) X_j(\mathbf{n} - \mathbf{i}) \quad \mathbf{0} \leqslant \mathbf{n} \leqslant \mathbf{N}, \tag{3.2}$$

where $G_1(\mathbf{n})=X_1(\mathbf{n})$ and $X_j(\mathbf{0})=1$, $j=1, \ldots, J$. The number of arithmetic operations required to compute $G(\mathbf{N})$ using (3.2) is on the order of

$$J\prod_{k=1}^{K}\frac{(N_k+2)(N_k+1)}{2}$$

which is typically much less than is required by a direct summation. If there are SSFR or IS queues in a network then (3.2) can be simplified and fewer operations are required. For example if all queues are SSFR then (3.2) simplifies to

$$G_j(\mathbf{n}) = G_{j-1}(\mathbf{n})+\sum_{k=1}^{K}\rho_{jk}G_{j-1}(\mathbf{n}-\mathbf{e}_k), \quad \mathbf{e}_k\leqslant\mathbf{n}\leqslant\mathbf{N}, \tag{3.3}$$

where $\mathbf{e}_k$ is a K-dimensional vector all of whose components are zero except the k-th component which equals one. Using (3.3) $G(\mathbf{N})$ can be computed with on the order of

$$JK\prod_{k=1}^{K}(N_k+1)$$

arithmetic operations. Further aspects of the Convolution algorithm are discussed in Chapter 3 of LAVENBERG [59], including the efficient computation of throughputs, mean queue lengths, queue length distributions and mean response times from normalizing constants as well as numerical issues. With regard to numerical issues, floating point range problems are inherent in the algorithm, although means were later proposed to at least partially alleviate these problems, e.g. LAM [56].

The Convolution algorithm was an important step in making product form queueing networks useful in practice and it was soon incorporated in queueing network modeling packages, e.g. REISER and SAUER [82]. The next major advance in product form queueing networks was the Mean Value Analysis (MVA) algorithm which appeared in REISER and LAVENBERG [81]. This algorithm allows throughputs, mean queue lengths and mean response times to be computed directly without computing normalizing constants. Let T_{jk} denote the mean service time and let θ_{jk} denote the relative visit ratio for chain k at queue j. Let $\lambda_{jk}(\mathbf{n})$, $L_{jk}(\mathbf{n})$ and $R_{jk}(\mathbf{n})$ denote respectively the throughput, mean queue length and mean response time for chain k at queue j if the network population vector is $\mathbf{n}=(n_1, \ldots, n_K)$. Let $\lambda_k(\mathbf{n})$ denote the chain k throughput at a designated queue that is visited by chain k and let the relative visit ratio at that queue be set to one so that $\lambda_{jk}(\mathbf{n})=\theta_{jk}\lambda_k(\mathbf{n})$. For a closed product form queueing network consisting of only SSFR queues the MVA algorithm is based on the following equations. For $j=1, \ldots, J$, $k=1, \ldots, K$ and $\mathbf{e}_k\leqslant\mathbf{n}\leqslant\mathbf{N}$

$$R_{jk}(\mathbf{n}) = T_{jk}\left[1+\sum_{m=1}^{K}L_{jm}(\mathbf{n}-\mathbf{e}_k)\right] \tag{3.4}$$

$$\lambda_k(\mathbf{n}) = \frac{n_k}{\sum_{j=1}^{J}\theta_{jk}R_{jk}(\mathbf{n})} \tag{3.5}$$

$$L_{jk}(\mathbf{n}) = \theta_{jk}\lambda_k(\mathbf{n})R_{jk}(\mathbf{n}). \tag{3.6}$$

If any queue, say queue j, is IS then (3.4) is replaced by $R_{jk}(\mathbf{n}) = T_{jk}$. Equation (3.6) is Little's equation applied to chain k jobs at queue j while (3.5) is obtained by summing (3.6) over all j. The key equation is the recursion in (3.4) which allows the mean response time for chain k at queue j to be computed in terms of the sum of the mean queue lengths for all chains at queue j for the network with one less customer in chain k. If queue j is FCFS then (3.4) is a direct consequence of the arrival theorem for closed product form queueing networks that appeared in LAVENBERG and REISER [60] and in SEVCIK and MITRANI [91]. One version of the arrival theorem states that the joint distribution of the number of jobs of each chain at each queue on arrival of a chain k job is identical to the joint distribution at an arbitrary time with one less chain k job in the network. Thus (3.4) states that the mean response time at an FCFS queue equals the mean service time plus the mean time to complete service for all jobs in the queue on arrival. We will see in Section 4 how the arrival theorem has provided a basis for MVA-like approximations for non-product form networks. For networks that consist of only SSFR and IS queues the MVA algorithm avoids the floating point range problems that are inherent in the Convolution algorithm. Furthermore, for such networks the computational requirements of MVA are of the same order as for Convolution. In REISER and LAVENBERG [81] networks with QLD queues were also considered and the MVA equations were extended to include a recursion that involves the queue length distribution at each QLD queue. (Hence they no longer involve mean values only.) However, numerical instability can occur when computing the probability of zero jobs at a QLD queue. A way of avoiding this numerical instability at the cost of increased computation can be found in Chapter 3 of LAVENBERG [59].

3.1.3. Approximate solution. It is a consequence of the recursions in (3.3) and (3.4) that the computational requirements of the Convolution and MVA algorithms for product form networks that consist of only SSFR queues grow at least as fast as

$$\prod_{k=1}^{K}(N_k+1).$$

This term can make the computational requirements prohibitive for networks with many chains or a few chains and many jobs per chain. A simple approximation which appeared in a conference abstract, SCHWEITZER [89], that breaks the recursion in (3.4) is

$$L_{jm}(\mathbf{n}-\mathbf{e}_k) = \begin{cases} L_{jm}(\mathbf{n}), & m \neq k, \\ \dfrac{n_k-1}{n_k}L_{jk}(\mathbf{n}), & m = k. \end{cases} \tag{3.7}$$

If this approximation is substituted into (3.4), then (3.4) along with (3.5) and

(3.6) form a set of simultaneous nonlinear equations in the unknowns $\lambda_k(\mathbf{n})$, $L_{jk}(\mathbf{n})$ and $R_{jk}(\mathbf{n})$. An early report on the extensive use of this approximation in computer system models can be found in Appendix B of BARD [5]. The nonlinear equations are solved iteratively by successive substitutions. In the hundreds of product form networks tested there were no convergence problems and the performance measures obtained by solving the equations were usually within five percent of the true values. It was not until later that the existence of a solution to these equations was established. Except for the single chain case, uniqueness and the convergence of iterative solution methods for these equations have yet to be established, nor have error bounds been obtained. Later examples were found in which this approximation yields large errors, e.g. over twenty percent, and new and more accurate approximations were proposed. We will discuss these more accurate approximations in Section 4. Nonetheless this approximation continues to be widely used in computer performance modeling.

3.2. Approximate solution of non-product form queueing networks

3.2.1. Decomposition. One major limitation of product form queueing networks for computer performance modeling applications is that they do not allow the explicit representation of contention for passive resources. The numerical solution of queueing network models of computer performance that are not product form networks is typically not feasible even if the usual Markovian assumptions are made. This is true for open networks such as the one in Fig. 3. It is also true for closed networks since the state space is typically very large due to the number of queues and/or chains and/or jobs in the network. Thus, one is typically forced to approximately solve non-product form queueing network models of computer performance. In Section 2.3 a method for approximately solving the model in Fig. 3 was presented. This is an example of a decomposition method for the approximate solution of certain Markovian non-product form queueing networks which we will call the 'flow equivalent' server method. For reasons to be mentioned later this method is mainly useful for single chain networks. Hence we will restrict our discussion to single chain networks. In decomposition by the flow equivalent server method a subnetwork is replaced by a server with exponentially distributed service times and queue length dependent service rates. The rates are obtained by solving for the throughput of the subnetwork in isolation once for each possible value of the number of jobs that can be in the subnetwork. The aggregated network consisting of this flow equivalent server and the rest of the original network is then solved. If the subnetwork in isolation and the aggregated network are product form networks then they can be efficiently solved. If they are not then they themselves can have state spaces that are too large to allow numerical solution. The flow equivalent server method is based on important results that appeared in the computer performance modeling literature which we now discuss.

In CHANDY ET AL. [17] the flow equivalent server method was shown to yield exact results for single chain product form networks. This result was called Norton's theorem for product form networks due to its analogy with Norton's theorem for electrical circuits in which a subsystem is replaced by a current source and parallel resistance that are equivalent to the original subsystem in terms of their effect on the rest of the system. Norton's theorem for product form networks suggests that the flow equivalent server method may yield fairly accurate approximations for networks that are almost product form networks. This is the case for the network in Fig. 3 under Markovian assumptions if M is large since it becomes a product form network in the limit as M tends to infinity.

A second justification for the use of this method can be found in COURTOIS [24] and COURTOIS [25]. Results from econometrics in SIMON and ANDO [92] on nearly completely decomposable systems were applied to the approximate analysis of closed Markovian queueing network models of computer system performance. Informally a continuous time Markov chain is nearly completely decomposable if, after identical permutations of its rows and columns, its transition rate matrix is almost block diagonal, i.e. the off block diagonal elements are small compared to the block diagonal elements. In such a case transitions between states within a block occur much more frequently than transitions between states in different blocks. The stationary distribution of the Markov chain can be approximated as follows. First each block submatrix is modified so that it is a legitimate transition rate matrix and the stationary distribution is computed for each such modified matrix. Then these within block stationary probabilities and the original off block diagonal elements are used to construct an approximate between block transition rate matrix for a Markov chain having one state for each block. The stationary distribution for this aggregated Markov chain is then computed and combined with the approximate within block stationary distributions to produce an approximation to the stationary distribution of the original Markov chain. The error in this approximation goes to zero in the limit as all of the original off block diagonal entries go to zero.

Decomposition based on the near complete decomposability of the transition rate matrix is in some cases equivalent to decomposition based on the flow equivalent server method. For example consider the closed queueing network shown in Fig. 5, which is a finite source version of the open network in Fig. 3. This is a more detailed model of a time sharing system than the machine repair model discussed in Section 2.2. If C, the mean number of times a job cycles through the subnetwork consisting of the processor and I/O devices, is large then transitions corresponding to jobs circulating within the subnetwork are much more frequent than transitions corresponding to jobs leaving the subnetwork and the Markov chain for the network is nearly completely decomposable. Each block corresponds to a set of states with a fixed number of jobs in the subnetwork and external queue and a state in the aggregated Markov chain is the number of such jobs. One possible set of modifications to the block submatrices corresponds to decomposition by the flow equivalent server method. Thus, near complete decomposability provides a justification for this decomposition when C is large. However, the flow equivalent server method has been found to provide reasonably accurate approximations in practice even if C is small.

We have restricted our discussion of the flow equivalent server method to single chain networks. Although a multiple chain version of the method is possible, it is usually not practical due to computational limitations, e.g. because the multiple chain aggregated network is not a product form network. Other issues that limit the application of the method to multiple chain networks are discussed in Chapter 6 of SAUER and CHANDY [85]. The flow equivalent server method is one of the few approximation methods for queueing network models of computer performance that has a theoretical basis. Therefore its practical limitation to single chain networks is unfortunate.

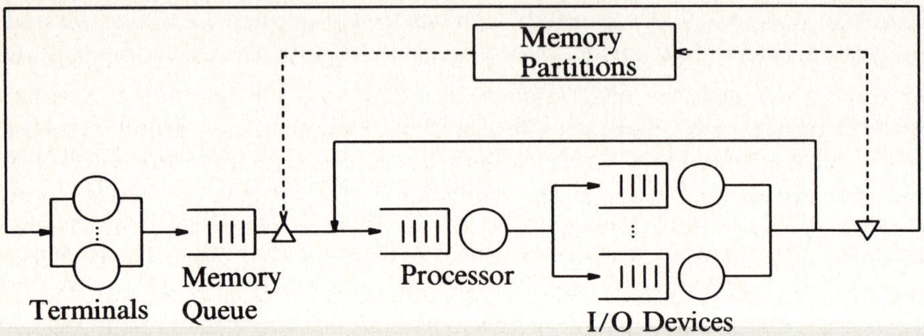

FIGURE 5. *Closed queueing network model of a multiprogramming time sharing system*

3.2.2. Other approximations. Queueing networks became widely used as computer performance models during the 1975-1980 time period. However, applying queueing networks to model a real system often requires a variety of approximations, many of them without a theoretical basis. Approximations were developed during this period to handle such features as non-exponential service times at FCFS queues, priority queueing disciplines and contention for passive resources in multiple chain networks, each of which does not allow a product form network representation. A survey can be found in CHANDY and SAUER [20]. A more accurate method for handling non-exponential service times in single chain networks than the ones discussed in CHANDY and SAUER [20] can be found in MARIE [64]. However, in practice computer performance modelers are usually less concerned with accurate distributional assumptions than with accurately representing structural features of the system being modeled such as priority queueing and contention for passive resources. This is so because experience has shown that incorrect structural assumptions can cause greater modeling errors than incorrect distributional assumptions, particularly for closed queueing networks which are much more widely used than open networks in computer performance modeling.

3.2.3. Applications. We now briefly discuss three successful applications of queueing networks to modeling real multiprogramming time sharing systems. All models are similar to but more complicated than the one shown in Fig. 5. In all three cases validation studies were carried out by comparing model predictions with measurements from a real system and the models were found to be sufficiently accurate for practical purposes. A single chain model of a CDC 6400 computer running a time sharing operating system developed at the University of Texas at Austin is described in BROWN ET AL. [12]. The main difference compared to the model in Fig. 5 is that the amount of memory required by a job is assumed to be random rather than fixed. The flow equivalent server approximation method was used and, in addition, an ad hoc approximation was developed to handle the probabilistic memory requirements. A multiple chain model of an IBM 370 computer running IBM's VM operating system is described in BARD [3,4]. A type of decomposition was used but due to the multiple chains the subnetwork that consists of the processor and I/O devices was replaced by an IS queue rather than by a flow equivalent server. An approximation similar to but less accurate than the one in (3.7) was used to approximately solve the multiple chain subnetwork. Later, as discussed in BARD [5] this approximation was replaced by the one in (3.7). Other approximations were also used. The approximations resulted in a set of simultaneous nonlinear equations relating the unknown performance measures. The equations were solved using successive substitutions. A multiple chain model of an IBM 370 computer running IBM's MVS operating system is discussed, but not described in detail, in BUZEN [16]. Again a variety of approximations were used. Both the VM and MVS models have been extensively validated and incorporated in performance modeling packages for these systems. These packages are used in capacity planning, i.e. in predicting the performance impact of workload growth in these systems and in evaluating the performance improvements due to various equipment upgrades proposed to handle the increased workloads.

3.3. I/O subsystem models

The time spent by a job waiting for and receiving I/O service from a rotating storage device such as a drum or a movable head disk can constitute a major portion of a job's response time in a computer system. Although there were some earlier efforts to model such devices the most interesting results began to be obtained during the 1975-1980 time period. We will discuss two such modeling efforts.

Drum-like devices have one fixed head per track and the heads are aligned. Such devices were modeled in FULLER and BASKETT [33]. Two types of drums were considered, file drums that store variable length records and paging drums that store fixed length records (pages) whose starting positions are aligned across the tracks. In addition two queueing disciplines were considered, FCFS and shortest latency time first (SLTF) in which the next record transferred is one which next passes under the heads. Both the FCFS and SLTF paging drums were modeled using the M/G/1 queue with server walking time that originated in SKINNER [93]. This is an interesting direct application of a result that appeared in the queueing literature to performance modeling. In addition a standard M/G/1 queue was used to model the FCFS file drum. However, the SLTF file drum could not be modeled using an existing queueing model. This is because the latency time (time from when the transfer of one record completes until the transfer of the next one begins) depends in a complicated way on the device's history. Models that ignore this dependency were developed for the SLTF file drum.

Queueing models of an I/O subsystem consisting of multiple movable head disk drives connected to a single channel were developed in WILHELM [108]. This work is of particular interest today due to the prevalence of disk storage devices and since the approach used has formed the basis for subsequent modeling work. A disk service consists of seek, latency and transfer operations, each of which is initiated by the channel which is either released once the operation starts or held throughout the operation. Thus the disks interfere with each other through their contention for the channel which is a passive resource in the sense discussed in Section 3.2. The seek time is the time to move the arm from its current track to the track containing the requested record. The channel is not used during the seek operation so that seeks on different disk drives can be overlapped with each other. The latency is the time for the beginning of the requested record to pass under the head. Two kinds of disk drives were modeled. One kind is typified by the IBM 2314 in which the channel is held during the latency operation. The other kind is typified by the newer IBM 3330 which uses rotational position sensing (RPS). With RPS each disk surface is divided into equal size sectors and the disk hardware senses the beginning of the sector where the requested record is stored without the use of the channel. Thus latencies on different disk drives can be overlapped with each other and with seeks on different drives. However the disk must reconnect to the channel for the transfer to occur and the channel is held throughout the transfer. If the channel is not free when the disk is ready to reconnect then the disk will have to wait for another complete rotation before trying to reconnect again. There can be several of these lost rotations due to so called RPS misses before the reconnect is successful.

The approach in WILHELM [108] is to consider the time from when the drive that stores the requested record becomes free until the transfer of the requested record is complete to be the service time for the drive. Each disk drive is then treated as an M/G/1 queue with this service time. We will discuss this approach only for the second kind of disk drive since it has better performance than the first kind and it is more prevalent in large computer systems today. The components of the service time for the drive are the two waiting times for the channel to become free to initiate the seek and later the latency operations (the times to initiate these operations are negligible and are assumed to be zero), the time for lost rotations due to RPS misses and the seek, latency and transfer times. These times are assumed to be statistically independent. Of the three times due to channel contention the largest one by far is the time due to lost rotations. This is computed by assuming that each attempt to reconnect to the channel is an independent Bernoulli trial with probability of success equal to one minus the channel utilization that is due to other drives. This utilization is known since the mean channel service time equals the known mean transfer time and with an open model the arrival rate to the channel is known. The results of this approximate analysis were compared to simulation results and found to be quite accurate except at high arrival rates in some cases. One source of inaccuracy in the approximation is the assumption that the probability of successful reconnect is the same on each successive attempt. Simulation results indicate that the probability of successful reconnect decreases on successive trials. Attempts to account for this can be found in ZAHORJAN ET AL. [111] and BARD [5]. A queueing model of a more complex I/O subsystem in which, among other things, a disk can be accessed via more than one channel was approximately analyzed in BARD [5]. So far we have only considered a model of an I/O subsystem in isolation. However, an I/O subsystem model is often used as a submodel in an overall system model. This is the case in BARD [5] where the I/O subsystem model was used in the VM model which we discussed in Section 3.2.3. The arrival rates to the I/O subsystem depend on the throughputs in the overall system and the throughputs in the overall system depend on the response times in the I/O subsystem. A type of decomposition was used along with successive substitutions to approximately solve the overall model.

4. POST-1980 DEVELOPMENTS

4.1. Product form queueing networks

The Convolution and MVA algorithms for the exact solution of multiple chain closed product form queueing networks become computationally intractable as the number of chains or the number of jobs per chain increases. Therefore interest quickly developed in the approximate analysis of such networks. The approximation in SCHWEITZER [89] for networks with SSFR and IS queues that we discussed in Section 3.1.3 (see (3.7)) is equivalent to the approximation that for each chain the mean fraction of its jobs in each queue does not change if there is one less job in the network. In reality these fractions do change. A method for estimating this change that yields a more accurate approximation than Schweitzer's can be found in CHANDY and NEUSE [19]. This approximation, called the Linearizer approximation, gives rise to $I+(I-1)K$ sets of simultaneous nonlinear equations to be solved, where K is the number of chains and typically I is two or three, rather than the single set of simultaneous nonlinear equations to be solved when Schweitzer's approximation is used. Each of these sets of equations is typically solved iteratively using successive substitutions. Thus there is a tradeoff between the Schweitzer and Linearizer approximations in terms of accuracy and cost. As with Schweitzer's approximation convergence has not been a problem in practice, although an example presented in ZAHORJAN ET AL. [110] demonstrates that convergence can be very slow with these and similar approximations. Performance measures computed using the Linearizer approximation have been found empirically to be very accurate in all the hundreds if not thousands of test cases considered with errors typically less than one percent and at most a few percent. Hence it can be used in practice with great confidence. The Linearizer approximation has been extended to networks with queue length dependent queues, e.g. KRZESINSKI and GREYLING [54]. However, there is much less empirical evidence about the accuracy of the extension.

One way in which closed queueing networks with many chains, at least dozens and perhaps hundreds, arise is as models of local area networks of computer systems in which jobs originating at one system can use the resources of other systems. In such models different chains are needed to represent jobs originating at different systems. The Linearizer approximation can become computationally expensive for such models. An approximation method that was motivated by the interest in such models can be found in DE SOUZA E SILVA ET AL. [27]. With this method the network is divided into subnetworks. Each subnetwork is solved with the rest of the network replaced by IS queues for some chains and by Poisson sources for the other chains. This is an inexact representation for the rest of the network. While a flow equivalent server representation of a portion of a network is exact for multiple chain product form networks, as shown in KRITZINGER ET AL. [51], such a representation would not reduce the computational cost of solving the network. The mean service times of the IS queues and the arrival rates of the Poisson sources depend on the performance measures for the rest of the network. The set of simultaneous nonlinear equations that results for the performance measures of the network is solved iteratively using successive substitutions. This approximation method was empirically found to be more accurate than Schweitzer's approximation and less costly, but less accurate, than the Linearizer approximation. It also applies to networks with multiple server queues.

Error bounds do not exist for the above approximations and the evidence of their accuracy is empirical. Nontrivial bounds have been derived for performance measures of closed multiple chain product form networks. We mention in particular the bounds based on asymptotic integral expansions in MCKENNA and MITRA [65] and MITRA and MCKENNA [68] and the bound hierarchies in EAGER and SEVCIK [29]. While these bounds can be computed relatively efficiently it appears that the computational effort may exceed that of the Linearizer approximation. Since the Linearizer approximation appears to be so accurate these bounds are less important in practice than they would otherwise be.

In addition to the above developments on approximations and bounds new algorithms for the exact solution of product form networks with many chains were developed. These include the tree convolution algorithm in LAM and LIEN [57] and tree MVA algorithms in TUCCI and SAUER [105] and HOYME ET AL. [41] for networks in which each chain visits only a small subset of the queues and the Recursion by Chain algorithm (RECAL) in CONWAY and GEORGANAS [23] and the Mean Value Analysis by Chain algorithm (MVAC) in CONWAY ET AL. [22] for networks with a large number of chains and few queues.

4.2. Approximate solution of non-product form queueing networks

4.2.1. Memory contention. Despite the important developments in the exact and approximate solution of queueing networks that we discussed in Sections 3.1 and 3.2 there were many shortcomings in the state of the art that existed at the beginning of the post-1980 period. One of the most significant was that the class of product form networks does not permit the explicit representation of contention for passive resources, in particular contention for memory. Decomposition by the flow equivalent server method was known to be an effective method for the approximate solution of single chain queueing networks that represent certain kinds of memory contention. However, the method has computational limitations when applied to multiple chain networks. We will next discuss advances made during the post-1980 period to at least partially overcome this shortcoming.

Consider a multiple chain version of the model shown in Fig. 5 in which the total memory space is partitioned among the chains so that at most M_k chain k jobs can be in memory simultaneously, i.e. at most M_k chain k jobs can be in the subnetwork that represents the processor and I/O devices. Using decomposition by the flow equivalent server method the subnetwork is replaced by a single server queue whose service rate for each chain is a chain dependent function of the number of jobs of each chain in the queue. The service rate functions are obtained by solving the subnetwork in isolation for the throughputs for each chain, once for each possible number of jobs of each chain in the subnetwork. The resulting multiple chain aggregated network is not a product form network and therefore is very costly to solve. (The method was applied to some two chain networks in SAUER [84].) Further approximations were developed in LAZOWSKA and ZAHORJAN [61] and BRANDWAJN [9] to avoid this computational limitation. A separate single chain aggregated network was constructed for each chain. The queue length dependent service rates of the flow equivalent server for chain k were obtained by solving the subnetwork in isolation, once for each possible number of chain k jobs in the subnetwork and with the number of jobs in the subnetwork for all other chains fixed at their mean values. (The two referenced papers use different methods of dealing with the possibly nonintegral values of the means.) Since these mean values are not known a set of simultaneous nonlinear equations arises that relates the various performance measures. These equations are solved iteratively using successive substitutions. Limited empirical evidence suggested that this approximation method is reasonably accurate. (Other approximations that appear to have similar accuracy were developed in THOMASIAN and BAY [97].) This method was extended in LAZOWSKA and ZAHORJAN [61], BRANDWAJN and MCCORMACK [10] and KRZESINSKI and TEUNISSEN [55] to handle memory constraints in which different chains share the same memory space, i.e. the memory space is not partitioned among the chains. However the approximations for shared memory constraints appear to be much less accurate than the approximations for partitioned memory constraints. Unfortunately shared memory constraints are common in computer systems. In THOMASIAN and BAY

[98] it was shown how the above approximations can be used to approximately analyze not only a single memory constrained system, but also a distributed system in which multiple memory constrained systems are interconnected by a communication network.

4.2.2. Priority queues. We next consider multiple chain queueing networks that violate the condition for product form by having one or more priority queues. Priority queueing disciplines are often used in computer systems to schedule jobs on a processor and can have a major impact on performance. A practically useful method for the approximate solution of networks with priority queues can be found in BRYANT ET AL. [13]. Earlier approximate solution methods for such networks are also reviewed there. Service times at priority queues are chain dependent and assumed to be exponentially distributed. The priorities are also chain dependent. Both nonpreemptive, also called head of line, and preemptive resume priority queues are allowed. Closed, open and mixed networks are considered. The approximation method is based on modifying the key recursion of Mean Value Analysis, given in (3.4), to approximately reflect the effect of the priority queueing discipline on the mean response times for a priority queue. This modification is in part based on assuming that the arrival theorem holds, which is an approximation for a non-product form network such as one with priority queues. Other approximations are also made. If all chains have equal priority at a priority queue then the queue reduces to an FCFS queue. In this case the modified recursion reduces to one used in REISER [79] to handle an FCFS queue with chain dependent mean service times. One thing that is particularly noteworthy about BRYANT ET AL. [13] is the systematic error study that was done to help establish conditions under which the approximation method can be expected to yield reasonably accurate results. A modification to the above method for the case of preemptive resume priority queues that can yield somewhat better accuracy can be found in VAN DOREMALEN ET AL. [28].

4.2.3. Lock contention. One of the major commercial uses of computer systems in the 1980s has been for transaction processing, e.g. airline reservation and banking systems. A transaction consists of a sequence of reads and/or writes of items in a data base interleaved with processing. If writes are allowed then some form of concurrency control is used to regulate the simultaneous access of multiple transactions to the same data item in order to guarantee correct transaction execution. Although many concurrency control methods have been proposed the most common method in practice is locking. In its most basic form there is one lock per data item and a transaction that wishes to access a data item first obtains the lock for that item. It releases the lock at some time after it completes its operations on the item, typically at the end of the transaction. In this way at most one transaction at a time can access and operate on a data item. In general there can be different types of locks for a data item depending on the operation to be performed, e.g. a read lock, which multiple transactions can share (also called a shared lock), and a write lock, which at most one transaction can use (also called an exclusive lock). The use of read and write locks allows multiple transactions which want only to read a data item to do so simultaneously. A transaction may request all of its locks at the beginning of the transaction, so called static locking, or it may request locks one at a time when the data item to be locked is referenced during transaction execution, so called dynamic locking. In a transaction processing system transactions contend not only for the memory, processor and I/O resources of the system, but they also contend for locks. It appears that lock contention does not have a significant impact on performance in current transaction processing systems, but it may have in the future. This is because with transaction processing systems running on faster processors and/or in parallel on more processors it is expected that greater numbers of transactions will attempt to simultaneously access the same data thereby causing increased lock contention.

A lock is a passive resource in the sense discussed in Section 2.3 and the exact solution of queueing network models that represent both lock contention and contention for hardware resources is not feasible. In order to simplify the problem of solving such models hardware contention is either ignored or it is included by representing the network of hardware resources, e.g. processor and I/O devices, by a flow equivalent server or other simple representation. Even with hardware contention ignored the only lock contention model that has been solved exactly is a model of static locking in which transactions that encounter lock contention do not queue but are discarded. The main performance measure that is solved for is the probability of lock contention. The analysis of this model in the case of exclusive locks only can be found in MITRA and WEINBERGER [69]. An extension to handle both shared and exclusive locks can be found in MITRA [67]. However, static locking without queueing is of little practical interest and the analysis method in the above papers does not appear to be extendable to other locking schemes. What is perhaps the most practically interesting approach to the approximate solution of lock contention models can be found in TAY ET AL. [95,96]. A very simple representation of hardware contention is used. In particular it is assumed that the mean execution time of a transaction in the absence of lock contention is a known function of the mean number of executing transactions. Several locking schemes are analyzed including static locking with queueing and dynamic locking with and without queueing. Simple approximations involving mean values are used in the analysis. Performance measures solved for include the probability of lock contention, transaction throughput and mean transaction response time. Comparisons with simulation results show the analytical results to be quite accurate. The analytical results are used to compare the performance of the various locking schemes and they reveal certain interesting phenomena such as thrashing due to data contention in which the throughput starts to decrease as a function of the number of transactions in the system when the number exceeds a certain value. Other papers on the approximate analysis of lock contention models include THOMASIAN and RYU [100], MORRIS and WONG [71,72] and RYU and THOMASIAN [83]. In these papers a flow equivalent server is used to represent the effect of hardware contention. First a model that represents lock contention only is approximately solved. Then the results of that analysis are combined with the flow equivalent server in an aggregated model to compute the transaction throughput and mean response time. (The approach in these papers is somewhat similar to the approach in POTIER and LEBLANC [78], although the model in this earlier paper is of much less interest than the models in the later papers.)

In the above papers very simple assumptions are made about the lock referencing characteristics of a transaction, e.g. if n locks are requested by a transaction then all subsets of size n are equally probable. (In TAY ET AL. [95,96] somewhat more general assumptions are handled.) There is no empirical evidence to support this kind of assumption nor have the results of the approximate analyses been compared to measurements of real systems. The simulation models used to test the accuracy of the approximate analyses make the same simple lock referencing assumptions. Thus, although existing lock contention models have provided some insight into the relative performance of different locking schemes and into certain performance phenomena, their ability to predict the performance of real systems has not been demonstrated.

4.2.4. Parallel processing. So far we have assumed that, with the exception of passive resources, a job cannot queue for or receive service from more than one resource at a time. Thus, a job receives its services sequentially at the active (non-passive) resources of the system. However, in some systems a job can receive some of its services in parallel. There are many examples. Some systems support the overlap of processor and I/O services for a job. This is one of the earliest examples of parallel processing. More recently some systems, e.g. the operating system for the Cray X-MP and IBM's MVS operating system, contain a multitasking feature that allows a job to spawn multiple tasks that can execute in parallel on a multiprocessor system. The spawning job waits until all of its spawned tasks have completed service before it resumes execution. We will call this fork/join parallelism. Another form of parallelism is exemplified by update propagation in Locus, a distributed operating system developed at UCLA that supports replicated files. When a job closes a file that it has updated, update propagation tasks are spawned on each machine that stores an old version of the file. In this case the spawning job does not wait and the spawned tasks receive service in parallel with the spawning job. We will call this asynchronous parallelism. While parallel processing of a job is possible on current computer systems, it is expected to become much more prevalent in the future. Some systems already exist and others are being built in which large numbers of processors are interconnected in order to provide parallel processing of a job. Examples of experimental parallel processing systems are the Cosmic Cube developed at Cal Tech, SEITZ [90], and the Research Parallel Processor Prototype (RP3) being developed at the IBM Research Center, PFISTER ET AL. [77]. A recent list of U.S. companies offering or building parallel processing systems can be found in FRENKEL [32].

A queueing network in which a job receives some of its services in parallel is not a product form network. An approximate solution method for single chain closed queueing network models that represent processor-I/O overlap is presented in TOWSLEY ET AL. [104]. We will focus on more recent developments in this section. The queueing networks whose approximate solution we will discuss would have been product form networks if there were no parallel processing of a job.

The case of asynchronous parallelism is considered in HEIDELBERGER and TRIVEDI [38] for single chain closed networks. The rate λ at which the spawned tasks are generated is a known constant times the unknown job throughput. The approximation is made that the spawned tasks are generated according to an independent Poisson process with arrival rate λ. The resulting network is a mixed product form network with one open and one closed chain whose closed chain throughput is a nonlinear function of the unknown open chain arrival rate λ. A simple method is used to iteratively solve the nonlinear equation that results. Necessary and sufficient conditions are obtained for the existence and uniqueness of the solution of this equation. The accuracy of the approximation is studied through extensive comparisons with simulations and found to be quite accurate in most cases. This approximation method can be extended in the obvious way to multiple chain closed networks. However the existence and uniqueness of the solution to the resulting simultaneous nonlinear equations is an open problem and the accuracy of the approximation has not been investigated.

The case of fork/join parallelism is considered in HEIDELBERGER and TRIVEDI [39] for single chain closed networks. It is assumed that each job spawns a fixed number of tasks that can be of different types. A job can be either active or waiting for its spawned tasks to complete service. A spawned task can be either active, waiting to be spawned or waiting for its sibling tasks to complete service. Two approximate solution methods are developed, the more accurate of which is a decomposition approximation. The state of the aggregated model is the number of active jobs and the number of active spawned tasks of each type and the state is assumed to evolve in time as a Markov chain. Approximate transition rates for this Markov chain are computed by solving a multiple chain closed product form queueing network with one chain for the active jobs and one chain for each type of the active spawned tasks. The network is solved once for each such possible population vector. (An approximation is also used to estimate whether any siblings of a spawned task are active when a spawned task completes service.) While this approximation method is empirically found to be highly accurate it becomes computationally expensive as the number of spawned tasks increases. Results are only presented for two and three spawned tasks. However parallel processing systems with much higher degrees of parallelism will become common in the future.

A more complex model of parallelism is considered in THOMASIAN and BAY [99]. A job is represented by a directed acyclic task graph whose edges indicate precedence relationships among tasks. All immediate predecessors of a task must complete service before the task can start and each task can be of a different type. An approximate solution method for a closed queueing network with a single such job is presented. The method is similar to the decomposition method in HEIDELBERGER and TRIVEDI [39]. The set of tasks that can be active simultaneously is assumed to evolve in time as a Markov chain whose approximate transition rates are obtained by solving a closed queueing network, once for each possible set of active tasks. This approximation method becomes computationally expensive for large task graphs. A comparison with simulation is only given for one example, a six task graph.

Fork/join parallelism is considered in NELSON and TANTAWI [73] in the case of an open queueing network consisting of Q identical parallel single server queues. Job arrivals are Poisson and each job spawns exactly Q tasks each of which joins a different one of the parallel queues. The job itself does not require any service, only its Q tasks do. All task service times are independent and identically distributed exponential random variables. A simple but very accurate closed form approximation is obtained for the mean job response time. In NELSON ET AL. [74] the case is considered where there is a single multiserver queue with Q servers instead of Q parallel single server queues. Recurrence relations are obtained to numerically compute the mean job response time. The mean response times for these two queueing systems with parallel processing are compared with each other and with the cases where a job's tasks must be served sequentially in these two queueing systems. The best performance is found to be obtained for the single queue parallel processing case.

4.2.5. Processor-memory interconnection. In a shared memory multiprocessor system multiple processors are connected via an interconnection network to multiple shared memory modules. In addition each processor can have its own private memory. A configuration that has frequently been modeled is the multiple bus configuration shown in Fig. 6. The throughput of jobs in this subsystem is limited to less than P times the throughput of a single processor, where P is the number of processors, due to memory and bus contention. A closed queueing network model of that configuration, using passive resources to represent the shared memory modules, is shown in Fig. 7. The number of jobs in the network is equal to the number of processors. A job executes continuously on a processor, accessing data in its private memory, until an access to shared memory is required. The job then queues for the particular shared memory module that contains the data to be accessed. This shared memory module is assumed to be chosen probabilistically. When the memory module is free it is instantaneously allocated to the job which then queues for a shared bus. When a bus is free the memory module and bus are then held for the memory access time after which they are instantaneously released and the job resumes execution on the processor. In this model the time to arbitrate and set up the processor-memory connection is assumed to be zero. Even if the processor execution times and memory access times are assumed to be exponentially distributed, this network is not a product form network due to the passive resource queues. Furthermore, if fixed sized units of data are accessed, e.g. pages, the memory access time is a constant.

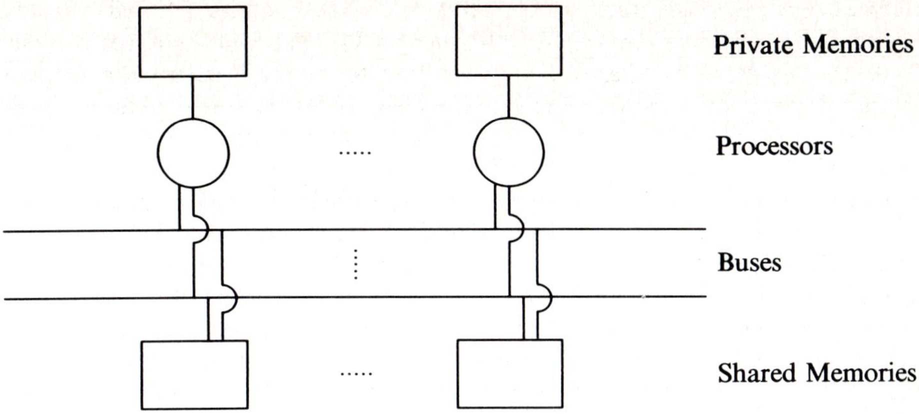

FIGURE 6. *Multiple bus configuration*

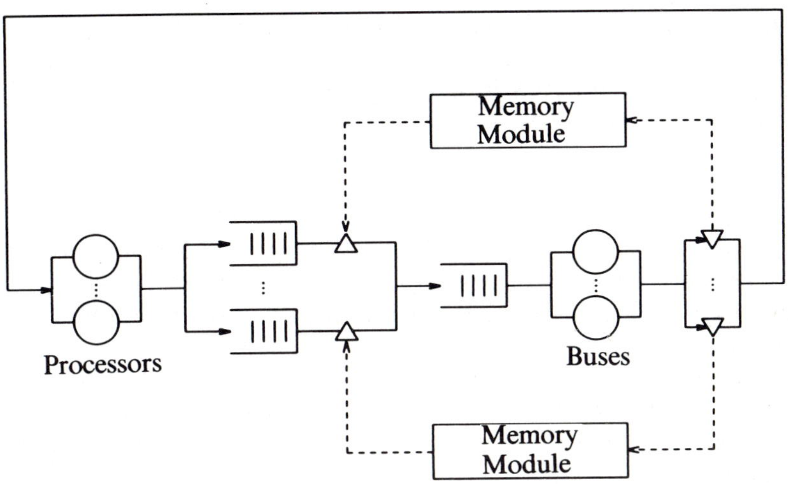

FIGURE 7. *Closed queueing network model of a multiple bus configuration*

Unless stated otherwise all jobs are assumed to belong to the same chain, i.e. they have the same processor execution time distribution, the same memory access time distribution and the same memory access probabilities. An exact solution for the mean number of busy processors is obtained in IRANI and ONYUKSEL [42] under the assumptions of exponential processor execution and memory access times and equal probabilities of accessing any shared memory module. The solution is obtained by exploiting symmetries in the resulting continuous time Markov chain. The solution is a closed form expression in terms of $P+1$ quantities that are related by a one step recursion. Several papers have developed very accurate approximation methods for the model in Fig. 7 under more general assumptions. In GOYAL and AGERWALA [36] the case of arbitrary processor execution time distribution, constant memory access time and equal memory access probabilities is considered. An approximation method developed in HOOGENDOORN [40] for the case of no interconnection network contention, equivalent to $B \geqslant M$ where B is the number of buses and M is the number of memory modules, is extended to the case $B < M$. In addition extensions are made to handle models that incorporate other features such as the effect of disk accesses that are caused by shared memory misses. In part this approximation method involves assuming that when a job encounters a busy memory module it immediately resumes execution on the processor. The case of unequal memory access probabilities is considered in TOWSLEY [102,103]. Two approximation methods are developed. One is a decomposition method in which the memory and bus subnetwork is replaced by a flow equivalent server. This method is applied to a discrete time model with constant memory access time in TOWSLEY [102]. The processor execution times are assumed to be geometrically distributed multiples of the memory access time. The method is applied to the case of exponential memory access times and general processor execution times in TOWSLEY [103]. In addition a second approximation method is developed in TOWSLEY [103] that involves removing the buses from the queueing network and accounting for the mean bus queueing delay by extending the mean memory access time by the unknown mean bus queueing delay. Given an estimate of the mean bus queueing delay, the resulting product form network can be solved for the distribution of the number of busy memory modules. This distribution can in turn be used to compute a new estimate of the mean bus queueing delay resulting in an iterative solution method. The case where not all jobs belong to the same chain can be handled using this method subject only to the computational limitations of solving multiple chain product form networks. Many other papers have considered the approximate solution of the model in Fig. 7 or variants of that model. We refer the reader to the papers we have discussed for further discussion and references.

Multistage banyan networks are a class of networks for interconnecting large numbers of processors and memory modules. They consist of a series of switching stages that provide a unique path from every network input port to every network output port. Each switching stage consists of multiple switches in parallel. Unless internal buffers are provided, if multiple inputs to a switch are destined for the same output port one will be transmitted and the others will be lost. A simple approximate expression for the mean response time for a buffer banyan network of square k by k switches is obtained in KRUSKAL and SNIR [52]. Infinite capacity buffers are assumed for each switch output port so that no inputs are lost. A discrete time queueing model is used. The numbers of arrivals to the network at different network input ports and at different discrete time instants are assumed to be independent and identically distributed Bernoulli random variables. Let p denote the probability of an arrival to a network input port at any discrete time instant. Then the numbers of arrivals at successive discrete time instants to any output queue of the first stage are binomial random variables with parameters p/k and k. This allows a simple exact expression to be obtained for the mean response time for the first stage. Unfortunately the numbers of arrivals at successive time instants at an output port of any stage other than the first are not independent and binomial. The approximate expression in KRUSKAL and SNIR [52] is obtained by making the independent binomial assumption for all output queues so that the mean response time for a K stage network is K times the mean response time for the first stage. Simulations show that the mean response for the first stage is less than that for any other stage. The use of this approximate expression in modeling the interconnection network of the RP3 parallel processing system being developed at the IBM Research Center is discussed in NORTON and PFISTER [76]. Results for more general arrival processes to the output queues of the first stage are obtained in KRUSKAL ET AL. [53]. In particular the number of arrivals at any discrete time instant can have a general distribution. Exact expressions are derived for the mean and variance of the response time for the first stage. In addition simulation results and the exact expressions for the first stage are combined to obtain approximate expressions for the response time statistics for stages other than the first and for the overall response time statistics.

4.2.6. Miscellaneous. In JACOBSON and LAZOWSKA [44] an approximate solution method is developed for closed multiple chain queueing networks that have population constrained subnetworks, such as can occur when there are passive resource queues in the network. The method is applied to networks that represent serialization delays in computer systems, e.g. multiple jobs trying to access a critical section of an operating system. Like many other approximate solution methods for non-product form networks this method is an iterative one that involves solving an interrelated collection of simplified networks. In each simplified network some of the resource contention in the original network is represented in an approximate way (many such ways are possible) so that the simplified network is a product form network. Some of the input parameter values for a simplified network depend on performance measures for the other simplified networks. Hence an iterative solution is required. A survey of such iterative methods can be found in DE SOUZA E SILVA ET AL. [26].

Further advances were made in the approximate solution of I/O subsystem models. Examples include BARD [6] and BRANDWAJN [11].

4.3. Load sharing

When computer systems, each having their own workload, are interconnected, e.g. via a local area network, the possibility exists of sharing the workload among the systems. Appropriate software must be provided to do this. An algorithm that is used to decide which jobs should be processed on which systems is called a load sharing algorithm. A load sharing algorithm can be based on static information, e.g. the mean service requirements of jobs, on dynamic information, e.g. the instantaneous queue lengths at each system's resources, or on a combination of both. To date only very simple queueing models have been used to analyze the effect of load sharing on system performance and much more work remains to be done. We will summarize some recent results. A taxonomy for load sharing algorithms and a summary of earlier results can be found in WANG and MORRIS [107].

Optimal static load sharing is considered in TANTAWI and TOWSLEY [94]. An open product form queueing network is used to model the interconnected systems. The processor and I/O resources of each system as well as the communication network that interconnects the systems can be represented in the model. The subnetworks that represent different computer systems need not be identical. When a job arrives at a computer system a probabilistic routing decision is made as to which system the job will be processed on, either the system at which the job arrived or another system. The job is processed in its entirety on the chosen system. A job that is transferred to another system for processing incurs a communication network delay, first to transfer the job and then to transfer the results. The processor overheads that occur when a job and its results are transferred are not represented in the model. The routing decision is based on the mean service requirements of the job at the resources of each system if the job were processed on that system and the mean communication delays if the job were transferred. It is assumed that a job can be processed on any system and that all jobs have the same mean service requirements and the same mean communication delays. The optimization problem considered is to determine the set of routing probabilities that minimizes the overall mean response time. Efficient algorithms are developed to compute the optimum probabilities.

While static load balancing has the capability of improving performance, as illustrated in the above work, the proper use of dynamic state information should yield even greater improvements in performance. However, it may be very costly to collect state dependent information and to keep this information consistent across systems and up to date. The possible benefits of using simple state dependent information in dynamic load sharing algorithms are investigated in EAGER ET AL. [30,31] and in LEE and TOWSLEY [62]. In these papers each system is represented as an identical M/M/1 queue in the absence of dynamic load sharing. (A more general arrival process is also considered in LEE and TOWSLEY [62].) Thus the load is optimally shared in the static sense. It is assumed that a job can be processed on any system and that a job is processed in its entirety on a single system. In EAGER ET AL. [30,31] the processor overhead required to transfer a job is represented but the communication network delay is not. In LEE and TOWSLEY [62] the processor overhead is not represented but the communication network delay is. The load sharing algorithms considered are of the threshold type, i.e. the decisions of whether or not to transfer a job and if so where to transfer it to, depend on whether certain queue lengths exceed a threshold. Several varieties of threshold algorithms are considered. For example, in one algorithm considered in EAGER ET AL. [30,31] a computer system tries to transfer a newly arrived job if its queue length including that job would exceed a threshold T. Up to L different systems are then probed independently and at random. The job is transferred to the first such system whose queue length including that job would not exceed *T*. If no such system is found the job is processed on the original system. This algorithm is found to yield very good performance compared to more complex algorithms and to yield substantially better performance than if dynamic load sharing were not used. Despite the simple queueing network model of the interconnected systems, i.e. identical M/M/1 queues, the model has not been solved exactly for the dynamic load sharing algorithms considered in these papers. In order to obtain an approximate solution it is assumed that the states of each queue evolve in time as independent Markov processes and the arrival process of transferred jobs at a system is Poisson.

The models used to analyze the effects of dynamic load sharing are overly simplistic in that each system is represented as a single server queue. Furthermore, it is assumed that a job is processed in its entirety on a single system. This may not be the case in a distributed database system. For example consider a distributed database system in which the database is partitioned among interconnected systems. A job (transaction) consists of a sequence of database requests interleaved with intervals of processor execution. A job can be started on any system but any database request issued by the job that references data on another system is sent to that system for processing with the results sent back to the system that originated the request. Thus the processing of a job is distributed among multiple systems. Dynamic load sharing algorithms for such systems are considered in
YU ET AL. [109]. It is assumed that all jobs arrive to a front end system that is used to route jobs to the other systems for processing according to the load sharing algorithm. The queueing network model used is more complex and realistic than in previous studies of dynamic load sharing. However simulation, rather than analysis, is used to estimate performance.

5. CONCLUSIONS

We have reviewed some of the major advances in the analysis of queueing models of computer system and subsystem performance. The emphasis has been on advances of practical importance. Among the most fundamental results are the exact solution of and computational algorithms for product form networks. However, as we have seen many important computer system features lead to models that are non-product form queueing networks. These features include memory contention, priority queues, lock contention, parallel processing and dynamic load sharing. Non-product form queueing networks also arise as models of I/O subsystems and processor-memory interconnection. A variety of approximate solution methods have been developed for these non-product form networks. These include decomposition based on the flow equivalent server method, approximate versions of the Mean Value Analysis algorithm and iterative methods. With some of these methods product form networks are solved at intermediate steps. The assessment of the accuracy of these approximation methods has typically been done by comparison with simulation results. This accuracy assessment was quite thorough and systematic for the approximation method for priority queues which we discussed in

Section 4.2.2 and less so for other methods. Some of these methods have been used to approximately solve models of large computer systems that have been incorporated in commercial computer performance modeling packages. Descriptions of early versions of two such models can be found in BARD [3,4] and BUZEN [16].

Some of the models that we have considered are too simplistic to be realistic representations of existing systems or possible future systems. This is for the most part true for models that represent lock contention, parallel processing or dynamic load sharing. These models have, however, been used to gain some insight into aspects of performance issues for systems that have these features. With the anticipated increase of database systems with high levels of concurrency, parallel processing systems and distributed systems we expect the development of more realistic models of such systems and their approximate solution to be a challenging area for future research.

ACKNOWLEDGEMENT

I would like to thank R. Nelson for his detailed comments on an earlier version of this paper.

REFERENCES

1. B. AVI-ITZHAK, D.P. HEYMAN (1973). Approximate queueing models for multiprogramming computer systems. *Oper. Res. 21,* 1212-1229.
2. A.D. BARBOUR (1976). Networks of queues and the method of stages. *Adv. in Appl. Probab. 8,* 584-591.
3. Y. BARD (1978). The VM/370 performance predictor. *Comput. Surveys 10,* 333-342.
4. Y. BARD (1978). An analytic model of the VM/370 system. *IBM J. Res. Develop. 22,* 498-508.
5. Y. BARD (1980). A model of shared DASD and multipathing. *Commun. ACM 23,* 564-572.
6. Y. BARD (1982). Modeling I/O systems with dynamic path selection, and general transmission networks. *Perform. Eval. Rev. 11, Proc. ACM SIGMETRICS Conf. Meas. Modeling Comput. Syst.,* 118-129.
7. F. BASKETT (1971). The dependence of computer system queues upon processing time distribution and central processor scheduling. *Proc. ACM SIGOPS Third Symposium on Operating Systems Principles,* 109-113.
8. F. BASKETT, K.M. CHANDY, R.R. MUNTZ, F. PALACIOS-GOMEZ (1975). Open, closed and mixed networks of queues with different classes of customers. *J. Assoc. Comput. Mach. 22,* 248-260.
9. A. BRANDWAJN (1982). Fast approximate solution of multiprogramming models. *Perform. Eval. Rev. 11, Proc. ACM SIGMETRICS Conf. Meas. Modeling Comput. Syst.,* 141-149.
10. A. BRANDWAJN, W.M. MCCORMACK (1984). Efficient approximation for models of multiprogramming with shared domains. *Perform. Eval. Rev. 12, Proc. ACM SIGMETRICS Conf. Meas. Modeling Comput.*

Syst., 186-194.

11. A. BRANDWAJN (1983). Models of DASD subsystems with multiple access paths: A throughput-driven approach. *IEEE Trans. Comput. C-32,* 451-463.
12. R.M. BROWN, J.C. BROWNE, K.M. CHANDY (1977). Memory management and response time. *Commun. ACM 20,* 153-165.
13. R.M. BRYANT, A.E. KRZESINSKI, M.S. LAKSHMI, K.M. CHANDY (1984). The MVA priority approximation. *ACM Trans. Comput. Syst. 2,* 335-359.
14. J.P. BUZEN (1971). *Queueing Network Models of Multiprogramming,* Ph.D. dissertation, Div. Eng. Appl. Phys., Harvard Univ., Cambridge, MA.
15. J.P. BUZEN (1973). Computational algorithms for closed queueing networks with exponential servers. *Commun. ACM 16,* 527-531.
16. J.P. BUZEN (1978). A queueing network model of MVS. *Comput. Surveys 10,* 319-331.
17. K.M. CHANDY, U. HERZOG, L. WOO (1975). Parametric analysis of queueing networks. *IBM. J. Res. Develop. 19,* 36-42.
18. K.M. CHANDY, J.H. HOWARD, JR., D.F. TOWSLEY (1977). Product form and local balance in queueing networks. *J. Assoc. Comput. Mach. 24,* 250-263.
19. K.M. CHANDY, D. NEUSE (1982). Linearizer: A heuristic algorithm for queueing network models of computer systems. *Commun. ACM 25,* 126-134.
20. K.M. CHANDY, C.H. SAUER (1978). Approximate methods for analysis of queueing network models of computer systems. *Comput. Surveys 10,* 263-280.
21. J.W. COHEN (1979). The multiple phase service network with generalized processor sharing. *Acta Informatica 12,* 245-284.
22. A.E. CONWAY, E. DE SOUZA E SILVA, S.S. LAVENBERG (1988). Mean value analysis by chain of product form queueing networks. *IEEE Trans. Comput.,* to appear.
23. A.E. CONWAY, N.D. GEORGANAS (1986). RECAL - A new efficient algorithm for the exact analysis of multiple chain queueing networks. *J. Assoc. Comput. Mach. 33,* 768-791.
24. P.J. COURTOIS (1975). Decomposability instabilities and saturation in multiprogramming systems. *Commun. ACM 18,* 371-377.
25. P.J. COURTOIS (1977). *Decomposability: Queueing and Computer System Applications,* Academic Press, New York.
26. E. DE SOUZA E SILVA, S.S. LAVENBERG, R.R. MUNTZ (1983). A perspective on iterative methods for the approximate analysis of closed queueing networks. G. IAZEOLLA, P.J. COURTOIS, A. HORDIJK (eds.). *Mathematical Computer Performance and Reliability,* North-Holland, Amsterdam, 225-244.
27. E. DE SOUZA E SILVA, S.S. LAVENBERG, R.R. MUNTZ (1986). A clustering approximation technique for queueing network models with a large

number of chains. *IEEE Trans. Comput. C-35,* 419-430.

28. J.B.M. VAN DOREMALEN, J. WESSELS, R. WIJBRANDS (1986). Approximate analysis of priority queueing networks. O.J. BOXMA, J.W. COHEN, H.C. TIJMS (eds.). *Teletraffic Analysis and Computer Performance Evaluation,* North-Holland, Amsterdam, 117-131.
29. D.L. EAGER, K. C. SEVCIK (1986). Bound hierarchies for multiple class queueing networks. *J. Assoc. Comput. Mach. 33,* 179-206.
30. D.L. EAGER, E.D. LAZOWSKA, J. ZAHORJAN (1986). Adaptive load sharing in homogeneous distributed systems. *IEEE Trans. Software Eng. SE*-12, 662-675.
31. D.L. EAGER, E.D. LAZOWSKA, J. ZAHORJAN (1986). A comparison of receiver-initiated and sender-initiated adaptive load sharing. *Performance Evaluation 6,* 53-68.
32. K.A. FRENKEL (1986). Evaluating two massively parallel machines. *Commun. ACM 29,* 752-758.
33. S.H. FULLER, F. BASKETT (1975). An analysis of drum storage units. *J. Assoc. Comput. Mach. 22,* 83-105.
34. D.P. GAVER (1967). Probability models for multiprogramming computer systems. *J. Assoc. Comput. Mach. 14,* 423-438.
35. W.J. GORDON, G.F. NEWELL (1967). Closed queueing systems with exponential servers. *Oper. Res. 15,* 254-265.
36. A. GOYAL, T. AGERWALA (1984). Performance analysis of future shared storage systems. *IBM J. Res. Develop. 28,* 95-108.
37. P. HEIDELBERGER, S.S. LAVENBERG (1984). Computer performance evaluation methodology. *IEEE Trans. Comput. C-33,* 1195-1220.
38. P. HEIDELBERGER, K.S. TRIVEDI (1982). Queueing network models for parallel processing with asynchronous tasks. *IEEE Trans. Comput. C-31,* 1099-1108.
39. P. HEIDELBERGER, K.S. TRIVEDI (1983). Analytic queueing models for programs with internal concurrency. *IEEE Trans. Comput. C-32,* 73-82.
40. C.M. HOOGENDOORN (1977). A general model for memory interference in multiprocessors. *IEEE Trans. Comput. C-26,* 990-1005.
41. K.P. HOYME, S.C. BRUELL, P.V. AFSHARI, R.Y. KAIN (1986). A tree-structured mean value analysis algorithm. *ACM Trans. Comput. Syst. 4,* 178-185.
42. K.B. IRANI, I.H. ONYUKSEL (1984). A closed-form solution for the performance analysis of multiple-bus multiprocessor systems. *IEEE Trans. Comput. C-33,* 1004-1012.
43. J.R. JACKSON (1963). Jobshop-like queueing systems. *Management Sci. 10,* 131-142.
44. P.A. JACOBSON, E.D. LAZOWSKA (1983). A reduction technique for evaluating queueing networks with serialization delays. A.K. AGRAWALA, S.K. TRIPATHI (eds.). *PERFORMANCE '83,* North-Holland, Amsterdam, 45-59.
45. F.P. KELLY (1975). Networks of queues with customers of different types. *J. Appl. Probab. 12,* 542-554.

46. F.P. KELLY (1976). Networks of queues. *Adv. in Appl. Probab. 8,* 416-432.
47. L. KLEINROCK (1964). Analysis of a time-shared processor. *Naval Research Logistics Quarterly 11,* 59-73.
48. L. KLEINROCK (1967). Time-shared systems: A theoretical treatment. *J. Assoc. Comp. Mach. 14,* 242-261.
49. L. KLEINROCK (1976). *Queueing Systems, Volume 2, Computer Applications,* New York, Wiley.
50. L. KLEINROCK (1988). Performance evaluation of distributed computer-communication systems. O.J. BOXMA, R. SYSKI (eds.). *Queueing Theory and its Applications - Liber Amicorum for J.W. Cohen,* North-Holland, Amsterdam.
51. P. KRITZINGER, S. VAN WYK, A. KRZESINSKI (1982). A generalization of Norton's theorem for multiclass queueing networks. *Performance Evaluation 2,* 98-107.
52. C.P. KRUSKAL, M. SNIR (1983). The performance of multistage interconnection networks for multiprocessors. *IEEE Trans. Comput. C-32,* 1091-1098.
53. C.P. KRUSKAL, M. SNIR, A. WEISS (1986). The distribution of waiting times in clocked multistage interconnection networks. *Proc. Intl. Conf. Parallel Processing,* 12-19.
54. A. KRZESINSKI, J. GREYLING (1984). Improved lineariser methods for queueing networks with queue dependent centres. *Perform. Eval. Rev. 12, Proc. ACM SIGMETRICS Conf. Meas. Modeling Comput. Syst.,* 41-51.
55. A. KRZESINSKI, P. TEUNISSEN (1985). Multiclass queueing networks with population constrained subnetworks. *Perform. Eval. Rev. 13, Proc. ACM SIGMETRICS Conf. Meas. Modeling Comput. Syst.,* 128-139.
56. S.S. LAM (1982). Dynamic scaling and growth behavior of queueing network normalization constants. *J. Assoc. Comput. Mach. 29,* 492-513.
57. S.S. LAM, Y.L. LIEN (1983). A tree convolution algorithm for the solution of queueing networks. *Commun. ACM 26,* 203-215.
58. E.R. LASSETTRE, A.L. SCHERR (1972). Modeling the performance of OS/360 time sharing option (TSO). W. FREIBERGER (ed.). *Statistical Performance Evaluation,* Academic Press, New York, 57-72.
59. S.S. LAVENBERG (ed.) (1983). *Computer Performance Modeling Handbook,* Academic Press, New York.
60. S.S. LAVENBERG, M. REISER (1980). Stationary state probabilities at arrival instants for closed queueing networks with multiple types of customers. *J. Appl. Probab. 17,* 1048-1061.
61. E.D. LAZOWSKA, J. ZAHORJAN (1982). Multiple class memory constrained queueing networks. *Perform. Eval. Rev. 11, Proc. ACM SIGMETRICS Conf. Meas. Modeling Comput. Syst.,* 130-140.
62. K.J. LEE, D. TOWSLEY (1986). A comparison of priority-based decentralized load balancing policies. *Perform. Eval. Rev. 14, Proc. PERFORMANCE '86 and ACM SIGMETRICS,* 70-77.

63. P.A.W. LEWIS, G.S. SHEDLER (1971). A cyclic-queue model of system overhead in multiprogrammed computer systems. *J. Assoc. Comput. Mach. 18,* 199-220.
64. R.A. MARIE (1979). An approximate analytical method for general queueing networks. *IEEE Trans. Software Eng. SE-5,* 530-538.
65. J. MCKENNA, D. MITRA (1982). Integral representations and asymptotic expansions for closed Markovian queueing networks: Normal usage. *Bell Syst. Tech. J. 61,* 661-683.
66. D. MERLE, D. POTIER, M. VERAN (1978). A tool for computer system performance analysis. D. FERRARI (ed.). *Performance of Computer Installations,* North-Holland, Amsterdam, 195-213.
67. D. MITRA (1985). Probabilistic models and asymptotic results for concurrent processing with exclusive and non-exclusive locks. *SIAM J. Comput. 14,* 1030-1051.
68. D. MITRA, J. MCKENNA (1986). Asymptotic expansions for closed Markovian networks with state dependent service rates. *J. Assoc. Comput. Mach. 33,* 568-592.
69. D. MITRA, P. J. WEINBERGER (1984). Probabilistic models of database locking: Solutions, computational algorithms, and asymptotics. *J. Assoc. Comput. Mach. 31,* 855-878.
70. C.G. MOORE, III (1971). *Network Models for Large-Scale Time-Sharing Systems,* Technical Report No. 71-1, Dept. of Ind. Eng., Univ. of Michigan, Ann Arbor, MI.
71. R.J.T. MORRIS, W.S. WONG (1985). Performance of concurrency control algorithms with non-exclusive access. E. GELENBE (ed.). *PERFORMANCE '84,* North-Holland, Amsterdam, 87-101.
72. R.J.T. MORRIS, W.S. WONG (1985). Performance analysis of locking and optimistic concurrency control algorithms. *Performance Evaluation 5,* 105-118.
73. R. NELSON, A.N. TANTAWI (1988). Approximate analysis of fork/join synchronization in parallel queues. *IEEE Trans. Comput.,* to appear.
74. R. NELSON, D. TOWSLEY, A.N. TANTAWI (1988). Performance analysis of parallel processing systems. *IEEE Trans. Software Eng.,* to appear.
75. D. NEUSE, K.M. CHANDY, J. MISRA, R. BERRY (1981). Simulation tools in performance evaluation. *CPEUG 81,* 331-334.
76. V.A. NORTON, G.F. PFISTER (1985). A methodology for predicting multiprocessor performance. *Proc. Intl. Conf. Parallel Processing,* 772-781.
77. G.F. PFISTER, W.C. BRANTLEY, D.A. GEORGE, S.L. HARVEY, W.J. KLEINFELDER, K.P. MCAULIFFE, E.A. MELTON, V.A. NORTON, J. WEISS (1985). The IBM research parallel processor prototype (RP3): introduction and architecture. *Proc. Intl. Conf. Parallel Processing,* 764-771.
78. D. POTIER, PH. LEBLANC (1980). Analysis of locking policies in database management systems. *Commun. ACM 23,* 584-593.
79. M. REISER (1979). A queueing network analysis of computer communication networks with window flow control. *IEEE Trans. Commun. COM-27,* 1199-1209.

80. M. REISER, H. KOBAYASHI (1975). Queueing networks with multiple closed chains: Theory and computational algorithms. *IBM J. Res. Develop. 19,* 283-294.
81. M. REISER, S.S. LAVENBERG (1980). Mean-value analysis of closed multichain queueing networks. *J. Assoc. Comput. Mach. 27,* 313-322.
82. M. REISER, C.H. SAUER (1978). Queueing networks models: Methods of solution and their program implementation. K.M. CHANDY, R.T. YEH (eds.). *Current Trends in Programming Methodology, Vol. III, Software Modeling and Its Impact on Performance,* Prentice-Hall, Englewood Cliffs, NJ.
83. I.K. RYU, A. THOMASIAN (1985). *Analysis of Database Performance with Dynamic Locking,* IBM Research Report RC 11428, Yorktown Heights, NY.
84. C.H. SAUER (1981). Approximate solution of queueing networks with simultaneous resource possession. *IBM J. Res. Develop. 25,* 894-903.
85. C.H. SAUER, K.M. CHANDY (1981). *Computer Systems Performance Modeling,* Prentice-Hall, Englewood Cliffs, NJ.
86. C.H. SAUER, E.A. MACNAIR, J.F. KUROSE (1982). The research queueing package: past, present and future. *Proc. National Computer Conference,* 273-280.
87. C.H. SAUER, M. REISER, E.A. MACNAIR (1977). RESQ - A package for solution of generalized queueing networks. *Proc. National Computer Conference,* 977-986.
88. A.L. SCHERR (1967). *An Analysis of Time-Shared Computer Systems,* MIT Press, Cambridge, MA.
89. P. SCHWEITZER (1979). Approximate analysis of multiclass closed networks of queues. *Proc. Int. Conf. Stochastic Control and Optimization,* Amsterdam, The Netherlands.
90. C.L. SEITZ (1985). The cosmic cube. *Commun. ACM 28,* 22-33.
91. K.C. SEVCIK, I. MITRANI (1981). The distribution of queueing network states at input and output instants. *J. Assoc. Comput. Mach. 28,* 358-371.
92. H.A. SIMON, A. ANDO (1961). Aggregation of variables in dynamic systems. *Econometrica 29,* 11-138.
93. C.E. SKINNER (1967). A priority queueing system with server-walking time. *Oper. Res. 15,* 278-285.
94. A.N. TANTAWI, D. TOWSLEY (1985). Optimal static load balancing in distributed systems. *J. Assoc. Comput. Mach. 32,* 445-465.
95. Y.C. TAY, N. GOODMAN, R. SURI (1985). Locking performance in centralized databases. *ACM Trans. Database Syst. 10,* 415-462.
96. Y.C. TAY, R. SURI, N. GOODMAN (1985). A mean value performance model for locking in databases: The no-waiting case. *J. Assoc. Comput. Mach. 32,* 618-651.
97. A. THOMASIAN, P.F. BAY (1984). Analysis of queueing network models with population size constraints and delayed blocked customers. *Perform. Eval. Rev. 12, Proc. ACM SIGMETRICS Conf. Meas. Modeling*

Comput. Syst., 202-216.

98. A. THOMASIAN, P.F. BAY (1985). Integrated performance models for distributed processing in computer communication networks. *IEEE Trans. Software Eng. SE-11*, 1203-1215.

99. A. THOMASIAN, P.F. BAY (1986). Analytic queueing network models for parallel processing of task systems. *IEEE Trans. Comput. C-35*, 1045-1054.

100. A. THOMASIAN, I.K. RYU (1983). A decomposition solution to the queueing network model of the centralized DBMS with static locking. *Perform. Eval. Rev., Proc. ACM SIGMETRICS Conf. Meas. Modeling Comput. Syst.*, 82-92.

101. D.F. TOWSLEY (1980). Queueing network models with state-dependent routing. *J. Assoc. Comput. Mach. 27*, 323-337.

102. D.F. TOWSLEY (1983). An approximate analysis of multiprocessor systems. *Perform. Eval. Rev., Proc. ACM SIGMETRICS Conf. Meas. Modeling Comput. Syst.*, 207-213.

103. D.F. TOWSLEY (1986). Approximate models of multiple bus multiprocessor systems. *IEEE Trans. Comput. C-35*, 220-228.

104. D.F. TOWSLEY, K.M. CHANDY, J.C. BROWNE (1978). Models for parallel processing within programs: Application to CPU:IO and IO:IO overlap. *Commun. ACM 21*, 821-830.

105. S. TUCCI, C.H. SAUER (1985). The tree MVA algorithm. *Performance Evaluation 5*, 187-196.

106. M. VERAN, D. POTIER (1985). QNAP2: A portable environment for queueing systems modeling. D. POTIER (ed.). *Modeling Techniques and Tools for Performance Analysis*, North-Holland, Amsterdam, 25-63.

107. Y-T. WANG, R.J.T. MORRIS (1985). Load sharing in distributed systems. *IEEE Trans. Comput. C-34*, 204-217.

108. N.C. WILHELM (1977). A general model for the performance of disk systems. *J. Assoc. Comput. Mach. 24*, 14-31.

109. P.S. YU, S. BALSAMO, Y-H. LEE (1986). Dynamic load sharing in distributed database systems. *Proc. Fall Joint Comput. Conf.*, 675-683.

110. J. ZAHORJAN, D.L. EAGER, H. SWEILLAM (1986). *Accuracy, Speed and Convergence of Approximate Mean Value Analysis*, Technical Report 86-08-07, Dept. of Computer Science, Univ. of Washington, Seattle, WA.

111. J. ZAHORJAN, J.N.P. HUME, K.C. SEVCIK (1978). A queueing model of a rotational position sensing disk system. *INFOR 16*, 199-216.

Markov Processes and Teletraffic

R. Syski
Department of Mathematics, University of Maryland
College Park, Maryland 20742, USA

After comments on probabilistic Potential Theory, the paper presents selected theoretical topics related to the structure of Markov processes, which find applications in queueing and in teletraffic. Emphasis is put on boundary problems (especially the Dirichlet problem) and on various decomposition theorems.

Note: This paper is dedicated to J.W. Cohen on his 65th birthday

1. Overview

Inspection of recent teletraffic literature indicates that applications of stochastic processes in teletraffic fall into two categories (which overlap to a great extent), namely:

(a) general theories concerning a class of systems or a class of problems;
(b) specific problems concerning an individual system.

Indeed, this duality is clearly visible when scanning proceedings of congresses and conferences, such as the prestigious International Teletraffic Congresses (ITC) and associated seminars, and numerous meetings on 'system analysis and performance'. In particular, Queueing Theory with its analysis of waiting and loss systems is the prime example of category (a), whereas investigations of specific systems of practical interest provide examples of category (b). It must be added that the majority of works belongs to group (b).

1.1. Theory and practice

Theoretical developments in stochastic processes as well as in statistical methods (inference on stochastic processes) lead to important structural theorems of great interest to theoreticians, but are frequently discarded by practitioners as 'academic exercises'. On the other hand, concrete technological developments present challenging new mathematical problems which, if solved at all, require new powerful tools. As these methods are too advanced and complicated, practitioners satisfy themselves often with rather crude but simpler techniques (sufficient as the first approximation). Hence, the proverbial gap between theory and practice grows wider and wider at an increasing pace.

There is another aspect of this phenomenon. Each of the theoretical disciplines develops rapidly for its own sake, and recent advances are accessible only to a few devoted specialists. This in turn makes it difficult for non-specialists and users to get acquainted with new methods and techniques.

This is the situation prevailing in Markov processes, regenerative processes, martingales, stochastic equations, and many other fields. For example, elementary facts such as birth-and-death processes, basic renewal theory, etc., are well-known to every traffic engineer, whereas Dirichlet problem, functionals, etc., are still not within the reach of a person interested in applications. And this is despite the fact that such theoretical concepts offer numerous advantages by providing foundations and better understanding. To be more specific, this gap may be illustrated by comparison of two well-known treatises on stochastic processes, namely that of KARLIN and TAYLOR [45] and that of DELLACHERIE and MEYER [20].

Fortunately, as it happens frequently, some theoretical structures developed by theoreticians prove their usefulness in a large variety of problems, and become widely accepted and used. The classical example is the concept of statistical equilibrium introduced by A.K. Erlang, and now in common use. The modern example is provided by the influence of martingale methods in teletraffic studies. There have been several attempts to put abstract theories to work on practical problems, and fortunately this trend continues, although rather slowly. It suffices to mention here the book of BRÉMAUD [6] on martingale techniques, BACCELLI and BRÉMAUD [1] on stationary queues, numerous books devoted to stochastic equations, such as CHUNG and WILLIAMS [9] or ELLIOTT [27], and books using martingale methods in statistical analysis, such as JACOBSEN [41] or SLUD [69].

1.2. Selection

To present the account of the influence of all these developments on teletraffic studies is too ambitious a task, and would require a book, instead of a mere paper. It is therefore necessary to restrict this review of 'stochastic processes in teletraffic' to a few selected topics, linked by a common thread. The field of teletraffic is a vast one, and branches in several directions. In this paper only aspects belonging to Queueing Theory will be discussed. Moreover, attention will be restricted to problems in category (a) mentioned above.

Foremost among stochastic processes which influence teletraffic studies are Markov processes, with all their ramifications and extensions. Although the Markov assumption is only approximate (and frowned upon by some investigators), it nevertheless offers the backbone structure on which more appropriate extensions can be built. Even in the Markov framework only selected topics of special interest will be mentioned. It is fortunate that the language of modern probabilistic Potential Theory offers a unifying treatment, simplifies discussion and provides considerable insight into the structure of the material presented here. This is especially true for Markovian queues, where topics such as boundary problems (especially the Dirichlet problem embracing busy periods, waiting times, absorption), main decomposition theorems (such as Riesz, Doob-Meyer, first entrance, last exit) and other potential theoretic concepts receive a traffic interpretation.

The approach taken in this paper, being purely subjective, is of course open to criticism for its choice of topics and perhaps for uneven presentation of basic ideas. Granted these shortcomings, the author can only say in defense that the topics chosen here are close to his heart; moreover, to all these topics Wim Cohen contributed significantly. Indeed, Wim's work provides this common thread around which this paper is organized.

> Let me take this opportunity to mention a few personal reminiscences. Wim and I met for the first time at the First International Teletraffic Congress in Copenhagen in 1955, although we were in correspondence for some time earlier. Since then we have been in touch sharing common interest in Queueing Theory and Teletraffic; see the book of personal recollections, edited by GANI [32]. Wim, a 'single server man', combines excellent mathematical knowledge with deep understanding of practical problems, a winning combination reflected in his work. It is my great pleasure to dedicate this paper to his honor.

2. GREAT POTENTIALS

2.1. Classical and modern

The theory of stochastic processes whose origin is traced back to statistical mechanics of Boltzmann, Maxwell and Gibbs, to the Fokker-Planck-Smoluchowski equation, to Brownian motion of Langevin and Wiener, and to pioneering work of Erlang on 'barred access', has today reached enormous heights of sophistication in its development, and has branched out in numerous directions, both in theoretical and applied fields. The progress has culminated in modern 'general theory of processes' and in extensive developments in special processes, especially Markov processes, martingales, regenerative processes, and others. These studies are couched in analytic and/or measure theoretic terms, and some of them found their way to many applications (for example, stochastic calculus). For a very comprehensive account of these developments, see, for example, the series of books by DELLACHERIE and MEYER [20], or books by CHUNG [8] and by WILLIAMS [84]; see also the encyclopedia article [81].

Classical (i.e., Newtonian) potential theory, developed through works of Gauss, Poisson, Laplace, Green, Dirichlet, Poincaré and others, has exerted tremendous impact on Mathematics and its applications, which continues up to the present day. It suffices to mention here boundary problems in partial differential equations associated with potentials in vector fields (e.g., electrostatic, gravitational). Analytic potential theory is a modern extension of the classical theory, based on work of Brelot, Deny, Cartan, F. and M. Riesz, and others. The theory deals with various kinds of potentials, (super-) harmonic functions and measures, special sets (thin, polar, etc.), the fine topology, Dirichlet problem, Martin boundary, capacity and energy, and various extensions and modifications. Of special interest are boundary problems in complex analysis.

A very readable account of analytic potential theory is given by HELMS [37] where further references can be found; see also the encyclopedia article [56]. An extensive treatment with historical notes is given in the recent book by DOOB [23]. For potential theory in complex analysis, see also [71]. For classical potential theory, the old books by KELLOG [49] and by MACMILLAN [55] still make interesting reading.

It was perhaps inevitable that these separate branches of Mathematics encountered a common point of interest. In this way the modern probabilistic potential theory was born. This theory deals with those aspects of stochastic processes (mostly Markov processes and martingales) which can be put in the framework of analytic potential theory in such a way that potential-theoretic concepts and problems have a probabilistic interpretation.

The origin of probabilistic potential theory goes back to early observations that the random walk may be used to solve the discrete analogue of the first boundary problem (see [59]) and to the independent discovery by Kac and Kakutani of a close analogy between properties of Brownian motion and the Newtonian potential [42], [43]. Indeed, the classical Dirichlet problem may be interpreted in terms of Brownian motion with generator given by the Laplacian.

The intimate relationship between potential theory and Markov processes was established in a series of papers by DOOB [21], [22] who identified (super-) martingale theory with the theory of (super-) harmonic functions and studied the probabilistic Dirichlet problem, and by HUNT [40] who developed the general potential theory of Markov processes (introducing excessive functions and measures, and stressing the role of balayage in connection with passage times). The work of FELLER [29] on boundaries induced by integro-differential operators (couched in analytic terms) provided another early link. Independently of these developments, KEILSON [47] introduced a concept of compensation, suggested by the boundary problems of the classical potential theory.

Connections between probabilistic and analytic potential theories are much deeper than formal analogies. On one hand, the potential-theoretic approach provided far-reaching unification for Markov processes and opened new research areas, and on the other hand, probabilistic methods helped to solve open problems in analytic potential theory. Potential theoretic notions are now indispensible notions in Markov processes and in the general theory of processes. In particular, of special interest both for theory and applications is the fact that the Dirichlet problem is just a problem of the first passage times in Markov processes.

Well-known expositions of probabilistic Potential Theory are given in the treatise by BLUMENTHAL and GETOOR [5] and in the voluminous work of DELLACHERIE and MEYER [20]. The recent treatise of DOOB [23], already mentioned, provides a comprehensive treatment, with an extensive list of references. The text by KEMENY et al. [50] gives a detailed account of potential theory for Markov chains; see also the book by REVUZ [63] and the survey by SYSKI [74].

2.2. Teletraffic and queueing

The whole field has received contributions from numerous workers, and it is impossible to characterize the recent developments in a brief outline. The following list of those works on teletraffic and queueing theories which use potential theory or related methods is incomplete and is presented here only as an indication of usefulness of potential theoretic ideas.

Although pioneering work of F. Pollaczek and of C. Palm was carried out from a different point of view, it admits a potential theoretic interpretation. In fact, Pollaczek used passage probabilities represented by contour integrals, and Palm's investigations of regenerative processes lead to a potential theoretic notion of a 'Palm measure'; see, for example, [33], [64], [1]. In his book on random walk, SPITZER [70] made extensive use of Potential Theory. In his study of passage times in Markov chains, KEMPERMAN [51] treated a waiting time problem as an absorption problem. Investigation of the ballot problem by TAKACS [83] and studies of the Wiener-Hopf technique by PRABHU [61] are clearly related conceptually to Potential Theory. As already noted, J. Keilson explored advantages of the Green function and introduced the compensation method (see Subsection 3.5.4 below). Similarly, the phase-type distributions introduced by M.F. Neuts have a potential-theoretic flavor (see again Subsection 3.5.4). The present author advocated the use of Potential Theory methods already in [72], and has continued to do so in his study of Markovian queues based on passage probabilities [79], [82]. Independent work on priority systems, based on entrance and exit times, was carried out by SCHASSBERGER [67]. Further references to special topics are mentioned later in Section 3 of this paper.

Of course, potential theoretic approaches extend beyond Markov processes. The intensive development of theory and applications of regenerative processes and point processes illustrates this trend. Among the vast literature in this field and its applications to queueing and traffic studies, one must mention here a series of papers by E. Çinlar on Markov additive processes [10], [11] and a survey paper [12]. A recent paper by KASPI and MAISONNEUVE [46] on regenerative systems is a major work in this area. The books on point processes and queueing by FRANKEN et al. [30] and by ROLSKI [64], the book by KALLENBERG [44] on random measures, and the treatment of point processes in teletraffic in terms of stochastic integrals developed by LE GALL [53], are also related to Potential Theory. Works on martingale methods by Brémaud and others have already been mentioned in Subsection 1.1. Further references are given in Section 3.

2.3. In Cohen's work

In the contribution of Wim Cohen to Probability Theory and its applications one can distinguish, from the point of view of Potential Theory, three aspects, namely:

i) passage times;
ii) regenerative processes;
iii) boundary problems.

In his fundamental book [15] Cohen treats the single server queue using methods of complex analysis (contour integrals) following Pollaczek. This approach, closely related to the Wiener-Hopf technique and renewal theory, may be interpreted in potential theoretic terms. Indeed, as noted by SPITZER in his book [70], p. 174, one deals here essentially with absorption problems in fluctuation theory; see also [73]. As Pollaczek himself asserted, his method transforms a probabilistic problem into a problem in complex analysis (integral equations).

In his monograph on regenerative processes [13] and in subsequent papers on queueing models, Cohen makes extensive use of stochastic mean value theorems. This concept bears close resemblance to a potential generated by an increasing process; see comments in Subsection 3.3.4.

His recent work on boundary value problems is obviously purely potential theoretic. As it will be discussed extensively at other places in this book, here only brief comments on comparison with Potential Theory described earlier are in order. It is remarkable that queueing problems can be treated as boundary value problems. The method is indeed powerful as Cohen and Boxma have shown in their monograph [18]. The Riemann-Hilbert problem concerns the investigation of a kernel of some functional equation for generating functions, and the solution of this equation. Thus, one considers analytic aspects, analogous to (two-dimensional) Analytic Potential Theory. Probabilistic Potential Theory deals rather with a stochastic process itself. Thus, one has here two levels of investigations. See Subsection 3.4.4 for further comments.

3. SELECTED TOPICS

This section presents several results from probabilistic Potential Theory which emphasize structural properties and whose use is widespread in applications. In particular, decomposition theorems provide useful representations of various quantities of interest, functionals become powerful tools in many applications, and boundary problems (such as the Dirichlet problem and Poisson equation) embrace a large variety of problems of practical interest.

Each subsection lists definitions and representative theorems, with references provided, and concludes with indications of applications.

3.1. Markov processes

It is convenient to introduce the terminology and notation to be used in the sequel, and to begin with a brief summary of properties of Markov processes. For details and for definitions of unexplained terms used here see the basic texts, such as BLUMENTHAL and GETOOR [5], CHUNG [8], DYNKIN [26] or WILLIAMS [84].

3.1.1. Consider a Markov process $X=(X_t, \mathfrak{F}_t, 0 \leqslant t < \infty)$ defined on a (complete) probability space $(\Omega, \mathfrak{F}, \mathbb{P})$ and taking values in a state space $(E, \mathfrak{E})$, where E is a (locally compact) metric space (in particular, a real line $\mathbb{R}$ or a countable set Π) and $\mathfrak{E}$ is its Borel field. The process (X_t) is adapted to the filtration $(\mathfrak{F}_t)$ satisfying the 'usual conditions' (right-continuity, augmentation). In the following, all stopping times, martingales, etc., are taken relative to the filtration ('history') $(\mathfrak{F}_t)$. For the purpose of this paper, it suffices to consider the 'natural' $\mathfrak{F}_t = \sigma(X_s, s \leqslant t)$ generated by the process. It is also assumed that the life time of the process is infinite. Furthermore, the process X is taken to be strongly Markov, quasi-left-continuous and cadlag (i.e., its paths are right-continuous with left limits). These properties characterize the Hunt process, but have been shown to hold for a Feller process (defined in analytic terms).

It is assumed that there exists a family of transition functions for a process X, taken to be time-homogeneous,

$$P_t(x,B) = \mathbb{P}(X_t \in B | X_0 = x), \quad x \in E, \quad B \in \mathfrak{E}.$$

The kernel $P_t(\cdot,\cdot)$ on $E \times \mathfrak{E}$ defines for each t the operator P_t acting on the space $\mathbb{B}$ of bounded measurable functions on E (and, dually, on the space $\mathbb{M}$ of bounded measures on $\mathfrak{E}$) by:

$$P_t f(x) = \int_E P_t(x,dy) f(y) = \mathbb{E}^x f(X_t),$$

$$\mu P_t(B) = \int_E \mu(dx) P_t(x,B) = \mathbb{P}^\mu(X_t \in B)$$

for $x \in E$, $B \in \mathfrak{E}$. For any $f \in \mathbb{B}$ and any $\mu \in \mathbb{M}$, the symbol μf denotes the inner product:

$$\mu f = \int_E \mu(dx) f(x).$$

In particular, if π is the initial distribution of X_0, then $\pi P_t f = \mathbb{E} f(X_t)$. The Markov property implies that the operators $(P_t, 0 \leqslant t < \infty)$ form a contraction semi-group:

$$P_{t+s} = P_t P_s, \quad t \geqslant 0, \quad s \geqslant 0 \quad \text{(Chapman-Kolmogorov equation)},$$

or equivalently in terms of the resolvent equation:

$$U^\alpha - U^\beta = (\beta - \alpha) U^\alpha U^\beta, \quad \alpha > 0, \quad \beta > 0,$$

with $U^\alpha U^\beta = U^\beta U^\alpha$. Here U^α is the Laplace transform of P_t, with kernel

$$U^{\alpha}(x,B) = \int_0^{\infty} e^{-\alpha t} P_t(x,B)dt$$

and (U^{α}, $\alpha>0$) is known as the resolvent (or α-potential) of the semi-group.

3.1.2. The full power of semi-group theory is available for the study of such Markov semi-groups. In addition to references mentioned above, see also LAMPERTI [52] and ETHIER and KURTZ [28] for the semi-group approach. The general theory of semi-groups is given in DUNFORD and SCHWARTZ [24] and HILLE and PHILLIPS [38]. Restricting attention to P_t acting on the function space $\mathbb{B}$, it suffices to mention here that with limit operations taken in the strong sense (supremum norm), the semi-group is usually assumed to be strongly continuous at $t=0$ (this implies continuity for all t). The infinitesimal generator Q of the semi-group is defined as the (strong) limit as $t \to 0$:

$$(P_t f - f)/t \to Qf.$$

The set of those f for which this limit exists is the domain of Q, denoted d(Q). Note that Q is in general an unbounded operator. It is further assumed that both the backward and the forward Kolmogorov equations for $P_t f$ hold.

In the resolvent form, continuity and differentiability conditions (at $t=0$) are expressed respectively by:

$$\alpha U^{\alpha} f \to f, \quad \alpha(\alpha U^{\alpha} f - f) \to Qf, \quad \text{as} \quad \alpha \to \infty,$$

and the Kolmogorov equations take the form:

$$\begin{aligned} \alpha U^{\alpha} f - f &= QU^{\alpha} f, \quad \text{(B)} \\ &= U^{\alpha} Qf, \quad \text{(F)} \end{aligned} \quad \text{for} \quad f \in \mathrm{d}(Q).$$

As strong continuity $P_t f \to f$, when $t \to 0$, means that $P_t f(x) \to f(x)$ pointwise uniformly in x (so it is too restrictive), weaker modes of limit operations are used. In particular, pointwise limits are of main interest.

3.1.3. For a stopping time T and $\alpha \geq 0$, define the operator P_T^{α} by:

$$P_T^{\alpha} f(x) = \mathbb{E}^x(e^{-\alpha T} f(X_T),\ T<\infty), \quad x \in E,$$

for any bounded (or positive finite) measurable function f on E. When $\alpha=0$, write P_T. Note that P_T^{α} is the extension of $P_t^{\alpha} = e^{-\alpha t} P_t$ (corresponding to a constant stopping time).

For any two stopping times T and S, the expression

$$T_S = S + T \circ \theta_S$$

is a stopping time (representing 'occurrence of event T after event S'), and

$$X(T_S) = X_T \circ \theta_S.$$

Here θ_S denotes the shift operator, namely the extension of the ordinary shift θ_s (for fixed s) whose effect on random variables is $X_t \circ \theta_s = X_{t+s}$.

The 'product theorem' asserts that:

$$P_S^\alpha P_T^\alpha = P_{T_s}^\alpha, \quad \alpha \geqslant 0.$$

Note that if both S and T are constant or if only T is constant, then the product theorem reduces to the semi-group property and the strong Markov property, respectively. If only S is constant (say, s), then $T_s = s + T \circ \theta_s$; a stopping time T is called a terminal time when $T_s = T$ whenever $T \geqslant s$.

The first hitting time of a (Borel) set $H \subset E$ is defined by

$$T_H = \inf(t: t > 0, X_t \in H)$$

where the infimum of the empty set is taken to be $+\infty$. Verification that T_H is a stopping time requires capacity theory. If $T = T_H$, write P_H^α for $P_{T_H}^\alpha$. The operator P_H^α is known as an *α-balayage*. The measure $P_H^\alpha(x, \cdot)$ is called the α-hitting distribution of H (starting at x) or an α-harmonic measure of H (relative to x). Thus, for $\alpha = 0$:

$$P_H(x, B) = \mathbb{P}^x(X(T_H) \in B, T_H < \infty).$$

The right-continuity of paths and the 0-1 law imply that $P_H^\alpha(x, \cdot)$ is concentrated on a set $H \cup H^r$, known as a fine closure of H, where H^r is the set of regular points of H, namely, those points x such that $\mathbb{P}^x(T_H = 0) = 1$.

3.1.4. With reference to a Markov process X, one has the following generalized definitions:

$$P_t f(X_s) = \mathbb{E}(f(X_{t+s}) | X_s), \quad U^\alpha f(X_s) = \int_0^\infty e^{-\alpha t} P_t f(X_s) dt,$$

$$Qf(X_s) = \lim_{h \to 0} \mathbb{E}(f(X_{h+s}) - f(X_s) | X_s)/h, \quad \text{a.e.}$$

which lead to the generalized Kolmogorov equations. In particular, the (F)-equation in its integrated form is:

$$\mathbb{E}(f(X_{t+s}) | X_s) - f(X_s) = \mathbb{E}\left[\int_s^{s+t} Qf(X_u) du \,\Big|\, X_s \right], \quad \text{a.e.}$$

for any $f \in \mathrm{d}(Q)$.

The following results are of importance. For proofs see DYNKIN [26], CHUNG [8] or WILLIAMS [84].

THEOREM 3.1.1. *For any bounded $f \in \mathrm{d}(Q)$, let*

$$M_t = f(X_t) - \int_0^t Qf(X_u) du, \quad t \geqslant 0.$$

Then $(M_t, \mathcal{F}_t, 0 \leqslant t < \infty)$ is a martingale with $\mathbb{E} M_t = \pi f$ (where π is the initial distribution of X_0).

From this theorem, one can deduce (using the optional sampling theorem) the basic Dynkin theorem:

THEOREM 3.1.2. *For any stopping time T and $\alpha>0$:*

$$U^\alpha f(x) = \mathbb{E}^x \int_0^T e^{-\alpha s} f(X_s)ds + P_T^\alpha U^\alpha f(x), \quad x \in E,$$

and if $\mathbb{E}^x T < \infty$, then for $\alpha=0$:

$$P_T f(x) - f(x) = \mathbb{E}^x \int_0^T Q f(X_u) du.$$

A Markov process $\tilde{X}$ killed at the first entrance time T_H to a set H is defined by

$$\begin{aligned} \tilde{X}_t &= X_t \quad \text{for } t<T_H \\ &= \Delta \quad \text{for } t \geqslant T_H \end{aligned}$$

where Δ is an adjoint state. This leads to taboo probabilities (i.e., transitions avoiding the taboo set H):

$$_HP_t(x,B) = \mathbb{P}^x(X_t \in B,\ t>T_H).$$

Writing $_HP_t$ and $_HU^\alpha$ for the corresponding sub-Markov semi-group and its resolvent, it follows from the Dynkin formula (applied to $T=T_H$) that one has the 'first entrance decomposition'

$$U^\alpha f(x) = {}_HU^\alpha f(x) + P^\alpha\, {}_HU^\alpha f(x), \quad x \in E, \quad \alpha \geqslant 0.$$

See Subsection 3.5 below.

3.1.5. Applications of Markov processes in Queueing and in Teletraffic are well-known and there is no need to elaborate on this subject. For the sake of illustration of the variety of applications, it suffices to mention here the waiting time process (W_n) in the Pollaczek and Lindley argument, the virtual waiting time in the M/G/1 queue, and of course Markovian queues (where Markov chains in continuous time prevail). See also COHEN [16] for historical comments on the birth-and-death process.

In contrast to diffusion processes (extensively treated in engineering applications) characterized by continuous sample functions (the Brownian motion being the prime example), many Markov processes occurring in applications have discontinuous paths with jumps and may have a deterministic component. The virtual waiting time is of this type, and so are various processes in storage models, insurance risk and dams. However, the best known example is provided by the Markov processes associated with the classical renewal process, namely the age $t-S_{N(t)}$ and the residual life $S_{N(t)+1}-t$, at time t (S_n and $N(t)$ have their classical meaning here). In this connection it is worthwhile to mention the famous Markov process called 'the uniform motion to the right' defined by $X_t = X_0 + vt$ for $t \geqslant 0$. The process moves deterministically to the right with constant 'velocity' $v>0$. For any bounded function f, one has now $P_t f(x) = f(x+vt)$, and the generator Q is the differential operator

$Qf(x)=vf'(x)$.

Topics such as first entrance decomposition (and its dual last exit decomposition), the concept of balayage, the Dynkin formula, etc., received considerable attention in applied literature. Recently, the so-called martingale problem has become a source of many important applications. This problem consists of finding the most general class of processes X, or of functions f, for which the family (M_t) in Theorem 3.1.1 is still a martingale. See the book of ETHIER and KURTZ [28] for a complete discussion, and the paper of ROSENKRANTZ [65] for queueing applications.

3.2. Riesz decompositions

These fundamental decomposition theorems, named after F. Riesz, are analogues of theorems from classical Potential Theory according to which a superharmonic function can be written as a sum of a harmonic and a potential component. Several different versions (not all equivalent) of these theorems are known, but here only the most common version is stated. Proofs can be found in references on Markov processes and Potential Theory, for example, DELLACHERIE-MEYER [20], DOOB [23], CHUNG [8], WILLIAMS [84].

3.2.1. Consider real-valued $\mathscr{E}$-measurable functions defined on the state space E, and write

$$P_t^\alpha = e^{-\alpha t}P_t \quad \text{for} \quad \alpha \geqslant 0.$$

A measurable non-negative function f is said to be α-supermean (or supermedian), relative to the semi-group (P_t), when (pointwise):

$$P_t^\alpha f \leqslant f \quad \text{for every} \quad t \geqslant 0,$$

(and α-invariant when equality holds). A function f is said to be α-excessive if it is α-supermean and:

$$P_t^\alpha f \to f, \quad \text{pointwise as} \quad t \to 0.$$

When $\alpha=0$, it will be dropped from notation, and the 0-excessive function is simply called excessive.

An excessive function f is said to be purely excessive when:

$$P_t f \to 0, \quad \text{pointwise as} \quad t \to \infty.$$

In terms of Laplace transforms, a function $f \geqslant 0$ is excessive iff:

$$\alpha U^\alpha f \leqslant f \quad \text{for all} \quad \alpha \geqslant 0 \quad \text{and} \quad \alpha U^\alpha f \to f \quad \text{as} \quad \alpha \to \infty;$$

and f is purely excessive iff it is excessive and $\alpha U^\alpha f \to 0$ as $\alpha \to 0$.

For a measurable function $a \geqslant 0$ on E, a function $U^\alpha a$ is called an α-potential of a charge a. A potential Ua is always purely excessive.

The following is the basic Riesz decomposition theorem: a finite excessive function f has a unique representation

$$f = p + h$$

where p is purely excessive and h is invariant. The idea of proof consists of verification that the (pointwise) limit

$$h = \lim_{\alpha\to 0} \alpha U^\alpha f$$

is invariant, and then that $p=f-h$ satisfies $\alpha U^\alpha p\to 0$ as $\alpha\to 0$.

3.2.2. This in turn leads to Riesz decompositions for supermartingales. The connection between Markov processes and martingales is established by:

THEOREM 3.2.1. *For a Markov process $X=(X_t, \mathfrak{F}_t)$ define*

$$S_t = f(X_t), \quad t\geqslant 0,$$

where f is a P_t-excessive function on E with $\pi f<\infty$ (where π is the initial distribution). Then, $S=(S_t, \mathfrak{F}_t, 0\leqslant t<\infty)$ is a positive supermartingale (a martingale if f is invariant), with $\mathbb{E}S_t\leqslant \pi f$. Moreover, S has a cadlag version.

By the supermartingale convergence theorem, a positive supermartingale always converges, and is closable, i.e., $(S_t, \mathfrak{F}_t, 0\leqslant t\leqslant\infty)$ is a supermartingale with $S_t\to S_\infty$ a.e., but

$$\lim_{t\to\infty} \mathbb{E}S_t \geqslant \mathbb{E}S_\infty$$

and equality holds iff (S_t) is uniformly integrable.

The term 'potential supermartingale' denotes a positive, right-continuous supermartingale $(S_t, \mathfrak{F}_t, 0\leqslant t<\infty)$ such that $\mathbb{E}S_t\to 0$ as $t\to\infty$. This implies that $S_t\to 0$ a.e., but the converse need not hold unless (S_t) is a uniformly integrable family.

With this preparation, one has two forms of the Riesz decomposition theorem:

THEOREM 3.2.2. *Let f be an excessive function with decomposition $f=p+h$. Then, a supermartingale S from Theorem 3.2.1 can be represented uniquely (up to equivalence) as the sum:*

$$f(X_t) = p(X_t)+h(X_t), \quad t\geqslant 0,$$

where $(p(X_t), \mathfrak{F}_t, 0\leqslant t<\infty)$ is a potential supermartingale with

$$\lim_{s\to\infty} \mathbb{E}(p(X_{t+s})|\mathfrak{F}_t) = 0 \quad \text{a.e.,} \quad t\geqslant 0,$$

and $(h(X_t), \mathfrak{F}_t, 0\leqslant t<\infty)$ is a martingale with

$$h(X_t) = \lim_{s\to\infty} \mathbb{E}(f(X_{t+s})|\mathfrak{F}_t) \quad \text{a.e.,} \quad t\geqslant 0.$$

Moreover, S and its components are cadlag.

Theorem 3.2.3. *A supermartingale S from Theorem 3.2.1 can be represented uniquely (up to equivalence) as the sum:*

$$S_t = V_t + M_t, \quad t \geqslant 0,$$

where $M_t = \mathbb{E}(S_\infty | \mathfrak{F}_t)$ *and* $(M_t,\ 0 \leqslant t < \infty)$ *is a uniformly integrable martingale, and where* $V_t = S_t - M_t$, $t \geqslant 0$, *form a supermartingale such that*

$$\lim_{t \to \infty} V_t = 0 \quad a.e.$$

In general, one has

$$h(X_t) \geqslant M_t, \quad p(X_t) \leqslant V_t,$$

and note that in the first decomposition it is the family $(p(X_t))$, and in the second one the family (M_t), which is uniformly integrable. When (S_t) is itself a uniformly integrable family, then both these decompositions coincide. This would be the case for bounded excessive functions.

3.2.3. Excessive functions and measures enjoy increasing visibility in applications. The following are some examples of this trend.

1. The typical example of an excessive function is provided by the 'first entry probability' defined as a balayage of constant function 1:

$$P_H 1(x) = \mathbb{P}^x(T_H < \infty)$$

   where T_H is the first entrance time to a set H. The function $P_H 1$ is the basic source of applications of the Dirichlet problem (see Subsection 3.4.2). More generally, if f is α-excessive, so is the balayage $P_H f$.
2. Characterization of excessive functions for a given Markov process constitutes an important problem. It is known that if a process is recurrent (in a specified sense), then all excessive functions are constant. This is the case of the Brownian motion of dimension 1 and 2; for dimension 3 and higher, all excessive functions are lower semi-continuous.
   Of special interest are calculations of excessive functions for Markov chains, in particular for birth-and-death processes. As an illustration, observe that any (positive) non-increasing function $f(i)$, $i \geqslant 0$ is excessive for a birth process.
3. A reference measure ξ on $\mathfrak{E}$ is a strictly positive, σ-finite and α-excessive measure such that the measure $U^\alpha(x, \cdot)$ for fixed x, is absolutely continuous with respect to ξ. Two Markov semi-groups (or two Markov processes X and $\hat{X}$) are said to be in duality relative to ξ, if there exists a non-negative function $v^\alpha(x,y)$ such that:

$$U^\alpha(x,A) = \int_A v^\alpha(x,y)\xi(dy), \quad \hat{U}^\alpha(A,y) = \int_A \xi(dx) v^\alpha(x,y)$$

   where U^α and $\hat{U}^\alpha$ are their respective resolvents. Thus, for a measurable function f, $U^\alpha f$ and $\hat{U}^\alpha f$ are functions given by:

$$U^\alpha f(x) = \int_E v^\alpha(x,y)\xi(dy) f(y), \quad \hat{U}^\alpha f(y) = \int_E f(x)\xi(dx) v^\alpha(x,y).$$

Similar relations hold for dual kernels $P_t(x,A)$ and $\hat{P}_t(A,y)$, $Q(x,A)$ and $\hat{Q}(A,y)$, etc. Conceptually, the notion of duality is associated with reversing (time) direction of a Markov process. For a full discussion of duality see [5].

For a function f its dual is a measure $\hat{f}$ defined by $\hat{f}(dy)=\xi(dy)f(y)$. An α-potential of a measure μ is a *function* $U^\alpha\mu$ defined as an α-potential of a function $\hat{\mu}$ (dual of μ):

$$U^\alpha\mu(x) = U^\alpha\hat{\mu}(x) = \int_E v^\alpha(x,y)\mu(dy).$$

In the case of a Markov chain with resolvent (u^α_{ij}), one has $u^\alpha_{ij}=v^\alpha(i,j)\xi(j)$ and the dual resolvent is given by:

$$\hat{u}^\alpha_{ji} = \xi(i)u^\alpha_{ij}/\xi(j).$$

Moreover, an α-potential of a measure μ is a function $U^\alpha\mu$ defined by:

$$U^\alpha\mu(i) = \sum_j v^\alpha(i,j)\mu(j).$$

The notion of duality has been used in Markov chain theory since early days, with the invariant distribution (e_j) in the role of measure ξ, with dual transition probabilities given by:

$$\hat{p}_{ij}(t) = e_j p_{ji}(t)/e_i.$$

Potential theoretic concepts relative to the dual are designated with the prefix co-, when applied to the direct semi-group. Thus a function f such that $\hat{P}_t f \leqslant f$ is co-excessive with reference to P_t. In particular, the reference measure ξ is both α-excessive and α-coexcessive, and its dual is a constant function 1.

4. Let μ be a finite measure on $\mathcal{E}$ with α-potential $U^\alpha\mu$. The inner product of a measure μ with a function $U^\alpha\mu$ (i.e. the integral of a potential with respect to a charge) is said to define an α-energy of a Markov process:

$$en^\alpha = \mu U^\alpha\mu = \int_E \mu(dx)U^\alpha\mu(x) = \int_E\int_E \mu(dx)v^\alpha(x,y)\mu(dy).$$

Note that if $a\geqslant 0$ is a function (a charge) and $\hat{a}$ its dual (a measure), so $U^\alpha\hat{a}=U^\alpha a=p^\alpha$, then energy is $en^\alpha=\hat{a}p^\alpha$. As defined, the concept of energy is the extension of the classical notion of energy in analytic potential theory which corresponds to a symmetric function $v^\alpha(x,y)$. Nevertheless, probabilistic energy offers interesting and challenging problems, and has received considerable attention recently; see [8], [20], [35], [76] and [78] where further references can be found. The notion of energy has not yet been fully employed in Queueing Theory, but it is believed that it may serve as a useful characterization of systems.

5. Excessive functions play an important role in optimization and control theory for Markov processes. There exists a voluminous literature devoted to this field and to its numerous applications to economics, management,

social sciences etc. (control of queues being of current interest). For the purpose of illustration, only a brief comment is made here; see [6], [27], [39] where further references may be found.

Let X be a Markov process, and let g be a non-negative function representing gain (cost, utility, etc.) associated with the process. Consider a function f defined by:

$$f(x) = \sup_T \mathbb{E}^x g(X_T)$$

where supremum is taken over all (finite) stopping times (relative to natural σ-fields of X). The main problem is to find the (optimal) stopping time T_0 such that the supremum is attained: $f(x)=\mathbb{E}^x g(X_{T_0})$. The function f is called the value function, and is clearly excessive.

3.3. Functionals

The notion of a functional of a Markov process is a powerful tool in theoretical considerations and its use in applications is widespread. However, special forms of functionals have been known in applied fields since early days. Here only some selected problems will be mentioned; for a full discussion, see BLUMENTHAL and GETOOR [5], DYNKIN [26], WILLIAMS [84]. Applications to Teletraffic were discussed in [78].

3.3.1. A family $A=(A_t,\ 0\leqslant t<\infty)$ of non-negative random variables is called an additive functional (AF) of a Markov process $X=(X_t,\ \mathcal{F}_t,\ 0\leqslant t<\infty)$ when

a) the paths of A are non-decreasing, right-continuous and $A_0=0$;
b) for each t, A_t is $\mathcal{F}_t$-measurable (adaptation);
c) for each t and s:

$$A_{t+s} = A_t + A_s \circ \theta_t$$

where θ_t is the shift operator.

A functional A is said to be continuous (CAF) if its paths are continuous; it is said to be (purely) discontinuous (DAF) if its paths are jump functions. A family (A_t) of integrable A_t satisfying conditions a) and b) only, is said to form an *increasing process*. An AF (or an increasing process) is said to be integrable if $\mathbb{E}A_\infty<\infty$.

For a non-negative measurable function f on E, the α-potential of f relative to an AF $A=(A_t)$ is defined by

$$U_A^\alpha f(x) = \mathbb{E}^x \int_0^\infty e^{-\alpha t} f(X_t)\,dA_t = \int_E u_A^\alpha(x,dy) f(y),$$

for $\alpha\geqslant 0$, $x\in E$, where $u_A^\alpha(x,dy)$ is the kernel of the potential operator U_A^α. When $f\equiv 1$, write $U_A^\alpha 1=u_A^\alpha$. The function u_A^α is called the α-potential of a functional A:

$$u_A^\alpha(x) = \mathbb{E}^x \int_0^\infty e^{-\alpha t}\,dA_t, \quad \alpha\geqslant 0, \quad x\in E.$$

When $\alpha=0$, it will be dropped from notation, and *the potential of A* is : $u_A(x)=\mathbb{E}^x A_\infty$.

For a functional A, the function of x defined by

$$F_x(t) = \mathbb{E}^x A_t$$

is called the *characteristic of A*.

An *integral functional* (IF) $A=(A_t)$ is defined by the (path) integral:

$$A_t = \int_0^t a(X_s)ds, \quad 0\leq t<\infty,$$

where a is a non-negative bounded function on E, called a charge. Thus, A is a CAF with

$$u_A^\alpha(x) = U^\alpha a(x), \quad F_x(t) = \int_0^t P_s a(x)ds.$$

A purely discontinuous functional (DAF) $A=(A_t)$ of the most commonly used form, is defined by

$$A_t = \sum_{0<s\leq t} g(X_{s-}, X_s), \quad t>0, \quad \text{and} \quad A_0=0,$$

where $g=g(\cdot,\cdot)$ is a non-negative bounded function on $E\times E$ such that $g(x,x)=0$ for all $x\in E$. To be of this form, A must be quasi-left continuous (have totally inaccessible jumps); see [2].

3.3.2. An integrable increasing process (or an AF) $A=(A_t)$ is predictable ('natural') iff for every bounded positive cadlag martingale $(Y_t, \mathfrak{F}_t)$, one has

$$\mathbb{E}\int_0^\infty Y_s dA_s = \mathbb{E}\int_0^\infty Y_{s-} dA_s.$$

For an integrable increasing process $A=(A_t)$, its *dual predictable projection* (also called a *compensator* of A) is a predictable increasing process $A^*=(A_t^*)$ defined by:

$$\mathbb{E}\int_0^\infty Z_t dA_t = \mathbb{E}\int_0^\infty Z_t dA_t^*$$

for every non-negative predictable process $Z=(Z_t)$. Two increasing processes A and B are said to be associated if they have the same dual predictable projection, and this is the case iff the process $A-B$ is a martingale. Thus, if A is an integrable increasing process and A^* its dual predictable projection, then $(A_t-A_t^*, \mathfrak{F}_t)$ is a martingale, and

$$\mathbb{E}(A_{t+s}-A_s|\mathfrak{F}_s) = \mathbb{E}(A_{t+s}^*-A_s^*|\mathfrak{F}_s), \quad t\geq 0, \quad s\geq 0, \quad \text{a.e.}$$

A potential $U=(U_t)$ generated by an integrable increasing process $A=(A_t)$ is defined (up to equivalence) by

$$U_t = \mathbb{E}(A_\infty | \mathcal{F}_t) - A_t, \quad t \geq 0.$$

It is easy to see that (U_t) is a potential supermartingale of class (D). Conversely, the basic Doob-Meyer decomposition theorem asserts that for any potential supermartingale of class (D) there exists a unique (up to equivalence) integrable and predictable increasing process which generates this potential. (If the potential is not of class (D), the martingale component must be replaced by a local martingale.) See [20] for a full discussion and definitions.

Returning to an increasing process A and its compensator A^*, it can be verified that A and A^* have the same characteristic $F_x(t)$ and the same α-potential:

$$\mathbb{E}^x \int_0^\infty e^{-\alpha t} dA_t = \mathbb{E}^x \int_0^\infty e^{-\alpha t} dA_t^*.$$

In particular, the common potential generated by A and A^* is:

$$U_t = \mathbb{E}(A_\infty | \mathcal{F}_t) - A_t = \mathbb{E}(A_\infty^* | \mathcal{F}_t) - A_t^*.$$

Writing (M_t) for the martingale $(A_t - A_t^*)$, the expression $A_t = M_t + A_t^*$, $t \geq 0$, is the Doob-Meyer decomposition of a submartingale (A_t); this form explains the term 'compensator' for the component A^*.

Finally, let $A = (A_t)$ be a DAF determined by a function g. The *Lévy system* for A is the pair (K, L) where K is a kernel on $E \times \mathbb{E}$ and L is a CAF such that the dual predictable projection A^* of A has the form:

$$A_t^* = \int_0^t Kg(X_s) dL_s, \quad t \geq 0,$$

where

$$Kg(X_s) = \int_E K(x, dy) g(x, y), \quad x \in E.$$

Hence, the potential and characteristic of a DAF can be expressed in terms of its Lévy system.

3.3.3. The following are some of the important results.

1. Extended Dynkin theorem. Let A be an AF with finite α-potential u_A^α. Then, for any stopping time T:

$$u_A^\alpha(x) = \mathbb{E}^x \int_0^T e^{-\alpha t} dA_t + P_T^\alpha u_A^\alpha(x), \quad x \in E, \quad \alpha \geq 0.$$

When A is an AF determined by a charge a, this reduces to the Dynkin theorem in Subsection 3.1.4.

2. Revuz measure. As shown in [62], for any AF A there exists a measure μ_A, called the Revuz measure, such that the α-potential of A coincides with the α-potential of this measure:

$$u_A^\alpha = U^\alpha \mu_A.$$

Here a potential of measure is defined in terms of a reference measure ξ, by duality, as

$$U^{\alpha}\mu_A(x) = \int_E v^{\alpha}(x,y)d\mu_A(y)$$

where $v^{\alpha}(x,y)$ is a function such that $U^{\alpha}(x,dy)=v^{\alpha}(x,y)\xi(dy)$; see Subsection 3.2.3.

3. Characteristic. It may be seen that a characteristic of a functional A satisfies:

$$F_x(t+s)-F_x(s) = P_sF_x(t), \quad x\in E, \quad t\geqslant 0, \quad s\geqslant 0,$$

and $p(x)=F_x(\infty)$ defines the purely excessive function, and

$$F_x(s) = p(x)-P_sp(x).$$

Conversely, if a given bounded function F satisfies the above relation (and some additional assumptions), then there exists a CAF such that its characteristic is given by F. This is an example of a class of theorems which provide conditions under which there exists a CAF whose potential u_A coincides with a given purely excessive function; see [5].

4. Random time change. Let A be an integral functional determined by a charge $a\geqslant 0$. The inverse of A is an increasing process (τ_t) defined by:

$$\tau_t = \inf(s : A_s>t),$$

and for each t, τ_t is a stopping time. Define now:

$$\tilde{X}_t = X(\tau_t), \quad t\geqslant 0.$$

The new process $\tilde{X}=(\tilde{X}_t)$ is called the time changed process, obtained from the original process $X=(X_t)$ by a random change determined by the CAF A. $\tilde{X}$ is shown to be Markov, and its life time is A_∞. For $t<A_\infty$, the process $\tilde{X}$ takes values in a measurable set $S\subset E$. Thus, roughly speaking, the process $\tilde{X}$ is obtained by watching the original Markov process X in S with change of 'a clock'.

Random time change is very useful and received considerable attention in the literature. For basic properties see [5], [20], [26]. In recent work, GLOVER [34] considered the 'raw time changes', GZYL [36] obtained the form of a generator, and MITRO [57] investigated discontinuous time changes. An application to Markovian queues is given in [79].

3.3.4. The following examples are of interest.

1. In the continuous parameter Markov chains occurring in Markovian queues, the Lévy formula expresses discontinuous functionals (referring to jumps) by their compensators. The Lévy system reduces to the generator q_{ij} (for $i\neq j$, and 0 for $i=j$) and to $A_t^*=t$. In particular, if A_t represents the number of jumps up to time t, so $g(i,j)=1$ for $i\neq j$, then

$$Qg(i) = \sum_{j\neq i} q_{ij} = q_i.$$

Hence, the α-potential is:

$$u_A^\alpha(i) = U^\alpha q_i.$$

2. Properties of passage times (like the first entrance to - or the last exit from - a set) may often be expressed in terms of functionals. An illustration of this approach, of interest in Markovian queues, is mentioned below in Subsection 3.4.3.
3. This example concerns an increasing process of the form:

$$A_t = \int_0^t a(V_s)ds, \quad t \geq 0,$$

where $a \geq 0$ and (V_t) is a regenerative process with finite duration T of the regenerative cycle. An argument analogous to that of the Dynkin theorem yields for the α-potential of A:

$$u_A^\alpha(x) = \mathbb{E}^x \int_0^T e^{-\alpha t} a(V_t)dt + \mathbb{E}^x(e^{-\alpha T})\mathbb{E} \int_0^\infty e^{-\alpha t} a(V_t)dt,$$

where use was made of the fact that T and V_{t+T} are independent, and V_t and V_{t+T} have the same distribution (in particular for $t=0$). Hence, one has:

$$\mathbb{E} \int_0^\infty e^{-\alpha t} a(V_t)dt = \frac{1}{1-\mathbb{E}(e^{-\alpha T})} \mathbb{E} \int_0^T e^{-\alpha t} a(V_t)dt,$$

so for the stationary process passage to the limit yields:

$$\mathbb{E}a(V_\infty) = (1/\mathbb{E}T)\mathbb{E} \int_0^T a(V_t)dt.$$

Taking $a(x)=e^{-cx}$, the above result expresses the mean value theorem used by J.W. Cohen in studies of waiting times (combined with an argument based on level crossing); see [13], [17], [14].

3.4. Dirichlet problem

This fundamental problem is the probabilistic analogue of the Dirichlet problem in classical Potential Theory and partial differential equations. Its importance for Markov processes is due to the fact that it embraces all first passage problems (through the concept of balayage). The problem received considerable attention in the literature; see DOOB [23], DYNKIN [26], BLIEDTNER and HANSEN [4] for general discussion, and CHUNG [8], DURRETT [25], FREIDLIN [31], PORT and STONE [60], WILLIAMS [84] for Brownian motion. Here only one version of the problem will be stated, followed by specialization to Markov chains as in SYSKI [79], [82].

3.4.1. A real valued measurable function f defined on a state space $(E, \mathcal{E})$ is said to be *superharmonic* on an open set $G \subset E$ for a Markov process $X=(X_t)$ if

$$P_{E-\bar{A}} f(x) \leqslant f(x), \quad x \in G,$$

for every open set A with compact closure $\bar{A} \subset G$. Here $P_{E-\bar{A}}$ is the balayage operator. With T denoting the first entrance time to $E-\bar{A}$, the above definition means:

$$\mathbb{E}^x\{f(X_T), T<\infty\} \leqslant f(x), \quad x \in G.$$

A function f is *harmonic* on G for X if equality holds in the above relations; this is the analogue of the 'mean value property' from classical theory:

$$P_T f(x) = f(x), \quad x \in G.$$

For a standard Markov process X, the set of all excessive functions coincides with the set of all positive superharmonic functions. However, a harmonic function need not be invariant, in general. If f is harmonic on G, then it follows that

$$Qf(x) = 0, \quad x \in G.$$

3.4.2. The (exterior) Dirichlet problem for a Markov process $X=(X_t)$ consists of finding a function f such that:
i) f is harmonic on G;
ii) $f = \phi$ on $E-G$;
iii) for regular points $x \in \partial G$:

$$\lim f(z) = \phi(x) \quad \text{as} \quad z \to x \quad \text{and} \quad z \in G,$$

where ϕ is a given 'boundary function', and ∂G is the (topological) boundary of G.

THEOREM 3.4.1. *Suppose that X is a strong Feller process, G an open set with compact closure and T the first entrance time to $E-G$. Let the boundary function ϕ be bounded and continuous on ∂G. Then, there exists a (minimal) bounded solution of the Dirichlet problem for X on G with the boundary function ϕ, given by*

$$f(x) = \mathbb{E}^x\{\phi(X_T), T<\infty\}.$$

The solution is unique when T is finite a.e.

A simpler version of the Dirichlet problem (appropriate for Markovian queues) is that for Markov chains with a discrete state space Π, stochastic transition matrix $P_t=(p_{ij}(t))$ and (conservative) generator $Q=(q_{ij})$. The problem consists of finding a non-negative bounded function f on Π such that

$$Qf(i) = 0 \quad \text{for} \quad i \in \Pi - H$$

$$f(i) = \phi(i) \quad \text{for} \quad i \in H$$

where H is a fixed subset, and ϕ is a given (non-negative bounded) function on the 'boundary' H.

THEOREM 3.4.2. *For a Markov chain X, let T be the first entrance time to a set H, and let* ϕ *be a given boundary function on H. Then, the (minimal) bounded solution of the Dirichlet problem on* $\Pi - H$ *for X, with the boundary function* ϕ, *is given by*

$$f(i) = \mathbb{E}^i(\phi(X_T), T<\infty), \quad i \in \Pi.$$

The solution is unique when T is finite a.e.

It may be worthwhile to outline briefly the basic ideas of the martingale proof of these theorems. First, the required function f is shown to be invariant for the stopped process $\tilde{X}=(\tilde{X}_t)$, where $\tilde{X}_t = X_{t\wedge T}$ for $t \geqslant 0$. Put $Y_t = f(\tilde{X}_t)$, $t \geqslant 0$. As f is bounded, the process (Y_t) is a non-negative uniformly integrable martingale. The martingale convergence theorem implies then that the martingale converges with probability one and has the usual representation:

$$Y_t \to Y_\infty \quad \text{a.e.,} \quad Y_t = \mathbb{E}(Y_\infty | X_t) \quad \text{a.e.,} \quad t \geqslant 0.$$

As the next step, it is verified that Y_∞ is an invariant random variable. This allows one to define f:

$$f(x) = \mathbb{E}(Y_\infty | \tilde{X}_0 = x).$$

The proof is completed by noting that $Y_\infty = \phi(X_T)$ when $T<\infty$.

Observe that taking the boundary function ϕ identically 1, one has:

$$f(x) = \mathbb{P}^x(T<\infty).$$

This is the first entry probability, already mentioned in Subsection 3.2.3. It should be observed that the unique solution of the Dirichlet problem is in fact the balayage of the boundary function, namely $P_H\phi$ (see Subsection 3.1.3).

3.4.3. The Dirichlet problem and its associated Poisson equation enjoy an important role in those applications where passage times are of interest. In Markovian queues, this involves busy periods, waiting times, absorption problems as well as other random times (such as overflow and output problems). The language of Potential Theory is most appropriate for this purpose, because it provides unifying and direct treatment. For exposition of these ideas in Markovian queues, see [79], [82].

1. For the purpose of illustration, the entrance probability

$$D_i = \mathbb{P}^i(T<\infty), \quad i \in \Pi,$$

where T is the first entrance time to a set $H \subset \Pi$, is the (minimal) solution of the Dirichlet problem, as already noted. The function D satisfies also the Poisson equation:

$$\begin{aligned} QD_i &= 0, \quad i \in \Pi - H \\ &= -b_i, \quad i \in H, \end{aligned}$$

where $b \geqslant 0$ is the 'escape rate' supported on H. Thus, D is excessive on Π, and its Riesz decomposition (for transient chains) takes the form:

$$D_i = Ub_i + s_i, \quad i \in \Pi,$$

where Ub is the purely excessive component which is also a potential of a charge, and s is invariant. Introducing the last exit time from a set H defined by

$$L = \sup\,(t : X_t \in H),$$

the above decomposition corresponds to the representation:

$$\mathbb{P}^i(0 < L < \infty) = Ub_i, \quad \mathbb{P}^i(L = \infty) = s_i,$$

so the component s is the probability of infinitely many visits to a set H. There is an interesting connection with functionals (as remarked in Subsection 3.3.4). Consider two integral functionals defined, respectively, by charges b and D:

$$B_t = \int_0^t b(X_s)ds, \quad D_t = \int_0^t D(X_s)ds.$$

Then, their characteristics and potentials are related to the distribution of L as follows:

$$\mathbb{E}^i B_t = \mathbb{P}^i(0 < L \leqslant t) \to \mathbb{P}^i(0 < L < \infty) = Ub_i = u_B(i),$$

$$\mathbb{E}^i D_t = \int_0^t \mathbb{P}^i(L > s)ds \to UD_i = u_D(i),$$

for $i \in \Pi$. See [82] for details.

2. It may be worthwhile to recall some well-known examples. In the classical waiting system M/M/s (with s servers) with a Markov chain (X_t) denoting the number of customers, the busy period T is just the first entrance time to the set $H = \{0, 1, \ldots, s-1\}$, and D_i is the probability that the busy period, originated at state $i \geqslant s$, is finite (the proper busy period corresponds to $i = s$). With the same H, but for a modified chain (Y_t) representing waiting customers, the first entrance time T is the waiting time and D_i is the probability that waiting time is finite.
Other well-known examples of this type are the classical gambler's ruin problem, the branching process, and other absorption problems, with D_i representing probability of absorption (extinction, etc.). In this type of problems, the set H may be split into subsets, and one considers probability of absorption in a particular subset (before entering others). This corresponds to the boundary function being the indicator of that subset. The classical overflow problem in a loss system is of this type, and the output problem may also be cast in this framework (by attaching an absorbing state).
It may be of interest to mention here that if X_t represents the number of

individuals at time t in the classical branching process, then the random variable

$$A_\infty = \int_0^\infty X_s ds$$

has the same distribution as the busy period in the M/M/1 queue with group input. This result constitutes a nice application of the random time change technique; see [79], [82]. As already mentioned earlier, the Potential Theory approach puts these diverse problems into the unifying framework of the Dirichlet problem.

3. The above discussion presents examples of probabilistic boundary problems, analogous to those in analytic Potential Theory. For a further discussion of boundary value problems solved by probabilistic methods, see DOOB [23], DURRETT [25], DYNKIN [26], FREIDLIN [31].

3.4.4. It is of great interest to comment on similarities between probabilistic boundary problems as described in this subsection and the boundary value problems considered by COHEN and BOXMA [18]. As already mentioned earlier in Subsection 2.3, two levels of analysis must be distinguished.

For illustrative purposes it is convenient to consider here a simple and well-known case of a one-dimensional random walk. This choice does not do justice to the amount of ingenuity involved in the analysis of two-dimensional random walks and in applications of the method to concrete cases. It suffices, however, to outline the main ideas for comparison purposes.

Consider a one-dimensional random walk (X_n, $n=0,1,...$) on the positive integers where

$$X_{n+1} = X_n + \xi_n \quad \text{for} \quad X_n \geq 1,$$

with 0 as an absorbing state. Here (ξ_n) is a family of independent identically distributed random variables, assuming values $-1,0,1,...,$ with distribution $b(k)$ for $k \geq -1$, and with generating function written in the form:

$$\psi(p) = p\mathbb{E}p^{\xi_n}, \quad 0 \leq p \leq 1.$$

Clearly, (X_n) is a Markov chain with one-step transition probabilities

$$p_{ij} = b(j-i), \quad j-i \geq -1, \quad i \geq 1, \quad p_{00} = 1,$$

and its n-step transition probabilities $p_{ij}^{(n)}$ coincide with the taboo probabilities ${}_Hp_{ij}^{(n)}$. Here, the taboo set is $H=(0)$, and $H^c=(1,2,...)$. Let $T=T_H$ be the first entrance time into H; note that $X_T=0$ on ($T<\infty$).

Introduce the notation (where conditioning is on the event $X_0=i$):

$$\Phi_i^{(n)}(p) = \mathbb{E}^i(p^{X_n}), \quad i \geq 1, \quad \text{with} \quad \Phi_i^{(0)}(p) = p^i,$$

$$\Phi_i(r,p) = \sum_{n=0}^\infty r^n \Phi_i^{(n)}(p), \quad \Phi_i(r,H) = \sum_{n=0}^\infty r^n p_{i0}^{(n)}, \quad 0 \leq r < 1,$$

and,

$$h_i^{(n)} = \mathbb{P}^i(X_T=0,\ T=n), \quad h_i(r) = \sum_{n=0}^{\infty} r^n h_i^{(n)}, \quad 0 \leq r < 1.$$

The usual argument yields the relation

$$(1-r)\Phi_i(r,H) = \mathbb{E}^i(r^T,\ T<\infty)$$

and the basic equation

$$[1-r\psi(p)/p]\Phi_i(r,p) = r[1-\psi(p)/p]\Phi_i(r,H)+p^i.$$

As this equation contains two unknown functions, the analysis of the 'kernel' $p-r\psi(p)$ is the starting point for the determination of a solution. In the present case, the characteristic equation

$$p-r\psi(p) = 0$$

has the unique real root $p=p(r)$ for $0 \leq r < 1$. It is then shown that both the LHS and RHS of the functional equation vanish at $p(r)$. This determines the unknown function

$$\Phi_i(r,H) = p^i(r)/(1-r).$$

Substitution of this expression into the functional equation determines the required generating function $\Phi_i(r,p)$.

Furthermore, one has now

$$h_i(r) = p^i(r).$$

So, in particular:

$$\mathbb{P}^i(T<\infty) = h_i(1) = p^i(1).$$

It is evident that $h_i = h_i(1)$ is the solution of the Dirichlet problem:

$$\sum_{j \geq i-1}^{\infty} p_{ij} h_j = h_i, \quad i \geq 1$$

$$h_0 = 1.$$

Solving it in the usual way, one finds again that $h_i = p^i$, where p is the solution of $p=\psi(p)$. Note that the case considered here refers to T being the busy period for the imbedded Markov chain in the M/G/1 queue.

In the general case of two-dimensional random walks, consideration of the kernel of the functional equation is the crucial and most difficult part of the analysis. Often solution of the functional equation can be reduced to solution of a Riemann problem, or a Riemann-Hilbert problem (which includes the Dirichlet problem as a special case). For a full discussion, see the monograph of COHEN and BOXMA [18] where further references may be found. See also the monograph by BLANC [3]; some probabilistic aspects are discussed in [80].

3.5. Decomposition theorems

In the study of Markovian systems, it is important to have representation theorems which relate the structure of two associated semi-groups. Typical examples are provided by the well-known 'first entrance' decomposition, by perturbation of semi-groups, and by the renewal equation. General references have already been given in Subsection 3.1, and a few specific references are provided below.

3.5.1. Let (P_t) be a Markov semi-group of transition operators on a space $\mathbb{B}=\mathbb{B}(E, \mathcal{E})$ of bounded measurable functions on a state space $(E, \mathcal{E})$. Here $P_0=I$ and $P_t 1=1$ for all t. Let $(\hat{P}_t)$ be a sub-Markov semi-group of transition operators on the same space $\mathbb{B}$, but with $\hat{P}_0 \leqslant I$ and $\hat{P}_0 1 \leqslant 1$. Denote by Q and $\hat{Q}$ the corresponding generators (in a specified sense), and similarly write U^α and $\hat{U}^\alpha$ for resolvents.

Suppose that operators $\hat{P}_t$ and P_t are related by the following *decomposition formula:*

$$P_t f = \hat{P}_t f + (K \star P)_t f, \quad f \in \mathbb{B}, \quad t \geqslant 0,$$

where K_t is a positive operator on $\mathbb{B}$ defined by a kernel $K_t(\cdot, \cdot)$ on $E \times \mathcal{E}$, and $\star$ denotes convolution:

$$(K \star P)_t = \int_0^t dK_s P_{t-s}, \quad t \geqslant 0,$$

with the integral taken over the closed interval $[0, t]$.

In applications, (K_t) is given and $(\hat{P}_t)$ is the 'original' semi-group, whereas (P_t) is the 'transformed' (or 'perturbed') semi-group, and the decomposition formula has the form of a Markov renewal equation.

It is more convenient to write the decomposition formula in terms of the resolvent:

$$U^\alpha = \hat{U}^\alpha + K^\alpha U^\alpha, \quad \alpha > 0,$$

where

$$K^\alpha = \int_0^\infty e^{-\alpha t} dK_t, \quad \lim_{\alpha \to \infty} K^\alpha = K_0.$$

Well-known examples of decomposition theorems are provided by Markov chains. With P_t being a matrix of transition probabilities $p_{ij}(t)$, one has

a) The (B)-equation in the integrated form:

$$p_{ij}(t) = \delta_{ij} e^{-q_i t} + \sum_k \int_0^t r_{ik}(s) p_{kj}(t-s) ds, \quad i, j \in \Pi,$$

where $\hat{P}_t$ is the diagonal matrix of exponential terms with $(\delta_{ij}) = I$ (the unit matrix), and the matrix $K_s = (r_{ik}(s))$ is such that its Laplace transform matrix K^α has elements $q_{ik}/(q_i + \alpha)$ for $i \neq k$, and 0 for $i = k$.

b) The first entrance decomposition:

$$p_{ij}(t) = {}_Hp_{ij}(t) + \sum_{k\in H} \int_0^t d_{ik}(s)p_{kj}(t-s)ds, \quad i,j\in\Pi,$$

where $\hat{P}_t = {}_HP_t$ is the matrix of taboo probabilities, and the matrix $K_s = (d_{ik}(s))$ is such that its Laplace transform matrix K^α is the α-balayage matrix $B^\alpha = (d^\alpha_{ik})$, denoted by P^α_H in Subsection 3.1.3.

3.5.2. Observe that

$$\hat{P}_0 + K_0 = P_0 = I, \quad \hat{P}_t 1 + \hat{K}_t 1 = P_t 1 = 1,$$

and that for the original semi-group one has:

$$\hat{P}_0\hat{P}_t = \hat{P}_t\hat{P}_0 = \hat{P}_t, \quad \hat{Q} = \lim_{t\to 0}(\hat{P}_t - \hat{P}_0)/t$$

with Kolmogorov equations:

$$\alpha\hat{U}^\alpha - \hat{P}_0 = \hat{Q}\hat{U}^\alpha, \quad (B)$$
$$= \hat{U}^\alpha\hat{Q}, \quad (F).$$

The following relations are listed here for convenience in applications. They can be formally deduced from definitions and properties of semi-groups; see the references for precise conditions and proofs.

As both $\hat{P}_t$ and P_t are assumed to be semi-groups, it follows that K_t must satisfy the condition:

$$\hat{P}_t(K_s - K_0) = K_{s+t} - K_t, \quad t\geqslant 0, \quad s\geqslant 0.$$

Using the (F)-equation for both semi-groups, one obtains a relation between their generators:

$$K^\alpha - K_0 = \hat{U}^\alpha(Q - \hat{Q}), \quad \alpha>0,$$

or

$$K_t - K_0 = \int_0^t \hat{P}_s(Q - \hat{Q})ds.$$

The next result indicates that a solution of the Markov renewal equation has essentially the form of a Riesz decomposition. The operator G^α defined here is the analogue of a renewal function.

THEOREM 3.5.1. *For $f\in\mathbb{B}$, the equation*

$$U^\alpha f = \hat{U}^\alpha f + K^\alpha U^\alpha f$$

has a solution $U^\alpha f$ iff the (strong) limit:

$$G^\alpha\hat{U}^\alpha f = \sum_{n=0}^{\infty}(K^\alpha)^n\hat{U}^\alpha f$$

exists. All solutions are of the form:

$$U^\alpha f = G^\alpha \hat{U}^\alpha f + W^\alpha f$$

where

$$W^\alpha f = \lim_{n\to\infty} (K^\alpha)^n U^\alpha f.$$

The solution is unique when

$$W^\alpha f = 0.$$

For a proof see papers by ÇINLAR [10], [11], and also, for a different approach, [24], [38]. A survey of operator equations was given in [66].

3.5.3. Examples a) and b) mentioned in Subsection 3.5.1 above are of great importance in the theory of Markov chains.

a) This example establishes the connection between a Markov chain and its associated jump chain (i.e., discrete parameter chain of jumps). With matrices:

$$U^\alpha = (u_{ij}^\alpha), \quad \hat{U}^\alpha = (\delta_{ij}/(q_i+\alpha)),$$

the relation (obtained by Feller in his early work):

$$U^\alpha = G^\alpha \hat{U}^\alpha$$

indicates that a minimal solution of Kolmogorov equations for transition probabilities is related to the structure of the jump chain. In fact, the formula:

$$K^\alpha - K_0 = \hat{U}^\alpha (Q - \hat{Q})$$

with $K_0=0$, $\hat{Q}=(-\delta_{ij}q_j)$, yields transition probabilities K^α for the α-jump chain, and G^α is the corresponding Green potential. See CHUNG [7], SYSKI [82]. In this sense, the Feller relation is equivalent to the Dynkin characteristic operator which in the present case coincides with the generator Q.

b) With H as a taboo set, the matrices (in a partitioned form) are:

$$K^\alpha = B^\alpha = \begin{pmatrix} 0 & B_{12}^\alpha \\ 0 & I \end{pmatrix}, \quad K_0 = \begin{pmatrix} 0 & 0 \\ 0 & I \end{pmatrix}, \quad \hat{U}^\alpha = {}_H U^\alpha = \begin{pmatrix} {}_H U_{11}^\alpha & 0 \\ 0 & 0 \end{pmatrix}.$$

Hence:

$$(B^\alpha)^n = B^\alpha \quad \text{for all} \quad n \geqslant 1, \quad B^\alpha \hat{U}^\alpha = 0,$$

so:

$$G^\alpha \hat{U}^\alpha = \hat{U}^\alpha, \quad W^\alpha = B^\alpha U^\alpha$$

and the solution of the Markov renewal equation reduces to the decomposition itself:

$$U^\alpha = {}_H U^\alpha + B^\alpha U^\alpha.$$

The relations stated earlier which involve K_t reduce now to the well-known relations involving the balayage matrix B_t. For example,

$$B^\alpha_{12} = {}_H U^\alpha_{11} Q_{12}.$$

Recall that the first entrance decomposition may be obtained from the Dynkin theorem applied to the first entrance time T to a set H.

c) This interesting example indicates that a Markov chain with all states absorbing may be 'revived' by some kind of perturbation to produce a 'live' chain. Thus,

$$\hat{P}_t = I, \quad \hat{Q} = 0, \quad \hat{U}^\alpha = I/\alpha.$$

Take $K_t = tA$ where A is a bounded matrix, so $K^\alpha = A/\alpha$. Evidently, $K_s - K_0 = K_{t+s} - K_t$ and

$$G^\alpha = \alpha(\alpha I - A)^{-1},$$

so the solution is:

$$U^\alpha = (\alpha I - A)^{-1}, \quad \text{or} \quad P_t = e^{At}.$$

This semi-group is probabilistically admissible when A is the generator Q (i.e. bounded, conservative).

3.5.4. It is apparent that the Markov renewal equation

$$U^\alpha = \hat{U}^\alpha + K^\alpha U^\alpha$$

can be written in the equivalent form which expresses U^α directly in terms of the original process:

$$\hat{P}_0 U^\alpha - \hat{U}^\alpha = \hat{U}^\alpha (Q - \hat{Q}) U^\alpha.$$

This relation is known as the 'second resolvent equation'; cf. HILLE and PHILLIPS [38] (where $\hat{P}_0 = I$).

In Perturbation Theory of semi-groups the requirement that the perturbation is 'sufficiently small' is usually expressed by some boundedness conditions imposed on $\hat{U}^\alpha(Q - \hat{Q})$. Then, the theory asserts the existence of U^α (corresponding to Q) satisfying the second resolvent equation, and given by a series

$$U^\alpha = \sum_{n=0}^{\infty} S^\alpha_n$$

where $S^\alpha_n = K^\alpha S^\alpha_{n-1}$ with $S^\alpha_0 = \hat{U}^\alpha$. For perturbation theory of semi-groups, see HILLE and PHILLIPS [38], DUNFORD and SCHWARTZ [24], ETHIER and KURTZ [28].

The 'compensation method' proposed by KEILSON [47], [48], constitutes in fact a special form of perturbation, carried out by an external replacement mechanism. For a full discussion, see [75] and [77] where applications to

Markovian queues are presented. A related method of phase-type distributions was introduced by NEUTS [58]. Explanation of these ideas in terms of the perturbation mechanism can be found in [77]. Other applications may be found in LEVIATAN [54] and in the recent paper by SIEGRIST [68] on revival processes; see also a paper by DAVIS [19] on piecewise-deterministic Markov processes.

REMARK. The Markov renewal equation considered here may be thought of as being associated with the first entrance times. The dual system, associated with the last exit time, takes the form:

$$U^\alpha = {}_H U^\alpha + U^\alpha C^\alpha$$

where C^α is the α-co-balayage. See [79], [82] for details.

ACKNOWLEDGEMENT
The author wishes to express his gratitude to the referees for their comments, which greatly improved the first version of the paper.

REFERENCES

1. F. BACCELLI, P. BREMAUD (1987). *Palm Probabilities and Stationary Queues,* Springer-Verlag, Berlin.
2. A. BENVENISTE, J. JACOD (1973). Systèmes de Lévy des processus de Markov. *Invent. Math. 21,* 183-198.
3. J.P.C. BLANC (1982). *Application of the Theory of Boundary Value Problems in the Analysis of a Queueing Model with Paired Services,* MC Tract 153, Centre for Mathematics and Computer Science, Amsterdam.
4. J. BLIEDTNER, W. HANSEN (1986). *Potential Theory (An Analytic and Probabilistic Approach to Balayage),* Springer-Verlag, Berlin.
5. R.M. BLUMENTHAL, R.K. GETOOR (1968). *Markov Processes and Potential Theory,* Academic Press, New York.
6. P. BRÉMAUD (1981). *Point Processes and Queues (Martingale Dynamics),* Springer-Verlag, Berlin.
7. K.L. CHUNG (1967). *Markov Chains with Stationary Transition Probabilities,* Springer-Verlag, Berlin (2nd edition).
8. K.L. CHUNG (1982). *Lectures from Markov Processes to Brownian Motion,* Springer-Verlag, Berlin.
9. K.L. CHUNG, R.J. WILLIAMS (1983). *Introduction to Stochastic Integration,* Birkhäuser, Basel.
10. E. ÇINLAR (1969). On semi-Markov processes on arbitrary spaces. *Proc. Camb. Phil. Soc. 66,* 381-392.
11. E. ÇINLAR (1972). Markov additive processes, I and II. *Z. Wahrsch. Verw. Gebiete 24,* 85-93, 94-121.
12. E. ÇINLAR (1975). Markov renewal theory: a survey. *Management Sci. 21,* 727-752.
13. J.W. COHEN (1976). *On Regenerative Processes in Queueing Theory,* Springer-Verlag, Berlin.
14. J.W. COHEN (1977). On up- and downcrossings. *J. Appl. Probab. 14,* 405-410.

15. J.W. COHEN (1982). *The Single Server Queue,* North-Holland, Amsterdam, 2nd edition.
16. J.W. COHEN (1985). Who introduced the birth-and-death technique? M. AKIYAMA (ed.). *Proc. ITC 11,* North-Holland, Amsterdam.
17. J.W. COHEN (1985). *On Renewal Processes with Discontinuous Sample Functions,* Preprint, Dept. Math., Univ. of Utrecht, nr. 401.
18. J.W. COHEN, O.J. BOXMA (1983). *Boundary Value Problems in Queueing System Analysis,* North-Holland, Amsterdam.
19. M.H.A. DAVIS (1984). Piecewise-deterministic Markov processes. *J. Roy. Statist. Soc. Ser. B46,* 353-388.
20. C. DELLACHERIE, P.A. MEYER (1978, 1982). *Probabilities and Potentials,* North-Holland, Amsterdam, Vol. A 1978, Vol. B 1982.
21. J.L. DOOB (1954). Semimartingales and subharmonic functions. *Trans. Amer. Math. Soc. 77,* 85-121.
22. J.L. DOOB (1956). Probability methods applied to the first boundary value problem. *Proc. 3rd Berkeley Symp. Math. Stat. Prob.,* 49-80, Univ. California Press.
23. J.L. DOOB (1984). *Classical Potential Theory and its Probabilistic Counterpart,* Springer-Verlag, Berlin.
24. N. DUNFORD, J.T. SCHWARTZ (1958). *Linear Operators,* J. Wiley (Interscience), New York.
25. R. DURRETT (1984). *Brownian Motion and Martingales in Analysis,* Wadsworth Advanced Books, Belmont.
26. E.B. DYNKIN (1965). *Markov Processes, Vol. I and II,* Springer-Verlag, Berlin.
27. R.J. ELLIOTT (1982). *Stochastic Calculus and Applications,* Springer-Verlag, Berlin.
28. S.N. ETHIER, TH.G. KURTZ (1986). *Markov Processes (Characterization and Convergence),* J. Wiley (Interscience), New York.
29. W. FELLER (1960). Some new connections between probability and classical analysis. *Proc. Intern. Congr. Math.,* 69-86. Cambridge Univ. Press.
30. P. FRANKEN, D. KÖNIG, U. ARNDT, V. SCHMIDT (1982). *Queues and Point Processes,* J. Wiley (Interscience), New York.
31. M. FREIDLIN (1985). *Functional Integration and Partial Differential Equations,* Princeton Univ. Press, Princeton.
32. J. GANI (ed.) (1986). *The Craft of Probabilistic Modelling,* Springer-Verlag, Berlin.
33. D. GEMAN, J. HOROWITZ (1975). Polar sets and Palm measures in the theory of flows. *Trans. Amer. Math. Soc. 208,* 141-159.
34. J. GLOVER (1981). Application of raw time-changes to Markov processes. *Ann. Prob. 9, 1019-1029.*
35. J. GLOVER (1983). *Topics in Energy and Potential Theory, Seminar on Stochastic Processes 1982,* 195-207, Birkhäuser, Basel.
36. H. GZYL (1980). Infinitesimal generators of time changed processes. *Ann. Prob. 8,* 716-726.

37. L.L. HELMS (1969). *Introduction to Potential Theory,* J. Wiley, New York.
38. E. HILLE, R.S. PHILLIPS (1957). Functional analysis and semi-groups. *Coll. Publ. Amer. Math. Soc.*
39. A. HORDIJK (1974). *Dynamic Programming and Markov Potential Theory,* MC Tract 52, Centre for Mathematics and Computer Science, Amsterdam (see also a review in *Technometrics 17* (1975), p. 514.)
40. G.A. HUNT. Markov processes and potentials. *Ill. J. Math.* I: *1,* 44-93, 1957; II: *1,* 316-369, 1957; III: *2,* 151-213, 1958.
41. M. JACOBSEN (1982). *Statistical Analysis of Counting Processes,* Springer-Verlag, Berlin.
42. M. KAC (1951). On some connections between probability theory and differential and integral equations. *Proc. 2nd Berkeley Symp. Math. Stat. Prob.,* 189-215, Univ. California Press.
43. S. KAKUTANI (1944). Two dimensional Brownian motion and harmonic functions. *Proc. Imp. Acad. Tokyo 20,* 706-714.
44. O. KALLENBERG (1976). *Random Measures,* Academic Press, New York.
45. S. KARLIN, H.M. TAYLOR (1975,1981). *A First Course in Stochastic Processes. A Second Course in Stochastic Processes,* Academic Press, New York.
46. H. KASPI, B. MAISONNEUVE (1986). *Regenerative Systems on the Real Line.* Submitted.
47. J. KEILSON (1965). *Green's Function Methods in Probability Theory,* Griffin, New York.
48. J. KEILSON (1979). *Markov Chain Models - Rarity and Exponentiality,* Springer-Verlag, Berlin.
49. O.D. KELLOG (1953). *Foundations of Potential Theory,* Dover, New York.
50. J.G. KEMENY, J.L. SNELL, A.W. KNAPP (1966). *Denumerable Markov Chains,* Van Nostrand, New York.
51. J.H.B. KEMPERMAN (1961). *The Passage Problem for a Stationary Markov Chain,* Univ. of Chicago Press, Chicago.
52. J. LAMPERTI (1977). *Stochastic Processes (A Survey of the Mathematical Theory),* Springer-Verlag, Berlin.
53. P. LE GALL (1974, 1978). Stochastic integral equations applied to telecommunication traffic without delay. *Stoch. Proc. Appl. 2,* 261-280, and *6,* 337-338.
54. T. LEVIATAN (1972). Perturbation of Markov processes. *J. Functional Analysis 10,* 309-325.
55. W.D. MACMILLAN (1958). *The Theory of the Potential,* Dover, New York.
56. MATHEMATICAL SOCIETY OF JAPAN (1977). *Encyclopedic Dictionary of Mathematics,* MIT Press, Boston.
57. J.B. MITRO (1986). A note on discontinuous time changes of Markov processes. *Stoch. Proc. Appl. 23,* 343-348.
58. M.F. NEUTS (1975). Probability distributions of phase type. *Liber Amicorum Prof. Emeritus H. Florin,* Univ. of Louvain, 173-206.
59. I.G. PETROVSKY (1954). *Partial Differential Equations,* J. Wiley, New York.

60. S.C. PORT, C.J. STONE (1978). *Brownian Motion and Classical Potential Theory,* Academic Press, New York.
61. N.U. PRABHU (1980). *Stochastic Storage Processes,* Springer-Verlag, Berlin.
62. D. REVUZ (1970). Mesures associées aux fonctionnelles additives de Markov. I. *Trans. Amer. Math. Soc. 148,* 501-531. II. *Z. Wahrsch. Verw. Gebiete 16,* 336-344.
63. D. REVUZ (1975). *Markov Chains,* North-Holland, Amsterdam.
64. T. ROLSKI (1981). *Stationary Random Processes Associated with Point Processes,* Springer-Verlag, Berlin.
65. W.A. ROSENKRANTZ (1981). Some martingales associated with queueing and storage processes. *Z. Wahrsch. Verw. Gebiete 58,* 205-222.
66. T.L. SAATY (1967). *Modern Nonlinear Equations,* McGraw-Hill, New York.
67. R. SCHASSBERGER (1972). On the work load process in a general preemptive resume priority queue. *J. Appl. Probab. 9,* 588-603.
68. K. SIEGRIST (1983). Harmonic functions and the Dirichlet problem for revival Markov processes. *Ann. Prob. 11,* 624-634.
69. E.V. SLUD. *Martingale Methods in Statistics.* Forthcoming book.
70. F. SPITZER (1964). *Principles of Random Walk,* Van Nostrand, New York.
71. M. SPIVAK (1965). *Calculus on Manifolds,* W.A. Benjamin, Inc., New York.
72. R. SYSKI (1965). Markovian queues. W.L. SMITH, W.E. WILKINSON (eds.). *Congestion Theory,* 170-227, Univ. of North Carolina Press, Chapel Hill.
73. R. SYSKI (1967). Pollaczek method in queueing theory. R. CRUON (ed.). *Queueing Theory,* 33-60, English University Press.
74. R. SYSKI (1973). Potential theory for Markov chains. A.T. BHARUCHA-REID (ed.). *Probabilistic Methods in Applied Mathematics Vol. 3,* 214-276.
75. R. SYSKI (1977). Perturbation models. *Stoch. Proc. Appl. 5,* 93-129.
76. R. SYSKI (1979). Energy of Markov chains. *Adv. Appl. Probab. 11,* 542-575.
77. R. SYSKI (1982). Phase-type distributions and perturbation model. *Applicationes Mathematicae XVII, 3,* 377-399.
78. R. SYSKI (1986). Markovian functionals in teletraffic analysis. O.J. BOXMA, J.W. COHEN, H.C. TIJMS (eds.). *Teletraffic Analysis and Computer Performance Evaluation,* 303-317, North-Holland, Amsterdam.
79. R. SYSKI (1986). *Introduction to Congestion Theory in Telephone Systems,* revised edition, North-Holland, Amsterdam.
80. R. SYSKI (1986). *Note on a Random Walk with Absorption,* Preprint Dept. Math., Univ. of Utrecht, nr. 420.
81. R. SYSKI (1988). Stochastic processes. *Encyclopedia of Statistical Sciences Vol. 8,* pp. 836-851. J. Wiley, New York.
82. R. SYSKI. *Passage Times in Markov Chains.* Book in preparation.
83. L. TAKACS (1967). *Combinatorial Methods in the Theory of Stochastic Processes,* J. Wiley, New York.
84. D. WILLIAMS (1979). *Diffusions, Markov Processes and Martingales,* J. Wiley, New York.

Computational Methods for Queueing Models: A Review

H.C. Tijms
Institute for Econometrics, Vrije Universiteit Amsterdam
P.O. Box 7161, 1007 MC Amsterdam, The Netherlands

This paper presents a survey of computational methods for basic queueing models with an isolated servicing node. We first discuss exact and approximate methods for the computation of the state and waiting-time probabilities in the single-server queue with general interarrival and service times. The approximate methods for complex queueing models are typically based on exact results for simpler related models and on asymptotic expansions. Finally, we give approximations for multi-server queues with Poisson arrivals.

1. Introduction

A remarkable change in queueing analysis took place about 1970. In the fifties and the sixties, the precomputer area, the emphasis was naturally on obtaining analytical results in a closed form. Recently queueing analysis has been much more directed to the goal of getting algorithms suited for (recursive) calculations on the computer. For example, the old and relatively simple 'birth-and-death' technique has seen a revival by the use of phase-type distributions and the ability of current computers to solve very large systems of linear equations. This basic technique goes at least back to Fry [13] and Molina [18], and was a powerful computational tool already in the early days of teletraffic analysis. Most of the new and important developments that took place in the seventies are nicely overviewed in the supplemented chapter of the second edition of Cohen's book 'The Single Server Queue' [5].

My own interest in algorithmic methods for queueing models was much stimulated by pleasant and useful discussions I had with Wim Cohen. It is therefore my great pleasure to give in this Liber Amicorum a survey on numerical methods for queueing models. In what follows I hope to make clear that algorithmic analysis of queueing models is more than getting numerical answers. The essence of computational probability is to find probabilistic ideas which make the computations transparent and natural.

In this survey I have made the choice to restrict my overview to basic queueing models with an isolated servicing node and not to discuss queueing networks. Even within this restriction I have to be selective. The paper is organized as follows. Section 2 discusses several basic algorithms for the computation of the state probabilities and the waiting-time probabilities for queueing models of the M/G/1 type. In Section 3 generally applicable algorithms and approximations for the GI/G/1 queue are given. The final Section 4 discusses approximations for multi-server queues.

2. Computational methods for M/G/1-type queueing models

In this section we discuss basic algorithms for the computation of the state probabilities and waiting-time probabilities.

2.1. The regenerative method for the state probabilities

The regenerative approach is a powerful computational tool that is generally applicable for the calculation of the state probabilities in a variety of queueing models of the M/G/1 type. This approach uses basic results from the theory of regenerative processes and a simple up- and downcrossings argument.

The regenerative approach is best explained via the standard M/G/1 queue having a Poisson arrival process with rate λ and a general probability distribution function $B(t)=P\{S\leq t\}$ for the service time S of a customer. It is assumed that the offered load $\rho=\lambda E(S)<1$. The equilibrium state probabilities p_j for $j=0,1,\ldots$ are defined by

$$p_j = \lim_{t\to\infty} P\{\text{there will be } j \text{ customers present a time } t \text{ from now}\}.$$

This limiting distribution exists and is independent of the initial state. Alternatively, the probability p_j can be interpreted as the long-run fraction of time the system will be in state j. Here state j corresponds to the situation that j customers are present. By the theory of regenerative processes (see e.g. COHEN [4]), we have the basic result

$$p_j = \frac{E(T_j)}{E(T)} \quad \text{for} \quad j=0,1,\ldots, \tag{2.1}$$

where

$T =$ the length of a cycle which is defined as the time elapsed between two arrivals finding the system empty;

$T_j =$ the amount of time that the process is in state j during one cycle.

Note that the epochs at which an arrival occurs and finds the system empty are regeneration epochs for the queue length process. The regenerative method is based on two fundamental relations between the expected values $E(T_j)$ and $E(N_k)$, where the random variable N_k is defined by

$N_k =$ the number of service completion epochs in one cycle at which k customers are left behind, $k=0,1,\ldots$.

The first relation is based on the up- and downcrossings argument that during each cycle the number of transitions out of the set of states $\{k+1, k+2,\ldots\}$ must be equal to the number of transitions into the set of states $\{k+1, k+2,\ldots\}$ for each k. The expected number of transitions out of the set $\{k+1, k+2,\ldots\}$ per cycle equals $E(N_k)$ by definition, while the expected number of transitions into the set $\{k+1, k+2, \ldots\}$ per cycle equals $\lambda E(T_k)$. For the latter result we use that the arrival process is a Poisson process and invoke a fundamental property of the Poisson process given in WOLFF [31]. Thus we obtain the relation

$$\lambda E(T_k) = E(N_k) \quad \text{for} \quad k=0,1,\ldots . \tag{2.2}$$

The second relation between $E(T_j)$ and $E(N_k)$ is obtained by splitting up the cycle in disjoint intervals via the service completion epochs and calculating $E(T_j)$ as the sum of the contributions of the disjoint intervals. Here we need again the memoryless property of the Poisson arrival process. Then, noting that during a service the queue size cannot decrease, we obtain as second relation

$$E(T_j) = A_{1j} + \sum_{k=1}^{j} E(N_k) A_{kj} \quad \text{for} \quad j=1,2,\ldots, \tag{2.3}$$

where the quantity A_{kj} is defined by

$A_{kj}=$ the expected amount of time that j customers are present during one service time at whose beginning k customers are present.

Note that $E(T_0)=1/\lambda$ and so, by (2.1), $E(T)=1/(\lambda p_0)$. Inserting (2.2) into (2.3), dividing both sides of the resulting equation by $E(T)$ and using (2.1), we obtain the recursion equation

$$p_j = \lambda A_{1j} p_0 + \sum_{k=1}^{j} \lambda A_{kj} p_k \quad \text{for} \quad j=1,2,\ldots . \tag{2.4}$$

Starting with p_0 $(=1-\rho)$, we can recursively compute $p_1, p_2, \ldots$. This recursive algorithm is numerically stable and is easy to apply. It remains to specify the constants A_{kj}. In general, we have the representation

$$A_{kj} = \int_0^\infty e^{-\lambda t} \frac{(\lambda t)^{j-k}}{(j-k)!} \{1-B(t)\} dt.$$

These integrals can efficiently be computed by numerical integration methods such as Gauss-Laguerre or Gauss-Legendre, cf. VAN HOORN [27]. For phase-type service time distributions the constants A_{kj} can also be computed from a simple recursion scheme, cf. also (2.8)-(2.9) below. For computational purposes it is important to note that A_{kj} depends on k and j only through $j-k$.

The regenerative approach having its origin in the paper by HORDIJK and TIJMS [14] is a flexible and generally applicable method to single-server queueing models in which the arrival process is Markovian. For example, the analysis can be straightforwardly extended to models with state-dependent Poisson arrivals (including the machine repairman model) and to variants of the M/G/1 queue such as models with server vacations, retrials or exceptional first services, see TIJMS [25].

As further illustration, consider the $M^X/G/1$ queue with batch arrivals. Let

$\{\beta_j, j \geq 1\}$ denote the batch size distribution. It is assumed that $\lambda E(B)E(S) < 1$, where the generic variable B denotes the batch size. The above analysis leading to the recursion scheme (2.4) needs only a modification with respect to the up- and downcrossings relation (2.2). This relation becomes now

$$\lambda \sum_{i=0}^{k} E(T_i) \sum_{m>k-i} \beta_m = E(N_k), \quad k=1,2,\ldots.$$

The recursion scheme for the state probabilities p_j is now given by

$$p_j = \lambda p_0 \sum_{m=1}^{j} \beta_m A_{mj} + \lambda \sum_{k=1}^{j} \left(\sum_{i=0}^{k} p_i \sum_{m>k-i} \beta_m\right) A_{kj}, \quad j=1,2,\ldots, \tag{2.5}$$

where $p_0 = 1-\rho$ with $\rho = \lambda E(B)E(S)$. The generating function $P(z) = \Sigma_{j=0}^{\infty} p_j z^j$ is explicitly given by

$$P(z) = (1-\rho)\frac{1-\lambda A(z)\{1-\beta(z)\}}{1-\lambda A(z)\{1-\beta(z)\}/(1-z)}, \tag{2.6}$$

where $\beta(z) = \Sigma_{j=1}^{\infty} \beta_j z^j$ and $A(z) = \int_0^{\infty} \{1-B(t)\} e^{-\lambda t(1-\beta(z))} dt$, and provides a tool to compute the state probabilities for special cases by using the fast Fourier transform method. For the recursion algorithm (2.5) the computation of the constants A_{kj} is not as simple as for the algorithm (2.4). However, an efficient recursion scheme for the constants A_{kj} can be given for the important case that the service time has a probability density of the form

$$b(t) = \sum_{i=1}^{r} q_i \mu^i \frac{t^{i-1}}{(i-1)!} e^{-\mu t}, \quad t \geq 0, \tag{2.7}$$

where $q_i \geq 0$ and $\Sigma_i q_i = 1$. That is, the service time distribution is a mixture of Erlangian distributions with the *same* scale parameters. Any probability distribution function with mass on $(0, \infty)$ can be arbitrarily closely approximated by a mixture of Erlangian distribution functions with the same scale parameters. This result due to SCHASSBERGER [20] is an extremely useful result in queueing analysis both for analytical and computational purposes. As a consequence of this result it is sufficient in many practical applications to consider service times that can be represented as a random sum of independent *exponentials* with the same means. For a service time with density (2.7) we have the useful interpretation that, with probability q_i, the service time is distributed as the sum of i independent exponential phases each with the same mean $1/\mu$. This interpretation together with the memoryless property of the exponential distribution enables us to compute the constants A_{kj} from

$$A_{kj} = \sum_{i=1}^{r} q_i \alpha_{j-k}^{(i)}, \tag{2.8}$$

where the numbers $\alpha_{j-k}^{(i)}$ are recursively computed from

$$\alpha_0^{(i)} = \frac{1}{\lambda+\mu} + \frac{\mu}{\lambda+\mu}\alpha_0^{(i-1)}, \quad 1 \leq i \leq r,$$

$$\alpha_n^{(i)} = \frac{\lambda}{\lambda+\mu}\sum_{j=1}^{n}\beta_j\alpha_{n-j}^{(i)} + \frac{\mu}{\lambda+\mu}\alpha_n^{(i-1)}, \quad n\geq 1,\ 1\leq i\leq r, \tag{2.9}$$

with $\alpha_n^{(0)}=0$ by convention. Here the quantity $\alpha_n^{(i)}$ should be interpreted as the expected amount of time that during a service time consisting of i exponential phases there are n additional customers present on top of the customers present at the beginning of that service.

For computational purposes it is important to point out that the recursive calculations in (2.5) may be stopped earlier by using the result that for j large enough the state probability exhibits a geometric behavior when the batch size distribution satisfies a weak regularity condition. This condition requires that the power series $\beta(x)=\Sigma_{j=1}^{\infty}\beta_j x^j$ has a convergence radius $R>1$ and that $\beta(x)\to\infty$ as $x\to R$. In words, the batch size distribution should not have an extremely long tail. Since the generating function $P(z)=\Sigma_{j=0}^{\infty}p_j z^j$ can be written as $F(z)/G(z)$ for functions $F(z)$ and $G(z)$ that are analytic in the region $1<|z|<R$ of the complex plane, we can apply partial fraction expansion to obtain for the service time density (2.7) that

$$\frac{p_j}{p_{j-1}} \approx \tau^{-1} \quad \text{for } j \text{ large enough}, \tag{2.10}$$

where τ is the smallest root on $(1,R)$ of the equation

$$1-x-\lambda[1-\beta(x)]\left[\sum_{i=1}^{r} q_i \sum_{j=0}^{i-1}\frac{\mu^j}{\{\lambda(1-\beta(x))+\mu\}^{j+1}}\right] = 0. \tag{2.11}$$

In practical applications it is not necessary to compute the constant τ from (2.11) on beforehand, but the constant τ can be estimated with any desired accuracy during the recursive computation of the p_j's. For nonlight traffic the asymptotic expansion (2.10) applies already for relatively small values of j. Using this empirical finding one may considerably reduce the computational effort of the recursion algorithm. The practical usefulness of asymptotic expansions in queueing analysis cannot be emphasized enough. Many other examples showing the practical importance of asymptotic expansions can be found in TIJMS [25], see also the discussion below.

The above analysis of the algorithm (2.5) assumed a service time density of the form (2.7). In certain applications a hyperexponential density $b(t)=\Sigma_{i=1}^{r}q_i\mu_i e^{-\mu_i t}$ may be more convenient (usually $r=2$). Then the constants A_{jk} are given by (2.8), where the constants $\alpha_n^{(i)}$ are now recursively computed from $\alpha_0^{(i)}=1/(\lambda+\mu_i)$ and $\alpha_n^{(i)}=[\lambda/(\lambda+\mu_i)]\Sigma_{j=1}^{n}\beta_j\alpha_{n-j}^{(i)}$, $n\geq 1$.

2.2. Computational methods for the waiting-time probabilities

In this subsection we continue with the $M^X/G/1$ queue. We discuss for this queueing system both exact methods and approximate methods for the computation of the waiting-time probabilities.

In the sequel it will be assumed that customers of different batches are served in order of arrival of the batches and that customers from a same batch are served in order of their (random) positions within that batch. We need the following preliminaries. Suppose that the customers of the batches are numbered as $n=1,2,\ldots$ in accordance with their order of service. By a standard result for backward recurrence times from renewal theory, we have

$$\lim_{n\to\infty} P\{\text{the } n\text{-th customer takes the } k\text{-th position in its batch}\}$$

$$= \frac{1}{E(B)}\sum_{j=k}^{\infty}\beta_j, \quad k=1,2,\ldots, \tag{2.12}$$

when we assume that the batch-size distribution $\{\beta_j\}$ is aperiodic (that is, there is no integer $d\geqslant 2$ such that $\Sigma_{n=1}^{\infty}\beta_{nd}=1$). This aperiodicity condition is not required for the following result. For each $k\geqslant 1$, the long-run fraction of customers taking the k-th position in their batch equals

$$\frac{1}{E(B)}\sum_{j=k}^{\infty}\beta_j \tag{2.13}$$

with probability 1. An intuitive proof of this result is as follows. Fix k and assume that a reward of 1 is received for each customer taking the k-th position in its batch. Then the above fraction can be interpreted as the long-run average reward per customer. The latter average is given by the expected reward per batch divided by the expected batch size. The expected reward per batch equals $\Sigma_{j\geqslant k}\beta_j$. Note from (2.13) that on the average a customer finds $\frac{1}{2}E(B^2)/E(B)-\frac{1}{2}$ customers from his batch in front of him.

Define now the random variable D_n as the delay in queue of the n-th customer, $n\geqslant 1$. Under the assumption that the batch-size distribution is aperiodic, the limit

$$W_q(x) = \lim_{n\to\infty} P\{D_n\leqslant x\} \tag{2.14}$$

exists for all x and represents a proper probability distribution, cf. COHEN [5]. Otherwise, define the waiting-time distribution function $W_q(x)$ by

$$W_q(x) = \lim_{n\to\infty}\frac{1}{n}\sum_{k=1}^{n} P\{D_k\leqslant x\}, \quad x\geqslant 0. \tag{2.15}$$

The latter limit always exists by the theory of regenerative processes. We have the unifying interpretation that, with probability 1, $W_q(x)$ gives the long-run fraction of customers who have a waiting time of no more than x.

Let us now consider the case that the service time density is given by (2.7). For the computation of the waiting-time probabilities, the computational usefulness of a service time representation (2.7) becomes particularly evident. Then, by representing the service time of a customer by a random number of independent exponential phases, the powerful technique of continuous-time Markov chain analysis can be used. Since each service phase has an exponential distribution with the same mean, the process describing the number of uncompleted service phases in the system is a continuous-time Markov chain. Denote by $\{f_j, j\geqslant 0\}$ the limiting distribution of this continuous-time Markov chain. It is important to note that this process has the feature that the value of its state cannot decrease with more than 1 at each transition. Then the state probabilities can be *recursively* computed by equating the rate at which the process leaves the *macrostate* of having at least j uncompleted service phases present to the rate at which the system enters that macrostate. This type of argument has a long history in teletraffic analysis, cf. also KOSTEN [16]. Thus we obtain the recursive balance equations

$$\mu f_j = \sum_{k=0}^{j-1} (\lambda \sum_{i \geq j-k} \nu_i) f_k, \quad j=1,2,\ldots, \tag{2.16}$$

where $f_0 = 1-\rho$ and ν_j denotes the probability that an arriving batch represents a total of j service phases. Obviously, $\nu_j = \sum_{n=1}^{j} \beta_n q_j^{n*}$ with q_j^{n*} denoting the probability that n customers represent a total of j service phases. In general the probabilities q_j^{n*} can be recursively computed from the convolution formula, but for the important case that only two q_j's are positive a binomial-type formula can be used.

The conditional waiting time of a customer who has just after arrival j uncompleted service phases in front of him is distributed as the sum of j independent exponentials with the same means and hence has an Erlang-j distribution. Denoting by z_j the long-run fraction of customers who just after arrival have j uncompleted service phases in front of them, we thus obtain

$$W_q(x) = 1 - \sum_{j=1}^{\infty} z_j \sum_{k=0}^{j-1} e^{-\mu x} \frac{(\mu x)^k}{k!} \tag{2.17}$$

$$= 1 - \sum_{k=0}^{\infty} e^{-\mu x} \frac{(\mu x)^k}{k!} \left\{ 1 - \sum_{j=0}^{k} z_j \right\}, \quad x \geq 0.$$

The latter representation for $W_q(x)$ is better suited for numerical calculations since it converges faster. The z_j's are easily expressed in the probabilities f_j. By the property *Poisson arrivals see time averages,* the equilibrium probability f_j has also the interpretation $f_j = \lim_{m \to \infty} P\{$there are j uncompleted service phases present just before the arrival of the m-th batch$\}$. Using this result together with (2.13), we obtain

$$z_j = \sum_{i=0}^{j} f_i \sum_{k=1}^{j-i+1} \eta_k q_{j-i}^{(k-1)*}, \quad j=0,1,\ldots, \tag{2.18}$$

where $\eta_k = [1/E(B)] \sum_{j \geq k} \beta_j$. The desired probabilities z_j can be directly computed from a recursive scheme. It is a matter of simple algebra to derive from (2.16) and (2.18) that

$$z_j = f_0 \sum_{k=1}^{j+1} \eta_k q_j^{(k-1)*} + \frac{\lambda}{\mu} \sum_{i=0}^{j-1} z_i (1 - \sum_{n=0}^{j-1-i} \nu_n), \quad j=1,2,\ldots, \tag{2.19}$$

where $z_0 = f_0/E(B)$ and $f_0 = 1-\rho$. For nonlight traffic the computational effort in the recursion scheme (2.19) can be reduced by using that, for some constant τ, $z_j/z_{j-1} \approx \tau$ for j large enough provided the batch size distribution has no extremely long tail. The constant τ can be estimated during the recursive computations.

Unlike the algorithm for the state probabilities, the algorithm for the waiting-time probabilities in the $M^X/G/1$ queue with hyperexponential services becomes more complicated than the corresponding algorithm for the case of Erlangian services. An exact analysis for the case of hyperexponential service requires a continuous-time Markov chain with two-dimensional states describing the number of customers present and the type of the exponential service in progress. The equilibrium state probabilities of this Markov chain are easily computed. However, the computation of the waiting-time distribution is more elaborate and involves the computation of the conditional waiting-time distributions as first-passage time distributions in a continuous-time Markov chain using the uniformization technique. We refer for details to EIKEBOOM and TIJMS [11].

For practical applications a more simple approach to the $M^X/G/1$ queue with phase-type service can be based on the asymptotic expansion

$$1-W_q(x) \approx \gamma e^{-\theta x} \quad \text{for } x \text{ large enough.} \tag{2.20}$$

This asymptotic expansion requires that the batch-size distribution has no extremely long tail. The constants γ and θ are easily computed for the case of phase-type service, see VAN OMMEREN [29]. For nonlight traffic the asymptotically exponential expansion (2.20) applies already for moderate values of x. Also, using the asymptotic expansion (2.20) and explicit expressions for the first two moments of the waiting time, the delay probability and the derivative of $W_q(x)$ at $x=0$, a practically useful approximation to $1-W_q(x)$ for all $x \geqslant 0$ can be obtained by a sum of three exponential functions, see VAN OMMEREN [30].

To conclude this section we discuss two-moment approximations for the $M^X/G/1$ queue when the squared coefficient of variation c_S^2 of the service time is not too large. Such approximations are possible in view of the empirical finding that for nonlight traffic the waiting-time probabilities in the $M^X/G/1$ queue are quite insensitive to more than the first two moments of the service time S provided c_S^2 is not too large, say $0 \leqslant c_S^2 \leqslant 2$. Thus for the $M^X/G/1$ queue with $0 < c_S^2 \leqslant 1$ the waiting-time probabilities can be approximately obtained by fitting a mixture of E_{k-1} and E_k distributions with the same scale parameters to the service time S by matching the first two moments (cf. appendix C in TIJMS [25]) and next applying the algorithm (2.17). In many practical applications only the probabilities in the tail of $1-W_q(x)$ are of interest. For the $M^X/G/1$ queue with $0 \leqslant c_S^2 \leqslant 2$ these tail probabilities (or, equivalently, the higher percentiles) can be accurately approximated by using the easily computed asymptotically exponential expansions of the waiting-time distributions for the particular cases of deterministic services and exponential services. This simple and practically useful approach will be discussed in more detail in Section 4.2 when considering the multi-server $M^X/G/c$ queue.

3. Computational methods for the GI/G/1 queue

This section presents algorithms and approximations for the waiting-time probabilities in the single-server GI/G/1 queue with service in order of arrival. We first discuss exact methods for two particular cases of phase-type services. Next we show that the exact results for phase-type services can be used to obtain accurate approximations for the case of general service by applying the powerful approach of linear interpolation with respect to the squared coefficient of variation of the service time.

3.1. *An embedded Markov chain algorithm for the* $GI/E_{1,\ldots,r}/1$ *queue*

In the $GI/E_{1,\ldots,r}/1$ queue the customers arrive according to a renewal process and the service time of a customer is distributed as a mixture of Erlangians and has as density

$$b(t) = \sum_{i=1}^{r} q_i \mu^i \frac{t^{i-1}}{(i-1)!} e^{-\mu t}, \quad t \geqslant 0, \tag{3.1}$$

where $q_i \geqslant 0$ and $\Sigma_i q_i = 1$. Recall that any service-time density can be arbitrarily closely approximated by a density of the form (3.1). It is assumed that the offered load is smaller than 1.

Denoting by D_n the delay in queue of the n-th customer, we wish to compute the stationary waiting-time distribution $W_q(x) = \lim_{n\to\infty} P\{D_n \leqslant x\}$ for $x \geqslant 0$ when service is in order of arrival. As in BUX [3] (cf. also [20]), we use the basic technique of embedded Markov chains together with the physical interpretation of the density (3.1). Since the service time of a customer is distributed as the random sum of independent exponential phases with the same means and the exponential distribution is memoryless, the embedded process $\{X_n\}$ defined by

X_n = the number of uncompleted service phases present just prior to the arrival of the n-th customer

is a discrete-time Markov chain. This Markov chain is aperiodic and positive recurrent and thus the limit $z_j = \lim_{n\to\infty} P\{X_n = j\}$ exists for all $j \geqslant 0$. Since the conditional waiting time of a customer finding upon arrival j phases present has an Erlang-j distribution, it follows that

$$W_q(x) = 1 - \sum_{j=1}^{\infty} z_j \sum_{k=0}^{j-1} e^{-\mu x} \frac{(\mu x)^k}{k!}, \quad x \geqslant 0. \tag{3.2}$$

Hence we have a useful algorithm for the waiting-time distribution provided the limiting probabilities z_j can be efficiently computed. As observed below (2.17), for numerical calculations it is recommended to rewrite (3.2) by interchanging the order of summation.

The probabilities z_j are the unique solution to the well-known balance equations

$$z_j = \sum_{i=0}^{\infty} p_{ij} z_i, \quad j = 0, 1, \ldots, \tag{3.3}$$

together with the normalization equation $\Sigma_{j=0}^{\infty} z_j = 1$. Here the p_{ij}'s are the one-step transition probabilities of the Markov chain $\{X_n\}$. Denoting by $A(t)$ the probability distribution function of the interarrival time, it is easy to see that for any $i \geqslant 0$,

$$p_{ij} = \sum_{k=1}^{r} q_k \int_0^{\infty} e^{-\mu t} \frac{(\mu t)^{i+k-j}}{(i+k-j)!} dA(t) \quad \text{for} \quad 1 \leqslant j \leqslant i+r, \tag{3.4}$$

while $p_{i0} = 1 - \Sigma_{j \geqslant 1} p_{ij}$. To solve numerically the linear equations for the equilibrium distribution of a Markov chain, iterative methods such as Gauss-Seidel or successive overrelaxation are highly recommended, see appendix D in TIJMS [25]. However, the system of linear equations (3.3) is infinite and has therefore to be truncated to a finite system of linear equations. A straightforward truncation is based on choosing a sufficiently large integer M such that $\Sigma_{j>M} z_j < \epsilon$ with ϵ some small accuracy number. For high traffic the truncation integer M will become very large which leads to excessive computing times. For the case of nonlight traffic there is a better alternative. We have the asymptotic result that

$$\frac{z_j}{z_{j-1}} \approx 1 - \frac{\theta}{\mu} \quad \text{for } j \text{ large enough,} \tag{3.5}$$

where θ is the unique solution on $(0, \mu)$ of the equation

$$\int_0^{\infty} e^{-\theta t} dA(t) \times \sum_{k=1}^{r} q_k \frac{\mu^k}{(\mu - \theta)^k} = 1. \tag{3.6}$$

A simple proof of (3.5) proceeds as follows. By general results in MILLER [17] for the batch-arrival queue with exponential service (note that the Markov chain $\{X_n\}$ can also be associated with a batch-arrival $GI^Q/M/1$ queue), we have that $z_j \approx \alpha \omega^j$ for j large enough for some $\alpha > 0$ and $0 < \omega < 1$. Substituting this asymptotic expansion into (3.2) and using the general result from FELLER [12] that $1 - W_q(x) \approx \gamma e^{-\theta x}$ for x large enough for some $\gamma > 0$ and θ given by (3.6), the relation (3.5) follows. In fact, using partial-fraction expansions for the generating function of the z_j's, it can be shown that

$$z_j = \sum_{k=1}^{s} \omega_k^j \sum_{i=0}^{m_k - 1} \alpha_{ik} i^j \quad \text{for all } j \geqslant 1,$$

where the ω_k's are the (complex) zeros of the function

$$1 - \left[\sum_{i=1}^{r} q_i z^{-i} \right] \int_0^{\infty} e^{-\mu(1-z)t} dA(t)$$

inside the unit circle $|z| = 1$ and m_k is the multiplicity of the root ω_k (the roots are all distinct for pure Erlangian service). This result and (3.2) show that $W_q(x)$ has the representation

$$1 - W_q(x) = \sum_{k=1}^{s} e^{-\theta_k x} \sum_{i=0}^{m_k - 1} \gamma_{ik} x^i, \quad x \geqslant 0,$$

where $\theta_k = \mu(1-\omega_k)$, cf. also results for the GI/K$_r$/1 queue in COHEN [5].

In view of the empirical fact that for nonlight traffic (3.5) holds already for moderate values of j, the result (3.5) can be used as follows for computational purposes. For an appropriately chosen integer N we replace z_j by $z_N(1-\theta/\mu)^{j-N}$ for $j>N$ in the linear equations for the z_j's. In this way we get a finite system of linear equations. For the case of *nonlight* traffic the integer N for which $z_j \approx z_N(1-\theta/\mu)^{j-N}$ accurately holds for $j \geqslant N$ will be typically much smaller than the earlier discussed truncation integer M. It is noted that when using an iterative method for solving the equations the chosen value for N can be dynamically adjusted during the iterative calculations. Also, from a computational point of view, it is important to point out that the probabilities p_{ij} for $j \neq 0$ depend on i and j only through the difference $i-j$, see (3.4). This simple observation reduces considerably the memory requirements for the computations. For deterministic or phase-type arrivals, the integrals in (3.4) reduce to explicit expressions, otherwise they can be calculated by Gauss-Laguerre or Gauss-Legendre integration. Finally, the following remark is in order. When applying an iterative method such as Gauss-Seidel of successive overrelaxation to solve the balance equations, it is always important to have an extra accuracy check to verify whether the stopping criterion of the iterative method has served its purposes. For the present model, we have that the long-run fraction of time the server is idle equals $1-\rho$. On the other hand, this long-run fraction equals the average amount of time the server is idle between two arrivals divided by the average time between two arrivals. Thus we have the accuracy check

$$1-\rho = \frac{1}{E(A)} \sum_{j=0}^{\infty} z_j \sum_{i=1}^{r} q_i \left[\int_0^{\infty} t \left\{ 1 - \sum_{k=0}^{i+j-1} e^{-\mu t} \frac{(\mu t)^k}{k!} \right\} dA(t) \right.$$

$$\left. - \frac{i+j}{\mu} \int_0^{\infty} (1 - \sum_{k=0}^{i+j} e^{-\mu t} \frac{(\mu t)^k}{k!}) dA(t) \right],$$

where $E(A)$ denotes the average interarrival time.

The above analysis can be straightforwardly extended to the GI^X/E$_{1,\ldots,r}$/1 queue with batch arrivals. The analysis applies as well to the finite capacity GI/E$_k$/1/N queueing system in which a customer finding upon arrival N other customers present is rejected. In the finite capacity model we restrict ourself to pure Erlangian service since otherwise the number of uncompleted service phases present does not unambiguously determine the number of customers present. Nevertheless the exact results for the GI/E$_k$/1/N queue can be used to calculate accurate approximations for several performance measures in the GI/G/1/N queue with a general service time S provided that c_S^2 is not too large. To do so, we consider the GI/E$_k$/1/N queue for two (or three) appropriately chosen values of k. For each GI/E$_k$/1/N queue considered we compute the exact value of the performance measure under consideration. Next the approximation for the GI/G/1/N queue is obtained by linear (or quadratic) interpolation of the exact results where the interpolation is based on the squared coefficient of variation of the service time, cf. formula (3.14) in Section 3.2. This approach can be used for performance measures such as the average queue size and the average waiting time per entering customer, but not for the waiting-time probabilities and the rejection probability. However, both the waiting-time percentiles and the minimal buffer size guaranteeing a prespecified rejection level allow for interpolation with respect to the squared coefficient of variation of the service time, cf. also NOBEL [19] and TIJMS [25].

To the end of this subsection we mention an alternative approach for the computation of certain long-run averages in the $GI/E_{1,\ldots,r}/1$ queue. This alternative approach is the (modified) value-iteration algorithm from Markov decision theory, cf. Chapter 3 in TIJMS [25]. Measures such as the delay probability and the average waiting time can be interpreted as an average cost by imposing an appropriate cost structure on the Markov chain $\{X_n\}$ describing the number of phases present just prior to an arrival. For the average waiting time, suppose that the Markov chain incurs a cost of i/μ in state i. Similarly, for the delay probability the Markov chain incurs a cost of 1 in state $i \neq 0$ and a cost of 0 otherwise; more generally, for the waiting-time probability $1-W_q(t_0)$ a cost of $\Sigma_{k=0}^{i-1} e^{-\mu t_0}(\mu t_0)^k/k!$ is associated with state i. The value-iteration algorithm does not require solving linear equations to compute the average cost but involves only recursive computations. Moreover, this approach has the important feature of providing at each iteration lower and upper bounds on the average cost. Thus the value-iteration approach may be used to calculate quick bounds for engineering purposes.

3.2. Approximations based on exact results for the $GI/C_2/1$ queue

The $GI/C_2/1$ queue denotes the particular $GI/G/1$ queue in which the service time S has Coxian-2 (C_2) density. That is,

$$S = \begin{cases} U_1 & \text{with probability } 1-b, \\ U_1+U_2 & \text{with probability } b, \end{cases}$$

for some $0 \leq b \leq 1$, where U_1 and U_2 are two independent exponentials with respective means $1/\mu_1$ and $1/\mu_2$. The class of C_2-densities coincides with the class of K_2-densities, where the probability density of a positive random variable is said to be a K_2-density when its Laplace transform is the ratio of a polynomial of at most degree 1 to a polynomial of degree 2. Any C_2-density has a squared coefficient of variation of at least 1/2. It is often convenient to fit a C_2-density to a positive random variable by matching its first two or its first three moments. Let S be a positive random variable with $c_S^2 \geq \frac{1}{2}$ and $m_i = E(S^i)$ denoting the i-th moment of S. If a three-moment fit to S by a C_2-density exists, the three parameters of this fit are given by

$$\mu_{1,2} = \tfrac{1}{2}(a_1 \pm \sqrt{a_1^2 - 4a_2}), \quad b = \frac{\mu_2}{\mu_1}(\mu_1 m_1 - 1), \tag{3.7}$$

where $a_1 = 1/m_1 + m_2 a_2/2m_1$ and $a_2 = (6m_1 - 3m_2/m_1)/[(3m_2^2/2m_1) - m_3]$.

An infinite number of C_2-densities can be fitted to S by matching only the first two moments. A very useful two-moment fit is the one with parameters

$$\mu_{1,2} = \frac{2}{m_1}\left[1 \pm \sqrt{\frac{c_S^2 - \frac{1}{2}}{c_S^2+1}}\right], \quad b = \frac{\mu_2}{\mu_1}(\mu_1 m_1 - 1). \tag{3.8}$$

This particular C_2-density has the same first three moments as a gamma density, cf. TIJMS [25].

For the $GI/C_2/1$ queue a tractable closed-form expression can be given for the stationary waiting-time distribution function $W_q(x)$, $x \geq 0$. The closed-form solution requires the computation of the two real zeros η_1 and η_2 with $0<\eta_1<\min(\mu_1,\mu_2)\leq\eta_2$ of the equation

$$x^2 - (\mu_1+\mu_2)x + \mu_1\mu_2 - \{\mu_1\mu_2 - (1-b)\mu_1 x\} \int_0^\infty e^{-xt} dA(t) = 0. \tag{3.9}$$

Here $A(t)$ is the probability distribution function of the interarrival time. It follows from the general relations (II.5.190) and (II.5.191) in COHEN [5] that

$$1 - W_q(x) = \sum_{i=1}^{2} a_i e^{-\eta_i x} \quad \text{for all} \quad x \geq 0, \tag{3.10}$$

and

$$W_q(0) = \frac{\eta_1\eta_2}{\mu_1\mu_2}, \quad \int_0^\infty x dW_q(x) = -\left[\frac{\mu_1+\mu_2}{\mu_1\mu_2}\right] + \frac{1}{\eta_1} + \frac{1}{\eta_2}. \tag{3.11}$$

Here the constants a_i are given by

$$a_1 = \frac{-\eta_1^2\eta_2 + \eta_1\eta_2(\mu_1+\mu_2) - \eta_2\mu_1\mu_2}{\mu_1\mu_2(\eta_1-\eta_2)}, \tag{3.12}$$

$$a_2 = \frac{\eta_1\eta_2^2 - \eta_1\eta_2(\mu_1+\mu_2) + \eta_1\mu_1\mu_2}{\mu_1\mu_2(\eta_1-\eta_2)}. \tag{3.13}$$

The practical usefulness of the exact result (3.10) should not be underestimated and is not restricted to C_2-service. The easily computed results for C_2-service enable us as well to calculate simple and accurate approximations for the $GI/G/1$ queue with a general service time S. For the case of $c_S^2 \geq \frac{1}{2}$, we suggest to fit a C_2-density to the service time S by using the three-moment fit (3.8) and to use the results (3.10)-(3.11) for this C_2-density. This will usually give excellent approximations provided the traffic load is not very low and c_S^2 is not very large. For the case of $c_S^2 < \frac{1}{2}$, we suggest to proceed as follows. To approximate a certain performance measure P in the $GI/G/1$ queue, the idea is to compute the exact values of the performance measure for two or three queueing systems of the $GI/C_2/1$ type and next to apply extrapolation with respect to the squared coefficient of variation of the service time. Each of the $GI/C_2/1$ models used should have the same interarrival-time distribution and the same mean service time as the original $GI/G/1$ model. The general form of the extrapolation formula is as follows. Suppose that the exact value $P(i)$ of the performance measure is known for the $GI/C_2/1$ queue with squared coefficient of variation $cv^2(i)$ for $i=1,\ldots,n$. Then the exact value of the performance measure for the GI/G/1 queue with cv^2 as squared coefficient of variation of the service time can be approximated by

$$P_{app}(cv^2) = \sum_{i=1}^{n} P(i) \prod_{\substack{k=1 \\ k\neq i}}^{n} \frac{cv^2 - cv^2(i)}{cv^2(i) - cv^2(k)}. \tag{3.14}$$

Usually $n=2$ (linear extrapolation) or $n=3$ (quadratic extrapolation). This approximation gives good results for performance measures such as the average waiting time and the waiting-time percentiles, provided the traffic load is not very small. It is an empirical finding that the extrapolation works for all waiting-time percentiles but not for the (small) waiting-time probabilities. In particular, extrapolation should not be used for the delay probability when the traffic load is very small. For nonlight traffic the delay probability may be approximated by using extrapolation. A good source for verifying experimentally the quality of approximations is the table book SEELEN ET AL. [21]. It should be pointed out that the approximation approach discussed above makes no approximation with respect to the interarrival-time distribution. This observation is important because performance measures are typically much more sensitive to the form of the interarrival-time distribution than to the form of the service time distribution, as is clearly demonstrated by the numerical results in Table 3.1. In many practical applications it suffices to use only the first two moments of the service time, provided c_S^2 is not too large; the range of c_S^2 depends to some extent on the variability in the interarrival time and on the traffic load. However, one should be more careful in using two-moment fits for the interarrival-time distribution. A more extensive discussion of this subject can be found in TIJMS [25].

In Table 3.1 we give some numerical results from which it can be verified that good approximations can be obtained by using exact results for the $GI/C_2/1$ queue in conjunction with (linear) extrapolation on the squared coefficient of variation of the service time.

For several values of c_A^2, c_S^2 and ρ, we give the exact values of the mean waiting time $E(W_q)$, the delay probability $\Pi_w = 1 - W_q(0)$ and the conditional waiting-time percentile $\eta(p)$ for $p=0.95$. Here c_A^2 and c_S^2 denote the squared coefficients of variation of the interarrival time A and the service time S. The conditional percentile $\eta(p)$ is defined by $P\{W_q > \eta(p) | W_q > 0\} = 1-p$, where W_q denotes the waiting time of a customer in the stationary situation. We vary c_A^2 as 0, 0.5 and 5 and c_S^2 as 0, 0.25, 0.50, 0.75, 1 and 2, where the value 0 corresponds to the deterministic distribution, the value 1/4 to the Erlang-4 distribution and the values 0.5, 0.75, 1 and 5 to C_2-distributions with the normalization (3.8). For $c_S^2=2$, we consider both the C_2-density with the gamma normalization (3.8) and the C_2-density with the normalization of balanced means (i.e. $p/\mu_1 = (1-p)/\mu_2$, where $p = 1 - b\mu_1/(\mu_1 - \mu_2)$). The offered load ρ is varied as $\rho=0.2$, 0.5 and 0.8. In all examples we have taken $E(S)=1$. The results for the particular case of the $H_2/D/1$ queue ($c_A^2=5$) are obtained by the algorithm in VAN HOORN [28]. The exact results for the $E_2/D/1$ queue are obtained by using that the $E_k/D/1$ queue is equivalent to an M/D/k queue, see also Subsection 4.3. The other exact results are calculated by using (3.10)-(3.13). The notation 5E-4 in Table 3.1 means 5×10^{-4}.

		$c_A^2=0$			$c_A^2=0.5$			$c_A^2=5$		
ρ	c_S^2	$E(W_q)$	Π_W	$\eta(.95)$	$E(W_q)$	Π_W	$\eta(.95)$	$E(W_q)$	Π_W	$\eta(.95)$
0.2	0	-	-	-	0.024	0.064	0.84	0.581	0.527	2.89
	0.25	0.000	3E-6	0.86	0.041	0.074	1.51	0.685	0.515	3.67
	0.50	0.000	5E-4	1.63	0.060	0.081	2.12	0.780	0.505	4.42
	0.75	0.002	0.003	2.42	0.081	0.088	2.72	0.869	0.496	5.15
	1	0.007	0.007	3.02	0.103	0.094	3.31	0.952	0.488	5.85
	2(g)	0.053	0.030	5.27	0.203	0.110	5.63	1.244	0.461	8.50
	2(b)	0.078	0.032	7.30	0.211	0.105	6.77	1.212	0.466	8.90
0.5	0	-	-	-	0.177	0.323	1.38	3.142	0.839	10.7
	0.25	0.017	0.047	1.07	0.281	0.344	2.26	3.320	0.826	11.7
	0.50	0.078	0.116	1.99	0.390	0.360	3.13	3.494	0.814	12.6
	0.75	0.160	0.165	2.90	0.503	0.372	3.99	3.664	0.803	13.6
	1	0.255	0.203	3.76	0.618	0.382	4.85	3.828	0.793	14.5
	2(g)	0.693	0.289	7.19	1.095	0.405	8.26	4.448	0.763	18.0
	2(b)	0.725	0.257	8.85	1.105	0.395	9.26	4.400	0.775	18.2
0.8	0	-	-	-	0.903	0.702	3.61	11.10	0.944	34.7
	0.25	0.319	0.446	2.08	1.381	0.716	5.59	11.61	0.940	36.6
	0.50	0.757	0.548	4.08	1.865	0.726	7.57	12.15	0.937	38.6
	0.75	1.217	0.598	6.08	2.353	0.734	9.54	12.67	0.933	40.6
	1	1.693	0.629	8.07	2.844	0.740	11.5	13.19	0.930	42.5
	2(g)	3.650	0.683	16.0	4.825	0.752	19.4	15.25	0.919	50.3
	2(b)	3.673	0.658	17.4	4.834	0.746	20.3	15.22	0.924	50.5

TABLE 3.1. *Numerical results for some GI/G/1 queues*

The above results can also be used to calculate an approximation to the average waiting time per customer in the general $GI^X/G/1$ queue with batch arrivals. This can be done by using the following result. Denoting by $E(W_q)$ the average waiting time per customer in the $GI^X/G/1$ queue and denoting by the generic variables B and S the number of customers in a batch and the service time of a customer, we have

$$E(W_q) = E(W_{q1})+\tfrac{1}{2}E(S)\left\{\frac{E(B^2)}{E(B)}-1\right\}. \tag{3.15}$$

Here the first term in the right-hand side of (3.15) denotes the average waiting time per 'supercustomer' in the single-arrival $GI/G_X/1$ queue in which the interarrival time is distributed as in the $GI^X/G/1$ queue and the service time is distributed as the total time needed to service all customers from one batch in the $GI^X/G/1$ queue. The second term in the right-hand side of (3.15) gives for the $GI^X/G/1$ queue the average amount of time per customer between the moment the batch of the customer arrives and the moment the customer enters service, cf. the relation (2.13) in Section 2. Using that the mean and the squared coefficient of variation of the service time U of a customer in the $GI/G_X/1$ queue are given by

$$E(U) = E(B)E(S) \quad \text{and} \quad c_U^2 = \frac{1}{E(B)}c_S^2 + c_B^2,$$

the average waiting time per customer in the $GI^X/G/1$ queue can be approximated by using the approach discussed above.

4. Approximations for multi-server queues

In this section we discuss some simple approximations for multi-server queues. For the general multi-server queue it is difficult to give exact algorithms that lend themselves to practical computations. Therefore in many situations one has to resort to approximate methods for calculating measures of system performance. Useful approximations to complex queueing systems are often obtained by using exact results for simpler related queueing systems. It will be seen again that asymptotic expansions and linear interpolation with respect to the squared coefficient of variation of the service time are powerful tools.

4.1. The M/G/c queue

In this section we consider the multi-server queue with c identical servers, Poisson arrivals at rate λ and a generally distributed service time S. It is assumed that $\rho = \lambda E(S)/c$ is smaller than 1. The M/G/c queue does not permit a tractable exact analysis with the exception of the special cases of the M/M/c queue and the M/D/c queue. A tractable analytical solution is even not possible for a simple measure as the average waiting time per customer. Nevertheless useful approximations for the waiting time can be obtained by having a closer look at the structure of the Pollaczek-Khintchine formula for the case of $c=1$ server. Denote by $E(W_q)$ the average waiting time per customer in the general M/G/c queue and let $E_{\exp}(W_q)$ and $E_{\det}(W_q)$ represent $E(W_q)$ for the special cases of exponential services and deterministic services with the same means $E(S)$. The Pollaczek-Khintchine formula for $E(W_q)$ for the case of $c=1$ server allows the following two representations,

$$E(W_q) = \tfrac{1}{2}(1+c_S^2)E_{\exp}(W_q), \tag{4.1}$$

and

$$E(W_q) = (1-c_S^2)E_{\det}(W_q) + c_S^2 E_{\exp}(W_q), \tag{4.2}$$

where c_S^2 denotes the squared coefficient of variation of the service time. For the case of $c>1$, the right-hand side of (4.1) can be used as a first-order approximation to $E(W_q)$. This is a practically useful approximation that is reasonably accurate provided c_S^2 is not too large. A more accurate approximation of a comparable simplicity is obtained by using the right-hand side of (4.2) in conjunction with the special-purpose approximation to $E_{\text{det}}(W_q)$,

$$E_{\text{det}}^{\text{app}}(W_q) = \tfrac{1}{2}\left[1+(1-\rho)(c-1)\frac{\{\sqrt{4+5c}-2\}}{16\rho c}\right]E_{\text{exp}}(W_q). \qquad (4.3)$$

This approximation was advocated by COSMETATOS [7] and is a quite accurate approximation to $E(W_q)$ provided c_S^2 is not too large, say $0 \leqslant c_S^2 \leqslant 2$. Unlike the M/G/1 queue, it is no longer true for the multi-server queue that the average waiting time is insensitive to more than the first two moments of the service time. The average waiting time becomes increasingly sensitive to the shape of the service time distribution as c_S^2 gets larger. The papers BOXMA ET AL. [2] and TIJMS ET AL. [24] give approximations of the form

$$E_{\text{app}}(W_q) = \tfrac{1}{2}f(1+c_S^2)E_{\text{exp}}(W_q), \qquad (4.4)$$

where f is some correction factor involving the service time distribution. Unlike the approximations based on (4.1)-(4.3), the latter approximations can also be used for larger values of c_S^2, see TIJMS [25] for details. This reference discusses also approximations for the state probabilities and the waiting-time probabilities. In particular, the delay probability $P\{W_q>0\}$ can be very well approximated by Erlang's delay probability $\Pi_W(\text{exp})$ for the M/M/c queue. We briefly discuss approximations to the waiting-time percentiles. It is convenient to do this for the conditional waiting-time percentiles $\eta(p)$ which are defined by $P\{W_q>\eta(p)|W_q>0\}=1-p$ for *all* $0<p<1$. Here W_q has the limiting distribution of the waiting time of a customer. Denoting by $\eta_{\text{exp}}(p)$ and $\eta_{\text{det}}(p)$ the corresponding percentiles for the special cases of the M/M/c queue and the M/D/c queue with the same means $E(S)$, an excellent approximation to $\eta(p)$ is given by

$$\eta_{\text{app}}(p) = (1-c_S^2)\eta_{\text{det}}(p)+c_S^2\eta_{\text{exp}}(p) \quad \text{for all } 0<p<1, \qquad (4.5)$$

provided that c_S^2 is not too large $(0 \leqslant c_S^2 \leqslant 2)$ and the traffic load is not too small (say, $\Pi_W(\text{exp}) \geqslant 0.2$). The percentile $\eta_{\text{exp}}(p)$ is trivial to compute, while a relatively simple algorithm can be given for the computation of $\eta_{\text{det}}(p)$, see TIJMS [25]. For the higher values of p the asymptotic expansion of the waiting-time distribution can be used to compute $\eta_{\text{det}}(p)$, see also Subsection 4.2 below. Also it is pointed out that for quick engineering calculations the first-order approximation $\tfrac{1}{2}(1+c_S^2)\eta_{\text{exp}}(p)$ to $\eta(p)$ can be used when p is close enough to 1 (say, $p \geqslant 1-\Pi_W(\text{exp})$), cf. TIJMS [26].

4.2. The $M^X/G/c$ queue

In the $M^X/G/c$ queue the customers arrive in batches rather than singly. The arrival process of batches is a Poisson process with rate λ, the service time S of a customer has a general distribution and the batch size B has a probability distribution $\{\beta_j, j \geq 1\}$. It is assumed that $\lambda E(B)E(S)/c$ is less than 1. Also, we make the technical assumption that the power series $\beta(x)=\Sigma_{j=1}^{\infty}\beta_j x^j$ has a convergence radius R larger than 1 and that $\beta(x)\rightarrow\infty$ as $x\rightarrow R$. This assumption is satisfied in cases of practical interest, e.g. when $\{\beta_j\}$ has finite support.

Supposing that customers from different batches are served in order of arrival and that customers from the same batch are served according to their random positions in the batch, denote by D_n the delay in queue of the n-th customer and let $W_q(x)=\lim_{n\rightarrow\infty}(1/n)\Sigma_{k=1}^{n}P\{D_k \leq x\}$ for $x \geq 0$. Also, denote by $\{p_j, j \geq 0\}$ the limiting distribution of the number of customers in the system at an arbitrary time. Using generating-function analysis and partial-fraction expansions, asymptotic expansions for $1-W_q(x)$ can be obtained for the particular cases of exponential and deterministic services. For the $M^X/M/c$ queue, it holds that (see CROMIE ET AL. [8]),

$$1-W_q(x) \approx \frac{(1/\rho)(\tau-1)\Sigma_{k=0}^{c-1}(c-k)p_k^{\exp}\tau^{k-c-1}}{(\lambda/\mu)[1-\tau(1-\tau)\beta'(\tau)-\beta(\tau)]}\, e^{-c\mu(1-1/\tau)x} \tag{4.6}$$

for x large enough, where $\mu=1/E(S)$ and τ is the smallest root on $(1,R)$ of the equation

$$c(1-x)-(\lambda/\mu)x\{1-\beta(x)\} = 0. \tag{4.7}$$

For the $M^X/D/c$ queue with the constant service time D, we have

$$1-W_q(x) \approx \gamma e^{-\lambda\{\beta(\tau)-1\}x} \quad \text{for } x \text{ large enough,} \tag{4.8}$$

with

$$\gamma = \frac{\{\beta(\tau)-1\}}{(\tau-1)^2 E(B)} \left\{c\tau^{c-1}-\lambda D\beta'(\tau)e^{-\lambda D(1-\beta(\tau))}\right\}^{-1} \sum_{j=0}^{c-1} p_j^{\det}(\tau^j-\tau^c), \tag{4.9}$$

where τ is the smallest root on $(1,R)$ of the equation

$$1-x^c e^{\lambda D(1-\beta(x))} = 0. \tag{4.10}$$

The state probabilities $p_j^{\exp}$ for the $M^X/M/c$ queue are computed from a recursive scheme like (2.16), while the state probabilities $p_j^{\det}$ are computed by solving a generalization of the equilibrium equations derived by CROMMELIN [9] for the M/D/c queue, see EIKEBOOM and TIJMS [11] for details. The first c state probabilities yield also the delay probability by

$$1-W_q(0) = 1-\sum_{j=0}^{c-1} p_j \sum_{i=1}^{c-j} \eta_i, \tag{4.11}$$

with $\eta_i=[1/E(B)]\Sigma_{k\geq i}\beta_k$, cf. (2.13) in Section 2.

Denote by $\eta(p)$ the p-th conditional waiting-time percentile for the $M^X/G/c$ queue. For the batch-arrival model the approximation based on interpolation of the percentiles for the particular cases of deterministic and exponential services works only for the higher percentiles. For p close enough to 1, a useful approximation to $\eta(p)$ is given by

$$\eta_{app}(p) = (1-c_S^2)\eta_{det}^{asy}(p)+c_S^2\eta_{exp}^{asy}(p), \tag{4.12}$$

provided that c_S^2 is not too large and the traffic load is not too small (that is, the probability that an arriving batch finds all servers busy should not be too small). Here $\eta_{det}^{asy}(p)$ and $\eta_{exp}^{asy}(p)$ are the asymptotic percentiles which are easily computed from (4.8) and (4.6) together with (4.11). As a rule of thumb, the approximation (4.12) can be used when $p \geqslant 1-\Pi_B(\exp)$ with $\Pi_B(\exp)=1-\Sigma_{j=0}^{c-1}p_j^{\exp}$.

c	ρ	p	constant batch-size				geometric batch-size			
			0.80	0.90	0.95	0.99	0.80	0.99	0.95	0.99
1	0.2	exa	2.927	3.945	4.995	7.458	5.756	8.122	10.49	15.98
		app	2.836	3.901	4.967	7.440	5.745	8.116	10.49	15.99
1	0.5	exa	5.107	7.170	9.231	14.02	9.044	12.84	16.64	25.45
		app	5.089	7.154	9.219	14.01	9.040	12.84	16.64	25.47
2	0.2	exa	1.369	1.897	2.431	3.661	2.989	4.172	5.355	8.101
		app	1.354	1.887	2.419	3.656	2.982	4.167	5.353	8.106
2	0.5	exa	2.531	3.561	4.592	6.985	4.600	6.498	8.395	12.80
		app	2.535	3.567	4.599	6.996	4.601	6.501	8.401	12.81
5	0.2	exa	0.621	0.845	1.063	1.560	1.298	1.773	2.246	3.345
		app	0.640	0.853	1.066	1.560	1.305	1.779	2.253	3.354
5	0.5	exa	1.063	1.476	1.889	2.846	1.898	2.657	3.417	5.179
		app	1.069	1.482	1.895	2.853	1.905	2.665	3.425	5.190
10	0.5	exa	0.553	0.764	0.971	1.451	0.980	1.360	1.740	2.622
		app	0.566	0.772	0.979	1.458	0.991	1.371	1.751	2.634
10	0.7	exa	0.923	1.295	1.667	2.530	1.547	2.181	2.815	4.287
		app	0.930	1.302	1.673	2.536	1.556	2.190	2.824	4.297

TABLE 4.1. *The conditional waiting-time percentiles $\eta(p)$ for the* $M^X/E_2/c$ *queue*

The average waiting time per customer can be accurately approximated by using an interpolation as in (4.2), see EIKEBOOM and TIJMS [11]. It should be pointed out that the simple first-order approximation (4.1) cannot be used for the batch-arrival model, as can be seen from formula (3.15). In Table 4.1 we give for the $M^X/E_2/c$ queue the exact and approximate values for the conditional waiting-time percentiles $\eta(p)$ for several values of c and p. The batch size B is constant or geometrically distributed with mean $E(B)=3$. In all cases the normalization $E(S)=1$ is used.

4.3. The GI/G/c queue

It seems obvious that the general GI/G/c queue offers enormous difficulties in getting practically useful results. Nevertheless, using powerful numerical techniques for solving large systems of linear equations for structured Markov chains, the continuous-time Markov chain method has proved to be extremely useful for analyzing the GI/G/c queue when the interarrival time and service time both have special phase-type distributions. By a detailed state description involving sufficient information about the number of customers present and the status of both the services in progress and the arrival in progress, it is possible to set up the equilibrium equations for the microstate probabilities. This system of linear equations possesses a structure enabling the application of special aggregation/disaggregation algorithms to solve the equations numerically. We refer for details of this approach to SEELEN [22] and TAKAHASHI and TAKAMI [23]. In this section we mention only some partial results obtained by simple approximation methods.

Numerical investigations reveal that in many practical GI/G/c queueing systems performance measures P such as the average queue length, the average waiting time and the (conditional) waiting-time percentiles may reasonably well be approximated by using the interpolation formula

$$P^{\text{app}}_{\text{GI/G/c}} = (1-c_S^2)P_{\text{GI/D/c}} + c_S^2 P_{\text{GI/M/c}} \tag{4.13}$$

provided c_S^2 is not too large. Here c_S^2 denotes the squared coefficient of variation of the service time and $P_{\text{GI/D/c}}$ and $P_{\text{GI/M/c}}$ denote the exact values of the performance measure under consideration for the special cases of the GI/D/c queue and the GI/M/c queue with the same mean service times. The approximation (4.13) takes into account the empirical finding that measures of system performance are typically much more sensitive to the shape of the interarrival-time distribution than to the shape of the service time distribution. The particular models of the GI/D/c queue and the GI/M/c queue allow for a relatively simple algorithmic analysis, see VAN HOORN [28] and COOPER [6]. The special case of the $E_k/D/c$ queueing model can be analyzed by solving the M/D/ck queueing model with the same server utilization, see Section 4.1 for a discussion of (approximate) results for the M/D/s queue. It is not difficult to show that these two models have the same stationary waiting-time distribution, see IVERSEN [15]. This result may be used as well to approximate measures of system performance in the GI/D/c queue by solving first the $E_k/D/c$ queue for two (or three) appropriately chosen values of k and by interpolating next the exact results with respect to the squared coefficient of variation of the interarrival time.

Much research remains still to be done for the general GI/G/c queue. In particular, tractable exact methods for the GI/C_2/c queue would be very useful. Some progress for this queueing model has been made by BERTSIMAS [1] and DE SMIT [10]. Exact results for the GI/C_2/c queue with $c_S^2 \geq ½$ together with extrapolation with respect to the squared coefficient of variation of the service time would provide a powerful method to get practically useful approximations for the general GI/G/c queue.

REFERENCES

1. D. BERTSIMAS (1988). An analytic approach to a general class of G/G/s queueing systems. To appear in *Oper. Res.*
2. O.J. BOXMA, J.W. COHEN, N. HUFFELS (1979). Approximations of the mean waiting time in an M/G/s queueing system. *Oper. Res. 27,* 1115-1127.
3. W. BUX (1979). Single server queues with general interarrival and phase type service time distributions. *Proceedings 9th ITC,* Torremolinos, paper 413.
4. J.W. COHEN (1976). *On Regenerative Processes in Queueing Theory,* Lecture Notes in Economics and Mathematical Systems, Vol. 121, Springer-Verlag, Berlin.
5. J.W. COHEN (1982). *The Single Server Queue,* 2nd ed., North-Holland, Amsterdam.
6. R.B. COOPER (1981). *Introduction to Queueing Theory,* Edward Arnold, London.
7. G.P. COSMETATOS (1975). Approximate explicit formulae for the average queueing time in the processes M/D/r and D/M/r. *INFOR 13,* 328-331.
8. M.V. CROMIE, M.L. CHAUDRY, W.K. GRASSMANN (1979). Further results for the queueing system M^X/M/c. *J. Oper. Res. Soc. 30,* 755-763.
9. C.D. CROMMELIN (1932). Delay probability formulae when the holding times are constant. *P.O. Elect. Engr. I. 25,* 41-50.
10. J.H.A. DE SMIT (1983). A numerical solution for the multi-server queue with hyperexponential service times. *Oper. Res. Letters 2,* 217-224.
11. A.M. EIKEBOOM, H.C. TIJMS (1987). Waiting time percentiles in the multi-server M^X/G/c queue with batch arrivals. *Prob. Engr. Inform. Sc. 1,* 75-96.
12. W. FELLER (1971). *An Introduction to Probability Theory and its Applications,* Vol. II, 2nd ed., Wiley, New York.
13. TH.C. FRY (1928). *Probability and its Engineering Uses,* D. van Nostrand, New York.
14. A. HORDIJK, H.C. TIJMS (1976). A simple proof of the equivalence of the limiting distributions of the continuous time and embedded process of the queue size in the M/G/1 queue. *Statistica Neerlandica 30,* 97-100.
15. V.B. IVERSEN (1983). Decomposition of an M/D/r.k queue into *k* E_k/D/r queues with FIFO. *Oper. Res. Letters 2,* 20-21.
16. L. KOSTEN (1973). *Stochastic Theory of Service Systems,* Pergamon Press, London.
17. R.G. MILLER (1959). A contribution to the theory of bulk queues. *J.*

Roy. Statist. Soc. Ser. B21, 320-337.

18. E.C. MOLINA (1927). Application of the theory of probability to telephone trunking problems. *The Bell Syst. Tech. J. 6,* 461-494.
19. R.D. NOBEL (1988). Practical approximations for finite-buffer queueing models with batch arrivals. To appear in *European J. Oper. Res.*
20. R. SCHASSBERGER (1973). *Warteschlangen,* Springer-Verlag, Berlin.
21. L.P. SEELEN, H.C. TIJMS, M.H. VAN HOORN (1985). *Tables for Multi-Server Queues,* North-Holland, Amsterdam.
22. L.P. SEELEN (1986). An algorithm for Ph/Ph/c queues. *European J. Oper. Res. 23,* 118-127.
23. Y. TAKAHASHI, Y. TAKAMI (1976). A numerical method for the steady-state probabilities of a GI/G/c queueing system in a general class. *J. Oper. Res. Soc. Japan 19,* 147-157.
24. H.C. TIJMS, M.H. VAN HOORN, A. FEDERGRUEN (1981). Approximations for the steady-state probabilities in the M/G/c queue. *Adv. in Appl. Probab. 13,* 186-206.
25. H.C. TIJMS (1986). *Stochastic Modelling and Analysis: A Computational Approach,* Wiley, Chichester.
26. H.C. TIJMS (1987). A simple and useful approximation to the waiting time distribution in the multi-server queue with priorities. G. IAZEOLLA, P.J. COURTOIS, O.J. BOXMA (eds.). *Computer Performance and Reliability,* North-Holland, Amsterdam, 161-169.
27. M.H. VAN HOORN (1984). *Algorithms and Approximations for Queueing Systems,* CWI Tract 8, CWI, Amsterdam.
28. M.H. VAN HOORN (1986). Numerical analysis of multi-server queues with deterministic services and special phase-type arrivals. *Zeitschrift für Oper. Res. 30,* 15-28.
29. J.C.W. VAN OMMEREN (1988). Exponential expansion for the tail of the waiting time probability in the single server queue with batch arrivals. To appear in *Adv. in Appl. Probab.*
30. J.C.W. VAN OMMEREN (1988). *Simple Approximations for the Batch Arrival $M^X/G/1$ Queue,* Research Report 1-88, Institute for Econometrics, Vrije Universiteit, Amsterdam.
31. R.W. WOLFF (1982). Poisson arrivals see time averages. *Oper. Res. 30,* 223-231.

PART II
THE SINGLE SERVER QUEUE

Approximations for the M/M/1 Busy-Period Distribution

Joseph Abate
AT&T Bell Laboratories
184 Liberty Corner Road, Warren, NJ 07060, USA

Ward Whitt
AT&T Bell Laboratories
600 Mountain Avenue, Murray Hill, NJ 07974, USA

This paper develops and evaluates relatively simple closed-form approximations for the M/M/1 busy-period cdf (cumulative distribution function). The first-order effect of the traffic intensity is identified and isolated by scaling. The principal approximations considered for the scaled busy-period cdf are: hyperexponential (H_2) approximations (mixtures of two exponentials) obtained by matching moments and derivatives of the cdf at the origin, asymptotic expansions as $t \to \infty$ and related approximations, approximants based on Widder's formula for Laplace transform inversion, and inverse Gaussian distributions. For small times, the hyperexponential approximation obtained by matching three derivatives at the origin performs best among these candidates. For larger times, an asymptotic normal approximation and an inverse Gaussian distribution perform remarkably well. Inverse Gaussian distributions arise naturally via diffusion approximations, but significantly better inverse Gaussian approximations can be obtained by making additional refinements, e.g., by matching moments or by matching the asymptotic exponential rate as $t \to \infty$. Overall, the results provide a better understanding of the M/M/1 model and a basis for developing approximations for busy-period cdf's in more general models and other quantities of interest in the M/M/1 model.

1. Introduction and summary

In the preface to the first edition of his fundamental book, *The Single Server Queue,* Cohen [15] describes the focus:

> 'The present book concentrates on the most basic model of queueing theory, i.e., the single server model. Its aim is twofold. Firstly a description of those mathematical techniques which have been proved to be the most fruitful for the investigation of queueing models, and secondly, an extensive analysis of the single server queue and its most important variants.'

Thus, the single server queue is of great interest for its own sake, as a model representing many queueing phenomena, as well as a means to illustrate useful techniques that can be applied to other stochastic models.

For these same reasons, we have also been studying the single-server queue

[1-6]. Our goal is to obtain simple approximations and a better understanding of the basic queueing models in order to facilitate practical engineering applications. In particular, our research has been directed toward developing simple approximations describing the transient behavior of the standard GI/G/1 model. We are trying to contribute theoretical insight to empirical investigations such as have been made by ODONI and ROTH [39]. Of course, we want to obtain good numbers, but more than numbers we want relatively simple formulas. The simple formulas help communicate understanding and help do further analysis, e.g., optimization and description of more complex models.

We also intend to apply our methods and results to analyze queues with time-dependent parameters, as in CLARK [14] and DUDA [20,21], but so far we have concentrated on the standard GI/G/1 model; we have been studying the transient behavior of a stationary model with general initial conditions. In fact, much of our work so far, including this paper, focuses on the most elementary M/M/1 special case. We hope to gain from such a narrow focus, just as Cohen does with *The Single Server Queue* (although some may think we are carrying a good thing to an extreme).

1.1. Simple approximations and asymptotic analysis

Developing simple approximations and a practical understanding is an important activity for the mathematician as well as the engineer. The possible benefits of mathematical insight applied in this direction are well illustrated by the approximations for the M/G/s queue developed by BOXMA, COHEN and HUFFELS [12].

In our analysis of transient behavior, a major role is played by asymptotic analysis. In particular, we exploit diffusion approximations resulting from heavy-traffic limit theorems [29]. The heavy-traffic limit theorems provide a scaling of time that is very useful. With appropriate scaling of space and time, the M/M/1 queue-length process approaches a nondegenerate limit as the traffic intensity ρ approaches 1. The limit at 1 is of course regulated or reflecting Brownian motion (RBM). With this scaling, we can thus regard RBM as the special M/M/1 model with $\rho=1$. We can thus obtain insight about RBM and M/M/1 from each other, as was done by COHEN and HOOGHIEMSTRA [16].

We are also interested in limit theorems showing how the time-dependent distribution approaches steady state, i.e., the theory of relaxation times as analyzed extensively by COHEN [15]. In agreement with NEWELL [38], we find that the first-order effect of the relaxation time is captured by the time scaling associated with the diffusion approximation. In agreement with ODONI and ROTH [39], we also find that the approximations following naturally from the relaxation-time limit theorems are not especially accurate. Of course, they do describe the important first-order effects, but they can be significantly improved. It appears that the region where the limit theorems related to the relaxation time are relatively accurate (within 10%) is usually considerably beyond the region of primary practical interest; see Table 3 of [1], Table 5 of [3], and Table 10[1] here. Fortunately (for applying steady-state results), the

1. All tables of this paper have been collected in Appendix A.

M/M/1 processes tend to be closer to their steady-state limits than the relaxation times predict at times of primary interest. (This summary judgement requires some qualification: It is correct for the quantities we have considered, such as the mean queue length starting empty or starting with relatively few customers, but it is not correct for the mean queue length starting with a great number of customers; then the approach to steady state is essentially linear.)

We not only investigate the quality of approximations of the transient behavior via relaxation-time limit theorems, but we also develop new approximations that are easy to apply and understand, and that are reasonably accurate. For example, mixtures of two exponentials (hyperexponential or H_2 distributions), which have three parameters, have proven to be very useful in our previous work [1-5]. Many M/M/1 quantities of interest turn out to be completely monotone (a general mixture of exponentials), which provides theoretical support for the H_2 approximations. Moreover, in regions of interest often one exponential dominates, so that we obtain a simple exponential approximation there. Then we have a very simple formula describing the time-dependent behavior of the queue.

1.2. The M/M/1 busy-period distribution

We continue this line of investigation here by investigating approximations for the M/M/1 busy-period distribution. It is well-known that the busy period plays a fundamental role in both the equilibrium distribution (e.g., via regenerative structure) and the transient behavior; see [4], [15] and [28]. In this paper, we consider several different approximations for the M/M/1 busy-period distribution and compare their accuracy. It turns out that the H_2 approximation obtained by matching three moments does not work as well for the busy-period distribution as it does for other M/M/1 quantities, so that we end up proposing different approximations. A specific approximation is proposed in Section 7, but additional insight is gained from seeing how all the candidates fare. The entire analysis illustrates, as others have discovered before, that the busy-period distribution is rather strange and difficult.

As indicated above, our motivation is not only to describe the M/M/1 model itself, but also to develop results and techniques applicable to more general models. In a sequel to this paper we apply the results here to develop approximations for the GI/G/1 busy-period distribution. We briefly discuss approximations for the GI/G/1 busy-period distribution in Section 8, but all numerical comparisons here are restricted to the M/M/1 model.

1.3. What is the busy-period distribution like?

Before discussing any formal results, it seems worthwhile to give a brief intuitive discussion of the behavior of the GI/G/1 busy-period distribution, because it has somewhat peculiar properties that are easy to understand after some thought. *The main observation is that the behavior for small time values is very different from the behavior for large time values.* The entire paper elaborates on this theme.

For small time values, the GI/G/1 busy-period distribution is quite sensitive to the specific interarrival-time and service-time distributions. For example, the probability that the GI/G/1 busy period coincides with one service time is $P(v<u)$ where v is a service time and u is an interarrival time. For very small time values, the busy-period cdf (cumulative distribution function) $B(t)$ is usually approximated very well by

$$B(t) \approx P(v \leqslant t, v<u), \tag{1.1}$$

which obviously depends strongly on the specific distributions for u and v. Moreover, for small time values the busy-period cdf does not depend critically on ρ. There is no dramatic change as ρ approaches 1 from below or even as ρ becomes greater than 1. Even if $\rho>1$, the busy period may be fairly short.

On the other hand, for large time values the busy-period distribution is primarily determined by a large number of convolutions of the interarrival-time and service-time distributions, so that central-limit-theorem behavior begins to play a central role. Thus, for large time values the busy-period distribution primarily depends on the interarrival-time and service-time distributions only through their first two moments, and approximations by RBM and other related quantities (normal and inverse Gaussian distributions) should be good. Moreover, for large time values the busy-period cdf obviously depends critically on ρ. As ρ approaches 1, the busy period can have a very fat tail. If $\rho>1$, then the busy period may not terminate at all.

The point we wish to emphasize is that the small-time and large-time behavior are quite different. Thus it should come as no surprise that we obtain useful results, both theorems and approximations, by treating these different regions separately. For example, we can describe the heavy-traffic behavior as $\rho \rightarrow 1$ in each of the separate time regions quite nicely, as we show below in Sections 2.6 and 2.7.

1.4. Literature review

There obviously is an enormous body of related literature, including the fundamental papers by KENDALL [33], BAILEY [8,9] and KARLIN and MCGREGOR [32]. For the most part, we refer to COHEN [15] for a map of the known world. With regard to previous work on *approximations* of busy-period distributions in the single-server queue, we cite RICE [40], RIORDAN [42, pp. 106-109], KOSTEN [34, pp. 42-45], HEYMAN [26] and DUDA [20,21]. We will discuss these previous approximations as they relate to our work.

1.5. Organization of this paper

We review supporting theory for the approximations in Section 2 and discuss the main approximations in Sections 3-7. We discuss hyperexponential approximations in Section 3, approximations related to asymptotic expansions in Section 4, approximations based on Widder's formula for Laplace transform inversion in Section 5, and inverse Gaussian approximations in Section 6. We present a final composite approximation in Section 7. We briefly discuss approximations for the GI/G/1 busy-period distribution in Section 8, and related approximations for the M/M/1 expected cumulative idle time up to time t in Section 9. Finally, we present conclusions in Section 10.

2. BACKGROUND

In this section we review supporting theory for our approximations. We draw on our previous M/M/1 papers [3-6], but we also make connections to the extensive literature.

2.1. The M/M/1 model with time scaling

Let $Q(t)$ represent the queue length (including the customer in service, if any) at time t in the M/M/1 model. Without loss of generality, let the service rate be 1, so that the arrival rate coincides with the traffic intensity ρ. Assume that $\rho<1$, so that the system is stable with $Q(t)$ converging in distribution to $Q(\infty)$ as $t\to\infty$, where $P(Q(\infty)=k)=(1-\rho)\rho^k$, $k\geqslant 0$.

As in [3-6], we further scale time by $(2\theta^2)^{-1}=2/(1-\rho)^2$; i.e., we consider $Q(2t/(1-\rho)^2)$ and let $P_{ij}(t)$ be the time-scaled transition function

$$P_{ij}(t) = P(Q(2t/(1-\rho)^2)=j\,|\,Q(0)=i). \tag{2.1}$$

The time scaling in (2.1) is a very important part of the story. Our goal is to describe the queue length process $Q(t)$, its transition function $P_{ij}(t)$ and the associated busy-period distribution. It is significant that the first-order effect of the single parameter ρ is captured by this time scaling. As discussed in Section 2.2 of [3], the time-scaling captures the heavy-traffic behavior as $\rho\to 1$. In particular, the family of processes $\{2^{-1}(1-\rho)Q(2t/(1-\rho)^2)\colon t\geqslant 0\}$ indexed by ρ converges to canonical regulated or reflecting Brownian motion (RBM), having drift coefficient -1 and diffusion coefficient 1, as $\rho\to 1$. Thus, with the time scaling, we can treat RBM as the nondegenerate M/M/1 queue with $\rho=1$. With the time scaling the M/M/1 processes for different ρ tend to fluctuate in about the same time scale; we remove the dominant ρ effect, so that we can more easily compare M/M/1 processes with different ρ. A simple practical consequence is that we can more meaningfully compare M/M/1 quantities for all possible traffic intensities ρ, $0\leqslant\rho\leqslant 1$, in the same table; see Table 1. In contrast, without the scaling, the interpretation of a given time t, depends more on ρ, so that one table including several different ρ is less meaningful; e.g., see HEYMAN [26]. (In fact, for the M/M/1 busy-period distribution, additional scaling is appropriate, as we will indicate in Section 2.7 below.)

For more general models, such as GI/G/1 or GI/G/m, it is useful to scale time so that *canonical* RBM also appears as the limit as $\rho\to 1$. As indicated in Section 2 of [1], it is easy to convert RBM with negative drift μ and variance coefficient ν to canonical RBM, so that it is easy to obtain the appropriate scaling; see Section 8. With such a scaling, we identify and isolate the first-order effect of the variability (as it differs from M/M/1) as well as the traffic intensity.

2.2. The busy-period distribution and related quantities

Here we are primarily interested in the busy-period distribution. Let $B(t)$ be the time-scaled busy-period cdf (cumulative distribution function); let $B^c(t) \equiv 1-B(t)$ be the complementary busy-period cdf; and let $b(t)$ be the density. Let $\hat{B}^c(s)$ and $\hat{b}(s)$ be the associated Laplace transforms. From KENDALL [33] or BAILEY [8,9], we know that

$$\hat{b}(s) \equiv \int_0^\infty e^{-st} b(t)dt = z_1(s) = [1-\theta+\theta^2 s-\theta\Psi(s)]/\rho,$$

$$\hat{B}^c(s) = 2\theta/[1+\theta s+\Psi(s)], \tag{2.2}$$

where

$$\begin{aligned} \theta &= (1-\rho)/2, & \Psi(s) &= [1+2(1-\theta)s+(\theta s)^2]^{1/2}, \\ r_1(s) &= \Psi+(1-\theta s), & r_2(s) &= \Psi-(1-\theta s), \\ \rho z_1 &= 1-\theta r_1, & \rho z_2 &= 1+\theta r_2, \\ r_1 r_2 &= 2s, \quad \rho z_1 z_2 = 1 & \text{and } \rho(1-z_1)(z_2-1) &= 2\theta^2 s. \end{aligned} \tag{2.3}$$

The functions $z_1 \equiv z_1(s)$ and $z_2 \equiv z_2(s)$ are the two roots of the basic quadratic equation $\rho z^2-(1+\rho+2\theta^2 s)z+1=0$. The Laplace transform $\hat{b}(s)$ is derived by KENDALL [33] by making connections to branching processes and by BAILEY [8,9] by considering a modified M/M/1 queue length process that is absorbed when it reaches the origin. As indicated in the proof of Theorem 3.1 of [4], we can also obtain $\hat{b}(s)$ directly from $\hat{P}_{10}(s)$ and $\hat{P}_{00}(s)$ using first principles, where $\hat{P}_{ij}(s)$ is the Laplace transform of $P_{ij}(t)$ in (2.1); in particular,

$$\hat{b}(s) = \hat{P}_{10}(s)/\hat{P}_{00}(s). \tag{2.4}$$

Indeed, the time-domain version of (2.4) is the starting point for Cohen's treatment in [15]; see (2.30) on p. 187.

Additional properties of the M/M/1 busy-period distribution appear in [3-5]. For example, Corollary 4.2.3 in [4] establishes an interesting connection between the probability of emptiness $P_{00}(t)$ and $B(t)$, namely,

$$P_{00}(t) = 1-\rho B(t), \quad t \geqslant 0, \tag{2.5}$$

which quickly yields expressions for the expected cumulative idle time in $[0, t]$, say $EI_0(t)$; the expected workload in the system at time t, $EW_0(t)$; and the expected queue length $EQ_0(t)$; all given that $Q(0)=0$ (this condition being indicated by the subscript), namely,

$$EI_0(t) = \int_0^t P_{00}(s)ds = \int_0^t [1-\rho B(u)]du,$$

$$EQ_0(t) = EW_0(t) = \rho t - t + EI_0(t); \tag{2.6}$$

see Corollary 4.2.6, the alternate proof of Corollary 5.2.1, and Theorem 8.1 of [4]. (In fact, (2.5) is an immediate consequence of the relation $\rho z_1 = 1-\theta r_1$ in (2.3), because $\hat{P}_{00}(s) = \theta r_1/s$ and $\hat{b}(s) = z_1$ by (2.6) of [4] and (2.2) above.) As indicated in Remark 4.1 after Corollary 4.2.3 in [4], (2.5) evidently has a long history, but it does not seem to have been given enough emphasis.

Relations (2.5) and (2.6) provide additional motivation for this paper because they show that other M/M/1 quantities of interest can be expressed in terms of the busy-period distribution. (For more on this, see [4-6] and Section 2.3 below.) Our approximations for $B(t)$ automatically yield associated approximations for these other M/M/1 quantities. We discuss some of these related approximations in Section 9.

In this regard, it is significant that the busy-period distribution coincides with the equilibrium waiting-time distribution in an M/M/1 queue with the last-come first-served (LCFS) discipline, as was first noted by RIORDAN [41]. This LCFS waiting-time distribution is useful for its own sake, but also because it provides an upper bound on the waiting-time distribution for a large class of service disciplines, as noted by VAULOT [47]; see p. 416 of HEYMAN and SOBEL [27]. Not only do the results here apply to this M/M/1-LCFS waiting-time distribution, but previous descriptions of the M/M/1-LCFS waiting-time distributions in VAULOT [47], RIORDAN [41] and pp. 106-109 of [42], and pp. 42-45 of KOSTEN [34] also apply here. Indeed, this previous work seems most closely related to our interest in relatively simple closed-form approximations. Vaulot, Riordan and Kosten discovered exact representations that are convenient for generating numbers as well as relatively simple closed-form approximations.

2.3. The stationary-excess relations

An important discovery in [3-5] is a connection between various M/M/1 quantities and the busy-period cdf $B(t)$ via the stationary-excess operator [51]. For any cdf $G(t)$ on the positive real line with mean m_1, the associated stationary-excess (or equilibrium-residual-life) cdf $G_e(t)$ is defined by

$$G_e(t) = m_1^{-1}\int_0^t [1-G(u)]du, \quad t \geq 0. \tag{2.7}$$

In [3] we focused on moment cdf's which are the (time-scaled) moments $E[Q(t)^k]$ under the condition that $Q(0)=0$, divided by their steady-state limits. Corollary 3.1.3 of [3] and Corollary 5.2.1 of [4] show that the first-moment cdf, denoted by $H_1(t)$, coincides with $B_e(t)$, the stationary-excess cdf associated with the busy-period cdf $B(t)$. In fact, this is just a restatement of the second relation in (2.6).

Theorem 1 of [5] shows that the correlation function of the *stationary* (time scaled) queue length process, denoted by $c_q(t)$, in turn coincides with $H_{1e}^c(t)$, the complementary stationary-excess cdf associated with the first-moment cdf $H_1(t)$. Thus, $c_q(t)$ is obtained from the busy-period $B(t)$ by applying the stationary-excess operator in (2.7) *twice.*

These stationary-excess relations have important implications for both theory and approximations. For example, the spectral representation and the asymptotic behavior for $B(t)$ thus easily extend to $H_1(t)$ and $c_q(t)$; see

Sections 2.8 and 4. Since the stationary-excess operator tends to be a smoothing operator [51], the functions $B^c(t)$, $H_1^c(t)$ and $c_q(t)$ tend to be successively better behaved. In particular, the stationary-excess relations help explain why it is harder to obtain good approximations for $B^c(t)$ than it is for $B_e^c(t)=H_1^c(t)$ or $c_q(t)$. DELBROUCK [18] and ERLANDER [22] previously considered $B_e(t)$ and noted that it has nice properties.

2.4. Numerical inversion of the Laplace transform

As first shown by KENDALL [33], the Laplace transform $\hat{b}(s)$ in (2.2) can be inverted (using pair 556.1 from CAMPBELL and FOSTER [13]) to obtain

$$b(t) = \frac{1}{t\sqrt{\rho}} e^{-t/\tau}[e^{-\nu} I_1(\nu)], \quad t \geqslant 0, \tag{2.8}$$

where $I_1(\nu)$ is a modified Bessel function of the first kind, e.g., p. 377 of ABRAMOWITZ and STEGUN [7],

$$\tau = \frac{(1+\sqrt{\rho})^2}{2} \quad \text{and} \quad \nu = t\theta^{-2}\sqrt{\rho}. \tag{2.9}$$

The parameter τ in (2.9) is the *time-scaled relaxation time* which describes the asymptotic behavior of (2.8) as $t \to \infty$. (Recall that $f(\nu) \sim g(\nu)$ means that $f(\nu)/g(\nu) \to 1$ as $\nu \to \infty$. Use 9.7.1 on p. 377 of [7] to see that $e^{-\nu} I_1(\nu) \sim (2\pi\nu)^{-1/2}$ as $\nu \to \infty$, so that $e^{-\nu} I_1(\nu) \sim \sqrt{t}$ as $t \to \infty$; i.e., $b(t) \sim Kt^{-3/2} e^{-t/\tau}$, so that $e^{-t/\tau}$ is the dominant term in (2.8) as $t \to \infty$; see Section 4. The unscaled relaxation time is $(1-\sqrt{\rho})^{-2}$, as given on p. 180 of COHEN [15]; τ is the time-scaled relaxation time, because $(1-\rho)^2 = (1-\sqrt{\rho})^2(1+\sqrt{\rho})^2$. Note that the dominant portion of the unscaled relaxation time for high ρ is in the time scaling.)

Asymptotic results related to the relaxation time can be used to generate approximations, but as in [1,3], we show that the natural approximations generated from these limits do not perform well for times of primarily practical interest. However, in Section 4 we show that appropriate refinements of the asymptotic results do perform well for times of practical interest.

We focus on the complementary cdf $B^c(t)$ instead of the density $b(t)$. Unfortunately, $\hat{B}^c(s)$ in (2.2) is not as easily inverted as $\hat{b}(s)$, so to obtain numerical values of $B^c(t)$ we numerically invert the Laplace transform $\hat{B}^c(s)$ in (2.2). Indeed, numerical transform inversion is not an unreasonable way to obtain numerical values for $b(t)$ in (2.8).

Given that so many queueing results are expressed in terms of Laplace transforms, it is surprising that relatively little attention has been given to numerical Laplace transform inversion in the applied probability literature. For example, queueing textbooks do not provide practical guidelines and procedures for numerically inverting Laplace transforms. However, as we indicated in Section 4.4 of [1], there are numerous techniques for the numerical inversion of Laplace transforms. As in [1], we use the Gaver-Stehfest procedure [23,43]. It yields good (but not exceptional) accuracy on a small computer (a personal computer with BASIC) with little programming effort. The

Gaver-Stehfest procedure was also employed and compared to other procedures by NANCE ET AL. [37]. For still other procedures plus comparisons, see DAVIES and MARTIN [17].

Our goal in numerical transform inversion is only to obtain limited accuracy. We are primarily interested in relatively simple analytical approximations that are convenient for practical engineering applications. If we actually wanted great numerical accuracy, we would use a different inversion technique. In fact, the best way to get numerical results is not to use transforms at all, but to exploit integral representations, as discussed in [6] and Section 2.8 below.

Numerical values of the complementary time-scaled busy-period cdf $B^c(t)$ obtained by applying the Gaver-Stehfest procedure are given in Table 1. (These values were checked by also applying an independent numerical inversion technique of JAGERMAN [30,31].) Table 1 shows that we have approximately the right time scale for times when $0.001 \leqslant B^c(t) \leqslant 0.2$, which we call the second regime [1]. (The second regime is the time interval where the process reaches steady state for practical purposes.) For example, consider $t=3.00$ where $B^c(t)=0.00378$ with $\rho=0.50$. *With the time scaling,* as ρ changes, the value of $B^c(3)$ changes quite slowly, until ρ gets very high. (This unusual behavior as $\rho \to 1$ is very different from our experience in [3] and [5]; we will discuss this anomaly further in Section 2.7.)

On the other hand, for very small times (which we call the first regime), $B^c(t)$ changes more slowly *when we omit the time scaling.* In Table 1 we remove the time scaling by considering the times $\theta^2/2$ and θ^2. The fact that the time scaling helps for larger times, but not for smaller times, reflects the special nature of the busy-period distribution discussed in Section 1.3.

2.5. Moments

While the M/M/1 busy-period distribution is somewhat inaccessible, its moments are readily available. Theorem 3.2 of [4] gives a basic recursion due to RIORDAN [41,42]: Let m_k be the k-th moment of the time-scaled busy-period distribution; then

$$m_{k+2} = (2k+1)(1-\theta)m_{k+1} - (k^2-1)\theta^2 m_k \tag{2.10}$$

for $m_0=1$ and $m_1=\theta$. (This result appears in a description of the M/M/1-LCFS waiting-time distribution.) Moreover, an explicit formula for the moments is given on p. 232 of TAKÁCS [44], namely,

$$m_{k+2} = \sum_{i=0}^{k} \frac{(k+i+2)!k!\theta^{k+1-i}(1-2\theta)^i}{(i+1)!2^{i+1}i!(k-i)!}. \tag{2.11}$$

The first five moments are

$$\begin{aligned} m_1 &= m_2=\theta, \quad m_3=3\theta(1-\theta), \quad m_4=15\theta[1-2\theta+\frac{4}{5}\theta^2], \\ m_5 &= 105\theta[1-3\theta+\frac{18}{7}\theta^2-\frac{4}{7}\theta^3]. \end{aligned} \tag{2.12}$$

From these moments we can obtain important insight into the busy-period distribution, especially its large-time behavior, and how it behaves as a function of $\theta \equiv (1-\rho)/2$ or ρ. First, the squared coefficient of variation (variance divided by the square of the mean) is

$$c_B^2 \equiv \frac{m_2 - m_1^2}{m_1^2} = \frac{\theta - \theta^2}{\theta^2} = \theta^{-1} - 1 = \frac{1+\rho}{1-\rho}, \tag{2.13}$$

which is independent of the time scale. Formula (2.13) shows that the distribution becomes highly variable as $\rho \to 1$; $c_B^2 \to \infty$ as $\rho \to 1$.

Moreover, the skewness and kurtosis also explode as $\rho \to 1$; i.e.,

$$\begin{aligned} \frac{m_3}{m_2^{3/2}} &= \frac{3\theta(1-\theta)}{\theta^{3/2}} = \frac{3(1-\theta)}{\sqrt{\theta}} = \frac{3\sqrt{2}(1+\rho)}{\sqrt{1-\rho}} \\ \frac{m_4}{m_2^2} &= \frac{15\theta[1-2\theta+\frac{4}{5}\theta^2]}{\theta^2} = \frac{15[1-2\theta+\frac{4}{5}\theta^2]}{\theta} = \frac{30\rho}{1-\rho} + 6(1-\rho). \end{aligned} \tag{2.14}$$

2.6. A power-series representation

In Theorem 9 of [5] we obtained a power-series representation for the complementary busy-period cdf in terms of the moments described in Section 2.5. In particular,

$$B^c(t) = 1 + \sum_{k=0}^{\infty} (-1)^{k+1} \frac{m_{k+2} t^{k+1}}{(k+2)!(k+1)!\theta^{2k+3}}. \tag{2.15}$$

We can use a few terms from (2.15) to get a good approximation for $B^c(t)$ for small t. However, the times t must be *relatively small before time scaling.* With $t = \alpha\theta^2$ (to undo the time scaling), the first three terms from (2.15) yield

$$\begin{aligned} B^c(\alpha\theta^2) &\approx 1 - \frac{m_2\alpha}{2\theta} + \frac{m_3\alpha^2}{12\theta} - \frac{m_4\alpha^3}{144\theta} \\ &\approx 1 - \frac{\alpha}{2} + \frac{(1+\rho)\alpha^2}{8} - \frac{(1+3\rho+\rho^2)\alpha^3}{48}. \end{aligned} \tag{2.16}$$

For example, for $\alpha = 1/2$, (2.16) becomes

$$B^c(\theta^2/2) \approx 0.779 + 0.023\rho - 0.0026\rho^2. \tag{2.17}$$

From Table 1, we see that (2.17) is very accurate, having a maximum relative percent error of about 0.1%. Similarly, when $\alpha = 1$, we obtain

$$B^c(\theta^2) \approx 0.604 + 0.063\rho - 0.0208\rho^2, \tag{2.18}$$

which yields a maximum relative percent error of about 1%.

In summary, (2.15) provides an alternative way to obtain numerical results and relatively simple approximations such as (2.16). However, the simple approximation (2.16) is only valid for relatively small times (small in the original time scaling).

2.7. Light and heavy traffic

Further insight into the busy-period distribution can be obtained by considering the limits as $\rho\to 0$ and $\rho\to 1$. As in Corollary 5.2.2(b) of [4], we can work with the Laplace transform to obtain the light-traffic limit. In particular, by the argument in [4],

$$\lim_{\rho\to 0} \hat{b}(s) = \frac{2}{2+s}; \tag{2.19}$$

so that

$$\lim_{\rho\to 0} B(t) = 1-e^{-2t}, \quad t\geqslant 0. \tag{2.20}$$

In other words, the busy-period distribution converges (weakly) to a simple exponential with mean $1/2$ as $\rho\to 0$. Indeed, this is obvious because the busy period obviously approaches a single service time as $\rho\to 0$. (The 2 appears because of the time scaling.) Moreover, this limit could be anticipated (and deduced) from the limit for the first-moment cdf $H_1(t)$ established in Corollary 5.2.2(b) of [4] plus the stationary-excess relation discussed in Section 2.2. (Empirically, note that the $\rho=0$ column of Table 1 here coincides with the $\rho=0$ column of Table 1 in [3].)

The heavy-traffic limiting behavior is more complicated. Unlike $H_1(t)$ in Corollary 5.2.2(a) of [4], $B^c(t)$ does *not* converge to a proper limit as $\rho\to 1$. However, Theorem 3.5 of [4] describes the limit as $\rho\to 1$. It is significant that, even with the time scaling, it involves an additional normalization by $\theta\equiv(1-\rho)/2$ as $\rho\to 1$. The limits for the normalized complementary cdf $B^c(t)$ and density $b(t)$ are

$$\begin{aligned}\lim_{\rho\to 1} 2(1-\rho)^{-1}B^c(t) &= \tilde{h}_1(t) = 2t^{-1/2}\phi(t^{1/2})-2[1-\Phi(t^{1/2})], \quad t\geqslant 0,\\ \lim_{\rho\to 1} 2(1-\rho)^{-1}b(t) &= (2\pi t^3)^{-1/2}e^{-t/2}, \quad t\geqslant 0,\end{aligned} \tag{2.21}$$

where $\Phi(t)$ is the cdf of a standard normal cdf (with mean 0 and variance 1) and $\phi(t)$ is its density. The limits in (2.21) are the density $\tilde{h}_1(t)$ of the first-moment cdf $\tilde{H}_1(t)$ for RBM (M/M/1 with $\rho=1$) as given in (4.4) of [1] and its derivative $\tilde{h}_1{}'(t)$. (We use the notation $\sim$ to designate the RBM case with $\rho=1$.) This is not too surprising, once we recognize that $\theta^{-1}B^c(t)$ on the left side of (2.21) is just the density of the stationary-excess cdf $B_e(t)\equiv H_1(t)$ for given ρ, as indicated in Section 2.3 above. Thus, (2.21) represents convergence of densities (i.e., $h_1(t)\to\tilde{h}_1(t)$ as $\rho\to 1$, a local limit theorem) and their derivatives, paralleling the convergence of distributions as $\rho\to 1$ established in Corollary 5.2.2(a) of [4].

By Theorem 1.3 of [1], the limit $\tilde{h}_1(t)$ in (2.21) also coincides with an exponential mixture of inverse Gaussian densities, i.e.,

$$\tilde{h}_1(t) = \int_0^\infty 2e^{-2x}f(t;x,0)dx, \tag{2.22}$$

where $f(t;x,0)$ is the density of an inverse Gaussian cdf $F(t;x,0)$, with

$$f(t;x,0) = \frac{x}{\sqrt{2\pi t^3}} \exp\left[-\frac{(x-t)^2}{2t}\right], \quad t \geqslant 0,$$
$$F(t;x,0) = \Phi\left[\frac{t-x}{\sqrt{t}}\right] + e^{2x}\Phi\left[\frac{-t-x}{\sqrt{t}}\right], \quad t \geqslant 0, \tag{2.23}$$

as in (1.5) and (1.6) of [1]; $f(t;x,0)$ is the density of the first-passage time from x to 0 for (canonical) RBM. Inverse Gaussian distributions arise naturally in diffusion approximations for first-passage-time distributions in queueing; e.g., see HEYMAN [26] and DUDA [20,21]. We will be considering the inverse Gaussian distribution further in Section 6.

The extra normalization by θ in (2.21) means that

$$B^c(t) \approx \frac{(1-\rho)}{2}\tilde{h}_1(t), \tag{2.24}$$

which converges to 0 as $\rho \to 1$, as can be seen by looking at the $\rho = 0.95$ and $\rho = 0.99$ columns of Table 1. A better scaling for the busy-period distribution is thus $\theta^{-1}B^c(t)$. (We are including the time scaling.) Note that $\theta^{-1}B^c(t)$ has proper limits both as $\rho \to 0$ and as $\rho \to 1$, just as $H_1(t)$ and $c_q(t)$ do after the time scaling alone. Part I of Table 2 displays the numerical values of $\theta^{-1}B^c(t)$ obtained by multiplying the values in Table 1 by θ^{-1}. Note that this doubly-scaled function is indeed even less sensitive to ρ. For example, for $t=1$, $\theta^{-1}B^c(t)$ decreases from 0.27 to 0.16 as ρ increases from 0 to 1. The remaining adjustment is not uniform over t though. In Table 2, $\theta^{-1}B^c(t)$ is increasing and then decreasing for $t=0.10$ and 0.25, strictly decreasing for $0.50 \leqslant t \leqslant 1.50$, and then strictly increasing for $t \geqslant 2.00$.

To a large extent, the behavior of $B^c(t)$ as a function of ρ is captured by the two scalings. What is left is shown in Part I of Table 2; the function $\theta^{-1}B^c(t)$ moves from an exponential at $\rho=0$ to $\tilde{h}_1(t)$ at $\rho=1$. A simple light-and-heavy-traffic-interpolation approximation based on this is the convex combination

$$\theta^{-1}B^c(t) \approx (1-w)2e^{-2t} + w\tilde{h}_1(t), \quad t \geqslant 0, \tag{2.25}$$

where $w \equiv w(\rho)$ is an increasing function of ρ such that $0 \leqslant w \leqslant 1$. For approximation (2.25) to perform reasonably well, it helps for $\theta^{-1}B^c(t)$ to be monotone. As we observed above, in Table 2 this is not true for $t \leqslant 0.25$, but it is true for $t \geqslant 0.5$. From Table 2, it is easy to see that (2.25) should perform quite well for $t \geqslant 0.25$. Moreover, from (2.12) here plus Corollary 1.3.4 of [1], we see that (2.25) with $w(\rho) = \rho$ yields the correct first three moments of $B(t)$. (It does not yield the correct fourth moment.) On the other hand, for very small t, (2.25) can not be good; e.g., $B^c(0)=1$ and $\tilde{h}_1(0)=\infty$. Nevertheless, (2.25) can be a pretty good approximation for $t \geqslant 0.25$. Experiments with a few weighting functions suggested the weight function $w(\rho)=\rho^x$ for $x=0.75$; this is displayed in Part II of Table 2. The performance is pretty good for t neither too small nor too large, e.g., for $0.50 \leqslant t \leqslant 4.00$. Having $x<1$ in this weighting function is consistent with the observation that the heavy-traffic limit

($\rho=1$) is more descriptive for ρ near 1 than the light traffic limit ($\rho=0$) is for ρ near 0. (Of course, the nice behavior near $\rho=1$ is achieved by the double scaling.) Approximation (2.25) is not very satisfying because, beyond the double scaling, it is ad hoc and because it evidently deteriorates for both small and large t. Some of our other approximations will do better in both regards.

2.8. The spectral representation

It is significant that the busy-period distribution can be represented as a mixture of exponential distributions, i.e., $B^c(t)$ is completely monotone; see Theorem 3.3 of [3]. An *explicit* representation as a mixture of exponentials follows from the spectral representation for $B^c(t)$. The spectral representation for the M/M/1 busy-period distribution was first given in (6.4) of KARLIN and MCGREGOR [32]; another derivation is given in [6]. With our time scaling, the spectral representation for the busy-period distribution has a remarkably simple form (taken from [6]), namely,

$$B^c(t) = \int_{\tau_1}^{\tau_2} e^{-t/y} w(y) dy, \quad t \geqslant 0, \tag{2.26}$$

where

$$w(y) = \frac{\theta\sqrt{(y-\tau_1)(\tau_2-y)}}{\pi\rho y^2}, \quad \tau_1 \leqslant y \leqslant \tau_2,$$
$$\tau_1 = \frac{1+\rho-2\sqrt{\rho}}{2} \quad \text{and} \quad \tau_2 = \frac{1+\rho+2\sqrt{\rho}}{2} = \frac{(1+\sqrt{\rho})^2}{2}; \tag{2.27}$$

i.e., the mixing distribution has a proper probability density $w(y)$, which is given in (2.27). The mixing density $w(y)$ in (2.27) is unimodal with mode

$$y_{\max} = \frac{3(1+\rho) - \sqrt{1+34\rho+\rho^2}}{4}. \tag{2.28}$$

The relaxation time appears as the upper limit of integration τ_2.

Note that (2.26) is an ideal representation for obtaining numerical results by numerical integration. Also the asymptotic behavior as $t \to \infty$ (to be discussed in Section 4) can easily be derived by applying Laplace's method to (2.26).

As noted in [6], corresponding simple spectral representations also hold for the complementary first-moment cdf $H_1^c(t)$ and the correlation function $c_q(t)$, by virtue of the stationary-excess relations discussed in Section 2.3; it is elementary to integrate with respect to t in (2.26). These spectral representations for $H_1(t)$ and $c_q(t)$ easily extend to RBM by taking the limit as $\rho \to 1$. The nice spectral representation for the correlation function of RBM was previously discovered by WOODSIDE ET AL. [53]. The resulting mixing distributions are all simple modifications of the beta distribution.

2.9. Connections to the unrestricted process

As indicated in Section 7 of [4] and as discussed by others before, one way to analyze the M/M/1 queue is to relate it to the unrestricted process on all the integers with transition function $Q_{ij}(t)$. For this process, a basic role is played by $Q_{00}(t)$ or the scaled version (with our time scaling)

$$\gamma_\rho(t) = \theta^{-1} Q_{00}(t) = \theta^{-1} e^{-t/\tau} e^{-\nu} I_0(\nu), \tag{2.29}$$

which is a bonafide probability density function; see (7.7) of [4]. From Corollary 7.2.2 of [4], we immediately obtain the representation

$$\theta^{-1} B^c(t) = \left[\frac{1+\rho}{2\rho}\right][2\gamma_\rho(t) - \gamma_{\rho e}(t)] + \frac{\theta^2}{\rho}\gamma_\rho'(t), \quad t \geq 0, \tag{2.30}$$

where $\gamma_{\rho e}(t)$ is the density of the associated stationary-excess cdf defined by (2.7). The heavy-traffic limit (2.21) follows directly from (2.30) as indicated in [4]. Moreover, all M/M/1 approximations could be generated from approximations for the basic quantity $\gamma_\rho(t)$ in (2.29) and its relatives $\gamma_{\rho e}(t)$ and $\gamma_\rho'(t)$.

2.10. A differential equation for the busy-period density

We conclude this section with a curiosity. We point out that the busy-period density $b(t)$ satisfies a second-order linear differential equation with monomial coefficients, namely,

$$\theta^2 t b''(t) + [2(1-\theta)t + 3\theta^2] b'(t) + [t + 3(1-\theta)] b(t) = 0, \tag{2.31}$$

with boundary conditions

$$b(0) = 1/2\theta^2 \quad \text{and} \quad b'(0) = -(1-\theta)/2\theta^4. \tag{2.32}$$

This follows from the Bessel function representation (2.8) and the differential equation for Bessel functions in 9.6.1 on p. 374 of [7].

3. HYPEREXPONENTIAL APPROXIMATIONS

In this section we consider hyperexponential (H_2) approximations for the M/M/1 complementary busy-period cdf $B^c(t)$. An H_2 distribution is a mixture of two exponentials; i.e., a complementary H_2 cdf is of the form

$$G^c(t) \equiv 1 - G(t) = pe^{-\lambda_1 t} + (1-p)e^{-\lambda_2 t}, \quad t \geq 0, \tag{3.1}$$

for $0 \leq p \leq 1$, $\lambda_1 \geq 0$ and $\lambda_2 \geq 0$, so that it is determined by the parameter triple $(p, \lambda_1, \lambda_2)$.

3.1. Motivation and methodology

We are motivated to consider H_2 approximations because we already have had considerable success using them to approximately describe the transient behavior of the M/M/1 queue and regulated Brownian motion (RBM) [1-5]. In particular, the first and second moment of the M/M/1 queue length at time t starting at 0 and the correlation functions of several stationary M/M/1 processes are well approximated (for t neither too small nor too large) by H_2

approximations obtained by matching the first three moments; see Tables 3 and 5 of [1], Tables 1-2, 5-7 of [3] and Tables 1-2, 6-9 of [5]. The qualification on time is somewhat vague and unpleasant, but as in Section 1.6 of [1], we contend that it is useful to consider three regimes: small time, medium time and large time, with the middle regime being when most M/M/1 processes reach steady state for most practical purposes, e.g., when the moments are within 0.1-15.0 percent of their steady-state limits. We evaluate our approximations with regard to their performance in the different regimes. Approximations that perform reasonably well in at least one regime seem worth considering, but we are primarily interested in the first two regimes; the third regime usually involves exceptionally large times that are of little practical interest.

As with many of the previous M/M/1 quantities studied, there is a strong theoretical basis for using an H_2 approximation for the busy-period distribution, because its density is completely monotone; i.e., the busy-period distribution *is* a mixture of exponentials, as indicated in Section 2.8. However, an H_2 fit still involves an approximation because the busy-period distribution is not a mixture of *two* exponentials, but uncountably many exponentials. Nevertheless, we might expect to capture the rather distinct small-time and large-time behavior of the busy-period distribution via the two exponential components.

An H_2 approximation for the busy-period cdf by moment matching is also attractive because the busy-period moments are readily available, as indicated in Section 2.5. Moreover, there are relatively straightforward schemes for expressing the H_2 parameter triple (p,λ_1,λ_2) in terms of the moment triple (m_1,m_2,m_3), i.e., the first three moments; see Section 5.1 of [1], especially (5.7) there. In fact, the three-moment H_2 approximation for the complementary busy-period cdf was previously developed by RIORDAN [42, p. 108] in his analysis of the waiting-time distribution in the M/M/1-LCFS system. (Riordan claims that the H_2 approximation matches the fourth moment too, but this seems to be incorrect.)

Moment matching usually is pretty accurate in describing the cdf for relatively large times, but not for small times. For smaller times, it is natural to match derivatives at the origin. As interest moves from times very near the origin to larger times, it is natural to shift from a three-derivative (3D) fit to a two-derivative, one-moment (2D, 1M) fit, to a one-derivative, two-moment (1D, 2M) fit, and eventually to a three-moment (3M) fit. Working with moments and derivatives at the origin, there are thus four candidate H_2 approximations we might consider. (We could of course obtain better numerical results by considering mixtures of more than two exponentials, but we would then lose the desired simplicity. Remember that the goal is not primarily numbers.)

It is significant that we can obtain all four of these H_2 approximations by the single three-moment matching procedure mentioned above. We obtain this simplification in fitting because the cases involving one or more derivatives are mapped into the three-moment case by the stationary-excess operator, as described in Section 7 and 8 of [5]. For a cdf $G(t)$ with a density $g(t)$, suppose that m_k is the k-th moment and d_k is the k-th derivative of $G(t)$ at $t=0$

(assume they exist); e.g., $d_1 = g(0)$. The stationary-excess cdf $G_e(t)$ associated with $G(t)$ is defined in (2.7). The key property is that the k-th moment m_{ek} and the k-th derivative (at $t=0$) d_{ek} of $G_e(t)$ are expressed in terms of the associated quantities of $G(t)$ by

$$m_{ek} = m_{k+1}/(k+1)m_1, \quad d_{e1} = 1/m_1 \quad \text{and} \quad d_{e(k+1)} = -d_{e1}d_k. \tag{3.2}$$

As a consequence, specifying the first i moments and first j derivatives of $G(t)$ is equivalent to specifying the first $(i-1)$ moments and first $(j+1)$ derivatives of $G_e(t)$. Moreover, the stationary-excess cdf associated with an H_2 cdf is another H_2 cdf with the same exponential parameters. Consequently, we only need to perform three-moment H_2 fits.

3.2. The H_2 formulas

Six H_2 approximations for the complementary busy-period cdf $B^c(t) \equiv 1 - B(t)$ are displayed in Table 3; they come from Table 4 of [5]. In addition to the four H_2 approximations based on derivatives and moments, we also display the simple one-moment exponential fit and the two-moment bounding H_2 distribution consisting of an atom at 0 plus a single exponential (the special case of (3.1) with $\lambda_1 = 0$). In a certain sense, all two-moment H_2 fits fall between these last two special cases; see [50].

It is significant for interpretation and further application, that we obtain all six H_2 approximations as relatively simple closed-form expressions of the traffic intensity ρ (the single M/M/1 parameter). Purely numerical results could be helpful too, but they would not provide nearly as much beyond numerical transform inversion of (2.2) or numerical integration of (2.26).

3.3. Performance for larger times (second regime)

Of course, it remains to describe how these various H_2 approximations perform. For relatively large times (e.g., $t \geqslant 1$ in scaled time), the quality of the approximations improves as we replace derivatives with moments; in particular, for larger times the three-moment H_2 fit is substantially better than the others. Table 4 compares the two leading candidates, (3M) and (1D, 2M), with the exact values obtained by numerical transform inversion in the case $\rho = 0.70$. (Other approximations appearing there will be discussed later.) Tables 5 and 6 display the (2D, 1M) H_2 approximation, along with other approximations to be discussed later, for the cases $\rho = 0.25$ and $\rho = 0.75$. For times away from zero with time scaling, replacing moments with derivatives clearly does not help.

To help put the H_2 three-moment approximation in perspective, we also compare it to HEYMAN'S [26] diffusion approximation for the M/M/1 busy-period distribution in Table 4. This table and other cases show that the new H_2 three-moment approximations offer order-of-magnitude improvements for times away from zero. Heyman's approximation is motivated by GAVER [24] and is related to our inverse Gaussian approximation in Section 6.

Table 4 shows that the three-moment H_2 approximation for the busy-period distribution performs pretty well, but it does not perform nearly as well as previous three-moment H_2 approximations for other M/M/1 quantities; cf. Tables 1-2, 6-7 of [3] and Tables 1-2 of [5]. In particular, the three-moment H_2 approximation deteriorates significantly as ρ increases. This is demonstrated by Table 7, which compares the three-moment approximations to the exact values for the cases $\rho=0.25$, $\rho=0.95$ and $\rho=0.99$. The relative percent errors (100[approx. − exact]/exact) are also given there. From Tables 4 and 7, we see that the (3M) H_2 approximation performs best in the second regime; the performance is not good for times that are very small or very large. (Indeed, it does not have the correct exponential rate as $t\to\infty$.) When ρ gets very high, the (3M) H_2 approximation really does not perform well at all. This observation motivated us to look for explanations and other approximations.

3.4. Performance for smaller times (first regime)

Of course, in the first regime, i.e., for very small time values (e.g., without time scaling), the derivatives do help. With time scaling, we achieve this by considering times of order θ^2, which undoes the time scaling. As indicated by Table 8, for such smaller times, the H_2 approximations based on three derivatives or two derivatives plus one moment perform very well. Moreover, for these times, the (3D) H_2 approximation is uniformly better than the (2D, 1M) approximation. As in Section 3.3, the quality of the approximation degrades somewhat as ρ increases.

In this time range, the relevant comparison is with the power series representation in Section 2.6. However, the three-derivative H_2 approximation performs better than a few terms from the power series. The power-series approximation, keeping terms up to t^3, is (in unscaled time)

$$\begin{aligned} B^c(t) &\approx 1-\frac{m_2}{2\theta}\left[\frac{t}{\theta^2}\right]+\frac{m_3}{12\theta}\left[\frac{t}{\theta^2}\right]^2-\frac{m_4}{144\theta}\left[\frac{t}{\theta^2}\right]^3 \\ &\approx 1-\frac{1}{2}\left[\frac{t}{\theta^2}\right]+\frac{(1+\rho)}{8}\left[\frac{t}{\theta^2}\right]^2-\frac{(1+3\rho+\rho^2)}{48}\left[\frac{t}{\theta^2}\right]^3. \end{aligned} \tag{3.3}$$

Three terms from the power series representation perform well only for times less than θ^2. For very large t, the truncated power series obviously performs very poorly.

When the three-derivative H_2 approximation is expanded in a power series, the first three terms necessarily coincide with the first three terms of the power series representation. In addition, the three-derivative H_2 approximation has approximately the correct shape overall. We thus conclude that the three-derivative H_2 approximation dominates the power-series representation, so that we do not consider the power-series representation further for approximations.

3.5. *Light and heavy traffic*

In Section 2.7 we described the light and heavy traffic behavior of the busy-period distribution. These limits as $\rho \to 0$ and $\rho \to 1$ provide useful insight into the performance of the H_2 approximations. First, since the busy-period distribution and its moments and derivatives converge to those of a simple exponential as $\rho \to 0$, the H_2 approximations are exact at $\rho = 0$ and obviously should tend to perform well as $\rho \to 0$. This explains why the results for $\rho = 0.25$ in Tables 5, 7 and 8 were so good.

On the other hand, the limiting behavior as $\rho \to 1$ is more complicated. Without time scaling, all percentiles of the cdf $B^c(t)$ converge to proper limits (the finite limits associated with $\rho = 1$, which for smaller times are well described by the power series representation in Section 2.5). However, with time scaling, the busy-period distribution converges to a proper limit as $\rho \to 1$ only if we do the additional normalization by θ in (2.21). Moreover, the limit as $\rho \to 1$ is not an H_2 distribution. Consequently, we should not be surprised to encounter difficulties with the H_2 approximation as $\rho \to 1$.

Indeed, the squared coefficient of variation of $B(t)$ given in (2.13) is $(1+\rho)/(1-\rho)$, which explodes as $\rho \to 1$; e.g., $c_B^2 = 39$ for $\rho = 0.95$. As explained in [49,50], high squared coefficients of variation present major problems with moment matching; the range of possible distributions consistent with the specified moments becomes very wide as c_B^2 increases. Thus, to obtain better approximations for high ρ, additional information about the true distribution is needed. In fact, additional information is available via the heavy-traffic limit and will be applied in later sections.

It is interesting to note that the exploding c_B^2 as $\rho \to 1$ in (2.13) for $B(t)$ is in distinct contrast to what occurs for the moment cdf's $H_k(t)$ in [3] and the M/M/1 correlation functions in [5]. For these other M/M/1 quantities, the squared coefficients of variation approached proper limits associated with RBM as $\rho \to 1$. This analysis helps explain why approximating the busy-period distribution is more complicated than approximating other M/M/1 quantities.

3.6. *Padé approximants*

As in Section 5.2 of [1], it is possible to generate an H_2 approximation directly from the Laplace transform $\hat{B}^c(s)$ of the complementary busy-period cdf $B^c(t)$. In particular, we can use a Padé approximant; see BAKER [10] and Section 16.4 of LUKE [36]. For this case, we rationalize $\Psi(s)$ in (2.3) by

$$\Psi(s) \equiv (1+2(1-\theta)s+\theta^2 s^2)^{1/2} \approx \frac{2+(1+2(1-\theta))s+\theta s^2}{2+s} \tag{3.4}$$

to get

$$\hat{B}^c(s) \approx \frac{\theta(2+s)}{2+2s+\theta s^2}, \tag{3.5}$$

which yields the three-moment H_2 fit discussed above.

3.7. The spectral representation

Further insight into the performance of the H_2 approximations is gained by considering the spectral representation, i.e., the explicit representation as a mixture of exponentials given in (2.26). For example, typical values for the case $\rho=0.75$ are given in Table 9. Note that the mixing density $w(y)$ in (2.27) has a sharp peak near the mode at 0.012 and very little mass at values near the upper limit τ_2 (the relaxation time). In this case, there is very little mass above $y=0.50$. On the other hand, the three-moment H_2 fit in this case is

$$B^c(t) \approx 0.933e^{-t/0.07}+0.0670e^{-t/0.93}, \quad t\geq 0. \tag{3.6}$$

From (3.6), we see that the H_2 approximation cannot be asymptotically correct as $t\to\infty$ because its relaxation time 0.93 does not agree with the true value 1.74. On the other hand, 0.93 is in the region above 0.50 where $w(y)$ has little mass. The low mass near τ_2 in $w(y)$ in (2.27) is a clear danger signal for approximation.

In contrast, the mixing densities for the first-moment-function complementary cdf $H_1^c(t)$ and the correlation function $c_q(t)$ studied in [3] and [5] are $(y/\theta)w(y)$ and $(2y^2/\theta)w(y)$, respectively, for $w(y)$ in (2.27). (These are easily obtained from using the stationary-excess operator, as mentioned in Section 2.8.) These other mixing densities clearly have much more mass near the upper end point τ_2, as is shown in Table 9. Indeed, the mixing density for the correlation function $c_q(t)$ is a symmetric beta distribution about $(1+\rho)/2$. Consequently, we should expect the H_2 approximations to perform better for $H_1(t)$ and $c_q(t)$ than for $B^c(t)$, as they do.

4. Asymptotic expansions

In this section we consider approximations associated with the exact asymptotic behavior as $t\to\infty$. The standard asymptotic expansion (three terms) for the complementary busy-period cdf $B^c(t)$ is

$$B^c(t) \sim \tau\theta L(t,\rho)\left[1-\frac{3}{t}\left(\frac{\tau}{2}\right)a_1+\frac{15}{t^2}\left(\frac{\tau}{2}\right)^2 a_2+O\left(\frac{1}{t^3}\right)\right], \tag{4.1}$$

where $\theta=(1-\rho)/2$ as in (2.3), $\tau=(1+\sqrt{\rho})^2/2$ as in (2.9), $a_1=1+\epsilon$, $a_2=1+\epsilon-\epsilon^2/2$, $\epsilon=\theta^2/4\tau\sqrt{\rho}$,

$$L(t,\rho) = (2\pi\rho^{3/2}t^3)^{-1/2}e^{-t/\tau}, \quad t\geq 0, \tag{4.2}$$

and $f(t)\sim g(t)$ means that $f(t)/g(t)\to 1$ as $t\to\infty$. The expansion (4.1) can be obtained directly from the Laplace transform $\hat{B}^c(s)$ in (2.2) by applying Heaviside's theorem; p. 254 of DOETSCH [19] or p. 165 of GNEDENKO and KOVALENKO [25]. Alternatively, it can be obtained from (2.8) using 9.7.1 of [7] plus integration by parts.

However, even with all three terms displayed in (4.1), we get quite poor numerical results for times of primary practical interest. For example, large relative percent errors at medium times ($2\leq t\leq 5$) are shown in Table 10 for the case $\rho=0.5$. (Other approximations yet to be discussed also appear there.)

Furthermore, for $\rho=0.75$ the relative percent error using (4.1) is 11%, 6% and 4% for the large times $t=8$, 10 and 12, respectively. (From Table 1, we see that $B^c(8)=0.000037$, so that $t=8$ is already quite a large time.) Thus, just as in Table 3 of [1] and Table 5 of [3], we find that the standard asymptotic theory related to the relaxation time is not very useful for generating good approximations. (Of course, relaxation times do give at least a rough indication.)

In Corollary 5.2.3 of [4] we produced another asymptotic expansion for $B^c(t)$ related to the heavy-traffic limit in (2.21), which we call the *asymptotic normal approximation.* In particular, in [4] we show that

$$\theta^{-1}B^c(t) \sim \frac{2}{\rho^3/4}\left[t^{-1/2}\phi(\sqrt{2t/\tau})-\sqrt{2/\tau}[1-\Phi(\sqrt{2t/\tau})]\right] \quad \text{as} \quad t\to\infty, \tag{4.3}$$

where again τ is given in (2.9), $\theta=(1-\rho)/2$, Φ is the standard normal cdf and ϕ is its density. Since $\tau\to2$ as $\rho\to1$, the right side of (4.3) approaches the heavy-traffic limit $\tilde{h}_1(t)$ in (2.21) as $\rho\to1$. Moreover, as shown in [4], the two asymptotic expansions for $B^c(t)$ in (4.1) and (4.3) are asymptotically equivalent as $t\to\infty$. In fact, (4.3) was previously developed for the M/M/1-LCFS waiting-time distribution; see p. 109 of RIORDAN [42] and p. 45 of KOSTEN [34].

It turns out that the asymptotic normal approximation for $\theta^{-1}B^c(t)$ in (4.3) performs significantly better than both the standard asymptotic expansion in (4.1) and the heavy-traffic limit $\tilde{h}_1(t)$ in (2.21). First, as indicated above, it coincides with the standard asymptotic expansion as $t\to\infty$ and the heavy-traffic limit as $\rho\to1$. However, (4.3) performs significantly better than the standard asymptotic expansion for $t<\infty$ and the heavy-traffic approximation for $\rho<1$.

Numerical comparisons with exact values are made for the asymptotic normal approximation in Tables 5, 6, 10 and 11. (Inverse Gaussian approximations will be discussed in Section 6.) Tables 5 and 6 make numerical comparisons for the cases $\rho=0.25$ and $\rho=0.75$, while Table 10 makes numerical comparisons for $\rho=0.50$. As should be expected, these tables show that the quality of the approximation improves as t increases for fixed ρ and as ρ increases for fixed t. Table 11 shows the values of t, as a function of ρ, where the relative percent error is less than 1%. For $\rho\geq0.5$, the asymptotic normal approximation is excellent in both the second and third regimes. In these regimes, it clearly dominates all the hyperexponential approximations discussed in Section 3. On the other hand, the asymptotic normal approximation for $B^c(t)$ is not good for small t; e.g., it need not even be a cdf. (An obvious refinement to any approximation for $B^c(t)$ is to replace values greater than 1 by 1.)

It is apparent from Tables 5, 6 and 10 that the asymptotic normal approximation is an upper bound on $B^c(t)$. However, we have yet to prove this.

5. Approximants from Widder's formula

In the general theory of Laplace transforms, there are techniques for generating relatively simple analytic approximate inversions called *approximants.* For example, in Section 3.6 we noted that our three-moment hyperexponential approximation for $B^c(t)$ can be obtained as a Padé approximant. In this section we briefly consider simple approximants obtained from Widder's inversion formula ([52], [30] or Section 3.1 of [17]), which represents a function $f(t)$ as the limit of successive derivatives of its Laplace transform $\hat{f}(s)$, namely,

$$f(t) = \lim_{n\to\infty} \frac{(-1)^n}{n!} \left[\frac{n+1}{t}\right]^{n+1} \hat{f}^{(n)}([n+1]/t), \tag{5.1}$$

where $\hat{f}^{(n)}(s)$ is the n-th derivative of $\hat{f}(s)$.

Formula (5.1) is not attractive for direct numerical inversion because it involves repeated differentiation, which leads to numerical instabilities, but it can be used to generate approximants. According to Davies and Martin [17], ter Haar [45] was the first to do this (in 1951), proposing essentially the 0-th term from (5.1),

$$f_0(t) \equiv t^{-1}\hat{f}(t^{-1}), \quad t \geqslant 0. \tag{5.2}$$

Jagerman [30,31] also proposed the ter Haar approximant and made a significant enhancement. In particular, Jagerman [30] observed that for a log-convex function we can determine the relaxation time constant from the singularity in the transform, so that we can force the approximant to have an exponential term with this time constant. The resulting Jagerman approximant is

$$f_\tau(t) \equiv \frac{e^{-t/\tau}}{(1-t/\tau)} f_0\left[\frac{t}{1-t/\tau}\right], \quad t \geqslant 0, \tag{5.3}$$

where $f_0(t)$ is the ter Haar approximant in (5.2) and τ is the relaxation time constant.

The ter Haar approximant (5.2) and the Jagerman approximant (5.3) are particularly appealing candidates because, as Jagerman [30, 31] shows, they inherit many of the structural properties of the original function $f(t)$. Thus, these approximants are clearly in the spirit of the simple closed-form expressions we are seeking.

Since the busy-period complementary cdf $B^c(t)$ is completely monotone, it is log-convex. Hence, it is natural to consider both the ter Haar and Jagerman approximants. In our time scale, the ter Haar approximant for $B^c(t)$ is

$$B_0^c(t) = \frac{2\theta}{\theta + t + \sqrt{\theta^2 + 2(1-\theta)t + t^2}}, \quad t \geqslant 0, \tag{5.4}$$

and the Jagerman approximant is

$$B_\alpha^c(t) = e^{-t/\tau} \frac{2\theta}{\theta + (1-\theta/\tau)t + \sqrt{\theta^2 + 2\sqrt{\rho}\, t}}, \quad t \geqslant 0, \tag{5.5}$$

where τ is the relaxation time in (2.9).

Even though the ter Haar approximant (5.4) inherits several properties of $B^c(t)$; e.g., $B_0^c(0)=1$, $B_0^c(\infty)=0$, and $B_0^c(t)$ is monotone and convex, the numerical results are not good, as can be seen from Table 4. Indeed it is obvious that (5.4) does not nearly decay quickly enough, because the exponential term is missing.

We might expect much better performance from (5.5) because it has the correct asymptotic exponential decay rate. Indeed, as $\rho\to 0$, (5.5) agrees with the exact result $B^c(t)=e^{-2t}$. However, (5.5) is not asymptotically correct as $t\to\infty$, i.e., we do *not* have $B^c(t)\sim B_\alpha^c(t)$ as $t\to\infty$. As $t\to\infty$, $B_\alpha^c(t)\sim Kt^{-1}e^{-t/\tau}$; there is a t in the denominator instead of the $t^{3/2}$ in (4.1) and (4.2), so $B_\alpha^c(t)$ also does not decay quickly enough, as Table 4 shows. Indeed (5.5) is certainly much better than (5.4), but (5.5) typically becomes greater than the standard asymptotic expansion for $2\leq t\leq 3$ (with time scaling). Thus, neither approximant performs well in the second and third regimes. In fact, even though (5.5) is asymptotically exact as $\rho\to 0$, it does not even perform well in the second regime for very small ρ such as $\rho=0.1$. The reason is that the order of the two limits $t\to\infty$ and $\rho\to 0$ matters.

From Sections 3 and 4 here plus our previous work [1-5], we know that for times of primary interest the actual exponential rate is substantially greater than the asymptotic exponential rate. Hence, we are motivated to force a bigger rate, as arises in the H_2 approximations in Section 3. Unfortunately, however, the Jagerman enhancement in [30] permits any smaller rate, but not any larger rate, so that we cannot improve the approximation in that way.

One approach to the poor performance of these approximants is to replace them by higher-order terms from the sequence (5.1), but we do not pursue this goal. With higher order terms, Theorem 8 of [30] can also be applied in various ways to match moments, but of course the resulting approximant gets more complicated. It is good to be aware that improved approximants by this general approach are possible, though. Indeed, JAGERMAN [30,31] uses (5.1) as a basis for creating a full numerical inversion procedure. To avoid the troublesome differentiation in (5.1), Jagerman applies complex analysis to obtain a Fourier-series type inversion procedure from (5.1). Since such higher-order approximants are not closed-form, they are perhaps best compared to other numerical inversion schemes and numerical integration. From Tables III-VI of [31], it is apparent that the higher-order approximants for $B^c(t)$ do not get accurate very quickly. It appears that $B^c(t)$ is a relatively difficult function to approximate by this general procedure.

6. Inverse Gaussian approximations

Inverse Gaussian approximations for the busy-period distribution arise naturally from diffusion approximations.

6.1. Heavy-traffic limits and RBM approximations

First, the inverse Gaussian (IG) cdf $F(t;x,0)$ in (2.23) is the cdf of the time required for canonical RBM to first reach 0 starting in x. Thus an IG cdf is a natural approximation for the cdf of a first passage time downward in a queue when we are using a diffusion (RBM) approximation for the queueing process. Indeed, an IG cdf arises as the heavy-traffic limit for queueing first passage times downward as $\rho \to 1$, as indicated in the proof of Corollary 3.4.1 of [4]. In particular, consider the time-scaled M/M/1 queue and let $T_n(\rho)$ be the first-passage time from state n to 0, as a function of ρ. (Of course, $T_n(\rho)$ is the sum of n independent busy periods.) Let $T_x(1)$ be the first-passage time from x to 0 in RBM (the M/M/1 queue with $\rho=1$) with IG cdf $F(t;x,0)$ in (2.23). Let $\Rightarrow$ denote convergence in distribution, as in BILLINGSLEY [11]. Then the standard heavy-traffic limit for $T_n(\rho)$ as $\rho \to 1$, obtained by applying the continuous mapping theorem (Theorem 5.1 of [11]) with the usual heavy-traffic limit for the queue-length process [29], is

$$T_{[x\theta^{-1}]}(\rho) \Rightarrow T_x(1) \quad \text{as} \quad \rho \to 1, \tag{6.1}$$

where $[x]$ is the greatest integer less than or equal to x. It is significant that essentially the same result holds for GI/G/m queues and many other models, so that the limit (6.1) suggests IG approximations for more general systems.

A difficulty with (6.1) for our purposes is that the starting state in the M/M/1 system with traffic intensity ρ is $[x\theta^{-1}]$, where

$$[x\theta^{-1}] = [2x/(1-\rho)] \to \infty \text{ as } \rho \to 1. \tag{6.2}$$

However, we can ignore difficulty (6.2), and suppose that

$$T_{[x\theta^{-1}]}(\rho) \approx T_x(1) \tag{6.3}$$

for fixed $x>0$ and $\rho<1$. We then set $x\theta^{-1}=1$ to make the left side equal to the desired $T_1(\rho)$; i.e., we approximate by

$$T_1(\rho) \approx T_\theta(1), \tag{6.4}$$

i.e., the first-passage time for canonical RBM starting in θ. We will come back to this later.

Instead of (6.1) and (6.4), we would probably prefer a limit directly for $T_1(\rho)$ as $\rho \to 1$. Fortunately, such a limit has been obtained, as indicated in Section 2.7, but the limit is not in the usual form, involving $[T_1(\rho)-a(\rho)]/b(\rho)$ for some functions of ρ, $a(\rho)$ and $b(\rho)$. In particular, (2.21)-(2.23) states that

$$\theta^{-1}B^c(t) = \frac{2}{1-\rho}P(T_1(\rho)>t) \to \tilde{h}_1(t) = \int_0^\infty 2e^{-2x} f(t;x,0)\,dx; \tag{6.5}$$

i.e., the limit is a mixture of IG *densities*, which seems to be not an especially convenient form.

6.2. Heuristic diffusion approximation

Diffusion process approximations can also be developed by 'direct fits,' without considering any heavy traffic limits; this is the approach of HEYMAN [26] and DUDA [20,21] for example. The idea is to apply some of the logic of the queue and some of the logic of the diffusion process. For example, Heyman works with the virtual waiting time process and approximates it by RBM. Since he works with the virtual waiting time process, the logic of the queue dictates that the initial level at the beginning of a busy period be the random amount determined by the service-time distribution. Thereafter, the process is assumed to evolve as RBM. Thus the approximation for the busy-period cdf is a mixture of IG cdf's, as in (6.5).

On the other hand, if we work with the queue-length process, as we do here, then with this procedure we would approximate the queue-length process by RBM but not randomize the initial position. We would thus approximate the busy-period cdf by the cdf of an IG distribution, without any randomization. Of course, with this procedure we still must specify the drift and diffusion coefficients of RBM and the initial state for the first passage time. The end result might then be (6.4). We mention that Heyman found that his approximation for the M/G/1 busy-period distribution worked best when the service-time distribution is deterministic, i.e., when no randomization was performed. As indicated in Table 4, Heyman's M/M/1 approximation does not perform very well. As discussed in [48], there seems to be a limit to what you can obtain by the diffusion logic alone. It is important to consider refinements based on additional properties such as moments.

6.3. Direct inverse Gaussian approximations

We propose using the heavy-traffic limit to motivate considering the IG cdf as an approximation for the busy-period cdf, but fitting the IG parameters directly. First, we note that there is a one-to-one correspondence between a two-parameter IG cdf and a two-parameter RBM representation. The first-passage time for RBM with drift coefficient $\mu<0$ and variance coefficient ν starting in state x has the complementary IG cdf

$$G^c(t) = \Phi^c\left[\frac{-\mu t - x}{\sqrt{\nu t}}\right] - e^{-\frac{2\mu x}{\nu}}\Phi^c\left[\frac{-\mu t + x}{\sqrt{\nu t}}\right], \quad t \geqslant 0. \tag{6.6}$$

However, there are only two basic parameters in (6.6). We can set $\mu=-1$ without loss of generality, and do, because the parameter triples (μ, ν, x) and $(-1, \nu/\mu^2, x/\mu)$ yield identical distributions in (6.6), as is easy to see. (We can also apply Proposition 2.1 of [1] for this.) The IG distribution with parameters $\mu=-1$, ν and x has moments

$$\begin{aligned} m_1 &= x, \quad m_2 = x(x+\nu), \quad m_3 = 3x\nu(x+\nu)+x^3, \\ m_{n+1} &= \sum_{i=0}^{n} \frac{(n+i)!x^{n+1-i}\nu^i}{(n-i)!2^i i!}; \end{aligned} \tag{6.7}$$

see p. 366 (12) of TWEEDIE [46].

A specific IG approximation is obtained by choosing x and v in (6.6). We choose x and v by matching moments to those of $B(t)$ in Section 2.4 and possibly by matching other properties. First, to match the mean, we set $x=\theta=(1-\rho)/2$. We then consider three cases of v:

$$\begin{aligned}
&\text{(i)} && v=1 \text{ (canonical RBM)} \\
&\text{(ii)} && v=1-\theta \text{ (all moments match up to order } \theta^3) \\
&\text{(iii)} && v=\frac{\tau}{2}\approx 1-\theta-\frac{\theta^2}{4}-\frac{\theta^3}{4} \text{ (proper exponential rate as } t\to\infty).
\end{aligned} \tag{6.8}$$

The first case in (6.8) with $v=1$ is just the heavy-traffic approximation (6.4) involving canonical RBM derived from (6.1). The second case, $v=1-\theta$, is obtained by matching the second moment, $\theta(\theta+v)=\theta$. From (6.5), we see that in this second case

$$m_{n+1} = (2n-1)!!(\theta-(n-1)\theta^2)+O(\theta^3), \tag{6.9}$$

where $(2n-1)!!=(2n-1)(2n-3)\cdots(3)1$, which together with (2.11) implies that *all* moments in (6.7) match the moments of $B(t)$ in Section 2.5 up to order $O(\theta^3)$ when $v=1-\theta$. Indeed, we can obtain the whole IG approximation directly from the moments of the busy period in this way. In particular, we can represent each busy-period moment by a power series in θ and look for a distribution that matches the moment sequence obtained by keeping the leading terms (matching terms up to order θ^3). From Section 2.5 and (6.7), we find that the IG distribution with $v=1-\theta$ solves this asymptotic moment problem. We thus should expect case (ii) to perform very well for higher traffic intensities, which it does.

The third case in (6.8), $v=\tau/2$, is just what is needed to give the IG distribution the proper exponential rate, as we would expect from (4.1) and (4.3); i.e., from (6.6) and (4.1), we easily obtain that

$$G^c(t) \sim A(\theta)B^c(t) \quad \text{as} \quad t\to\infty, \tag{6.10}$$

where

$$\begin{aligned}
A(\theta) &= \frac{(\tau/2)^{-1/2}e^{2\theta/\tau}}{\rho^{-3/4}} \\
&= \frac{\rho^{3/4}(1+\sqrt{\rho})^2e^{2(1-\rho)/(1+\sqrt{\rho})^2}}{4} \to 1 \quad \text{as} \quad \rho\to 1.
\end{aligned} \tag{6.11}$$

The third case also matches all moments up to order $O(\theta^2)$.

In fact, the corresponding densities also have the almost asymptotic property (6.10). Indeed all the approximations for the complementary cdf $B^c(t)$ yield corresponding approximations for the density $b(t)$ by differentiating, but in general the quality of approximations can deteriorate drastically upon differentiation. However, the IG density $g(t)$ in case (iii) of (6.8) is also good for $b(t)$. Indeed,

$$g(t) \sim A(\theta)b(t) \quad \text{as} \quad t\to\infty, \tag{6.12}$$

for $A(\theta)$ in (6.11). In fact, we can make the IG approximations for $b(t)$ and $B^c(t)$ asymptotically correct by dividing $g(t)$ and $G^c(t)$ by $A(\theta)$ in (6.11). However, then $g(t)$ is no longer a probability density function and $G^c(t)$ is no longer a complementary cdf. This modification gives the best numbers for t not too small, though; see Section 6.5 below.

It is furthermore significant that all three IG cdf's are asymptotically correct in heavy traffic, in the strong sense that

$$\lim_{\theta\to 0} \theta^{-1}[B^c(t)-G^c(t)] = 0, \tag{6.13}$$

i.e., in addition to the heavy-traffic limit theorem for $B^c(t)$ in (2.21),

$$\lim_{\theta\to 0} \theta^{-1}G^c(t) = \tilde{h}_1(t). \tag{6.14}$$

Indeed, the limit is true for the density

$$\lim_{\theta\to 0} \theta^{-1}g(t) = \tilde{h}_1{}'(t)=-L(t, 1) \tag{6.15}$$

for $L(t, \rho)$ in (4.2). The limit in (6.15) is easily established by differentiating (6.6) and taking the limit.

We remark that we do not yet have any direct probabilistic explanation why we should change the variance coefficient ν to the second or third candidate in (6.8) in the context of RBM approximations for the queue.

6.4. Numerical comparisons

Numerical comparisons of the IG approximations with exact values for the M/M/1 complementary busy period cdf $B^c(t)$ and other approximations appear in Tables 5, 6, 10, 11 and 12. Table 10 compares the three IG approximations with the asymptotic approximations in Section 4 for the case $\rho=0.50$. The relative percent errors are displayed in Table 10.

From Tables 4 and 10, we see that for the first IG approximation in (6.8) involving canonical RBM with $\nu=1$ performs better than Heyman's diffusion approximation using the random initial conditions, but not quite as well as the three-moment H_2 approximation. More importantly, this first IG approximation is clearly dominated by the other two IG approximations with $\nu=1-\theta$ and $\nu=\tau/2$. The second and third IG approximations perform very well except for very small t. There is relatively little difference between these two approximations when ρ is not too small (e.g., $\rho\geqslant 0.50$) because

$$\tau/2=(1+\sqrt{\rho})^2/4 = (1+\rho+2\sqrt{\rho})/4 \quad\text{and}\quad 1-\theta=(1+\rho)/2. \tag{6.16}$$

The third IG approximation may be preferred because it is nearly asymptotically correct as $t\to\infty$; it is good for the third regime as well as the second regime.

Tables 5 and 6 show that the asymptotic normal approximation in (4.3) is better than the second IG approximation with $\nu=1-\theta$ for very large t, but the IG approximations are better for small t. Tables 5 and 6 also show that the quality of the IG approximation with $\nu=1-\theta$ improves with increasing ρ; it is much better for $\rho=0.75$ than at $\rho=0.25$. Table 11 gives the range for which the relative percent error is less than one percent. For $\rho \geqslant 0.5$, the second and third IG approximations perform very well in the second regime; the third IG approximation also performs well in the third regime.

We have seen that the IG distribution is not a good approximation for very small times. This should not be surprising because the heavy-traffic limit (6.1) focuses on relatively large times. Indeed, the IG density $f(t;x,0)$ is 0 and increasing at $t=0$, while the true M/M/1 busy-period density is positive and strictly decreasing. However, the IG distribution does describe the busy-period distribution remarkably well when we consider larger times.

6.5. *Density approximations*

From (2.21) and (6.12), we should expect that the IG *density,* using (iii) in (6.8), is a good approximation for the M/M/1 busy period *density* $b(t)$ for t sufficiently large, and indeed it is.

However, we have just noted that $g(0)=0$, while $b(0)=1/2\theta^2$. In fact, all derivatives of g at 0 are 0. In contrast, $b'(0)=-(1-\theta)/2\theta^4$; see (2.32). Hence, for small t the IG approximation is very bad. It is natural to ask how large t must be for this bad initial match to disappear. This is partly answered by properties of the IG density. From p. 365 of TWEEDIE [46] we see that the IG density is unimodal with mode at (in our time scale)

$$t_{\text{mode}} = \theta\left\{\left[1+\frac{9\tau^2}{16\theta^2}\right]^{1/2} - \frac{3\tau}{4\theta}\right\} \approx \frac{2\theta^2}{3\tau} \approx \frac{\theta^2}{3}, \tag{6.17}$$

so that the initial period where the IG density has the wrong slope is brief. In fact, numerical evidence indicates that the IG density $g(t)$ is a good approximation for $t \geqslant 2\theta^2$, i.e., after about one service time. Comparisons between the IG density $g(t)$ and the busy-period density $b(t)$ for the cases $\rho=0.50$ and 0.75 appear in Tables 13 and 14.

The *relative percent error,*

$$\text{RE} \equiv \frac{g(t)-b(t)}{b(t)} \times 100, \tag{6.18}$$

is also given in Tables 13 and 14. Interestingly, the relative percent error quite rapidly approaches the *asymptotic relative percent error,*

$$\text{ARE} \equiv \lim_{t\to\infty} \text{RE} = 100(A(\theta)-1) \tag{6.19}$$

where $A(\theta)$ is given in (6.11). Values of the ARE appear in Table 15. The IG approximation in (iii) of (6.8) truly does exceptionally well as long as ρ and t are not too small.

7. A COMPOSITE APPROXIMATION

None of the approximations considered so far performs well for all t and ρ. However, it is easy to form a composite approximation that does. For this purpose, we suggest using the third inverse Gaussian (IG) approximation with

$\nu=\tau/2=(1+\rho+2\sqrt{\rho})/4$ from Section 6 when $t \geq t^*(\rho)$ and the three-derivative (3D) H_2 approximation from Section 3 when $t < t^*(\rho)$, where

$$t^*(\rho) = \frac{3\theta^2}{2\rho} = \frac{3(1-\rho)^2}{8\rho}. \tag{7.1}$$

From our previous analysis, we see that the composite IG-3D approximation is asymptotically exact as $\rho \to 0$, $\rho \to 1$, and as $t \to 0$. By (6.10)-(6.12), it is nearly asymptotically exact as $t \to \infty$, and can be modified to be asymptotically exact, as indicated in Section 6.

Numerical comparisons supporting the composite IG-3D approximation appear in Table 12. The relative percent error is given for the cases $\rho=0.25$, 0.50, 0.75 and 0.85. For $\rho \geq 0.50$, the relative percent error is evidently never greater than 2.8%.

8. Extensions to the GI/G/1 model

The principal approximations considered in this paper apply equally well to the M/G/1 queue with general service-time distribution, because the moments and derivatives of the M/G/1 busy-period cdf $B(t)$ are readily available. In particular, for the M/G/1 queue we propose the IG approximation in Section 6 based on a two-moment match for the second regime and the IG-3D composite in Section 7 overall.

With the identical time scaling, the first two moments of the M/G/1 busy period are $m_1=\theta$ and $m_2=\theta(c_s^2+1)/2$, where c_s^2 is the squared coefficient of variation (variance divided by the square of the mean) of the service-time distribution. The resulting IG approximation obtained from matching the first two moments is thus (6.6) with $x=\theta$, just as for M/M/1, and

$$\nu = \frac{c_s^2+1}{2} - \theta = \frac{c_s^2+\rho}{2}. \tag{8.1}$$

In fact, however, for the GI/G/1 queue we would scale time by $(c_a^2+c_s^2)/(1-\rho)^2$, where c_a^2 is the squared coefficient of variation of the interarrival time. This is the proper scaling to yield canonical RBM in the limit as $\rho \to 1$; see [29] and Section 2 of [1]. For the M/G/1 queue, with time scaled by $(1+c_s^2)/(1-\rho)^2$, the first two moments of the busy period again coincide, yielding

$$m_1 = m_2 = \frac{1-\rho}{1+c_s^2}. \tag{8.2}$$

With this time scaling and the same approximation procedure, (ii) in (6.8), we thus set $x=x(x+\nu)=m_1$ to obtain the IG approximation (6.6) with

$$x = \frac{1-\rho}{1+c_s^2} \quad \text{and} \quad \nu = 1-x. \tag{8.3}$$

For GI/G/1 with the time scaling $(c_a^2+c_s^2)/(1-\rho)^2$, we do not have the busy-period moments in such a convenient form. Nevertheless, it is natural to consider the IG approximation in (6.6) with

$$x = \frac{1-\rho}{c_a^2+c_s^2} \quad \text{and} \quad \nu = 1-x, \tag{8.4}$$

provided that $x<1$. Note that (8.4) reduces to (8.3) when $c_a^2=1$ and (8.3) reduces to the second M/M/1 case in (6.8) with $\nu=1-\theta$ when $c_s^2=1$. Also note that this IG approximation differs from what appears in (6.4) and DUDA [20,21] because $\nu<1$ in (8.4). Table 10 shows the improvement for the M/M/1 case. We intend to discuss approximations for the GI/G/1 busy-period distribution and make numerical comparisons in a subsequent paper.

9. RELATED QUANTITIES OF INTEREST

Approximations for the M/M/1 complementary busy-period cdf $B^c(t)$ immediately translate into approximations for the M/M/1-LCFS waiting-time cdf due to [41,42] and the M/M/1 probability of emptiness $P_{00}(t)$ due to (2.5). From the probability of emptiness, we then get the expected cumulative idle time $EI_0(t)$ and the mean queue length $EQ_0(t)$ via (2.6). Equivalently, from [3] we know that the normalized mean queue length $H_1(t)$ is just

$$H_1(t) = B_e(t) = (1-\rho)\int_0^t B^c(u)du, \quad t\geq 0. \tag{9.1}$$

Consequently, any approximation for $B^c(t)$ translates into an approximation for $H_1(t)$ through an integration. However, the relatively simple direct H_2 approximation for $H_1(t)$ developed in [3] still seems preferable. (Direct H_2 approximations are compared to H_2 approximations obtained from integration in [5].)

The approximation for $H_1(t)$ in [3] then can be used to obtain an approximation for $EI_0(t)$ via (2.6), namely,

$$EI_0(t) = (1-\rho)t+EQ_0(t) = (1-\rho)t+\frac{\rho}{1-\rho}+\frac{\rho}{1-\rho}H_1^c(t). \tag{9.2}$$

In fact, KUMAR and WONG [35] propose the linear portion of (9.2) as an approximation (and bound) for $EI_0(t)$; i.e., they use (9.2) with $H_1^c(t)\approx 0$, and show that it captures the essential behavior. This linear approximation can also be motivated by the fact (Theorem 9.3 of [4]), that $EI_0(t)$ coincides with a constant (the mean service time) multiplied by the renewal function associated with the busy-period distribution. Renewal functions are well-known to be asymptotically linear.

10. CONCLUSIONS

We have investigated several closed-form approximations for the M/M/1 complementary busy-period cdf $B^c(t)$. For larger times, the best approximations are the asymptotic normal approximation in (4.3) and the inverse Gaussian (IG) distribution in (6.6) with $\nu=1-\theta$ or $\nu=\tau/2$ (cases (ii) and (iii) in (6.8)). The light-and-heavy-traffic interpolation in (2.25) also performs well. For small times, the three-derivative (3D) hyperexponential (H_2) approximation in Table 3 performs very well. For very small times the power series representation in Section 2.6 is also good. A single composite approximation based on (3D-H_2) and (IG) is proposed in Section 7.

The three-moment (3M) H_2 approximations, which worked so well in [1-5], do not perform as well for the M/M/1 busy period (Section 3). Simple diffusion approximations as proposed by HEYMAN [26] and DUDA [20,21] do not perform especially well, as shown in Table 4, but our proposed inverse Gaussian approximations build on these diffusion approximations. Approximants based on Widder's formula (Section 5) also do not perform very well here, but these are very general procedures, which in some circumstances are excellent.

The M/M/1 complementary busy-period cdf has proven to be difficult to approximate. As illustrated by the relatively poor performance of the (3M) H_2 approximations, it is not enough to match three moments. As illustrated by the poor performance of the three-term asymptotic expansion in (4.1), it is not nearly enough to capture the asymptotic behavior as $t \to \infty$. As illustrated by the relatively poor performance of unrefined diffusion approximations in [20,21,26], it is not enough to consider the standard heavy-traffic limits and diffusion approximations. However, by combining several of these properties, it is possible to obtain quite good approximations. The positive approximation results provide a basis for treating busy periods in more general models (Section 8) and other M/M/1 quantities (Section 9).

ACKNOWLEDGMENT

We thank D. L. Jagerman, A. Kumar and W. S. Wong for helpful discussions about their work and for suggesting that we consider approximants based on Widder's formula (Section 5).

REFERENCES

1. J. ABATE, W. WHITT (1987). Transient behavior of regulated Brownian motion, I: starting at the origin. *Adv. in Appl. Probab. 19,* 560-598.
2. J. ABATE, W. WHITT (1987). Transient behavior of regulated Brownian motion, II: non-zero initial conditions. *Adv. in Appl. Probab. 19,* 599-631.
3. J. ABATE, W. WHITT (1987). Transient behavior of the M/M/1 queue starting at the origin. *Queueing Systems 2,* 41-65.
4. J. ABATE, W. WHITT (1988). Transient behavior of the M/M/1 queue via Laplace transforms. *Adv. in Appl. Probab. 20,* to appear.
5. J. ABATE, W. WHITT (1988). The correlation functions of RBM and M/M/1. *Stochastic Models,* to appear.
6. J. ABATE, W. WHITT (1989). Simple spectral representations for the M/M/1 queue. *Queueing Systems,* to appear.
7. M. ABRAMOWITZ, I.A. STEGUN (eds.) (1982). *Handbook of Mathematical Functions,* Dover, New York.
8. N.T.J. BAILEY (1954). A continuous time treatment of a simple queue using generating functions. *J. Roy. Statist. Soc. B 16,* 288-291.
9. N.T.J. BAILEY (1957). Some further results in the non-equilibrium theory of a simple queue. *J. Roy. Statist. Soc. B 19,* 326-333.

10. G.A. BAKER, JR. (1975). *Essentials of Padé Approximants,* Academic Press, New York.
11. P. BILLINGSLEY (1968). *Convergence of Probability Measures,* Wiley, New York.
12. O.J. BOXMA, J.W. COHEN, N. HUFFELS (1979). Approximations of the mean waiting time in an M/G/s queueing system. *Oper. Res. 27,* 1115-1127.
13. G.A. CAMPBELL, R.M. FOSTER (1948). *Fourier Integrals for Practical Application,* Van Nostrand, New York.
14. G.M. CLARK (1981). Use of Polya distributions in approximate solution to nonstationary M/M/s queues. *Commun. ACM 24,* 206-217.
15. J.W. COHEN (1982). *The Single Server Queue,* 2nd ed., North-Holland, Amsterdam.
16. J.W. COHEN, G. HOOGHIEMSTRA (1981). Brownian excursion, the M/M/1 queue and their occupation times. *Math. Oper. Res. 6,* 608-629.
17. B. DAVIES, B. MARTIN (1979). Numerical inversion of the Laplace transform. *J. Comp. Phys. 33,* 1-32.
18. L.E.N. DELBROUCK (1976). Approximations for certain congestion functions in single server queueing systems. *Eighth Int. Teletraffic Congress, 233-1-5.*
19. G. DOETSCH (1974). *Introduction to the Theory and Application of the Laplace Transformation,* Springer-Verlag, New York.
20. A. DUDA (1983). Transient diffusion approximation for some queueing systems. *Perf. Eval. Review 12,* 118-128.
21. A. DUDA (1986). Diffusion approximations for time-dependent queueing systems. *IEEE J. Sel. Areas Commun. SAC-4,* 905-918.
22. S. ERLANDER (1965). The remaining busy period for a single server queue with Poisson input. *Oper. Res. 13,* 734-746.
23. D.P. GAVER, JR. (1966). Observing stochastic processes and approximate transform inversion. *Oper. Res. 14,* 444-459.
24. D.P. GAVER, JR. (1968). Diffusion approximation and models for certain congestion problems. *J. Appl. Probab. 5,* 607-623.
25. B.V. GNEDENKO, I.N. KOVALENKO (1968). *Introduction to Queueing Theory,* Israel Scientific Translations Ltd., Jerusalem.
26. D.P. HEYMAN (1974). An approximation for the busy period of the M/G/1 queue using a diffusion model. *J. Appl. Probab. 11,* 159-169.
27. D.P. HEYMAN, M.J. SOBEL (1982). *Stochastic Models in Operations Research,* Vol. I, McGraw Hill, New York.
28. P. HOKSTAD (1979). On the relationship of the transient behavior of a general queueing model to its idle and busy period distributions. *Math. Operationsforsch. Statist. Ser. Optimization 10,* 421-429.
29. D.L. IGLEHART, W. WHITT (1970). Multiple channel queues in heavy traffic, II: sequences, networks and batches. *Adv. in Appl. Probab. 2,* 355-369.
30. D.L. JAGERMAN (1978). An inversion technique for the Laplace transform with applications to approximation. *The Bell System Tech. J. 57,* 669-710.

31. D.L. JAGERMAN (1982). An inversion technique for the Laplace transform. *The Bell System Tech. J. 61,* 1995-2002.
32. S. KARLIN, J. MCGREGOR (1958). Many server queueing processes with Poisson input and exponential service times. *Pacific J. Math. 8,* 87-118.
33. D.G. KENDALL (1951). Some problems in the theory of queues. *J. Roy. Statist. Soc. B 13,* 151-171.
34. L. KOSTEN (1973). *Stochastic Theory of Service Systems,* Pergamon Press, New York.
35. A. KUMAR, W.S. WONG (1987). *Some Mean Value Formulas for the Transient M/M/1 Queue,* unpublished paper, AT&T Bell Laboratories, Holmdel, NJ.
36. Y.L. LUKE (1969). *The Special Functions and their Approximants,* Academic Press, New York.
37. R.E. NANCE, U.N. BHAT, B.G. CLAYBROOK (1972). Busy period analysis of a time-sharing system: transform inversion. *J. Assoc. Comput. Mach. 19,* 453-462.
38. G.F. NEWELL (1982). *Applications of Queueing Theory,* 2nd ed., Chapman and Hall, London.
39. A.R. ODONI, E. ROTH (1983). An empirical investigation of the transient behavior of stationary queueing systems. *Oper. Res. 31,* 432-455.
40. S.O. RICE (1962). Single server systems — II. Busy periods. *The Bell System Tech. J. 41,* 279-310.
41. J. RIORDAN (1961). Delays for last-come first-served service and the busy period. *The Bell System Tech. J. 40,* 785-793.
42. J. RIORDAN (1962). *Stochastic Service Systems,* Wiley, New York.
43. H. STEHFEST (1970). Algorithm 368. Numerical inversion of Laplace transforms. *Commun. ACM 13,* 47-49 [erratum *13,* 624].
44. L. TAKÁCS (1967). *Combinatorial Methods in the Theory of Stochastic Processes,* Wiley, New York.
45. D. TER HAAR (1951). An easy approximate method of determining the relaxation spectrum of a viscoelastic material. *J. Polymer Sci. 6,* 247-250.
46. M.C.K. TWEEDIE (1957). Statistical properties of inverse Gaussian distributions I. *Ann. Math. Statist. 28,* 362-377.
47. E. VAULOT (1954). Délais d'attente des appels téléphoniques dans l'ordre inverse de leur arrivée. *C. R. Acad. Sci.* (Paris) *238,* 1188-1189.
48. W. WHITT (1982). Refining diffusion approximations for queues. *Oper. Res. Letters 1,* 165-169.
49. W. WHITT (1984). On approximations for queues, I: extremal distributions. *AT&T Bell Lab. Tech. J. 63,* 115-138.
50. W. WHITT (1984). On approximations for queues, III: mixtures of exponential distributions. *AT&T Bell Lab. Tech. J. 63,* 163-175.
51. W. WHITT (1985). The renewal-process stationary-excess operator. *J. Appl. Probab. 22,* 156-167.
52. D.V. WIDDER (1934). The inversion of the Laplace integral and the

related moment problem. *Trans. Amer. Math. Soc. 36,* 107-200.
53. C.M. Woodside, B. Pagurek, G.F. Newell (1980). A diffusion approximation for correlation in queues. *J. Appl. Probab. 17,* 1033-1047.

Appendix A

time	values of the traffic intensity							
t	$\rho = 0.00$	$\rho = 0.10$	$\rho = 0.25$	$\rho = 0.50$	$\rho = 0.75$	$\rho = 0.85$	$\rho = 0.95$	$\rho = 0.99$
$\theta^2/2$	.7788	.7812	.7848	.7905	.7961	.7983	.8004	.8012
θ^2	.6065	.6140	.6249	.6422	.6584	.6646	.6707	.6731
0.10	.8187	.7836	.7116	.5158	.2394	.1356	.0425	.00829
0.25	.6065	.5495	.4501	.2676	.1143	.0646	.0203	.00398
0.50	.3679	.3128	.2344	.1303	.0563	.0321	.0102	.00201
0.75	.2231	.1837	.1348	.0764	.0340	.0196	.00628	.00124
1.00	.1353	.1109	.0827	.0489	.0224	.0130	.00422	.00084
1.50	.0498	.0432	.0352	.0229	.0112	.00664	.00219	.00044
2.00	.0183	.0180	.0165	.0119	.00619	.00374	.00125	.00025
3.00	.00248	.00358	.00423	.00378	.00221	.00139	.000480	.000097
4.00	.000335	.000796	.00122	.00135	.000888	.000577	.000206	.000042
5.00	.000045	.000189	.000380	.000514	.000381	.000256	.000094	.000020
6.00	.000006	.000047	.000123	.000205	.000171	.000119	.000045	.000009
8.00	.000000	.000003	.000014	.000036	.000037	.000028	.000011	.000002

TABLE 1. *Numerical values of the time-scaled M/M/1 complementary busy-period cdf $B^c(t)$ obtained by Laplace transform inversion of (2.2).*

Part I: exact values of $\theta^{-1} B^c(t)$								
time	values of the traffic intensity							
t	$\rho = 0.00$	$\rho = 0.10$	$\rho = 0.25$	$\rho = 0.50$	$\rho = 0.75$	$\rho = 0.85$	$\rho = 0.95$	$\rho = 0.99$
0.10	1.64	1.74	1.90	2.06	1.92	1.81	1.70	1.66
0.25	1.21	1.22	1.20	1.07	0.91	0.86	0.81	0.80
0.50	0.74	0.70	0.63	0.52	0.45	0.43	0.41	0.40
0.75	0.44	0.41	0.36	0.31	0.27	0.26	0.25	0.25
1.00	0.27	0.25	0.22	0.20	0.18	0.17	0.17	0.16
1.50	0.100	0.096	0.094	0.092	0.089	0.089	0.088	0.088
2.00	0.036	0.040	0.044	0.048	0.050	0.050	0.050	0.050
3.00	0.0049	0.008	0.011	0.015	0.018	0.019	0.019	0.020
4.00	0.00067	0.002	0.003	0.005	0.007	0.008	0.008	0.008
5.00	0.00009	0.00044	0.0010	0.0020	0.0030	0.0035	0.004	0.004
6.00	0.000012	0.00011	0.00035	0.0008	0.0014	0.0016	0.002	0.002
8.00	0.00000	0.000007	0.00004	0.00014	0.0003	0.0004	0.0005	0.0005

Part II: approximate values of $\theta^{-1} B^c(t)$					
time	values of the traffic intensity				
t	$\rho = 0.10$	$\rho = 0.25$	$\rho = 0.50$	$\rho = 0.75$	$\rho = 0.95$
0.10	1.64	1.64	1.64	1.65	1.65
0.25	1.14	1.06	0.96	0.87	0.81
0.50	0.68	0.62	0.54	0.46	0.41
0.75	0.41	0.38	0.33	0.29	0.25
1.00	0.25	0.23	0.21	0.19	0.17
1.50	0.097	0.095	0.092	0.089	0.088
2.00	0.039	0.041	0.045	0.048	0.050
3.00	0.0075	0.010	0.014	0.017	0.019
4.00	0.0021	0.0034	0.0053	0.0070	0.0082
5.00	0.00078	0.0014	0.0024	0.0032	0.0038
6.00	0.00035	0.00068	0.0011	0.0015	0.0018
8.00	0.000087	0.00017	0.00029	0.00039	0.00047

TABLE 2. *Part I: numerical values of the doubly-scaled M/M/1 complementary busy-period cdf,* $\theta^{-1} B^c(t)$*, obtained by multiplying the values in Table 1 by* $\theta^{-1} = 2/(1-\rho)$*;*
Part II: light-and-heavy-traffic interpolation $(1-w)2e^{-2t} + w\tilde{h}_1(t)$ *in (2.25) with* $w(\rho) = \rho^x$ *for* $x = 0.75$.

type of approximation	$B^c(t) \approx pe^{-\lambda_1 t} + (1-p)e^{-\lambda_2 t}$		
	p	λ_1^{-1}	λ_2^{-1}
One-moment exponential fit	0	—	$\theta = \frac{1-\rho}{2}$
Two-moment bounding H_2 (atom at 0 plus exponential)	$(1-2\theta) = \rho$	∞ $(\lambda_1 = 0)$	$\frac{m_1(1+c^2)}{2} = \frac{1}{2}$
Three derivatives $g''(0), g'(0), g(0)$	$\frac{1 - \sqrt{\rho/(\rho+4)}}{2}$	$\frac{2\theta^2 p}{1-p}$	$\frac{2\theta^2(1-p)}{p}$
One moment and two derivatives $g'(0), g(0), m_1$	$\frac{1}{2}$	$\theta(1-\sqrt{\rho})$	$\theta(1+\sqrt{\rho})$
Two moments and a derivative $g(0), m_1, m_2$	$\frac{1+\sqrt{\rho/(4-3\rho)}}{2}$	$\frac{\theta(1-p)}{p}$	$\frac{\theta p}{1-p}$
Three moments m_1, m_2, m_3	λ_2^{-1}	$\frac{1-\sqrt{\rho}}{2}$	$\frac{1+\sqrt{\rho}}{2}$

TABLE 3. *Hyperexponential (H_2) approximations for the time-scaled $M/M/1$ complementary busy-period cdf $B^c(t)$.*

time t	exact transform inversion	H_2 Fits: three moments	H_2 Fits: two moments and one deriv.	HEYMAN'S [26] diffusion approx.	approximants from Widder's formula (5.1): TER HAAR [45]	approximants from Widder's formula (5.1): JAGERMAN [30]
.09	.317	.38	.24	.30	.45	.44
.18	.188	.17	.15	.23	.32	.31
.27	.132	.095	.13	.19	.26	.24
.36	.0994	.066	.11	.17	.22	.19
.45	.0782	.054	.094	.15	.19	.16
.90	.0321	.031	.045	.12	.12	.08
1.00	.0273	.027	.038		.11	.07
1.25	.0189	.021	.026		.09	.05
1.50	.0135	.016	.017		.08	.04
2.00	.0074	.0093	.0075		.06	.02
3.00	.0026	.0031	.0015		.04	.01
4.00	.0010	.0011	.0003		.03	.004
5.00	.0004	.0004	.0001			.002
6.00	.0002	.0001	.0000			.001

TABLE 4. *A comparison of five approximations of the time-scaled M/M/1 complementary busy-period cdf* $B^c(t)$ *with exact values obtained from Laplace transform inversion: the case of* $\rho = 0.70$.

time t	exact transform inversion	H_2 fit one moment, two derivatives	inverse Gaussian (6.6), $\nu = 1-\theta$	asymptotic normal (4.3)
$\theta^2/4 = 0.035$	.884	.884	.98	3.2
.05	.840	.840	.94	2.5
.10	.712	.712	.77	1.5
.25	.450	.452	.435	.64
.50	.234	.240	.216	.28
.75	.135	.141	.125	.15
1.00	.0827	.0869	.0785	.090
1.50	.0352	.0349	.0350	.0372
2.00	.0165	.014	.0172	.0172
3.00	.00424	.0024	.00486	.00436
4.00	.00123	.0004	.00154	.00125
6.00	.000125	.00005	.000187	.000125

TABLE 5. *A comparison of approximations of the time-scaled $M/M/1$ complementary busy-period cdf $B^c(t)$ with exact values obtained by Laplace transform inversion: the case of $\rho = 0.25$.*

time t	exact transform inversion	H_2 fit one moment, two derivatives	inverse Gaussian (6.6), $\nu = 1-\theta$	asymptotic normal (4.3)
$\theta^2/4 = 0.0039$	.888	.888	.96	1.8
.05	.378	.43	.372	.40
.10	.239	.33	.236	.25
.25	.114	.17	.113	.116
.50	.0563	.059	.0560	.0568
.75	.0340	.020	.0338	.0342
1.0	.0224	.0069	.0224	.0225
1.5	.0112	.0008	.0112	.0112
2.0	.00619		.00620	.00620
3.0	.00221		.00222	.00222
4.0	.000887		.000895	.000889
6.0	.000171		.000173	.000171

TABLE 6. *Same comparison as Table 5: the case of $\rho = 0.75$.*

time t	$\rho = 0.25$			$\rho = 0.95$			$\rho = 0.99$		
	exact transform inversion	three-moment H_2 fit	relative percent error	exact transform inversion	three-moment H_2 fit	relative percent error	exact transform inversion	three-moment H_2 fit	relative percent error
0.10	0.712	0.722	+ 1.4	0.0425	0.0118	—72.1	0.0083	0.0023	—72.3
0.25	0.450	0.455	+ 1.1	0.0203	0.0098	—50.0	0.0040	0.0020	—50.0
0.50	0.234	0.230	— 1.7	0.0102	0.0076	—20.0	0.0020	0.0015	—25.0
0.75	0.135	0.129	— 4.5	0.0063	0.0059	— 6.3	0.0012	0.0012	— 4.8
1.00	0.0827	0.0796	— 3.7	0.0042	0.0046	+ 9.5	0.0008	0.00092	+ 9.5
1.50	0.0352	0.0357	+ 1.4	0.00219	0.0028	+27.3	0.00044	0.00056	+27.3
2.00	0.0165	0.0176	+ 6.7	0.00129	0.0017	+30.8	0.00025	0.00034	+36.0
3.00	0.0042	0.0046	+ 8.0	0.00048	0.00061	+27.1	0.00010	0.00012	+27.8
4.00	0.00123	0.00121	— 1.6	0.00021	0.00022	+ 4.8	0.00004	0.000045	+ 7.1
5.00	0.00038	0.00032	—16.8	0.00009	0.00008	—14.9	0.00002	0.000017	—15.0
6.00	0.00013	0.000084	—38.5	0.00005	0.00002	—35.6	0.00001	0.000006	—35.8
8.00	0.000014	0.0000058	—57.2	0.000011	0.000004	—63.7	0.000002	0.0000008	—66.7

TABLE 7. *A comparison of the three-moment H_2 approximation of the time-scaled M/M/1 complementary busy-period cdf $B^c(t)$ with exact values obtained from Laplace transform inversion: the cases of $\rho = 0.25$, 0.95 and 0.99.*

time t	$\rho = .25$		$\rho = .50$		$\rho = .75$	
	(2D, 1M)	(3D)	(2D, 1M)	(3D)	(2D, 1M)	(3D)
.5 θ^2	+0.0	—0.0	+ 0.0	—0.0	+ 0.1	—0.0
1.0 θ^2	+0.1	—0.0	+ 0.1	—0.0	+ 0.9	—0.1
1.5 θ^2	+0.4	—0.1	+ 1.1	—0.2	+ 2.7	—0.4
2.0 θ^2	+0.7	—0.2	+ 2.4	—0.7	+ 5.1	—1.2
2.5 θ^2	+1.2	—0.4	+ 4.1	—1.5	+ 8.3	—2.5
3.0 θ^2	+1.8	—1.1	+ 5.9	—2.4	+12.0	—4.3
4.0 θ^2	+3.0	—2.5	+10.1	—6.3	+19.8	—9.4

TABLE 8. *The small-time performance of two hyperexponential fits: two derivatives and one moment (2D, 1M) and three derivatives (3D). The values displayed are relative percent error, i.e., 100× (Approx. − Exact)/Exact.*

time y	mixing density $w(y)$ in $B^c(t)$ (2.27)	in $H_1^c(t)$ [3]	in $c_q(t)$ [5]
$\tau_1 = 0.009$	0.00	0.00	0.00
0.010	22.1	1.77	0.04
$y_{max} = 0.012$	26.6	2.55	0.06
0.015	24.0	2.88	0.09
0.030	11.2	2.69	0.16
0.100	2.05	1.64	0.33
0.50	0.17	0.68	0.68
1.00	0.05	0.40	0.80
1.50	0.01	0.12	3.60
$\tau_2 = 1.74$	0.00	0.00	0.0

TABLE 9. *Values of the mixing density $w(y)$ in (2.8) in the spectral representation of the busy-period complementary cdf $B^c(t)$ for the case $\rho = 0.75$. Also given are the corresponding mixing densities for the first-moment function complementary cdf $H_1^c(t)$ and the correlation function $c_q(t)$.*

time	inverse Gaussian approximations (6.6) and (6.8)			asymptotic	standard asymptotic
t	$\nu = 1$	$\nu = 1-\theta$	$\nu = \tau/2$	normal (4.3)	expansion (3 terms) (4.1)
0.016	+6.2	+9.6	+9.8	+149.	
0.05	—4.3	+5.0	+6.0	+51.	
0.10	—10.7	—0.8	+0.3	+24.	
0.25	—11.2	—3.7	—2.7	+8.2	
0.50	—5.1	—2.3	—2.7	+4.3	
0.75	+1.0	—2.1	—2.5	+3.0	
1.00	+7.2	—1.4	—2.5	+2.3	
1.50	+21.	+0.0	—2.2	+1.7	
2.00	+34.	+1.7	—2.2	+1.3	+252.
3.00	+63.	+3.4	—2.1	+0.8	+84.
4.00		+5.2	—2.0	+0.7	+39.
5.00		+8.0	—1.9	+0.6	+21.

TABLE 10. *A comparison of five approximations of the time-scaled $M/M/1$ complementary busy-period cdf $B^c(t)$ for the case $\rho = 0.50$. The displayed values are relative percent error, 100 (Approx. − Exact)/Exact.*

traffic intensity ρ	inverse Gaussian $(v = 1 - \theta)$	asymptotic normal
0.25	1.4 - 1.6	5.5 - ∞
0.50	1.1 - 1.9	2.5 - ∞
0.75	0.18 - 4.4	0.45 - ∞
0.85	0.07 - 12	0.11 - ∞
0.95	0.01 - ∞	0.01 - ∞

TABLE 11. *Time range where the relative percent error is less than one percent for the asymptotic normal approximation in (4.3) and the inverse Gaussian approximation in (6.6) with $x = \theta$ and $v = 1 - \theta$.*

time t	traffic intensity							
	$\rho = .25$		$\rho = .50$		$\rho = .75$		$\rho = .85$	
	3D	IG	3D	IG	3D	IG	3D	IG
1 θ^2	—0.0		—0.0		—0.1		—0.1	
2 θ^2	—0.2		—0.7		—1.2	—0.8	—1.5	—0.7
3 θ^2	—1.1		—2.4	—2.4		—1.2		—1.0
4 θ^2	—2.5			—2.7		—1.2		—1.0
5 θ^2	—4.7			—2.8		—1.2		—1.0
6 θ^2	—8.1	—8.3		—2.8		—1.1		—0.8
1.0		—8.5		—2.5		—0.4		—0.2
1.5		—8.5		—2.2		—0.4		—0.2
2.0		—8.5		—2.2		—0.4		—0.1
3.0		—8.7		—2.1		—0.4		—0.1
4.0		—8.6		—2.0		—0.3		—0.1
5.0		—8.4		—1.9		—0.3		—0.1

TABLE 12. *Relative percent errors, 100 (Approx. — Exact)/Exact, for the composite IG-3D approximation in Section 7 and its inverse Gaussian (IG) and three-derivative (3D) H_2 approximation components.*

time t	busy-period density $b(t)$	IG density $g(t)$	relative percent error RE in (6.18)
0.00	128.0	0.00	
0.01	19.0	25.0	
0.02	12.1	13.8	
0.03	8.28	8.65	
0.04	5.98	6.02	+0.8
0.05	4.52	4.48	—0.8
0.10	1.71	1.68	—1.7
0.50	0.1295	0.1286	—0.7
1.00	0.0346	0.0344	—0.5
2.00	0.00691	0.00689	—0.4
4.00	0.000776	0.000773	—0.4
6.00	0.0001340	0.0001336	—0.3
8.00	0.0000276	0.0000275	—0.3

TABLE 13. *A comparison between the inverse Gaussian (IG) density $g(t)$ using (iii) in (6.8) and the M/M/1 busy-period density $b(t)$ in the case* $\rho = 0.75$.

time t	busy-period density $b(t)$	IG density $g(t)$	relative percent error RE in (6.18)
0.00	8.000	0.000	
0.01	7.106	2.243	
0.02	6.333	6.726	
0.03	5.662	7.432	
0.04	5.078	6.852	
0.06	4.123	5.261	
0.08	3.388	4.030	
0.10	2.816	3.161	+12.4
0.20	1.301	1.295	—0.4
0.40	0.458	0.444	—3.0
0.80	0.225	0.219	—2.9
1.0	0.0816	0.0794	—2.6
2.0	0.0148	0.0144	—2.3
3.0	0.00407	0.00399	—2.1
4.0	0.00134	0.00131	—2.0
5.0	0.000482	0.000473	—2.0
6.0	0.000185	0.000181	—2.0
7.0	0.0000739	0.0000724	—2.0
8.0	0.0000305	0.0000299	—1.9
9.0	0.0000129	0.0000126	—1.9

TABLE 14. *A comparison between the inverse Gaussian density $g(t)$ using (iii) in (6.8) and the M/M/1 busy-period density $b(t)$ in the case* $\rho = 0.50$.

traffic intensity ρ	asymptotic relative percent error ARE in (6.19)
0.10	—23.6
0.15	—15.9
0.20	—11.3
0.25	—8.2
0.30	—6.0
0.35	—4.48
0.40	—3.33
0.45	—2.47
0.50	—1.82
0.55	—1.32
0.60	—0.947
0.65	—0.660
0.70	—0.443
0.75	—0.283
0.80	—0.167
0.85	—0.087
0.90	—0.0359
0.95	—0.0084

TABLE 15. *The asymptotic relative percent error for the inverse Gaussian (IG) density approximation, using (iii) in (6.8), of the M/M/1 busy-period density.*

Storage of the Single-Server Queue

E. G. Coffman, Jr., I. Mitrani[1]
AT&T Bell Laboratories
Murray Hill, New Jersey 07974, USA

Efficient storage and maintenance of queues is a fundamental problem of computer science. In this paper, we study a stochastic model of linear storage devices with the M/M/1 queue as the underlying process of arrivals and departures. We present probabilistic bounds on the wasted space under the FIFO and processor-shared service disciplines and a large class of practical allocation policies.

As a contribution to a *liber amicorum* in honor of J.W. Cohen, we feel that our work here builds on the fundamental contributions of Cohen in two broad areas: computer system modeling, and the methodology of queueing theory. In the first instance, we refer to extensions to combined service and storage models, and in the second, to the asymptotic methods for problems where explicit formulas seem too much to expect.

1. Introduction

In many queueing applications a comprehensive study of performance entails not only an analysis of the number in queue and waiting times, but also an analysis of the locations and arrangement of queued items. In this paper we consider applications modeled by the following assumptions:

i) the waiting area is an unbounded, linear array of cells available for the storage of items;
ii) an item is placed in a cell at the time of its arrival, and remains there until it is served and departs.

Within this set-up, models are distinguished by their service discipline and storage policy. By the latter term we mean the decision rule that selects the empty cells in which new arrivals are placed. As an example, Fig. 1 defines the single-server system analyzed in Section 2, with FIFO service and a storage policy that appends new arrivals to the right of the queue; when the queue is empty an arrival is placed in the first, i.e. leftmost, cell. The arrival and departure processes are those of the M/M/1 queue, with the usual parameters λ and μ, respectively. The following sections also treat the processor-sharing (PS) service discipline and the often used first-fit storage policy (an arrival is placed

1. On leave from the University of Newcastle-upon-Tyne, Newcastle-upon-Tyne NE1 7RU, England.

in the leftmost unoccupied cell). Such models arise in studies of dynamic storage allocation and the general problems of queue storage and management in computer systems. KNUTH [9] has a full discussion of these applications.

The processes of interest include Q_t, the queue-length process, R_t, the index of the rightmost occupied cell at time t, and $W_t = R_t - Q_t$, the number of unoccupied cells or 'wasted space' to the left of the cell indexed by R_t. We define $W_t = R_t = 0$ if $Q_t = 0$. Past research on the analysis of R_t and W_t has been confined to the first-fit storage policy and arrival and departure processes defined by the M/M/∞ queue. The earliest results appeared over 50 years ago in the work of KOSTEN [10]. Explicit results for the stationary distribution of R_t can be found in the more recent work of COFFMAN ET AL. [2,4]; asymptotics for $E[R_t]$ were derived by ALDOUS [1]. NEWELL [11] recently published a monograph on this model.

In this paper, we concentrate on the apparently more difficult, single-server (M/M/1) queue. While Q_t is a well understood Markov process (e.g., see COHEN [6]), neither R_t nor W_t are Markov processes in the models we study. However, in the simpler models (e.g., the one in Fig. 1), (R_t, Q_t) and (W_t, Q_t) are tractable bivariate Markov processes.

We are primarily concerned with stationary behavior. When we omit the subscript t from our process notation, we refer to a random variable with the stationary distribution, assuming that one exists. We let $\{p_i\}$ and $\{q_i\}$ denote the distributions of W and Q, respectively, and we let $\{p_{ij}\}$ denote the joint distribution of (W, Q). We define the generating function

$$F(u, v) = \sum_{i \geq 0} \sum_{j \geq 0} p_{ij} u^i v^j, \tag{1.1}$$

so that $F(u, 1)$ and $F(1, v)$ generate the p_i's and q_i's, respectively. In all of the models studied in this paper, standard theory provides that, so long as $\rho = \lambda / \mu < 1$, the above distributions exist and are unique. In the case of Q, we have [6]

$$q_i = (1-\rho)\rho^i, \quad i \geq 0. \tag{1.2}$$

It is convenient to extend our probability notation to negative index values, but to define such probabilities as zero. Thus, $p_i \equiv 0$ and $q_i \equiv 0$ if $i < 0$, and $p_{ij} \equiv 0$ if either $i < 0$ or $j < 0$.

The theme of our results is the stationary behavior of the wasted-space process, W_t, so throughout the remainder of the paper we assume $\rho < 1$. In our first model, shown in Fig. 1, we find that the convenience of keeping the storage and arrival orders the same comes at the cost of a wasted space, W, which is stochastically quite large. In Sections 3 and 4, we consider more efficient storage policies that place new arrivals only into wasted cells, i.e., empty cells to the left of the rightmost occupied cell. Sections 3 and 4 derive performance bounds on such policies and Section 5 presents asymptotic results for the first-fit rule under both the PS and FIFO service disciplines.

2. A FIFO QUEUE

We consider the FIFO model in Fig. 1, where a new arrival is placed in the empty cell just beyond the last occupied cell (or in the first cell if the queue is empty). Our primary objective is the distribution $\{p_i\}$ of W; a method similar to the one below can be applied to the analysis of R as well.

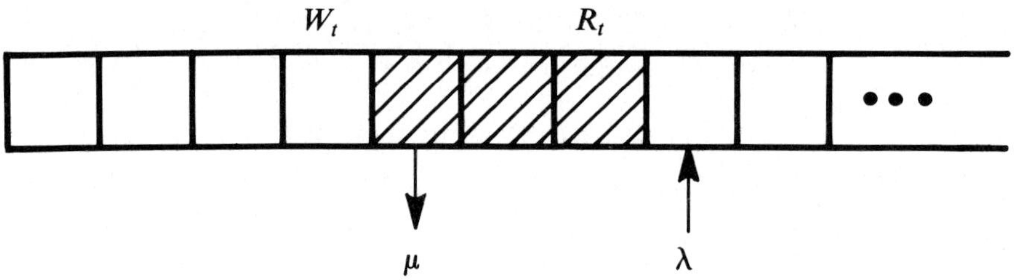

FIGURE 1. *Storage of a FIFO queue*

While W_t is not Markov, it can be analyzed as a marginal of the bivariate Markov process (W_t, Q_t), with the stationary probabilities p_{ij}. Since by definition of the model, $Q_t=0$ implies $W_t=0$, we have immediately

$$p_{00} = q_0=(1-\rho), \quad p_{i0}=0, \quad i\neq 0. \tag{2.1}$$

For the general case, the following birth-and-death equation is easily derived:

$$p_{ij} = \frac{1}{1+\rho}\, p_{i-1,j+1}+\frac{\rho}{1+\rho}\, p_{i,j-1}, \quad j\neq 0. \tag{2.2}$$

Introducing the generating function $F(u, v)$ in (1.1), we obtain by means of the usual manipulations

$$F(u, v) = (1-\rho)\frac{u-(1+\rho)v+uvK(u)}{\rho v^2-(1+\rho)v+u}, \tag{2.3}$$

where $K(u)=\frac{1}{1-\rho}\sum_{i\geq 0} p_{i1}u^i$ is an unknown function.

Now $F(u, v)$ is analytic for u and v each in the unit disk, so wherever the denominator in (2.3) has a root in this region, so also must the numerator. An easy analysis shows that for each u in the unit disk, there is exactly one v in the unit disk for which the denominator vanishes, and that this value is given by

$$v(u) = \frac{1+\rho-\sqrt{(1+\rho)^2-4\rho u}}{2\rho}. \tag{2.4}$$

Substituting (2.4) for v in the numerator of (2.3) and setting the numerator to 0, then yields

$$K(u) = \frac{(1+\rho)v(u)-u}{uv(u)}. \tag{2.5}$$

Substituting (2.5) into (2.3) and then setting $v=1$ produces, after some simplification, the desired generating function for the p_i's:

$$F(u,\, 1) = (1-\rho)\,\frac{u(1-v(u))}{(1-u)v(u)}. \tag{2.6}$$

At this point, moments can be found in the usual way. For example, one finds $E(W)=(\rho/(1-\rho))^2$.

In [5], rather complicated but explicit formulas are found for p_{ij} and p_i. However, for computing moments, the generating function derived here appears to be more convenient.

3. An upper bound model

In the remainder of the paper, we study storage policies that reclaim wasted space by placing newly arriving items into wasted cells, whenever possible. This is not possible when $W_t=0$ at the time of arrival; in these cases all disciplines considered place the new arrival into cell R_t+1. Storage policies having these properties are said to be *thrifty*. In this section, we define a model with a thrifty storage policy, whose wasted space W is stochastically at least as large as any other model with a thrifty storage policy.

We define the Leftmost-Preemptive-Resume (or LPR) model as follows. Whenever the queue is not empty, the leftmost of the items present is being served. At an arrival instant, the action taken depends on whether there are any wasted cells or not. In the latter case, the incoming item simply goes in the next available cell (cell R_t+1) and waits there. In the former, it is placed in the rightmost wasted cell, preempting the current service. Interrupted services are eventually resumed from the point of interruption. Fig. 2 illustrates the LPR model. In contrast to a first-fit system, we note that LPR retains the advantage of the model in Section 2 whereby stored items appear in consecutive cells.

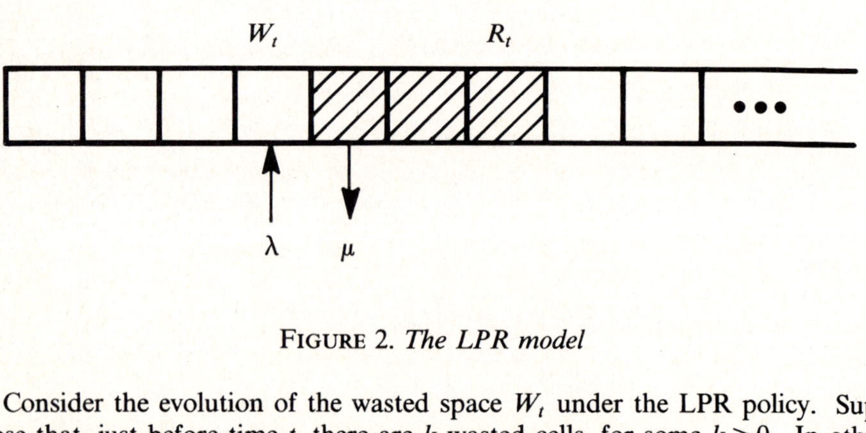

FIGURE 2. *The LPR model*

Consider the evolution of the wasted space W_t under the LPR policy. Suppose that, just before time t, there are k wasted cells, for some $k>0$. In other words, $W_{t^-}=k$. If an arrival occurs at time t, then one of these cells becomes occupied, leading to $W_t=k-1$. Of course, the same thing would happen under an arbitrary thrifty discipline. If, on the other hand, a departure occurs at time t, then unless the queue is left empty, i.e., $Q_{t^-}=1$, a new wasted cell is always created: $W_t=k+1$. Under other thrifty policies, when $Q_{t^-}>0$, the wasted space may increase by one, remain the same or decrease by varying amounts, depending on whether an arrival or departure took place, whether or not the rightmost item departed in the case of a departure, and whether or not $W_{t^-}=0$.

It should be clear from the above that, within the class of thrifty policies for the single-server queue, LPR achieves the largest wasted space. That is, for any given sequence of interarrival and service times, the wasted space under LPR, W_t^{LPR}, and the wasted space under any other thrifty policy, W_t^A, satisfy

$$W_t^A \leqslant W_t^{\mathrm{LPR}}, \quad t\geqslant 0. \tag{3.1}$$

Note that LPR is not unique in possessing this extremal property. Any thrifty policy which ensures that the rightmost item is served only when there are no others in the system has essentially the same behavior as LPR.

Together with (2.1), the distribution of (W,Q) is determined from the following balance equations:

$$\begin{aligned} p_{ij} &= \frac{\rho}{1+\rho}\, p_{i+1,j-1}+\frac{1}{1+\rho}\, p_{i-1,j+1}; \quad i\geqslant 1, \quad j\geqslant 2, \\ p_{0j} &= \frac{\rho}{1+\rho}\, (p_{1,j-1}+p_{0,j-1}); \quad j\geqslant 2, \\ p_{i1} &= \frac{1}{1+\rho}\, p_{i-1,2}; \quad i\geqslant 1, \\ p_{01} &= \frac{\rho}{1+\rho}\, p_{00}=\frac{1}{1+\rho}\sum_{i=0}^{\infty} p_{i1}. \end{aligned} \tag{3.2}$$

It is convenient to introduce the modified generating function

$$G(u, v) = \sum_{i\geqslant 0}\sum_{j\geqslant 1} p_{ij}u^i v^{j-1} \tag{3.3}$$

(this definition accounts more explicitly for (2.1)). In terms of $G(u, v)$, the set of recurrences in (3.2) reduces to

$$[(1+\rho)uv-\rho v^2-u^2]G(u, v) = uvG(1, 0)-u^2G(u, 0)-\rho v^2(1-u)G(0, v). \tag{3.4}$$

The right-hand side of (3.4) contains a constant, $G(1, 0)$, and two unknown functions, $G(u, 0)$ and $G(0, v)$. From the known marginal distribution of the number of items in the queue (which can also be derived by setting $u=1$ in (3.4)), we easily obtain $G(1,0)=\rho(1-\rho)$. To determine the two unknown functions, note that the coefficient of $G(u, v)$ in (3.4) has a particularly simple factorization:

$$(1+\rho)uv-\rho v^2-u^2 = -(u-v)(u-\rho v). \tag{3.5}$$

Since $G(u, v)$ is finite for all u and v in the unit disk, (3.5) implies that the right-hand side of (3.4) vanishes when $u=v$, and also when $u=\rho v$ ($|v|\leqslant 1$).

This yields two equations which, after simplifying, can be written as

$$\rho(1-\rho)-G(v,0) = \rho(1-v)G(0,v),$$
$$\rho(1-\rho)-\rho G(\rho v,0) = (1-\rho v)G(0,v). \tag{3.6}$$

These equations allow $G(0,v)$ to be eliminated, leaving a relation between $G(v,0)$ and $G(\rho v,0)$. That relation can be written in a simpler form by introducing the function

$$H(u) = \frac{\rho(1-\rho) - G(u,0)}{1-u}.$$

Then we have

$$H(u) = \frac{\rho^2(1-\rho)^2}{1-\rho u} + \rho^2 H(\rho u). \tag{3.7}$$

From (3.7), a series expansion for $H(u)$ is obtained:

$$H(u) = \rho^2(1-\rho)^2 \sum_{i=0}^{\infty} \frac{(\rho u)^i}{1-\rho^{i+2}}. \tag{3.8}$$

Both unknown functions are now determined by (3.6) and (3.8), so various performance measures can be derived. Suppose, for instance, that we are interested in the expected wasted space, $E(W)$. This is given by

$$E(W) = \frac{\partial G(1,1)}{\partial u} = \frac{2H(1)+H'(1)-\rho^2}{1-\rho}. \tag{3.9}$$

For arbitrary values of ρ, there are no closed-form expressions for the series representing $H(1)$ and $H'(1)$. However, one can easily establish the following limiting result:

$$\lim_{\rho\to 1} [E(W)(1-\rho)] = \frac{\pi^2}{6} - 1. \tag{3.10}$$

In other words, as the system approaches saturation, the growth of the average wasted space is proportional to that of the average number of items.

4. A LOWER BOUND PROCESSOR-SHARING MODEL

The Rightmost-Preemptive-Resume (RPR) model provides a trivial lower bound to go along with the LPR upper bound for thrifty policies. Indeed, by always serving the rightmost item, with preemptions at arrival times, we ensure that $W_t=0$ for all $t \geqslant 0$. (Note that RPR is equivalent to the preemptive LIFO service discipline.)

To obtain desired properties of waiting time distributions, it is common to implement service disciplines like FIFO or PS, which are independent of the ordering of items in storage. In these cases the most natural storage policy is first-fit: at the time of an arrival, the storage cells are scanned from left to right and the new item is placed in the first available cell. Unfortunately, exact analyses of the first-fit, FIFO and PS models are still open problems. However, asymptotic, heavy-traffic results are available, as shown in the next section.

In the remainder of this section, we define and analyze a model whose wasted space is stochastically no worse than any model with PS service and a thrifty storage policy. It turns out, as shown in Section 5, that the performance of the first-fit PS model is much closer to that of the lower bound model in this section than it is to the upper bound (LPR) results of the previous section.

To describe the lower bound model in simplest terms, we temporarily dispense with our ground rule that waiting items can not be moved. Take a model with PS service and any thrifty storage policy, and modify it as follows. Whenever the rightmost item departs, storage is compacted; i.e., if $n \geqslant 1$ waiting items remain, then these items are moved so as to occupy cells $1, 2, \ldots, n$. Thus, the wasted space just after such a departure jumps to 0.

Call the resulting model, the Periodic Compaction (PC) model. Clearly, changing the storage policy of the original model has no effect on wasted space, so long as we stay within the class of thrifty policies. It is easy to see that for all $x, t \geqslant 0$, $\Pr\{W_t^{PC} \leqslant x\} \geqslant \Pr\{W_t^A \leqslant x\}$, where A is any (non-compacting) model with PS service and a thrifty storage policy.

Together with (2.1), the following set of balance equations determines the steady-state joint probabilities, p_{ij}, of i wasted cells and j items in storage:

$$
\begin{aligned}
p_{ij} &= \frac{\rho}{1+\rho} p_{i+1,j-1} + \frac{1}{1+\rho} \frac{j}{j+1} p_{i-1,j+1}; \quad i \geqslant 1, \quad j \geqslant 2 \\
p_{0j} &= \frac{\rho}{1+\rho} (p_{0,j-1} + p_{1,j-1}) + \frac{1}{1+\rho} \frac{1}{j+1} \sum_{i=0}^{\infty} p_{i,j+1}; \quad j \geqslant 2 \\
p_{i1} &= \frac{1}{2} \frac{1}{1+\rho} p_{i-1,2}; \quad i \geqslant 1 \\
p_{01} &= \frac{\rho}{1+\rho} p_{00} + \frac{1}{2} \frac{1}{1+\rho} \sum_{i=0}^{\infty} p_{i2} = \frac{1}{1+\rho} \sum_{i=0}^{\infty} \left[p_{i1} + \frac{1}{2} p_{i2} \right].
\end{aligned}
\tag{4.1}
$$

Introducing the modified generating function, $G(u, v)$, in (3.3), this set is reduced to the single equation

$$
\begin{aligned}
(v-u)\left[v - \frac{u}{\rho}\right] \frac{\partial G(u, v)}{\partial v} &= \left[2\frac{1+\rho}{\rho} u - 3v\right] G(u, v) \\
&+ (1-u)\left[3vG(0, v) + v^2 \frac{\partial G(0, v)}{\partial v}\right] - (1-\rho)u\left[\frac{\rho}{1-\rho v} + 2\right].
\end{aligned}
\tag{4.2}
$$

Unfortunately, (4.2) has so far resisted attempts to solve it analytically. It does, however, allow us to reduce the problem of determining the average number of wasted cells to that of finding the value of $G(0, v)$ and its derivative at $v = 1$. More precisely, it follows from (4.2) that

$$
E(W) = \frac{\partial G(1,1)}{\partial u} = \frac{\rho}{2-\rho} \left[3G(0,1) + \frac{\partial G(0,1)}{\partial v} - 2\rho\right]. \tag{4.3}
$$

It turns out that one can find a solution to the set of equations (4.1) of the form (see [2])

$$p_{ij} = a_{i+j} b_j; \quad i \geqslant 0; \quad j \geqslant 1, \tag{4.4}$$

where the b_j's satisfy the recurrence

$$\begin{aligned} b_1 &= 1, b_2 = 2(1+\rho), \\ b_i &= \frac{\rho}{1+\rho} b_{i-1} + \frac{1}{1+\rho} \frac{i}{i+1} b_{i+1}; \quad i \geqslant 2, \end{aligned} \tag{4.5}$$

and where

$$a_m = (1-\rho) \left[\frac{\rho^m}{b_m} - \frac{\rho^{m+1}}{b_{m+1}} \right]; \quad m \geqslant 1. \tag{4.6}$$

Hence, the probabilities p_{0j} which are implicit in (4.3) are given by

$$p_{0j} = a_j b_j = (1-\rho)\rho^j \left[1 - \rho \frac{b_j}{b_{j+1}} \right]; \quad j \geqslant 1. \tag{4.7}$$

Relations (4.5) and (4.7) suggest a simple numerical procedure for computing $E(W)$ for arbitrary values of ρ. It can be shown that $E(W) \sim \frac{1}{2} \sqrt{\frac{\pi}{1-\rho}}$ as $\rho \to 1$ (see [2]).

5. Asymptotic estimates

As pointed out earlier, a direct analysis of Markov processes modeling the first-fit, FIFO and PS storage queues appears to be quite difficult. However, important information on heavy-traffic behavior (as $\rho \to 1$) can still be obtained through an asymptotic analysis. As a convenient normalization for this purpose, we take $\lambda = 1 - 1/N$ and $\mu = 1$, and let $1 < N < \infty$ be the index of traffic intensity. The dependence on N is made explicit in the notation, $W_t(N)$. Note that, with the above values of λ and μ, we have $E(Q) = N-1$, so asymptotics in N will also be asymptotics in the mean queue length.

A typical expected-value bound takes one of the forms $E[W(N)] = O(f_N)$, $\Omega(f_N)$ or $\Theta(f_N)$, where f_N is some increasing function of N. The second of these is simply the converse of the first; i.e., $E[W(N)] = \Omega(f_N)$, if there exists a constant c such that $E[W(N)] \geqslant c f_N$ for all N sufficiently large. We write the third, $E[W(N)] = \Theta(f_N)$, if both $E[W(N)] = \Omega(f_N)$ and $E[W(N)] = O(f_N)$ hold.

As an illustration of the types of results that can be proved, we consider the first-fit PS discipline.

Theorem 1. *In the first-fit PS system, we have for all t sufficiently large*

$$E[W_t(N)] = \Theta\left[\sqrt{N} \log N\right]. \tag{5.1}$$

Proof sketch. We discuss below the principal steps of the argument. The missing details are largely routine and can be found in [3].

We consider the $O\left[\sqrt{N\log N}\right]$ upper bound first, and write

$$E[W_t(N)] = \sum_{i\geq 1}\Pr\{W_t(N)\geq i\}\leq\sqrt{N\log N}\sum_{k\geq 0}\Pr\{W_t(N)\geq k\sqrt{N\log N}\}. \quad (5.2)$$

Thus, to prove the upper bound in (5.1),R
it is enough to show that there exists a constant $\epsilon>0$ such that

$$\Pr\{W_t(N)\geq k\sqrt{N\log N}\}=O\left[\frac{1}{k^{1+\epsilon}}\right]. \quad (5.3)$$

(We emphasize that ϵ and the constants hidden in the notation $O\left[\frac{1}{k^{1+\epsilon}}\right]$ must be independent of N.) Although tighter bounds are possible, we show only that (5.3) holds for all $\epsilon<1/3$; this restriction simplifies the presentation.

Our approach is to analyze the events that can occur in any suitably long interval $[t-d, t]$, with the only known property of the storage state at time $t-d$ being that Q_{t-d} has the stationary distribution. For this approach to succeed, d must be chosen sufficiently large that, with high probability, the items in storage at time t consist solely of the new arrivals in $[t-d, t]$.

The argument begins with the easily verified fact that $W_t(N)\geq k\sqrt{N\log N}$, $t\geq d$, only if

A. at least one item in storage at time $t-d$ remains in storage at time t, or
B. the *variation* (or *range*) of the process Q_τ during $[t-d, t]$ is at least $k\sqrt{N\log N}$ (i.e., the magnitude of the difference in the maximum and minimum points visited by Q_τ during $[t-d, t]$ is at least $k\sqrt{N\log N}$).

Thus, for $t\geq d$,

$$\Pr\{W_t(N)\geq k\sqrt{N\log N}\}\leq\Pr\{A\}+\Pr\{B\}. \quad (5.4)$$

We now analyze the events A and B separately.

In order for $\Pr\{A\}$ to be sufficiently small, we require that d be of the form $Nk^{2/3}(c_1\log N+c_2\log k)$. (It will be clear later that the exponent 2/3 here is tied to our choice $0<\epsilon<1/3$ for (5.3). Convenient choices for the constants $c_1>0$ and $c_2>0$ will be made shortly.) The arguments below are independent of $t\geq d$, so for convenience, we take $t=d$. We let m_t and v_t denote, respectively, the maximum point and the variation of Q_τ over $0\leq\tau\leq t$.

Conditioned on $m_t\leq l$, the probability that any given item remains in storage throughout $[0, t]$ is at most $[1+o(1)](1-1/l)^t$ as $N\to\infty$. Based on this observation, it is convenient to take $c_1=2$ and $c_2=4$, so we can write

$$\Pr\{A\}\leq 2[1+o(1)]Nk^{2/3}\left[1-\frac{1}{2Nk^{2/3}}\right]^{2Nk^{2/3}(\log N+2\log k)}$$
$$+\Pr\{m_t>2Nk^{2/3}\}, \quad (5.5)$$

as $N \to \infty$. Simplifying (5.5) and substituting into (5.4), we have for any fixed N,

$$\Pr\{W_t(N) \geq k\sqrt{N \log N}\} \leq \Pr\{m_t > 2Nk^{2/3}\} + \Pr\{v_t > k\sqrt{N \log N}\} + O\left(\frac{1}{k^{4/3}}\right). \tag{5.6}$$

Next, introduce the bound,

$$\Pr\{m_t > 2Nk^{2/3}\} \leq \Pr\{m_t > 2Nk^{2/3} \mid Q_0 \leq Nk^{2/3}\} + \Pr\{Q_0 > Nk^{2/3}\}. \tag{5.7}$$

Since m_t increases stochastically as Q_0 increases, we can replace the inequality $Q_0 \leq Nk^{2/3}$ in (5.7) by $Q_0 = Nk^{2/3}$. By (1.2) with $\rho = 1 - 1/N$, we have $\Pr\{Q_0 > Nk^{2/3}\} = O(e^{-k^{2/3}})$, so (5.7) can be rewritten

$$\Pr\{m_t > 2Nk^{2/3}\} \leq \Pr\{m_t > 2Nk^{2/3} \mid Q_0 = Nk^{2/3}\} + O(e^{-k^{2/3}}). \tag{5.8}$$

By a similar argument, (5.8) also applies with the event $m_t > 2Nk^{2/3}$ replaced by $v_t > k\sqrt{N \log N}$, so after substitution into (5.6), we get

$$\Pr\{W_t(N) \geq k\sqrt{N \log N}\} \leq \Pr\{m_t > 2Nk^{2/3} \mid Q_0 = Nk^{2/3}\} + \Pr\{v_t > k\sqrt{N \log N} \mid Q_0 = Nk^{2/3}\} + O\left(\frac{1}{k^{4/3}}\right). \tag{5.9}$$

To bound the first probability on the right of (5.9), we can easily show that m_t increases stochastically with N. Thus,

$$\Pr\{m_t > 2Nk^{2/3} \mid Q_0 = Nk^{2/3}\} \leq \Pr\{m_t^* > 2Nk^{2/3} \mid Q_0 = Nk^{2/3}\}, \tag{5.10}$$

where m_t^* is the maximum over $[0, t]$ of the symmetric process Q_τ^* corresponding to Q_τ; for Q_τ^* we have $\lambda = \mu = 1$ and an event rate of 2.

The problem of estimating the probability on the right of (5.10) can now be reduced to the analysis of a simple (unconstrained, symmetric) random walk. Let ζ_i, $0 \leq i \leq n$, denote this random walk, where $\zeta_0 = Nk^{2/3}$ and the number of steps is taken to be

$$n = \left\lfloor 4Nk^{2/3}(\log N + 2\log k)\right\rfloor,$$

i.e., $n = \lfloor 2t \rfloor$ is the integer part of the expected number of events of Q_τ^* in $[0, t]$. We leave the details to the interested reader (see [3]), but it is easy to argue heuristically that our probability estimates based on ζ_i, $0 \leq i \leq n$, can be taken as probability estimates based on Q_τ^*, $0 \leq \tau \leq t$ (to within constants hidden by our notation). Clearly, if the boundary is removed from Q_τ^*, then the values of the resulting free process at its first n event epochs are stochastically equal to ζ_i, $1 \leq i \leq n$. But with $Q_0^* = Nk^{2/3}$ and k large, it is easily verified that the probability of a visit to the origin by Q_τ^* in $[0, t]$, $t = 2Nk^{2/3}(\log N + 2\log k)$, is very small — indeed, sufficiently small that we can expect Q_τ^* to behave essentially as an unconstrained process in $[0, t]$. Furthermore, the number of events of Q_τ^* in this interval is a Poisson distributed random variable whose coefficient of variation tends to 0 as $k \to \infty$. Hence, we can also expect a negligible change in probability estimates when $m_t^* > 2Nk^{2/3}$ is conditioned on the number of events being fixed at the integer part of the mean, $n = \lfloor 2t \rfloor$.

Thus, for the probability on the right of (5.10), we seek an estimate of $\Pr\{\max_{0 \leq i \leq n} \zeta_i > 2Nk^{2/3}\}$. By a classical result for simple random walks (see [7], p. 89), we can relate the maximum value over n steps to the value at the n-th step by

$$\Pr\{\max_{0 \leq i \leq n} \zeta_i > 2Nk^{2/3}\} \leq 2\Pr\{\zeta_n > 2Nk^{2/3}\}. \tag{5.11}$$

We can now apply standard Chernoff bounds for ζ_n, which is manifestly a sum of i.i.d. bounded random variables (see [8]). On doing so, we find easily that $\Pr\{\zeta_n > 2Nk^{2/3}\}$ tends to 0, as $k \to \infty$, much faster than $1/k^a$ for any $a > 0$. From (5.9)-(5.11), we conclude that

$$\Pr\{W_t(N) \geq k\sqrt{N \log N}\} \leq \Pr\{v_t > k\sqrt{N \log N} \mid Q_0 = Nk^{2/3}\} + O\left(\frac{1}{k^{4/3}}\right). \tag{5.12}$$

Turning now to estimates of the variation, it is not difficult to show first that

$$\Pr\{v_t > k\sqrt{N \log N} \mid Q_0 = Nk^{2/3}\} \leq \Pr\{v_t^{**} > k\sqrt{N \log N} \mid Q_0^{**} = Nk^{2/3}\},$$

where v_t^{**} is the variation over $[0, t]$ of the unconstrained process Q_τ^{**} corresponding to Q_τ with its boundary conditions removed. We now argue as before that, without affecting the probability estimate, we can fix the number of events of Q_τ^{**} in $[0, t]$ at the integer part of the mean

$$n = \left\lfloor 2(2 - 1/N)Nk^{2/3}(\log N + 2\log k) \right\rfloor.$$

Instead of Q_τ^{**}, $0 \leq \tau \leq t$, we then analyze the corresponding, asymmetric and unconstrained random walk δ_i, $0 \leq i \leq n$, with $\delta_0 = Q_0^{**} = Nk^{2/3}$; i.e., the jumps $+1$ and -1 have probabilities $(1 - 1/N)/(2 - 1/N)$ and $1/(2 - 1/N)$, respectively.

Accordingly, let $\operatorname{var}_{0 \leq i \leq n} \delta_i$ denote the variation of δ_i over $0 \leq i \leq n$, and note that $E(\delta_i) = Nk^{2/3} - i/(2N - 1)$. We observe that the event $\operatorname{var}_{0 \leq i \leq n} \delta_i > x$ occurs only if there exists an i $(0 \leq i \leq n)$ such that

$$|\delta_i - E(\delta_i)| > \frac{x}{2} - \frac{i}{2N - 1}.$$

But by Kolmogorov's inequality (see [7], p. 234), we have

$$\Pr\{\exists i\ (0 \leq i \leq n)\colon |\delta_i - E(\delta_i)| > \alpha\sigma(\delta_n)\} \leq \frac{1}{\alpha^2}, \tag{5.13}$$

where $\sigma(\delta_n)$ denotes the standard deviation of δ_n. Thus, if we put

$$\alpha\sigma(\delta_n) = \frac{k\sqrt{N \log N}}{2} - \frac{i}{2N - 1},$$

and use $i \leqslant n$ along with

$$\sigma(\delta_n) \sim \sqrt{n} = \sqrt{2(2-1/N)Nk^{2/3}(\log N + 2\log k)}, \quad as \quad n \to \infty,$$

then we find that

$$\alpha = \Omega\left(\frac{k^{2/3}}{\sqrt{\log k}}\right), \tag{5.14}$$

where, as before, the constants hidden in this notation are independent of N. Hence, we obtain from (5.12)-(5.14), for any $\epsilon < 1/3$, that

$$\Pr\{\operatorname*{var}_{0 \leqslant i \leqslant n} \delta_i > k\sqrt{N \log N}\} = O\left(\frac{1}{k^{1+\epsilon}}\right). \tag{5.15}$$

From (5.12) and (5.15) we conclude that (5.3) holds for any $\epsilon < 1/3$, thus completing the upper bound argument.

For the lower bound, $E[W_t(N)] = \Omega(\sqrt{N \log N})$ for all t sufficiently large, consider an arbitrary interval $[t_0, t]$ with $t - t_0 = \frac{1}{2}N \log N$. (The coefficient 1/2 was chosen for simplicity; the argument below can be made to work for any coefficient in (0,1).) Clearly, it is enough to show that, with a probability bounded from below by a positive constant independent of N for all N sufficiently large, the following events occur jointly:

A. $Q_{t_0} \geqslant 2N$,

B. one of the rightmost $\sqrt{N}$ items in storage at time t_0 is still in storage at time t, and

C. $Q_{t_0} - Q_t \geqslant \sqrt{N \log N}$.

By (1.2), $\Pr\{A\} \sim e^{-2}$ as $N \to \infty$. Given event A, we have by our earlier analysis of the variation of Q_τ that $Q_\tau \geqslant N$, $t_0 \leqslant \tau \leqslant t$, with a probability that tends to 1 as $N \to \infty$. Hence, the instantaneous departure rate of any item in storage during $[t_0, t]$ is at most $1/N$ with a probability that tends to 1 as $N \to \infty$. But given this last event, the probability that all $\sqrt{N}$ rightmost items at time t_0 depart in $(t_0, t]$ is at most

$$[1+o(1)]\left(1 - \left(1 - \frac{1}{N}\right)^{\frac{1}{2}N \log N}\right)^{\sqrt{N}} \sim \left(1 - \frac{1}{\sqrt{N}}\right)^{\sqrt{N}} \sim e^{-1} \quad \text{as } N \to \infty. \tag{5.15}$$

Thus, the probability that at least one of the rightmost $\sqrt{N}$ items at t_0 remains at t tends to at least $1 - e^{-1}$, so that as $N \to \infty$,

$$\Pr\{A,B\} = \Pr\{B \mid A\}\Pr\{A\} \geqslant [1+o(1)](1-e^{-1})e^{-2}. \tag{5.16}$$

Now consider the event C conditioned on events A and B. It is easy to see that the conditioning provided by event B can be ignored, since this event is restricted to a fraction ($\leqslant 1/\sqrt{N}$) of the occupied cells which tends to 0 as $N \to \infty$; i.e., $\Pr\{C \mid A, B\}$ and $\Pr\{C \mid A\}$ tend to the same limit as $N \to \infty$. But for $\Pr\{C \mid A\}$, we need only appeal to the normal limit law for the displacement in a simple random walk of $\lfloor 2(t-t_0) \rfloor = \lfloor N \log N \rfloor$ steps (see our discussion of the approximations in the upper-bound argument). Thus,

$$\Pr\left\{ \frac{Q_{t_0} - Q_t}{\sqrt{N \log N}} \geq x \,\middle|\, Q_{t_0} \geq 2N \right\} \sim 1 - \Phi(x), \quad \text{as} \quad N \to \infty,$$

where $\Phi(\cdot)$ is the normal distribution function with mean 0 and unit variance. Hence, as $N \to \infty$,

$$\Pr\{C \mid A, B\} \sim \Pr\{Q_{t_0} - Q_t \geq \sqrt{N \log N} \mid Q_{t_0} \geq 2N\} = [1 + o(1)][1 - \Phi(1)],$$

which, together with (5.16), proves the lower bound. □

Similar methods can be applied to the first-fit, FIFO system. It is interesting to find that expected wasted space is less in this system than in the first-fit, PS system.

THEOREM 2. *In the first-fit, FIFO system, we have for all t sufficiently large* [3]

$$E[W_t(N)] = \Theta\left(\sqrt{N}\right).$$

REFERENCES

1. D. ALDOUS (1986). Some interesting processes arising as heavy traffic limits in a M/M/∞ storage process. *Stoch. Proc. Appl. 22*, 291-313.
2. E.G. COFFMAN, JR., L. FLATTO, C. KNESSL, I. MITRANI, L.A. SHEPP (1988). Stochastic models of queue storage. *Probability in Engineering and the Informational Sciences* (in press).
3. E.G. COFFMAN, JR., L. FLATTO, F.T. LEIGHTON (1988). *First-Fit Allocation of Queues: Tight Asymptotic Bounds on Expected Wasted Space,* AT&T Bell Laboratories, Murray Hill, NJ 07974 (to appear).
4. E.G. COFFMAN, JR., T.T. KADOTA, L.A. SHEPP (1985). A stochastic model of storage fragmentation. *SIAM J. Comput. 14,* 416-425.
5. E.G. COFFMAN, JR., A.C. MCKELLAR (1968). On the motion of a Markov queue in random access storage. *IEEE Trans. Comput. C-17,* 600-603.
6. J.W. COHEN (1969). *The Single Server Queue,* John Wiley & Sons, New York.
7. W. FELLER (1968). *An Introduction to Probability Theory and its Applications, Vol. I,* 3rd edition, John Wiley & Sons, New York.
8. W. HOEFFDING (1963). Probability inequalities for sums of bounded random variables. *Amer. Stat. Assoc. J. 58,* 13-30.
9. D.E. KNUTH (1973). *Fundamental Algorithms,* 2nd edition, Addison-Wesley, Reading, MA.
10. L. KOSTEN (1937). Uber Sperrungswahrscheinlichkeiten bei Staffelschaltungen. *Electra Nachrichten-Technik 14,* 5-12.
11. G.F. NEWELL (1984). *The M/M/∞ Service System with Ranked Servers in Heavy Traffic,* Springer-Verlag, New York.

Sojourn Time in an M/G/1 Queue with Bernoulli Feedback

B. T. Doshi, J. S. Kaufman
AT&T Bell Laboratories
Holmdel, New Jersey 07733, USA

Single class and multi class versions of an M/G/1 queue with Bernoulli feedback are studied. The analysis focuses on obtaining the sojourn time of a tagged customer on successive passes through the queue. Transform expressions for the distribution of sojourn time are obtained in the single class case and mean sojourn time expressions are obtained in the multi class case. Numerical examples are presented which illustrate different facets of the models.

1. Introduction

In this paper we study the sojourn time of a tagged customer who has just completed its m-th pass ($m = 1,2,\ldots$) in an M/G/1 queue with Bernoulli feedback. The analysis explicitly accounts for the dependence between successive sojourn times (sojourn times of the tagged customer in successive passes through the queue). The primary result obtained in Section 2 is the joint Laplace Stieltjes transform of the m successive sojourn times ($m = 1,2,\ldots$). This transform expression is numerically inverted to generate a variety of results for the sojourn time distribution after m passes. In particular, we contrast the exact analysis with an approximate analysis which assumes that successive sojourn times are independent.

In Section 3 we generalize the model to multiple job classes, each class of which has a distinct service time distribution and feedback parameter. The method used in Section 2 to obtain the joint transform of successive sojourn times generalizes easily, but leads to a term which we are unable to evaluate exactly - even in the case of exponential service time distributions. Although several approximate approaches suggest themselves, this is not pursued here. Instead, we content ourselves with exact results for the mean sojourn time of successive passes - and after m passes - for each class, in the case of generally distributed service times.

The method used in Section 2 to obtain the Laplace Stieltjes transform of the sojourn time after m passes is not new. The essential observations are evident in the classical paper [9] due to Takács. However, in [9] Takács focuses primarily on obtaining the distribution of (total) sojourn time, largely ignoring the question of the distribution of sojourn time after m passes, $m=1,2,\ldots$. Thus, our results in Section 2 can be viewed as a variation on a theme presented in [9]. DISNEY ET AL. [4] and FONTANA and DIAZ BERZOSA [5] also studied the queue-length and waiting-time distributions of the M/G/1 queue with Bernoulli feedback in some detail. The results we present in Section 2 are similar to those obtained earlier by LAM and SHANKAR [8] and more recently by VAN DEN BERG, BOXMA and GROENENDIJK [2] and VAN DEN BERG and BOXMA [3]. In [8], customers depart after a random number of passes determined by an arbitrary discrete distribution having finite support. In [2], all customers depart after a fixed number of passes and in [3] a far more general feedback scheme is analyzed which subsumes both the Bernoulli feedback model as well as the fixed number of passes model in [2]. The results in [2,3,8] are, however, valid only when the service time during each pass is exponentially distributed. The multiclass results in Section 3 appear to be new. The multiclass distributional result in Section 3.1 while incomplete, provides a reasonable point of departure for approximation schemes. The exact mean sojourn time results obtained in Sections 3.2-3.4 nicely complement known single class results.

2. THE SINGLE CLASS MODEL

The model under consideration is shown in Fig. 1. Customers arrive at mean rate λ according to a Poisson process, are served in first-in-first-out order and after receiving an amount of service drawn from a general distribution $F(\cdot)$ having mean μ^{-1}, depart from the system with probability $1-p$ or are fed back to rejoin the end of the queue with probability p. Consider a tagged customer and denote the successive sojourn times of this customer by $Y_1, Y_2, \ldots$:

$$Y_i = \text{sojourn time of customer on } i\text{-th pass through the queue.}$$

Also let

$$N_{i-1} = \text{number in system at the beginning of the } i\text{-th pass.}$$

Then we may write:

$$Y_i = x_{i1} + \sum_{j=2}^{N_{i-1}} x_{ij} + s_i, \quad i=1,2,\ldots, \tag{2.1}$$

where x_{i1} is the remaining service time of the customer in service, $x_{i2},\ldots$ are the service times of the customers waiting when the tagged customer begins its i-th pass and s_i denotes the service time of the tagged customer on its i-th pass. $x_{i2}, x_{i3},\ldots$ are i.i.d. with distribution function $F(\cdot)$ for $i\geqslant 1$. Note that x_{i1} has distribution function $F(\cdot)$ for $i\geqslant 2$ but for $i=1$, x_{i1} has distribution function $F(\cdot)$ and is independent of N_0 only if $F(\cdot)$ is exponential. We consider two cases:

case 1: s_i has distribution $F(\cdot)$, independent of $\{x_{ij}\}_{j=1}^{N_{i-1}}$ and $\{s_j\}_{j\neq i}$

case 2: s_i, $i=1,2,\ldots$ is a prespecified sequence.

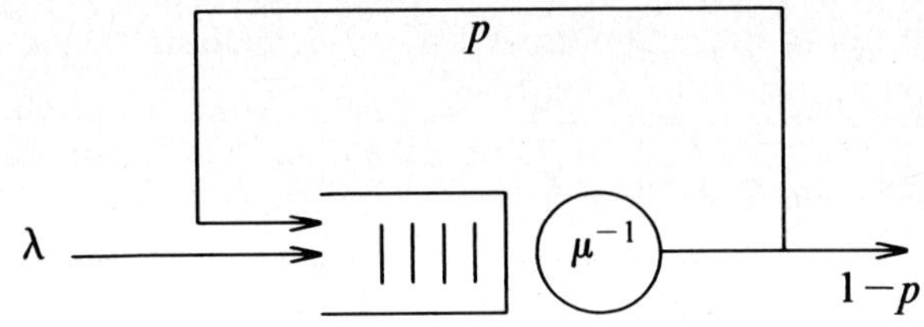

FIGURE 1. *M/G/1 queue with Bernoulli feedback*

Our analysis proceeds assuming case 1, but we will comment on the slight modification needed to accommodate case 2.

Denote the joint Laplace Stieltjes transform of $Y_1,\ldots,Y_m$ by $\phi_m(\underline{\theta})$,

$$\phi_m(\underline{\theta}) = E\{e^{-(\theta_1 Y_1+\theta_2 Y_2+\cdots+\theta_m Y_m)}\}. \tag{2.2}$$

The basic idea in evaluating $\phi_m(\theta)$ is to condition successively on $(N_0, N_1,\ldots,N_{i-1}; Y_1, Y_2,\ldots,Y_i)$ beginning with $i=m-1$ $(i=m-1,\ldots,1)$ and use the fact that N_i (which determines the distribution Y_{i+1}) is conditionally independent of $(N_0,\ldots,N_{i-2}; Y_1,\ldots, Y_{i-1})$ given (N_{i-1}, Y_i).

Thus we condition on $(N_0,\ldots,N_{m-2}; Y_1,\ldots, Y_{m-1})$ and obtain

$$\phi_m(\underline{\theta}) = E\left\{e^{-(\theta_1 Y_1+\cdots+\theta_{m-1}Y_{m-1})}\, E\left\{e^{-\theta_m\left[\sum_{j=1}^{N_{m-1}} x_{mj}+s_m\right]} \,|N_{m-2}, Y_{m-1}\right\}\right\},$$

where

$$E\left\{e^{-\theta_m\left[\sum_{j=1}^{N_{m-1}} x_{mj}+s_m\right]} \,|N_{m-2}, Y_{m-1}\right\} = F^*(\theta_m)\, E\left\{[F^*(\theta_m)]^{N_{m-1}} \,|N_{m-2}, Y_{m-1}\right\}$$

and $F^*(\cdot)$ denotes the Laplace Stieltjes transform of $F(\cdot)$.

But N_i, $i=1,2,\ldots$ is the sum of two components N_i^p, N_i^f where N_i^f is the number of customers fed back at the end of the tagged customer's i-th sojourn from among these present (N_{i-1}) at the beginning of its i-th sojourn and N_i^p is the number of Poisson arrivals during Y_i. Thus we have

$$E\left\{F^*(\theta_m)^{N_{m-1}} \,|N_{m-2}, Y_{m-1}\right\} = [(1-p)+pF^*(\theta_m)]^{N_{m-2}} e^{-\lambda(1-F^*(\theta_m))Y_{m-1}}$$

since N_i^f, N_i^p are conditionally independent given (N_{i-1}, Y_i).

After this first conditioning step we obtain:

$$\phi_m(\underline{\theta}) = F^*(\theta_m)E\left\{e^{-(\theta_1 Y_1+\cdots+\theta_{m-2}Y_{m-2})}[1-p+pF^*(\theta_m)]^{N_{m-2}} \times e^{-[\theta_{m-1}+\lambda(1-F^*(\theta_m))]Y_{m-1}}\right\}. \tag{2.3}$$

Now, we can condition on $(N_0,\ldots, N_{m-3}; Y_1,\ldots, Y_{m-2})$, repeat the above arguments, and continue in this fashion for $m-1$ steps. Notation is a burden until one notices the following recursive parameterization:

$$z_m(\underline{\theta}) = 1$$
$$\psi_m(\underline{\theta}) = \theta_m + \lambda(1 - z_m(\underline{\theta}))$$
$$\vdots$$
$$z_i(\underline{\theta}) = (1-p+pz_{i+1}(\underline{\theta}))F^*(\psi_{i+1}(\underline{\theta}))$$
$$\psi_i(\underline{\theta}) = \theta_i + \lambda(1 - z_i(\underline{\theta}))$$
$$\vdots$$
$$z_0(\underline{\theta}) = (1-p+pz_1(\underline{\theta}))F^*(\psi_1(\underline{\theta}))$$

With this notation, and suppressing $\underline{\theta}$ in $z_i(\underline{\theta})$ and $\psi_i(\underline{\theta})$, (2.3) takes the form

$$\phi_m(\underline{\theta}) = F^*(\psi_m)E\left\{e^{-(\theta_1 Y_1 + \cdots + \theta_{m-2}Y_{m-2})}(1-p+pz_{m-1})^{N_{m-2}}e^{-\psi_{m-1}Y_{m-1}}\right\}. \tag{2.4}$$

After $m-1$ steps, we obtain

$$\phi_m(\underline{\theta}) = \prod_{i=2}^{m} F^*(\psi_i)E\left\{(1-p+pz_1)^{N_0}e^{-\psi_1 Y_1}\right\}. \tag{2.5}$$

Now

$$Y_1 = x_{11} + \sum_{j=2}^{N_0} x_{1j} + s_1$$

where x_{11} is the residual service time of the customer in service as seen by the tagged customer upon arrival to the system. Thus, we have

$$E\{(1-p+pz_1)^{N_0}e^{-\psi_1 Y_1}\} = (1-\rho)F^*(\psi_1) + \rho E\{z_0^{N_0}e^{-\psi_1 R} \mid N_0 > 0\}$$

where $R = x_{11}$, $z_0 = (1-p+pz_1)F^*(\psi_1)$ and $\rho = \dfrac{\lambda\mu^{-1}}{1-p}$. In [9] Takács uses a result by WISHART [10] to obtain $E\{z_0^{N_0}e^{-\psi_1 R} \mid N_0 > 0\}$ for the M/G/1 queue with Bernoulli feedback. Using this result, we obtain

$$E\{(1-p+pz_1)^{N_0}e^{-\psi_1 Y_1}\} = (1-\rho)F^*(\psi_1) + \tag{2.6}$$
$$(1-\rho)\lambda z_0(1-z_0)\frac{[F^*(\psi_1) - F^*(\lambda(1-z_0))]}{[z_0 - (1-p+pz_0)F^*(\lambda(1-z_0))](\psi_1 - \lambda + \lambda z_0)}.$$

In the special case of exponentially distributed service times, (2.6) reduces to $F^*(\psi_1)E\{z_0^{N_0}\}$ and $E\{z_0^{N_0}\} = \dfrac{1-\rho}{1-\rho z_0}$ so (2.5) can be written:

$$\phi_m(\underline{\theta}) = \left[\prod_{i=1}^{m} F^*(\psi_i)\right]\frac{1-\rho}{1-\rho z_0}. \tag{2.7}$$

If the service times of the tagged customer on successive passes form a specified sequence $s_1, s_2, \ldots$ (rather than random variables drawn from the service time distribution $F(\cdot)$), it is easy to see that the only change in (2.7) is to replace $F^*(\psi_i)$ by $e^{-\psi_i s_i}$, $i=1,2,\ldots,m$.

Before considering numerical results obtained via transform inversion, we examine the special case $m=2$ with exponentially distributed service times for which explicit closed form results are easily obtained. We assume that the service time of the tagged customer is also exponentially distributed. It is easy to verify that in this case, we obtain:

$$z_2(\underline{\theta})=1, \quad \psi_2(\underline{\theta})=\theta_2, \quad F^*(\psi_2)=\frac{\mu}{\theta_2+\mu},$$

$$z_1(\underline{\theta}) = \frac{\mu}{\theta_2+\mu}, \quad \psi_1(\underline{\theta})=\theta_1+\frac{\lambda\theta_2}{\theta_2+\mu}, \quad F^*(\psi_1)=\frac{\mu(\theta_2+\mu)}{(\theta_1+\mu)(\theta_2+\mu)+\lambda\theta_2},$$

$$z_0(\underline{\theta}) = \frac{\mu((1-p)\theta_2+\mu)}{(\theta_1+\mu)(\theta_2+\mu)+\lambda\theta_2},$$

and therefore (2.7) simplifies to yield

$$\phi_2(\theta_1,\theta_2) = \frac{\mu^2(1-\rho)}{(\theta_1+\mu)(\theta_2+\mu)-\rho\mu^2} \tag{2.8}$$

where $\rho=\dfrac{\lambda\mu^{-1}}{1-p}$.

From (2.8) it is obvious that the density of the sojourn time for each of the two passes is exponential with parameter $\mu(1-\rho)$. Of course, this is obvious since the model in question is a BCMP [1] network. If we denote the density of the sojourn time of a tagged customer after completing its second pass by $f_2(\cdot)$, then from (2.8) we have

$$\phi_2(\theta,\theta) = E\{e^{-\theta(Y_1+Y_2)}\}=\frac{\mu(1-\rho)}{2\sqrt{\rho}}\left[\frac{1}{\theta+\alpha_1}-\frac{1}{\theta+\alpha_2}\right]$$

where $\alpha_1 = \mu(1-\sqrt{\rho})$, $\alpha_2=\mu(1+\sqrt{\rho})$, and therefore

$$f_2(t) = \frac{\mu(1-\rho)}{2\sqrt{\rho}}\{e^{-\mu(1-\sqrt{\rho})t}-e^{-\mu(1+\sqrt{\rho})t}\}, \quad t\geqslant 0.$$

If $\bar{F}_2(t)$ denotes the complementary distribution of Y_1+Y_2 we have

$$\bar{F}_2(t) = \frac{1+\sqrt{\rho}}{2\sqrt{\rho}}e^{-\mu(1-\sqrt{\rho})t}-\frac{1-\sqrt{\rho}}{2\sqrt{\rho}}e^{-\mu(1+\sqrt{\rho})t}, \quad t\geqslant 0. \tag{2.9}$$

By way of contrast, the complementary distribution of Y_1+Y_2 *assuming* that Y_1 and Y_2 are independent is given by

$$\bar{F}_2(t) = (1+\mu(1-\rho)t)e^{-\mu(1-\rho)t}, \quad t\geqslant 0. \tag{2.10}$$

It is not surprising to find that the exact distribution of Y_1+Y_2 has a longer 'tail' than the distribution obtained by assuming independent passes since the two successive passes are positively correlated. Although intuitively obvious, this can be verified by calculating the correlation coefficient γ between Y_1 and Y_2. Thus

$$E(Y_1 Y_2) = \frac{\partial^2}{\partial\theta_1\partial\theta_2}\phi_2(\theta_1,\theta_2)\,|_{\theta_1=\theta_2=0} = \frac{(1+\rho)}{(1-\rho)^2}\mu^{-2},$$

and since $E(Y_1)=E(Y_2)=\frac{\mu^{-1}}{1-\rho}$ we obtain

$$\gamma=\rho.$$

It is interesting to note that the Bernoulli feedback parameter p enters into the distribution of Y_1+Y_2 only via the server utilization ρ. Thus, as long as the server utilization remains constant, the distribution after the 2nd pass is independent of the average number of passes. The numerical examples in Figures[1] 5 and 6 show that this is a peculiarity for the case $m=2$.

Fig. 2 contrasts equations (2.9) and (2.10), using parameters $\lambda=.25$, $\mu^{-1}=1.0$ and $p=0.5$. The mean sojourn time of customers in this M/M/1 system with Bernoulli feedback is 4 (the mean number of passes a customer makes is 2, and the average sojourn time per pass is two). The curve labelled 'exact' is the complementary distribution of the sojourn time of a tagged customer which has just completed its second pass. The curve labelled 'independent passes' is the complementary distribution of a random variable obtained by convolving the sojourn time distribution in an M/M/1 system having parameters $\rho=.50$, $\mu^{-1}=1.0$ with itself. As asserted earlier, the exact distribution of Y_1+Y_2 has a longer tail than the independent pass approximation. Figures 3 and 4 show a similar comparison for tagged customers completing their 4th and 8th pass respectively. The parameters chosen in Figures 2-4 are such that the server utilization and the mean sojourn time of the tagged customer at the completion of its m-th pass (respectively 2,4,8) is identical (server utilization = 0.50, mean sojourn time = 4). As expected, the difference between the tails of the exact distribution and the 'independent pass' approximation increases with m. The Bernoulli parameter p has been chosen so that $m=(1-p)^{-1}$ in Figures 2-4. Fig. 5 illustrates the relative insensitivity of the sojourn time distribution of a tagged customer upon completing its 4th pass to p (e.g., to the mean number of passes). In this figure the server utilization is held constant at 0.50 by varying λ ($\lambda=1-p$). Fig. 6 is similar to Fig. 5 but with m increased to 16 and the server utilization again held constant at 0.50 ($\lambda=4(1-p)$). In Figures 7 and 8 we plot the complementary sojourn time distribution after 4 and 32 passes, respectively, for deterministic, exponential and hyperexponential service time distributions. The latter has density $f(t)=2\mu d^2 e^{-2\mu dt}+2\mu(1-d)^2 e^{-2\mu(1-d)t}$, where $d=.113$, which corresponds to a coefficient of variation equal to 2. The mean service time obtained after m passes ($m=4,32$) is held constant at 4, the server utilization is .50 and p is chosen so that $m=(1-p)^{-1}$. Note that the behavior evident in Figure 8 is not surprising. Our model should tend to the M/D/1 processor sharing model

1. Fig.2 - Fig. 11 have been collected at the end of this paper.

when $m \to \infty$ and $1/\mu \to 0$ with m/μ held constant. The numerical inversion technique used in Figures 2-8 is due to JAGERMAN [6,7].

3. MULTI CLASS GENERALIZATIONS

Consider an M/G/1 Bernoulli feedback model with K classes: class k has Bernoulli parameter p_k, service time distribution $F_k(\cdot)$ and Poisson arrival rate λ_k, $k=1,\ldots,K$. In the context of this model we discuss two different multi class problems. The first problem is the extent to which the approach in Section 2 can be generalized. In particular, in Section 3.1 we show that the details in Section 2 generalize easily, but the resulting Laplace Stieltjes transform of successive sojourn times of a class k customer depends on a term which we cannot evaluate. Although a variety of approximations may be suggested, this approach is not pursued. In Sections 3.2-3.4 we narrow our scope considerably and show that the mean number of class k customers in system and the mean sojourn time of a class k customer during its i-th pass (and its cumulative mean sojourn time after completing i passes) can be computed. Explicit closed form results are obtained for the two class case.

3.1. Laplace Stieltjes transform of successive sojourn times

Let Y_i^k denote the i-th sojourn time of a tagged class k customer and let $N_{j,0}^k$, $j=1,\ldots,K$ denote the number of class j customers in system as seen by a class k Poisson arrival (of course $N_{j,0}^1 = N_{j,0}^2 = \ldots = N_{j,0}^K$). More generally $N_{j,i-1}^k$ denotes the number of class j customers in system as seen by a class k customer when beginning its i-th pass through the queue. The basic observation is that for any $k=1,\ldots,K$ the sequence $(N_{j,i-1}^k, Y_i^k)$, $i=1,2,\ldots$ is Markovian for any $j=1,\ldots,K$. Thus, by focusing on a class k customer we can compute

$$\phi_m^k(\underline{\theta}) = E\{e^{-(\theta_1 Y_1^k + \cdots + \theta_m Y_m^k)}\}, \quad m=1,2,\ldots, \tag{3.1.1}$$

by successively conditioning on $(\underline{N}_0^k,\ldots,\underline{N}_{i-1}^k; Y_1^k,\ldots,Y_i^k)$ in the order $i=m-1$ to $i=1$ where

$$\underline{N}_j^k = (N_{1,j}^k,\ldots,N_{K,j}^k).$$

Proceeding in this manner - which is perfectly analogous to that in Section 2 - we obtain

$$\phi_m^k(\underline{\theta}) = \prod_{i=2}^{m-1} F_k^*(\psi_i) E\{\prod_{j=1}^{K}[1-p_j+p_j z_1^j]^{N_{j,0}} e^{-\psi_1 Y_1^k}\}, \tag{3.1.2}$$

where

$$z_m^j = 1$$

$$\psi_m = \theta_m + \sum_{l=1}^{K} \lambda_l (1 - z_m^l)$$

$$\vdots$$

$$z_i^j = ((1-p_j)+p_j z_{i+1}^j)F_j^*(\psi_{i+1})$$

$$\psi_i = \theta_i + \sum_{l=1}^{K} \lambda_l (1-z_i^l)$$

$$\vdots$$

$$z_0^j = (1-p_j+p_j z_1^j)F_j^*(\psi_1)$$

for $j=1,\ldots,K$.

If $F_i(t)=1-e^{-\mu_i t}$, $t \geq 0$, then (3.1.2) simplifies to

$$\phi_m^k(\underline{\theta})=\prod_{i=1}^{m} F_k^*(\psi_i) E\{\prod_{i=1}^{K}(z_0^i)^{N_{i,0}}\}. \tag{3.1.3}$$

The joint generating function of $N_{1,0},\ldots,N_{K,0}$ is unknown except in the special (and uninteresting) case where $\mu_1 = \ldots = \mu_K$ and the $\{p_k\}$ are arbitrary - in which case the joint generating function factors and we obtain

$$E\{\prod_{i=1}^{K}(z_0^i)^{N_{i,0}}\} = \prod_{i=1}^{K} \frac{(1-\rho_i)}{(1-\rho_i z_0^i)}.$$

Approximation schemes which use (3.1.2) or (3.1.3) as points of departure may prove useful.

3.2. Mean sojourn time - general formulation

In this section we show how the mean sojourn time of each of the K classes can be calculated for any given pass. In Sections 3.3-3.4 closed form results are obtained for the two class case.

Let $N_{i,j}^k$ denote the mean number of class i customers in the system observed by a class k customer at the start of its $(j+1)$-st pass. Let b_k and b_{k2} denote, respectively, the first and the second moment of the service time of a class k customer. Let S_j^k denote the mean sojourn time of a class k customer during its j-th pass ($j=1,2,\ldots$). Also, let S^k denote the mean total sojourn time of a randomly selected class k customer. Thus,

$$S^k = \sum_{j=1}^{\infty} S_j^k p_k^{j-1}. \tag{3.2.1}$$

Note that, for $j \geq 1$, $i=1,\ldots,K$, $k=1,\ldots,K$,

$$N_{i,j}^k = p_i N_{i,j-1}^k + \lambda_i S_j^k. \tag{3.2.2}$$

Also, for $j \geq 2$, $k=1,\ldots,K$,

$$S_j^k = \sum_{i=1}^{K} N_{i,j-1}^k b_i + b_k. \tag{3.2.3}$$

Finally,

$$S_1^k = \sum_{i=1}^{K} N_{i,0}^k b_i + b_k + \sum_{i=1}^{K} \rho_i \left(\frac{b_{i2}}{2b_i} - b_i\right), \tag{3.2.4}$$

where $\rho_i = \dfrac{\lambda_i b_i}{1-p_i}$. Let $R_i = \rho_i(\dfrac{b_{i2}}{2b_i} - b_i) = b_i\rho_i(\dfrac{b_{i2}}{2b_i^2} - 1)$, and $R = \sum_{i=1}^{K} R_i$.

Note that $N_{i,0}^k$ is independent of k. From equations (3.2.2)-(3.2.4), we get

$$S_j^k = \sum_{i=1}^{K} N_{i,0} p_i^{j-1} b_i + b_k + R\delta_{\{j=1\}} + \sum_{i=1}^{K} \sum_{m=1}^{j-1} S_m^k \lambda_i b_i p_i^{j-1-m}, \tag{3.2.5}$$

where $\delta_I = 1$ or 0 depending on whether I is true or false. Equation (3.2.5) can be derived directly by looking at the components contributing to the j-th sojourn time. In particular, b_k is the service time of the class k customer during its j-th pass; $N_{i,0} p_i^{j-1}$ is the mean number of class i customers who were present when the tagged class k customer arrived and are still present when it begins its j-th pass; $S_m^k \lambda_i p_i^{j-1-m}$ is the number of class i customers arriving during the tagged class k customer's m-th pass ($m<j$) and still present when the tagged customer begins its j-th pass. These class i customers contribute, on the average, b_i to the waiting time of the tagged class k customer on its j-th pass. Finally, R is the correction term required for the first pass to account for the fact that the customer in service, if any, when the tagged class k customer freshly arrives, contributes the forward recurrence time of its service time.

Now let

$$S^k(z) = \sum_{j=1}^{\infty} S_j^k z^j, \quad k=1,\ldots,K, \tag{3.2.6}$$

and

$$N_i^k(z) = \sum_{j=0}^{\infty} N_{i,j}^k z^j, \quad k=1,\ldots,K; \quad i=1,\ldots,K. \tag{3.2.7}$$

Then, from (3.2.2), we get

$$N_i^k(z) = N_{i,0} + p_i z N_i^k(z) + \lambda_i S^k(z),$$

or

$$N_i^k(z) = \frac{N_{i,0} + \lambda_i S^k(z)}{1-p_i z}. \tag{3.2.8}$$

Also, from (3.2.5), we get

$$S^k(z) = \sum_{i=1}^{K} \frac{N_{i,0} b_i z}{1-p_i z} + \frac{b_k z}{1-z} + Rz + zS^k(z) \sum_{i=1}^{K} \frac{\hat{\rho}_i}{1-p_i z},$$

or

$$S^k(z) = \frac{z}{1 - \sum_{i=1}^{K} \frac{\hat{\rho}_i z}{1-p_i z}} \left[\sum_{i=1}^{K} \frac{N_{i,0} b_i}{1-p_i z} + \frac{b_k}{1-z} + R \right], \tag{3.2.9}$$

where $\hat{\rho}_i = \lambda_i b_i$, $i = 1, \ldots, K$.

Next, note that by Little's law,

$$S^k = \sum_{j=1}^{\infty} S_j^k p_k^{j-1} = \frac{S^k(p_k)}{p_k} = \frac{N_{k0}}{\lambda_k}, \quad k = 1, \ldots, K.$$

Thus, from (3.2.9) we get for $k = 1, \ldots, K$,

$$N_{k,0} = \frac{S^k(p_k)\lambda_k}{p_k} = \frac{\lambda_k}{1 - \sum_{i=1}^{K} \frac{\hat{\rho}_i p_k}{1 - p_i p_k}} \left[\sum_{i=1}^{K} \frac{N_{i,0} b_i}{1 - p_i p_k} + \frac{b_k}{1 - p_k} + R \right]. \tag{3.2.10}$$

Equation (3.2.10) provides K linear equations in K unknowns $\{N_{k,0}, k = 1, 2, \ldots, K\}$. Once these are determined, S^k, $S^k(z)$, $N_i^k(z)$ are readily obtained from (3.2.8) and (3.2.9). $S^k(z)$ and $N_i^k(z)$ can then be numerically or analytically inverted to obtain $S_j^k, N_{i,j}^k$ for $j \geq 1$. We now proceed to do this for $K = 2$.

3.3. Mean total sojourn time and number in system for the two class case

For $K = 2$, equation (3.2.9) reduces to

$$\frac{S^1(z)}{z} = \frac{N_{10} b_1 (1 - p_2 z) + N_{20} b_2 (1 - p_1 z) + \left[\frac{b_1}{1 - z} + R \right] (1 - p_1 z)(1 - p_2 z)}{D(z)} \tag{3.3.1}$$

$$\frac{S^2(z)}{z} = \frac{N_{10} b_1 (1 - p_2 z) + N_{20} b_2 (1 - p_1 z) + \left[\frac{b_2}{1 - z} + R \right] (1 - p_1 z)(1 - p_2 z)}{D(z)} \tag{3.3.2}$$

where

$$\begin{aligned} D(z) &= 1 - (p_1 + p_2 + \hat{\rho}_1 + \hat{\rho}_2) z + (p_1 p_2 + p_1 \hat{\rho}_2 + p_2 \hat{\rho}_1) z^2 \\ &= (p_1 p_2 + p_1 \hat{\rho}_2 + p_2 \hat{\rho}_1)(z - z_1)(z - z_2), \end{aligned} \tag{3.3.3}$$

and

$$z_1, z_2 = \frac{(p_1 + p_2 + \hat{\rho}_1 + \hat{\rho}_2) \pm \sqrt{(p_1 - p_2 + \hat{\rho}_1 - \hat{\rho}_2)^2 - 4 \hat{\rho}_1 \hat{\rho}_2}}{2(p_1 p_2 + p_1 \hat{\rho}_2 + p_2 \hat{\rho}_1)}. \tag{3.3.4}$$

Equation (3.2.10) gives

$$\frac{N_{10}}{\lambda_1} = \frac{N_{10} b_1 (1 - p_1 p_2) + N_{20} b_2 (1 - p_1^2) + \left[\frac{b_1}{1 - p_1} + R \right] (1 - p_1^2)(1 - p_1 p_2)}{D(p_1)} \tag{3.3.5}$$

and

$$\frac{N_{20}}{\lambda_2} = \frac{N_{20} b_2 (1 - p_1 p_2) + N_{10} b_1 (1 - p_2^2) + \left[\frac{b_2}{1 - p_2} + R \right] (1 - p_1 p_2)(1 - p_2^2)}{D(p_2)} \tag{3.3.6}$$

where $D(\cdot)$ is given by (3.3.3).

Solving (3.3.5) and (3.3.6) for N_{10} and N_{20}, we get, after considerable simplification,

$$N_{10} = \frac{\rho_1}{1-\rho}+\frac{\rho_2}{1-\rho}\frac{\lambda_1(b_2-b_1)}{1-p_1\rho_2-p_2\rho_1-p_1p_2(1-\rho)} + \frac{\lambda_1 R[1-p_2\rho-p_1p_2(1-\rho)]}{(1-\rho)(1-p_1\rho_2-p_2\rho_1-p_1p_2(1-\rho))}, \tag{3.3.7}$$

and

$$N_{20} = \frac{\rho_2}{1-\rho}+\frac{\rho_1}{1-\rho}\frac{\lambda_2(b_1-b_2)}{1-p_1\rho_2-p_2\rho_1-p_1p_2(1-\rho)} + \frac{\lambda_2 R[1-p_1\rho-p_1p_2(1-\rho)]}{(1-\rho)(1-p_1\rho_2-p_2\rho_1-p_1p_2(1-\rho))},$$

where $\rho = \rho_1+\rho_2$. Also, using Little's Law we have

$$S^1 = \frac{b_1}{(1-p_1)(1-\rho)}+\frac{\rho_2}{1-\rho}\frac{b_2-b_1}{1-p_1\rho_2-p_2\rho_1-p_1p_2(1-\rho)} + \frac{R[1-p_2\rho-p_1p_2(1-\rho)]}{(1-\rho)(1-p_1\rho_2-p_2\rho_1-p_1p_2(1-\rho))}, \tag{3.3.8}$$

and

$$S^2 = \frac{b_2}{(1-p_2)(1-\rho)}+\frac{\rho_1}{1-\rho}\frac{b_1-b_2}{1-p_1\rho_2-p_2\rho_1-p_1p_2(1-\rho)} + \frac{R[1-p_1\rho-p_1p_2(1-\rho)]}{(1-\rho)(1-p_1\rho_2-p_2\rho_1-p_1p_2(1-\rho))}.$$

We now look at some special cases:

(i) symmetric feedback: $p_1=p_2=p$.

In this case,

$$1-p_1\rho-p_1p_2(1-\rho) = (1-p)(1+p(1-\rho)).$$

Thus,

$$N_{10} = \frac{\rho_1}{1-\rho}+\frac{\rho_1\rho_2}{1-\rho}\frac{\frac{b_2}{b_1}-1}{1+p(1-\rho)}+\frac{\lambda_1 R}{1-\rho}, \tag{3.3.9}$$

$$N_{20} = \frac{\rho_2}{1-\rho}+\frac{\rho_1\rho_2}{1-\rho}\frac{\frac{b_1}{b_2}-1}{1+p(1-\rho)}+\frac{\lambda_2 R}{1-\rho}, \tag{3.3.10}$$

and

$$N_{10}+N_{20} = \frac{\rho}{1-\rho}+\frac{\rho_1\rho_2}{1-\rho}\frac{\frac{b_2}{b_1}+\frac{b_1}{b_2}-2}{1+p(1-\rho)}+\frac{(\lambda_1+\lambda_2)R}{1-\rho}. \tag{3.3.11}$$

(ii) $p_1=p_2=p,\ b_1=b_2=b,\ b_{12}=b_{22}=b_{02}$

This is a special case of (i). In this case,

$$R=b_1\rho_1(\frac{b_{12}}{2b_1^2}-1)+b_2\rho_2(\frac{b_{22}}{2b_2^2}-1) = \rho b(\frac{b_{02}}{2b^2}-1).$$

Thus, from (3.3.9) - (3.3.11), we get

$$N_{10} = \frac{\rho_1}{1-\rho}+\frac{\rho}{1-\rho}\hat{\rho}_1(\frac{b_{02}}{2b^2}-1),$$

$$N_{20} = \frac{\rho_2}{1-\rho}+\frac{\rho}{1-\rho}\hat{\rho}_2(\frac{b_{02}}{2b^2}-1),$$

$$N_{10}+N_{20} = \frac{\rho}{1-\rho}(1+\hat{\rho}(\frac{b_{02}}{2b^2}-1))=\frac{\rho}{1-\rho}(1+\rho(1-p)(\frac{b_{02}}{2b^2}-1)). \tag{3.3.12}$$

Note that, in this case, there is no real class distinction and the total number in system should behave like that in a single class M/G/1 queue with Bernoulli feedback or, equivalently, in a simple M/G/1 queue without feedback but with service time corresponding to the geometric number of service times in the queue with feedback. If we let $G=\Sigma_{j=1}^{\infty}(1-p)p^{j-1}F^{*(j)}$, then it is easy to verify that (3.3.12) gives the correct expression for the mean queue length in an M/G/1 queue with arrival rate $\lambda=\lambda_1+\lambda_2$ and service time distribution G.

(iii) $R=0$, b_i and p_i arbitrary.
This corresponds to a coefficient of variation of 1 for both classes. In particular, this includes the case where the service time distribution during each pass is exponential. Let $b_{it}=\dfrac{b_i}{1-p_i}$ denote the total mean service time of a class i customer, $i=1,2$. Then (3.3.7) can be rewritten as

$$N_{10} = \frac{\rho_1}{1-\rho}+\frac{\rho_1\rho_2}{1-\rho}\frac{\frac{b_{2t}}{b_{1t}}(1-p_2)-(1-p_1)}{1-p_1\rho_2-p_2\rho_1-p_1p_2(1-\rho)}, \tag{3.3.13}$$

$$N_{20} = \frac{\rho_2}{1-\rho}+\frac{\rho_1\rho_2}{1-\rho}\frac{\frac{b_{1t}}{b_{2t}}(1-p_1)-(1-p_2)}{1-p_1\rho_2-p_2\rho_1-p_1p_2(1-\rho)}, \tag{3.3.14}$$

and

$$N_{10}+N_{20} = \frac{\rho}{1-\rho}+\frac{\rho_1\rho_2}{1-\rho}\frac{(1-p_1)(\frac{b_{1t}}{b_{2t}}-1)+(1-p_2)(\frac{b_{2t}}{b_{1t}}-1)}{1-p_1\rho_2-p_2\rho_1-p_1p_2(1-\rho)}. \tag{3.3.15}$$

If, in addition, $b_{1t}=b_{2t}$, then (3.3.15) gives

$$N_{10}+N_{20} = \frac{\rho}{1-\rho}. \tag{3.3.16}$$

If the individual passes have exponential service times, the total service time of a customer is also exponentially distributed. When the mean total service time is independent of the customer class, each class has the same exponential distribution for total service time. This is also the distribution of the remaining total service time of a fed back customer, irrespective of its class and the number of passes it has gone through. It is clear that in this case, the entire distribution of the total number in the system in steady state should behave like that in an M/M/1 queue. Equation (3.3.16) shows that the mean value of the total number in the system is the same as in an M/M/1 queue without feedback as long as the coefficient of variation of the service time is 1 and the mean total service time is the same for both classes of customers.

(iv) $R=0$, $b_1=b_2=b$, arbitrary p_1 and p_2.
From equation (3.3.7), we get

$$N_{10} = \frac{\rho_1}{1-\rho}, \quad N_{20}=\frac{\rho_2}{1-\rho}. \tag{3.3.17}$$

If the service time per pass is exponentially distributed, then $b_1=b_2$ implies a product form network and the expressions in (3.3.17) follow. However, (3.3.17) is valid for any service time distributions having the same mean and coefficient of variation equal to 1.

3.4. Mean sojourn time per pass for the two class case
In this section we explicitly evaluate the mean sojourn time S_j^1 and S_j^2 for the j-th pass of class 1 and class 2 customers, respectively. Our starting point is the set of generating functions ($S^1(z)$ and $S^2(z)$) given by Equations (3.3.1) and (3.3.2). We write (3.3.1) as

$$\frac{S^1(z)}{z} = \frac{N_{10}b_1(1-p_2z)}{D(z)}+\frac{N_{20}b_2(1-p_1z)}{D(z)}+\frac{b_1(1-p_1z)(1-p_2z)}{(1-z)D(z)} + \frac{R(1-p_1z)(1-p_2z)}{D(z)}. \tag{3.4.1}$$

Recall that

$$D(z) = (p_1p_2+p_1\hat{\rho}_2+p_2\hat{\rho}_1)(z-z_1)(z-z_2), \tag{3.4.2}$$

where z_1 and z_2 are given by (3.3.4). Simple algebra shows that

$$(1-z_1)(1-z_2) = \frac{(1-\rho)(1-p_1)(1-p_2)}{p_1p_2+p_1\hat{\rho}_2+p_2\hat{\rho}_1}, \tag{3.4.3}$$

and

$$z_1z_2 = \frac{1}{p_1p_2+p_1\hat{\rho}_2+p_2\hat{\rho}_1}. \tag{3.4.4}$$

(3.4.3) and (3.4.4) imply that $z_1>1$, $z_2>1$ as long as $\rho<1$.
Since the two roots of $D(z)=0$ are explicitly known, each term on the

right-hand side of (3.4.1) can be inverted explicitly to give $S_j^1, j \geq 1$. This simple but tedious algebra results in

$$S_j^1 = \frac{b_1}{1-\rho} + \frac{Rp_1p_2}{p_1p_2+p_1\hat{\rho}_2+p_2\hat{\rho}_1}\delta_{\{j=1\}}$$
$$+\frac{z_1z_2}{z_1-z_2}\Bigg[N_{10}b_1(1-z_2p_2)+N_{20}b_2(1-z_2p_1)$$
$$+\left\{\frac{b_1}{1-z_2}+R\right\}(1-z_2p_1)(1-z_2p_2)\Bigg]\frac{1}{z_2^j}$$
$$-\frac{z_1z_2}{z_1-z_2}\Bigg[N_{10}b_1(1-z_1p_2)+N_{20}b_2(1-z_1p_1)$$
$$+\left\{\frac{b_1}{1-z_1}+R\right\}(1-z_1p_1)(1-z_1p_2)\Bigg]\frac{1}{z_1^j}. \qquad (3.4.5)$$

Equivalently,

$$S_j^1 = b_1\Bigg[\frac{1}{1-\rho}+\frac{z_1z_2(1-z_2p_1)(1-z_2p_2)}{(z_1-z_2)(1-z_2)}\frac{1}{z_2^j}$$
$$-\frac{z_1z_2(1-z_1p_1)(1-z_1p_2)}{(z_1-z_2)(1-z_1)}\frac{1}{z_1^j}\Bigg]$$
$$+R\Bigg[p_1p_2z_1z_2\delta_{\{j=1\}}+\frac{z_1z_2(1-z_2p_1)(1-z_2p_2)}{z_1-z_2}\frac{1}{z_2^j}$$
$$-\frac{z_1z_2(1-z_1p_1)(1-z_1p_2)}{z_1-z_2}\frac{1}{z_1^j}\Bigg]$$
$$+N_{10}b_1\left[\frac{z_1z_2(1-z_2p_2)}{z_1-z_2}\frac{1}{z_2^j}-\frac{z_1z_2(1-z_1p_2)}{z_1-z_2}\frac{1}{z_1^j}\right]$$
$$+N_{20}b_2\left[\frac{z_1z_2(1-z_2p_1)}{z_1-z_2}\frac{1}{z_2^j}-\frac{z_1z_2(1-z_1p_1)}{z_1-z_2}\frac{1}{z_1^j}\right]. \qquad (3.4.6)$$

Expressions for S_j^2 follow readily by interchanging the subscripts on the right-hand side of (3.4.5) and (3.4.6).

From (3.4.5) - (3.4.6) and the corresponding expressions for S_j^2, we observe the following:

(i) $$\lim_{j\to\infty} S_j^i = \frac{b_i}{1-\rho},\ i=1,2;$$

(ii) $$S_j^1 - S_j^2 = (b_1 - b_2)\times$$

$$\left[\frac{1}{1-\rho} + \frac{z_1 z_2}{z_1 - z_2}\left[\frac{(1-z_2p_1)(1-z_2p_2)}{(1-z_2)z_2^j} - \frac{(1-z_1p_1)(1-z_1p_2)}{(1-z_1)z_1^j}\right]\right].$$

This equation shows that if $b_1 = b_2$ but $b_{12} \neq b_{22}$, both classes have the same mean sojourn time for any given pass. Another consequence is that $S_1^1 - S_1^2 = b_1 - b_2$. Finally, it is easy to show that the quantity inside the bracket in the above equation is a positive increasing function of j, being 1 at $j=1$ and $\frac{1}{1-\rho}$ at $j=\infty$. Thus $\frac{S_j^1 - S_j^2}{b_1 - b_2}$ is an increasing function of j. As the example in Figure 10 illustrates, S_j^i, $i=1,2$ need not be monotone in j.

We conclude this section by looking at the special case where $p_1 = p_2 = p$. In this case,

$$z_1 = \frac{1}{p} > 1,$$

$$z_2 = \frac{1}{p+\rho(1-p)} > 1.$$

Thus, from (3.4.6),

$$S_j^1 = \frac{b_1}{1-\rho}\left[1 - \rho\frac{1}{z_2^{j-1}}\right] + R\left[pz_2\delta_{\{j=1\}} + \rho(1-p)\frac{1}{z_2^{j-2}}\right]$$

$$+ N_{10}b_1\left[\frac{1}{z_2^{j-1}}\right] + N_{20}b_2\left[\frac{1}{z_2^{j-1}}\right]$$

$$= \frac{b_1}{1-\rho} + \frac{Rp\delta_{\{j=1\}}}{p+\rho(1-p)}$$

$$+ \left[N_{10}b_1 + N_{20}b_2 + \frac{R\rho(1-p)}{p+\rho(1-p)} - \frac{b_1\rho}{1-\rho}\right]\frac{1}{z_2^{j-1}}.$$

From equations (3.3.9) and (3.3.10), we get

$$b_1N_{10} + b_2N_{20} = \frac{b_1\rho_1}{1-\rho} + \frac{b_2\rho_2}{1-\rho} + \frac{\rho(1-p)R}{1-\rho}.$$

Thus,

$$S_j^1 = \frac{b_1}{1-\rho} + \frac{Rp\delta_{\{j=1\}}}{p+\rho(1-p)}$$

$$+ \left[\frac{\rho_2}{1-\rho}(b_2 - b_1) + \frac{R\rho(1-p)(1+p-\rho p)}{(1-\rho)(p+\rho(1-p))}\right]\frac{1}{z_2^{j-1}}$$

$$= \frac{b_1}{1-\rho} + \frac{Rp\delta_{\{j=1\}}}{p+\rho(1-p)}$$

$$+(p+\rho(1-p))^{j-1}\left[\frac{\rho_2}{1-\rho}(b_2-b_1)+\frac{R(1-p)(1+p-p\rho)}{p+\rho(1-p)}\frac{\rho}{1-\rho}\right],$$

and

$$S_j^2 = \frac{b_2}{1-\rho}+\frac{Rp\delta_{\{j=1\}}}{p+\rho(1-p)}$$
$$+(p+\rho(1-p))^{j-1}\left[\frac{\rho_1}{1-\rho}(b_1-b_2)+\frac{R(1-p)(1+p-p\rho)}{p+\rho(1-p)}\frac{\rho}{1-\rho}\right].$$

When $R=0$, these reduce to

$$S_j^1 = \frac{b_1}{1-\rho}+(p+\rho(1-p))^{j-1}\frac{\rho_2}{1-\rho}(b_2-b_1),$$

$$S_j^2 = \frac{b_2}{1-\rho}+(p+\rho(1-p))^{j-1}\frac{\rho_1}{1-\rho}(b_1-b_2).$$

Figures 9-11 illustrate the behavior of S_j^i, $i=1,2$ for the parameters: $b_1=0.1$, $b_2=0.2$, $\rho_1=0.2$, $\rho_2=0.4$, $p_1=p_2=0.5$ and $R=0$, 0.03 and 0.1, respectively.

4. Concluding remarks

The results presented in this paper can be generalized in a number of directions. Thus, with only minor changes, the results in Sections 2 and 3 can be generalized to include pass dependent feedback. Such a pass dependent feedback model has recently been studied in [3] and indeed helps to place our results and those in [2] in perspective. Similarly, the model can be generalized to include pass dependent service time distributions as well as a priority structure - a version of the latter can be found in [5]. Finally, we remark that by using Jagerman's technique [6,7], inverting the transform expressions obtained in Section 2 to obtain distributions of interest, was essentially as simple (and as accurate) as obtaining the more traditional, but far less informative first and second moments.

References

1. F. Baskett, K.M. Chandy, R.R. Muntz, F.G. Palacios (1975). Open, closed and mixed networks of queues with different classes of customers. *J. Assoc. Comput. Mach. 22,* 248-260.
2. J. L. van den Berg, O. J. Boxma, W. P. Groenendijk (1987). Sojourn times in the M/G/1 queue with deterministic feedback. Submitted to *Stochastic Models.*
3. J. L. van den Berg, O. J. Boxma (1987). *Sojourn Times in Feedback Queues,* Report OS-R8710, Centre for Mathematics and Computer Science, Amsterdam, The Netherlands.
4. R. L. Disney, D. König, V. Schmidt (1984). Stationary queue-length and waiting-time distributions in single-server feedback queues. *Adv. in Appl. Probab. 16,* 432-446.
5. B. Fontana, C. Diaz Berzosa (1985). M/G/1 queue with N-priorities

and feedback: joint queue-length distributions and response time distribution for any particular sequence. M. AKIYAMA (ed.). *Proc. ITC 11*, North-Holland, Amsterdam, 452-458.
6. D. L. JAGERMAN (1978). An inversion technique for the Laplace transform with application to approximation. *The Bell System Tech. J. 57*, 669-710.
7. D. L. JAGERMAN (1982). An inversion technique for the Laplace transform. *The Bell System Tech. J. 61*, 1995-2002.
8. S. S. LAM, A. U. SHANKAR (1981). A derivation of response time distributions for a multi-class feedback queueing system. *Performance Evaluation 1*, 48-61.
9. L. TAKÁCS (1963). A single server queue with feedback. *The Bell Syst. Tech. J. 42*, 505-519.
10. D. M. G. WISHART (1960). Queueing systems in which the discipline is last-come-first-served. *Oper. Res. 8*, 591-599.

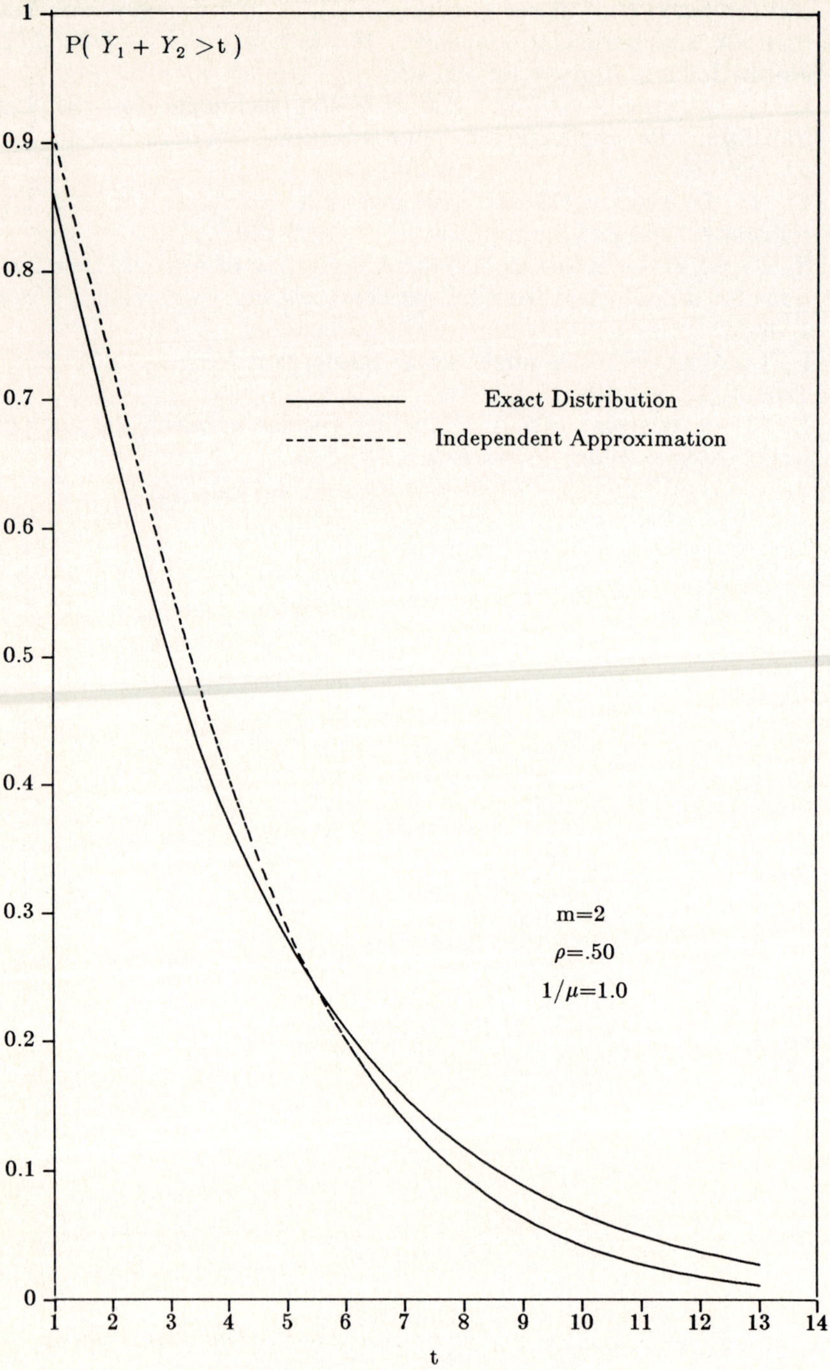

FIGURE 2. *Contrasting the exact and approximate distributions*

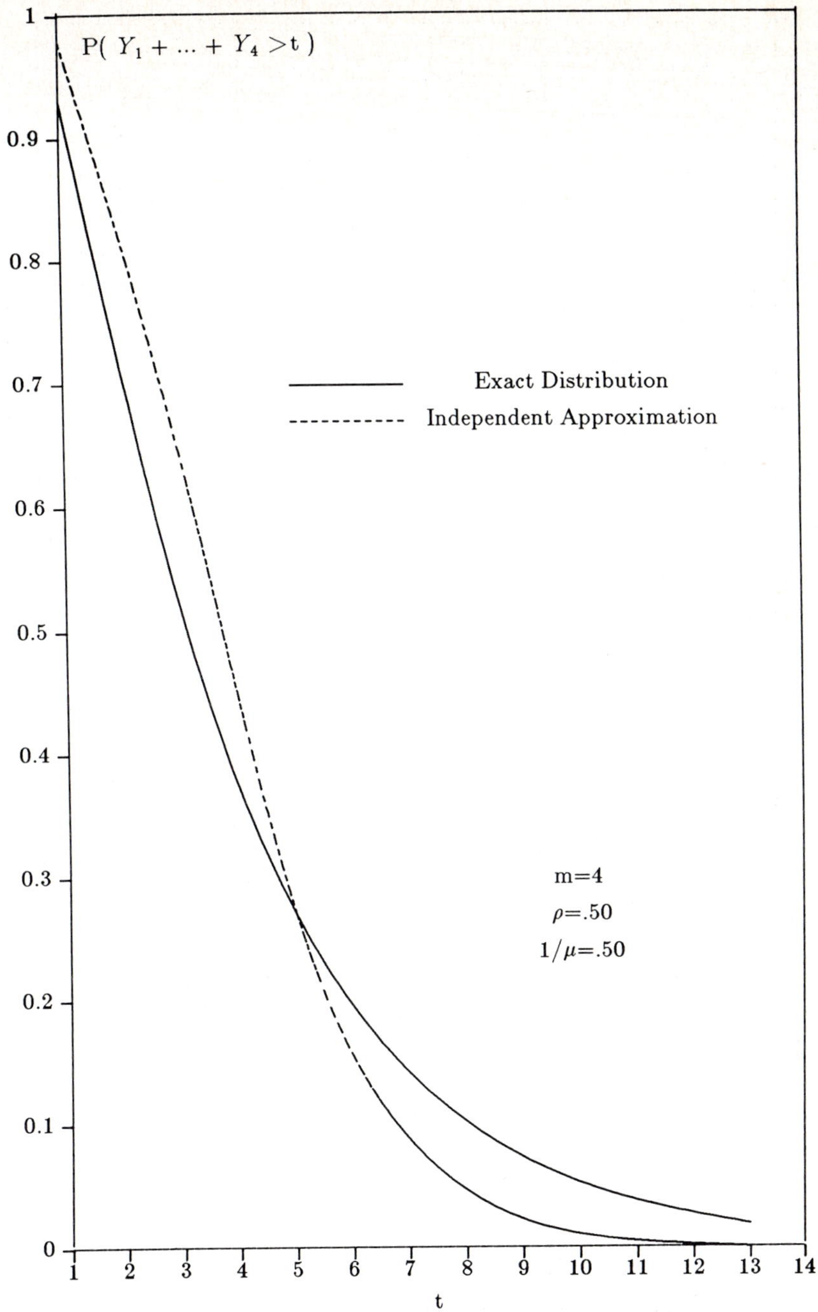

FIGURE 3. *Contrasting the exact and approximate distributions*

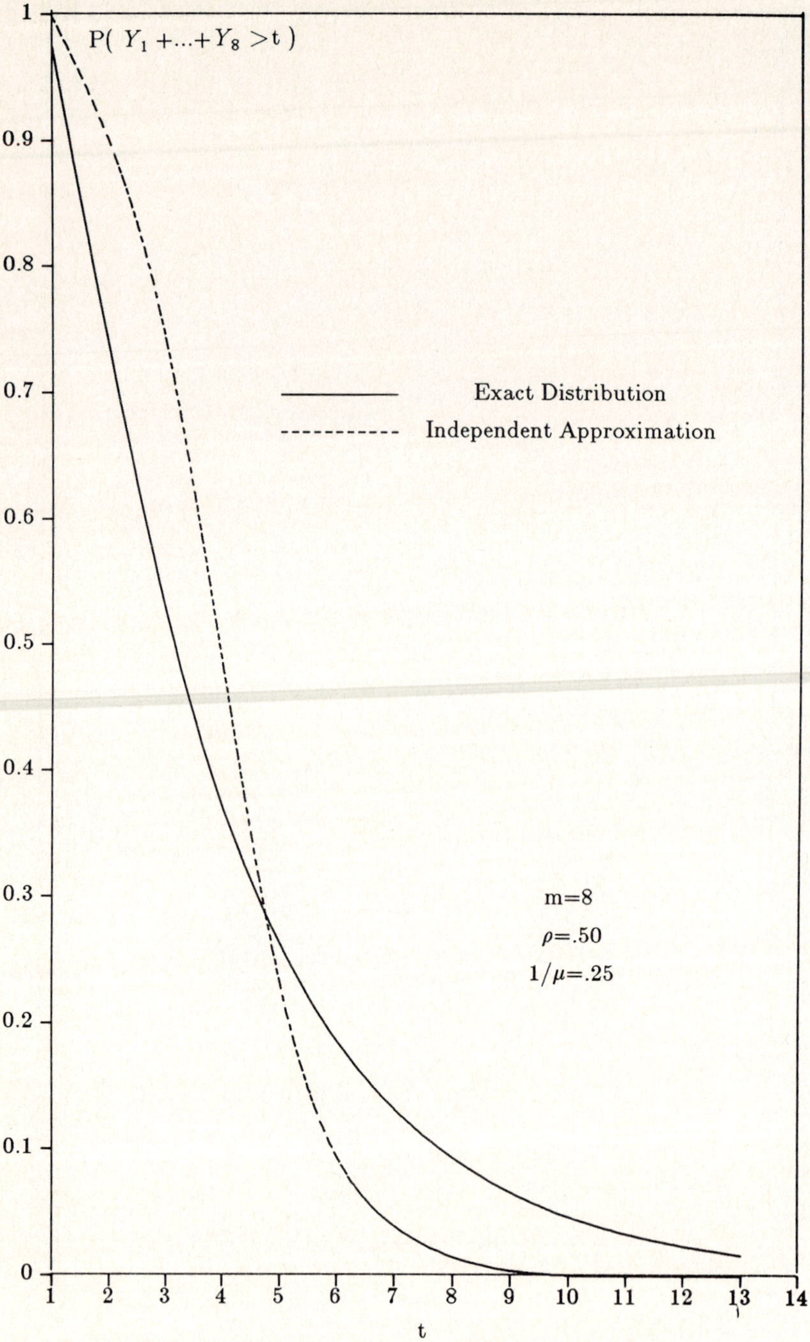

FIGURE 4. *Contrasting the exact and approximate distributions*

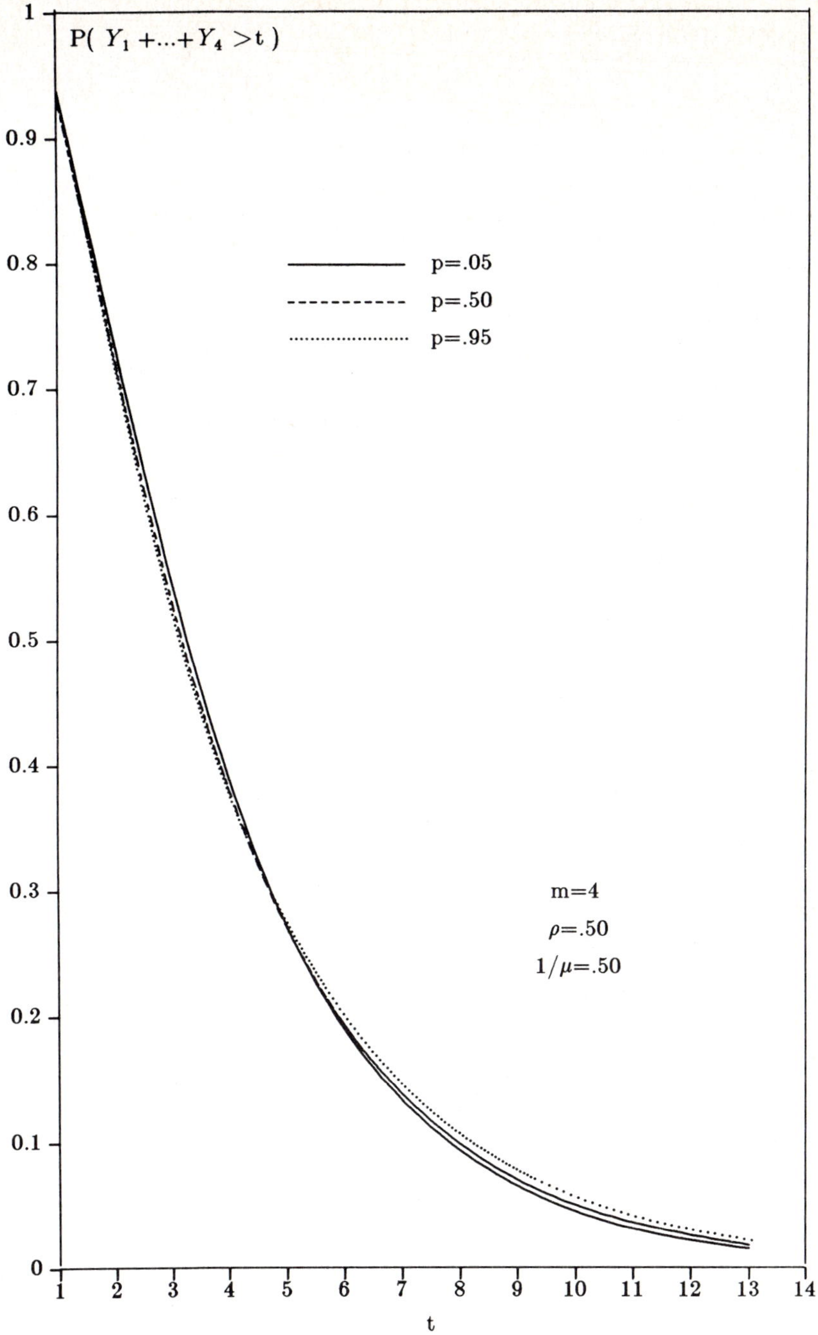

FIGURE 5. *Sensitivity to p with server occupancy fixed*

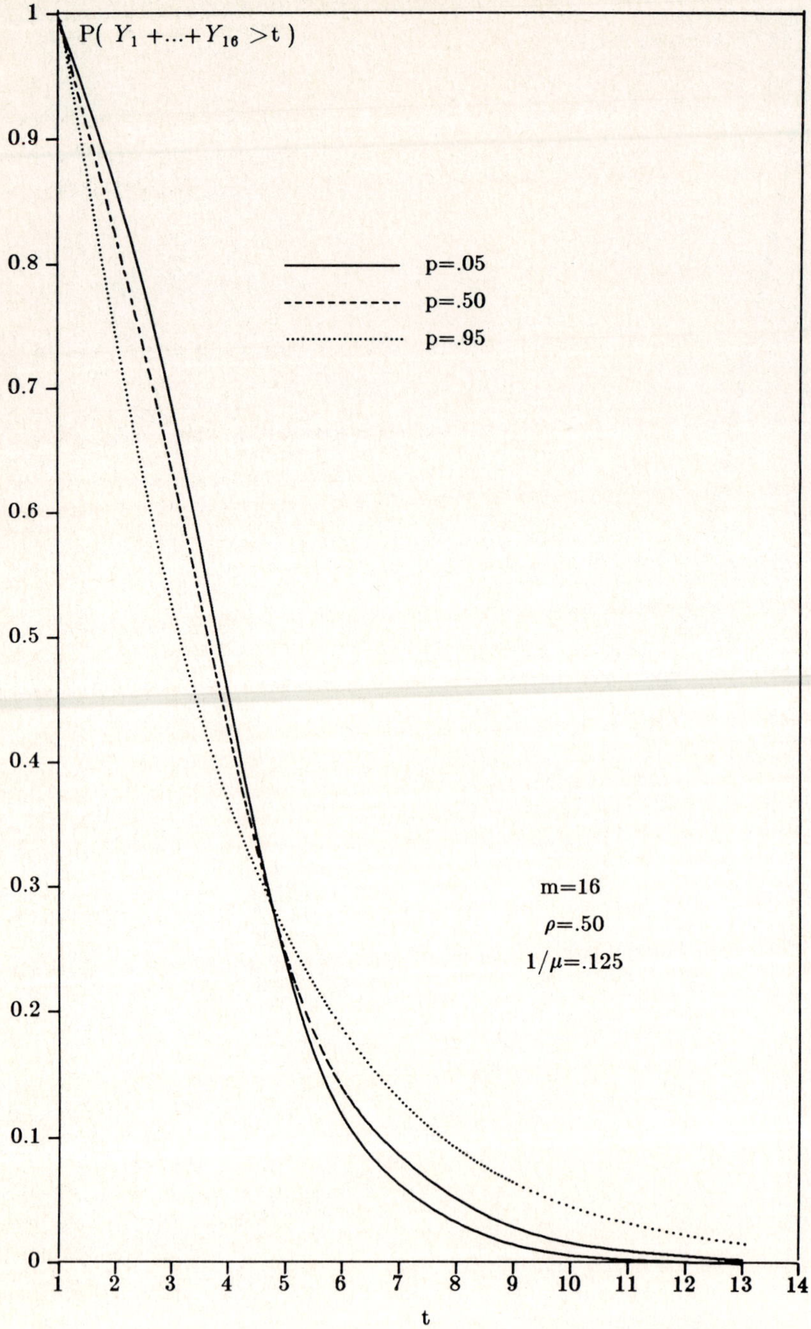

FIGURE 6. *Sensitivity to p with server occupancy fixed*

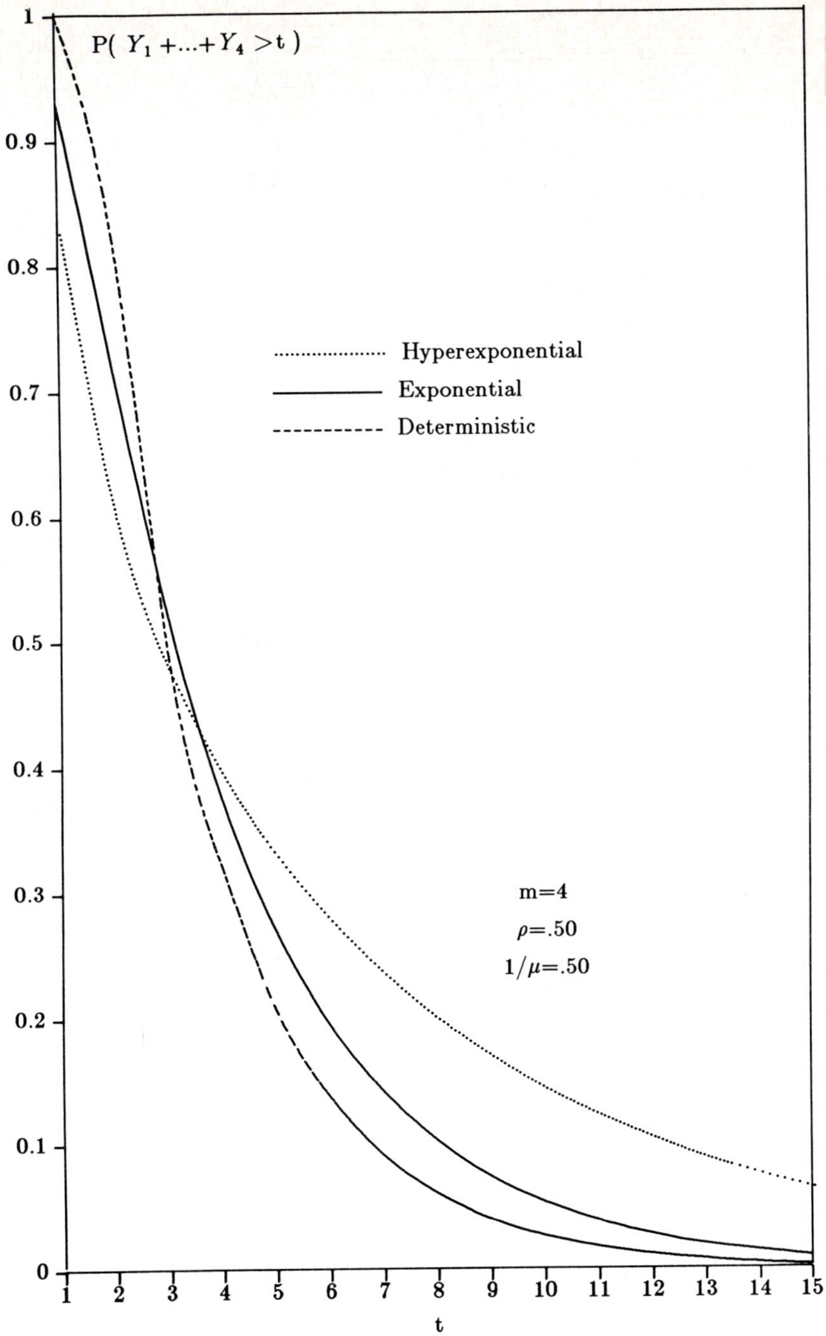

FIGURE 7. *Hyperexponential, exponential and deterministic distributions*

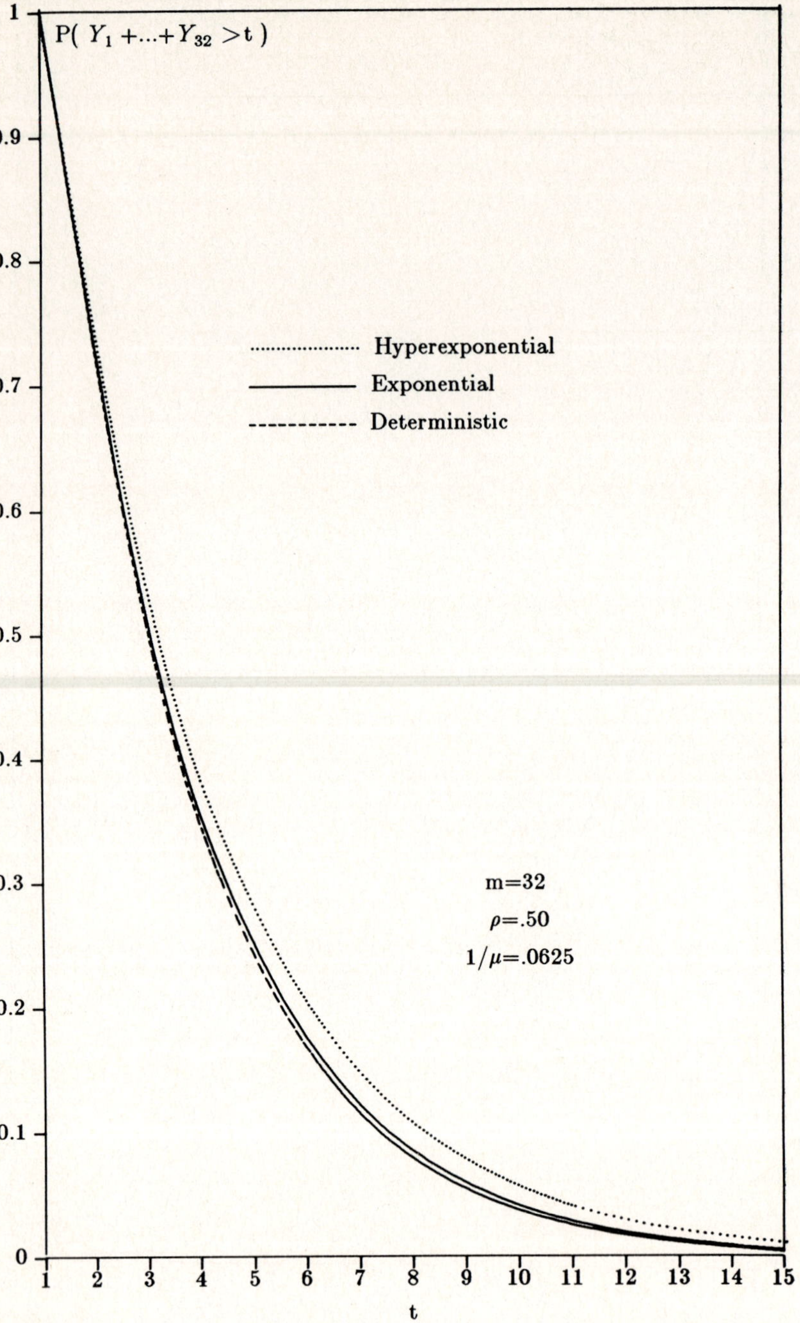

FIGURE 8. *Hyperexponential, exponential and deterministic distributions*

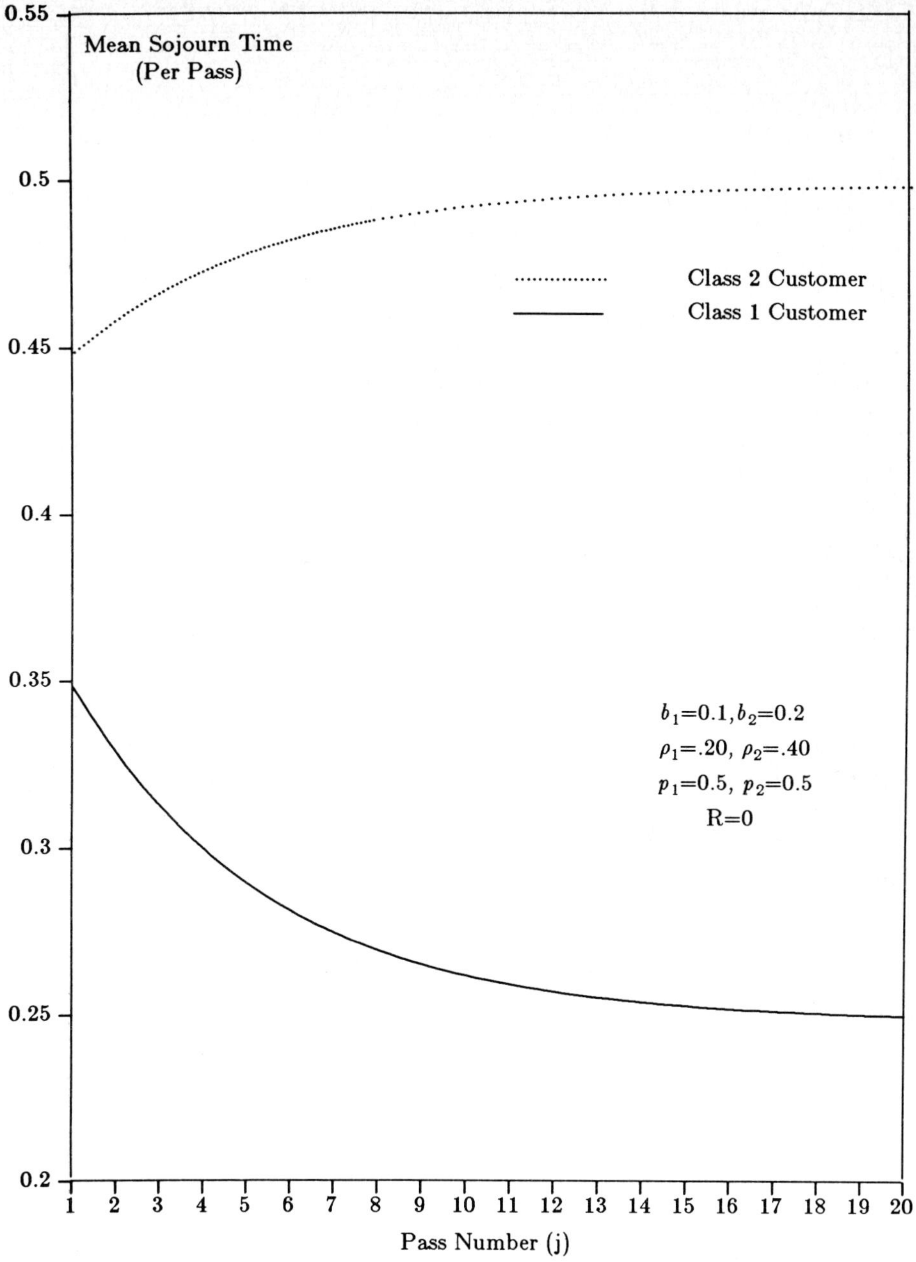

FIGURE 9. *Mean sojourn time per pass*

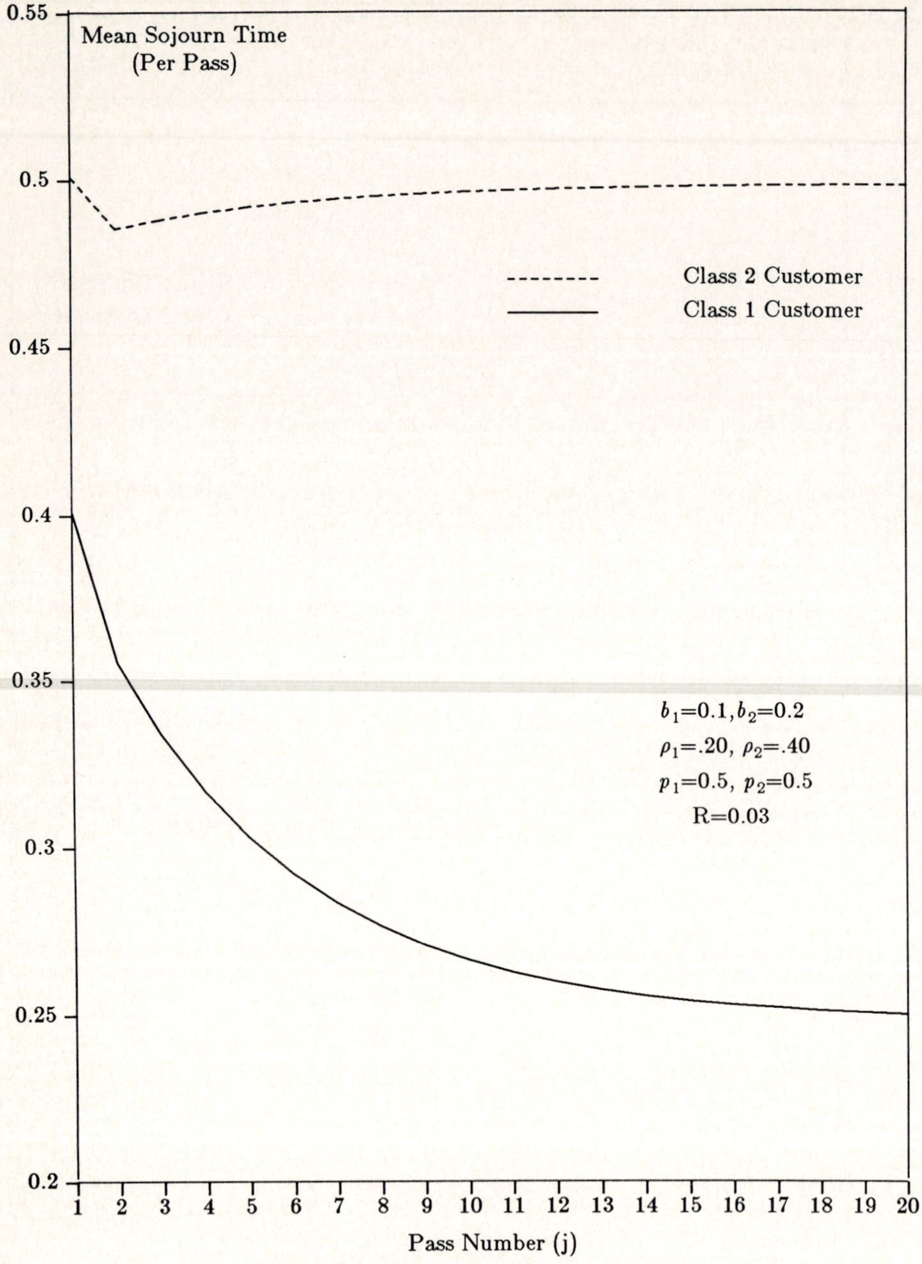

FIGURE 10. *Mean sojourn time per pass*

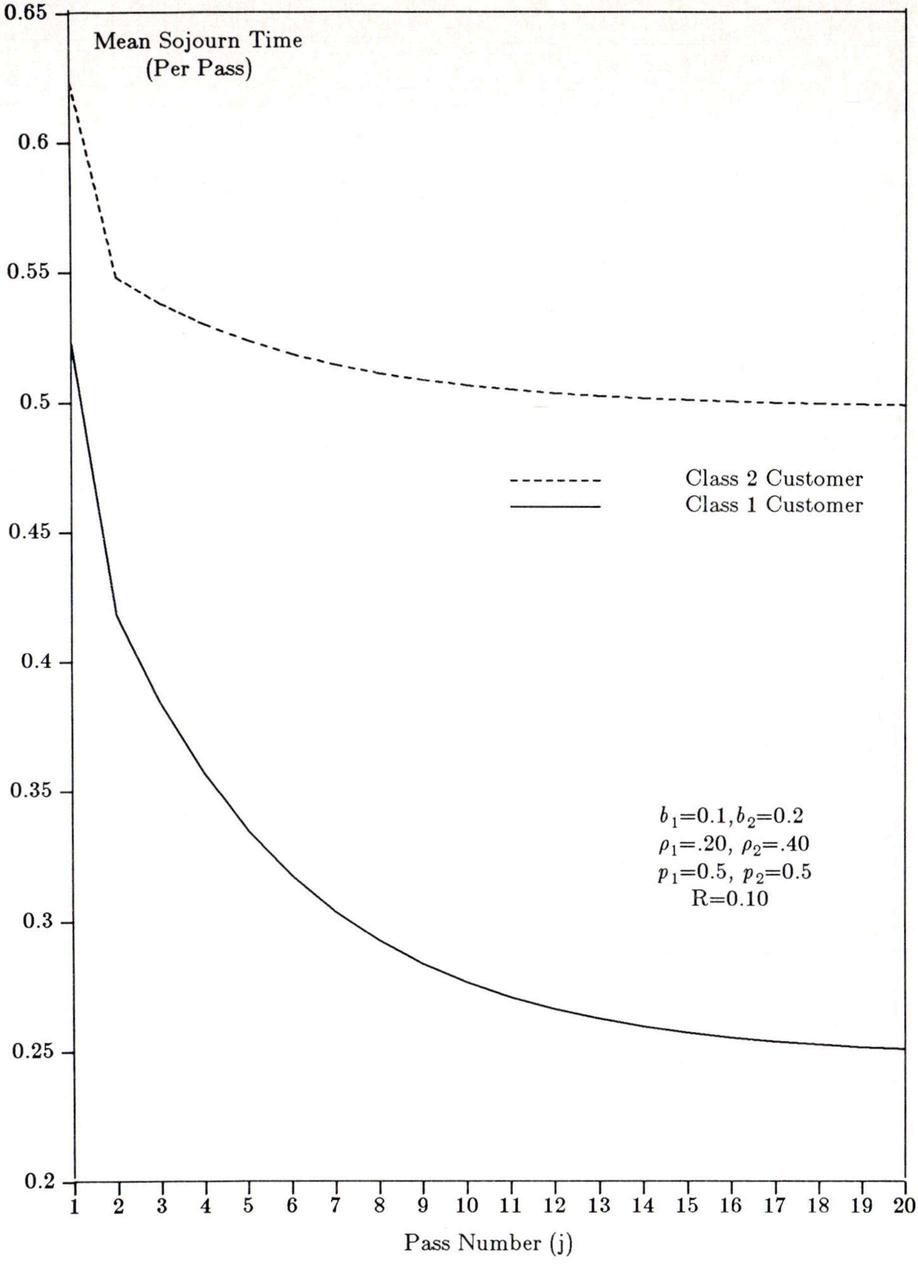

FIGURE 11. *Mean sojourn time per pass*

PART III
ANALYTIC METHODS
IN
QUEUEING THEORY

On the Relaxation Times of Open Queueing Networks

J.P.C. Blanc
Tilburg University, Faculty of Economics
P.O. Box 90153, 5000 LE Tilburg, The Netherlands

The relaxation times of open queueing networks with two single server stations, Poisson arrival streams, exponential service times and random transitions are determined with the aid of a powerful method for the analysis of queueing models with two waiting lines. This method makes use of the theory of boundary value problems for regular functions. A conjecture concerning the relaxation times of networks with arbitrary numbers of stations is made on the basis of the results for networks with two stations.

1. Introduction

Networks of waiting lines are widely used for modeling computer communication systems, manufacturing jobshops, etc. The study of queueing networks has often been restricted to steady-state characteristics. Most analytic results have been obtained for networks with a product-form joint queue length distribution, cf. [1,11]. However, many circumstances can be found in which knowledge of equilibrium characteristics of a system is not sufficient, for instance when the system starts operating in a position which does not reflect steady-state behavior and when the time the system needs to reach equilibrium is not negligible compared with the time the system is operating under unchanged conditions. Most studies on time-dependent behavior of queueing systems are of numerical or approximative character, cf. [10,13,15]. The exact analysis of time-dependent queueing phenomena is generally very complicated and leads to rather intractable expressions; see, e.g., [7, II.2.1], for the M/M/1 system. If one is not interested in a complete description of the evolution of a system over time, but if one mainly looks for a measure of the time a system needs to reach equilibrium, the concept of relaxation time will be useful. The relaxation time of a queueing system is defined to be the inverse of the exponent which dominates the asymptotic behavior in the long run of the queueing characteristics. The relaxation time of the GI/G/1 queueing system has been studied in [7, III.7.3]. In [8] and [9] a technique has been developed for the determination of the joint queue length distribution for systems with two waiting lines. This technique can also be used to study time-dependent phenomena, in particular to determine the relaxation time of such systems, cf. [3,4].

The present paper is concerned with the study of the relaxation times of exponential open queueing networks with two single server stations. It extends results of [4], in which a system with two service stations in series was studied, to a network with random transitions. Part of the results which will be proved in the present paper have already been announced (in a slightly incorrect form) in [5] and [6]. The latter paper contains a review of results on relaxation times of queueing systems.

2. Description of the network

Consider the following open queueing network with K, $K>1$, single server stations. Jobs arrive from outside the network according to a Poisson process with mean interarrival time α. A job enters the network at station k with probability c_k, $k=1,\ldots,K$, $c_1+\cdots+c_K=1$. Jobs receive at station k an amount of service which is negative exponentially distributed with mean β_k, $k=1,\ldots,K$. At each node jobs are served according to some work-conserving discipline, and successive service times are independent. At the instant that the service of a job at station k is completed, this job goes instantaneously to station j with probability q_{kj}, or it leaves the network with probability $1-\Sigma_j q_{kj}$ $(k,j=1,\ldots,K)$. For the network to be open it is necessary and sufficient that the matrix $Q=(q_{kj})$ does not possess eigenvalues equal to one. Then each job visits with probability one a finite number of stations and leaves the network afterwards.

First we shall discuss some classical results for the class of networks described above. Let λ_j, $j=1,\ldots,K$, be the solution of the set of linear equations:

$$\lambda_j = c_j/\alpha + \sum_{k=1}^{K} \lambda_k q_{kj}, \quad j=1,\ldots,K. \tag{2.1}$$

This set of equations possesses a unique solution because the network is open. It has been proved in [11] that if the traffic intensities of all the stations are bounded by one, i.e. if

$$\rho_j := \lambda_j \beta_j < 1, \quad j=1,\ldots,K, \tag{2.2}$$

then the network is stable and λ_j represents the rate of arrivals (external and internal) to station j, $j=1,\ldots,K$; and then the stationary joint queue length distribution possesses the following product-form: for n_j, $x_j=0,1,\ldots$, $j=1,\ldots,K$,

$$\begin{aligned}&\lim_{t\to\infty} \Pr\{N_j(t)=n_j,\ j=1,\ldots,K \mid N_j(0)=x_j,\ j=1,\ldots,K\}\\ &= \prod_{j=1}^{K}(1-\rho_j)\rho_j^{n_j};\end{aligned} \tag{2.3}$$

here $N_j(t)$ denotes the number of jobs at node j at time t, $t\geq 0$, $j=1,\ldots,K$. The transitions of a tagged job through the network are described by a Markov chain with K transient states (the nodes of the network) and one absorbing state (the exterior of the network). Let m_j be the mean number of visits to node j by a job which visits node j at least once, $j=1,\ldots,K$. Then from standard Markov chain theory it follows that

$$m_j = |I-Q|_{jj}/|I-Q|, \quad j=1,\ldots,K; \tag{2.4}$$

here I is the $K\times K$ identity matrix, $|I-Q|$ is the determinant of the matrix $I-Q$, and $|I-Q|_{jj}$ is the determinant of the submatrix of $I-Q$ obtained by deleting the j-th row and column.

For the understanding of the structure of the relaxation time of a stable open queueing network it is important to note that the traffic intensities at the different stations can be decomposed as a product of two factors in the following way (as an alternative to (2.2)):

$$\rho_j := (\lambda_j/m_j)(m_j\beta_j), \quad j=1,\ldots,K; \tag{2.5}$$

here λ_j/m_j can be interpreted as the arrival rate at station j of jobs which visit station j for the first time (possibly after having visited several other stations), and $m_j\beta_j$ stands for the mean total amount of service received at station j by a job which visited station j at least once.

The present paper is mainly concerned with networks with two nodes ($K=2$); see Fig. 1. In this case the solution of (2.1) reads

$$\lambda_1 = \alpha^{-1}[1-q_{11}]^{-1}[c_1+c_2r_2]/[1-r_1r_2],$$
$$\lambda_2 = \alpha^{-1}[1-q_{22}]^{-1}[c_1r_1+c_2]/[1-r_1r_2], \tag{2.6}$$

with

$$r_1 := q_{12}/(1-q_{11}), \quad r_2 := q_{21}/(1-q_{22}). \tag{2.7}$$

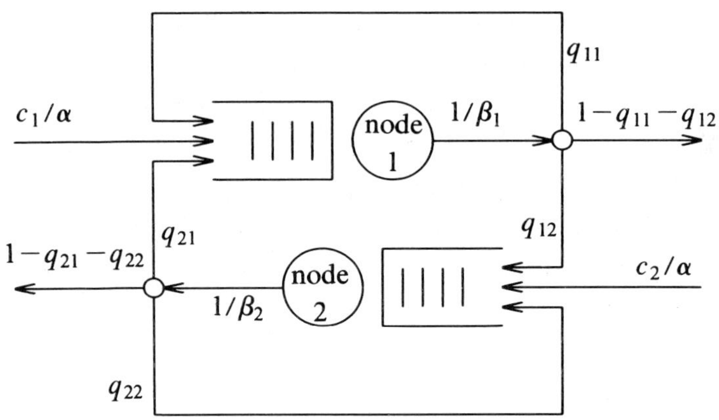

FIGURE 1. *General Jackson network with two nodes*

Introducing the notation

$$b_j := [1-q_{jj}]^{-1}\beta_j/\alpha, \quad j=1,2, \tag{2.8}$$

it is obtained from (2.2), (2.6), (2.4), that in the case $K=2$,

$$\rho_1 = b_1[c_1+c_2r_2]/[1-r_1r_2], \quad \rho_2 = b_2[c_1r_1+c_2]/[1-r_1r_2], \tag{2.9}$$

$$m_j \beta_j = \alpha b_j / [1 - r_1 r_2], \quad j = 1, 2. \tag{2.10}$$

The above notations will also be used in case a network is not stable, but then they do not have the usual interpretations. To avoid trivialities some restrictions are placed on the parameters of the network. For the network to be open it is required that $q_{11} < 1$, $q_{22} < 1$, and $q_{12} + q_{11} < 1$ or $q_{21} + q_{22} < 1$. Further, we want the network to be connected, i.e. $q_{12} > 0$ or $q_{21} > 0$. The nodes will be numbered such that $q_{12} > 0$ and $q_{21} + q_{22} < 1$ (the nodes can be renumbered, if $q_{12} = 0$ and $0 < q_{21} < 1 - q_{22}$). Finally, we want each node to receive traffic; this requires that $c_1 > 0$ or $q_{21} > 0$. The foregoing choices imply that, cf. (2.7),

$$r_2 < 1 \wedge r_1 > 0 \wedge (c_1 > 0 \vee r_2 > 0); \tag{2.11}$$

see also Fig. 2.

These conditions allow to number the nodes of the network according to the following inequality (which is essential for the proof of Lemma 5.4):

$$c_1 r_1 (1 - r_2) \geqslant c_2 r_2 (1 - r_1). \tag{2.12}$$

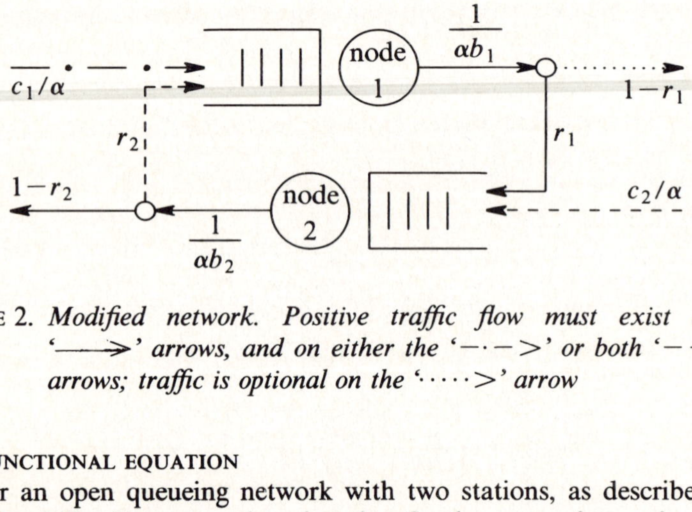

FIGURE 2. *Modified network. Positive traffic flow must exist on the '——>' arrows, and on either the '—·—>' or both '- - - >' arrows; traffic is optional on the '· · · · >' arrow*

3. THE FUNCTIONAL EQUATION

Define for an open queueing network with two stations, as described in Section 2, the following generating function/Laplace transform: for $\mathrm{Re}\,\phi > 0$, $|z_1| \leqslant 1$, $|z_2| \leqslant 1$, $x = (x_1, x_2)$, $x_1, x_2 = 0, 1, \ldots$,

$$\Omega_x(\phi; z_1, z_2) := \int_0^\infty e^{-\phi t} E\{z_1^{N_1(t)} z_2^{N_2(t)} | N_1(0) = x_1, N_2(0) = x_2\} dt. \tag{3.1}$$

PROPOSITION 3.1. *The transform $\Omega_x(\phi; z_1, z_2)$ satisfies the following functional equation: for* $\mathrm{Re}\,\phi > 0$, $|z_1| \leqslant 1$, $|z_2| \leqslant 1$,

$$\begin{aligned} &K(\phi; z_1, z_2) \Omega_x(\phi; z_1, z_2) \\ &= \alpha z_1^{x_1} z_2^{x_2} + H_1(z_1, z_2) \Omega_x(\phi; 0, z_2) + H_2(z_1, z_2) \Omega_x(\phi; z_1, 0); \end{aligned} \tag{3.2}$$

here the kernel of the functional equation is defined as:

$$K(\phi; z_1, z_2) := \alpha\phi + 1 - c_1 z_1 - c_2 z_2 + H_1(z_1, z_2) + H_2(z_1, z_2); \tag{3.3}$$

further,

$$H_1(z_1, z_2) := [z_1 - 1 - r_1(z_2 - 1)]/(b_1 z_1), \tag{3.4}$$

$$H_2(z_1, z_2) := [z_2 - 1 - r_2(z_1 - 1)]/(b_2 z_2). \tag{3.5}$$

PROOF. Functional equation (3.2) can be derived in a standard way from the Kolmogorov equations for the state probabilities of the queueing network. □

Note that equation (3.2) depends on the transition matrix Q and the mean service times β_1 and β_2 only through the four parameters r_1, r_2, b_1 and b_2, cf. (2.7), (2.8). A standard observation from equation (3.2) is that, cf. e.g. [9],

$$H_1(z_1, z_2)\Omega_x(\phi; 0, z_2) + H_2(z_1, z_2)\Omega_x(\phi; z_1, 0) + \alpha z_1^{x_1} z_2^{x_2} = 0, \tag{3.6}$$

for all triples $(\phi; z_1, z_2)$ for which

$$K(\phi; z_1, z_2) = 0, \quad \mathrm{Re}\,\phi > 0, \quad |z_1| \leqslant 1, \quad |z_2| \leqslant 1, \tag{3.7}$$

i.e. for which the kernel vanishes and the transform $\Omega_x(\phi; z_1, z_2)$ is known to be finite. Therefore, the zeros of the kernel will be studied in detail in Section 4; then, in Section 5, relation (3.6) will be used to formulate a boundary value problem for the function $\Omega_x(\phi; 0, z_2)$.

4. ANALYSIS OF THE KERNEL

This section is concerned with the study of the zeros of the kernel $K(\phi; z_1, z_2)$ of functional equation (3.2). The results of this section will be used to formulate a Riemann-Hilbert boundary value problem for the unknown function $\Omega_x(\phi; 0, z_2)$ in Section 5, and to determine the relaxation times of the queueing networks in Section 6.

From (3.3)-(3.5) it is seen that $z_1 z_2 K(\phi; z_1, z_2)$ is a quadratic function of z_1 as well as of z_2. Solving the equation

$$z_1 z_2 K(\phi; z_1, z_2) = 0, \tag{4.1}$$

as equation in z_2 leads to:

$$z_2 = \zeta(\phi; z_1), \tag{4.2}$$

with

$$\zeta(\phi; z_1) := \frac{A(\phi) - c_1 z_1 - (1 - r_1)/(b_1 z_1) \pm \sqrt{D(\phi; z_1)}}{2(c_2 + r_1/(b_1 z_1))}, \tag{4.3}$$

$$\begin{aligned} D(\phi; z_1) := \; & [A(\phi) - c_1 z_1 - (1 - r_1)/(b_1 z_1)]^2 \\ & - 4[c_2 + r_1/(b_1 z_1)][1 - r_2 + r_2 z_1]/b_2, \end{aligned} \tag{4.4}$$

$$A(\phi) := \alpha\phi + 1 + (1/b_1) + (1/b_2); \tag{4.5}$$

$\zeta(\phi; z_1)$ is a two-valued function of z_1 of which the limiting values at 0 and at ∞ are defined by

$$\zeta(\phi; 0) = \{0, -(1-r_1)/r_1\}, \quad \zeta(\phi; \infty) = \{\infty, -r_2/(c_1 b_2)\}. \tag{4.6}$$

The zeros of $D(\phi; z_1)$ will be described by a function $z_1 = y(\phi)$, which satisfies the following relation, cf. (4.4),

$$\begin{aligned} &[A(\phi) - c_1 y(\phi) - (1-r_1)/(b_1 y(\phi))]^2 \\ &= 4[c_2 + r_1/(b_1 y(\phi))][1 - r_2 + r_2 y(\phi)]/b_2. \end{aligned} \tag{4.7}$$

We consider the above functions to be defined on the extended complex plane $\mathbb{C}^* = \mathbb{C} \cup \{\infty\}$. The properties of these functions are summarized below.

PROPOSITION 4.1. *The set of zeros of the kernel $K(\phi; z_1, z_2)$ is completely described by relation (4.2). The function $\zeta(\phi; z_1)$ defined in (4.3)-(4.6) is a two-valued analytic function. The set of branch points of the function $\zeta(\phi; z_1)$ is completely described by the relation $z_1 = y(\phi)$, where $y(\phi)$ is a four-valued analytic function, implicitly defined by (4.7), which possesses six (first order) branch points (possibly coinciding).*

PROOF. Most of the properties are obvious from the definitions (4.3)-(4.7) and (3.3)-(3.5). Only the property that the function $y(\phi)$ possesses six branch points needs some explanation. The function $y(\phi)$ is defined as zero $y = y(\phi)$ of the function $D(\phi; y)$. Hence, it has branch points when $(\partial/\partial y) D(\phi; y)$ vanishes at $y = y(\phi)$, i.e., cf. (4.4),

$$\begin{aligned} &[A(\phi) - c_1 y(\phi) - (1-r_1)/(b_1 y(\phi))][(1-r_1)/(b_1 y^2(\phi)) - c_1] \\ &= 2[c_2 r_2 - r_1(1-r_2)/(b_1 y^2(\phi))]/b_2. \end{aligned} \tag{4.8}$$

Combining (4.7) and (4.8) shows that at its branch points the function $y(\phi)$ takes values y which are solutions of the equation

$$\begin{aligned} &[c_2 r_2 y - r_1(1-r_2)/(b_1 y)]^2/b_2 \\ &= [(1-r_1)/(b_1 y) - c_1 y]^2 [c_2 + r_1/(b_1 y)][1 - r_2 + r_2 y]. \end{aligned} \tag{4.9}$$

This equation has six roots in general (with the understanding that roots may be located at infinity or at the origin, see Remark 4.2 below). Each value of y determines one value of ϕ through (4.8), (4.5), except when a root y of (4.9) satisfies, cf. (4.8),

$$y = \pm\sqrt{(1-r_1)/(b_1 c_1)}. \tag{4.10}$$

The values of y in (4.10) can only be roots of (4.9) if for these values also the left-hand side of (4.9) vanishes. This is only possible when

$$c_1 r_1/(1-r_1) = c_2 r_2/(1-r_2). \tag{4.11}$$

In this special case the two values of y in (4.10) are both double roots of (4.9), while the other two roots satisfy

$$[r_1(1-r_2)/(1-r_1)]^2/b_2 = [c_2+r_1/(b_1y)][1-r_2+r_2y]. \tag{4.12}$$

Equation (4.8) is in this case equivalent to the following two equations:

$$\begin{aligned} y^2(\phi) &= (1-r_1)/(b_1c_1), \\ A(\phi)-c_1y(\phi)-(1-r_1)/(b_1y(\phi)) &= -2r_1(1-r_2)/((1-r_1)b_2). \end{aligned} \tag{4.13}$$

Each of the two roots of equation (4.12) determines one branch point of the function $y(\phi)$ through the second equation of (4.13), while each of the two values in (4.10) determines two branch points of the function $y(\phi)$ through equation (4.7). Hence, the function $y(\phi)$ possesses six branch points also when (4.11) holds. □

REMARK 4.1. Note that relation (4.11) determines the subclass of reversible networks with two nodes, cf. [12]. The six branch points of the function $y(\phi)$ can be explicitly evaluated in this case. The positive value in (4.10) leads with (4.7) to:

$$A(\phi) = 2[\sqrt{c_1(1-r_1)/b_1} \pm \{\sqrt{c_2(1-r_2)/b_2} + \sqrt{r_1r_2/(b_1b_2)}\}]; \tag{4.14}$$

the negative value in (4.10) leads with (4.7) to:

$$A(\phi) = 2[-\sqrt{c_1(1-r_1)/b_1} \pm \{\sqrt{c_2(1-r_2)/b_2} - \sqrt{r_1r_2/(b_1b_2)}\}]; \tag{4.15}$$

the two roots of (4.12) both lead with the second equation of (4.13) to:

$$\begin{aligned} A(\phi) = \; & -r_2(1-r_1)/(b_1(1-r_2))-r_1(1-r_2)/(b_2(1-r_1)) \\ & -(1-r_1)(1-r_2)/(r_1(1-r_2)+r_2(1-r_1)). \end{aligned} \tag{4.16}$$

Note that the six branch points of the function $y(\phi)$ are all real in this special case, that at the single ϕ value in (4.16) there are two branch points (two couples of branches of $y(\phi)$ meet at - in general - different y values, which are not necessarily real, cf. (4.12)), and that the corresponding y values coincide for the two ϕ values in (4.14) and similarly for those in (4.15). The largest branch point will be denoted by ϕ_0; from (4.14)-(4.16) it is readily seen that

$$A(\phi_0) = 2[\sqrt{c_1(1-r_1)/b_1} + \sqrt{c_2(1-r_2)/b_2} + \sqrt{r_1r_2/(b_1b_2)}]. \tag{4.17}$$

REMARK 4.2. When $c_1=0$, $c_2=0$ or $r_2=0$, a root of (4.9) is located at infinity; when $r_1=1$ a root of (4.9) is located at the origin. Each root of (4.9) at infinity or at the origin (see also (2.11)) leads to a branch point of the function $y(\phi)$ at infinity by (4.8). For instance, three branch points are located at infinity in the case of two queues in series: $r_1=1$, $c_2=0$, $r_2=0$, cf. [4]. When $c_1=0$ and $r_1=1$ the network is reversible, cf. (4.11); in this case the two branch points given by (4.16) are located at infinity.

For the study of the relaxation time it is important to know which of the six branch points of the function $y(\phi)$ has the largest real part. This branch point, which will turn out to be real, will be denoted by ϕ_0. In the case that (4.11) holds, the value of ϕ_0 is explicitly given by (4.17). In the sequel we shall characterise ϕ_0 in the general case. First we shall prove a preliminary result.

LEMMA 4.1. *For ϕ real, $\phi>0$, the four branches $y_j(\phi)$, $j=1,\ldots,4$, of the function $y(\phi)$, defined by (4.7), are real, and they can be numbered such that for all ϕ, $\phi>0$, holds:*

$$0\leqslant y_1(\phi)<y_2(\phi)<1<y_3(\phi)<y_4(\phi)\leqslant\infty. \tag{4.18}$$

For all ϕ, $\phi>0$, $y_1(\phi)>0$ if $r_1<1$, $y_1(\phi)=0$ if $r_1=1$, and $y_4(\phi)<\infty$ if $c_1>0$, $y_4(\phi)=\infty$ if $c_1=0$.

PROOF. Let ϕ be real, $\phi>0$, and let $c_1>0$, $r_1<1$. It is readily verified that $D(\phi;z_1)>0$ as $z_1\downarrow 0$, for $z_1=1$, and as $z_1\rightarrow\infty$, cf. (4.4), (4.5). On the other hand, the function $A(\phi)-c_1z_1-(1-r_1)/(b_1z_1)$ is negative as $z_1\downarrow 0$ and as $z_1\rightarrow\infty$, but it is positive at $z_1=1$; hence, this function has one zero on the interval (0,1) and one zero on the interval $(1,\infty)$. Clearly, $D(\phi;z_1)<0$ at these two points. This proves the assertions in the case $c_1>0$, $r_1<1$. The cases $c_1=0$ and/or $r_1=1$ are left to the reader (see also the proof of Proposition 4.1). □

In the sequel the branches of the function $y(\phi)$ will be assumed to be numbered according to (4.18).

PROPOSITION 4.2. *There exists a unique branch point ϕ_0 of the function $y(\phi)$ with the property that its real part is larger than the real parts of the other branch points of $y(\phi)$ in the finite plane; ϕ_0 is real and given by*

$$A(\phi_0)=\min_{y>0}\{c_1y+(1-r_1)/(b_1y)+2\sqrt{[c_2+r_1/(b_1y)][1-r_2+r_2y]/b_2}\}. \tag{4.19}$$

Moreover, the four branches of the analytic function $y(\phi)$ are monotonous functions for real ϕ, $\phi>\phi_0$, and satisfy (cf. Lemma 4.1),

$$0\leqslant y_1(\phi)<y_2(\phi)<y_3(\phi)<y_4(\phi)\leqslant\infty,\quad \phi>\phi_0, \tag{4.20}$$

$$0\leqslant y_1(\phi_0)<y_2(\phi_0)=y_3(\phi_0)<y_4(\phi_0)\leqslant\infty. \tag{4.21}$$

PROOF. First we note that the inverse of the four-valued analytic function $y(\phi)$ is a two-valued analytic function $\phi(y)$, cf. (4.7), which is given by

$$A(\phi(y))=c_1y+(1-r_1)/(b_1y)\pm 2\sqrt{[c_2+r_1/(b_1y)][1-r_2+r_2y]/b_2}. \tag{4.22}$$

It is readily verified that the positive branch in (4.22) possesses a positive minimum for y on the interval $(0,\infty)$. By comparing the graphs of the functions $[A(\phi)-c_1y-(1-r_1)/(b_1y)]^2$ and $4[c_2+r_1/(b_1y)][1-r_2+r_2y]/b_2$ on the real y-axis (the first one for different real values of ϕ) it is found that the function $y(\phi)$ must have at least four real branch points (some of which may be located at infinity, cf. Remark 4.2) and that the one defined by (4.19) satisfies the inequality (see also (4.17))

$$A(\phi_0) \geqslant 2[\sqrt{c_1(1-r_1)/b_1} + \sqrt{c_2(1-r_2)/b_2} + \sqrt{r_1 r_2/(b_1 b_2)}], \tag{4.23}$$

while $A(\phi)$ is smaller than this lower bound for $A(\phi_0)$ for the other three. When the remaining two branch points of $y(\phi)$ are real, cf., e.g., (4.16), they must also be smaller than ϕ_0. When they are not real, we strongly conjecture (but could not generally prove) that their real part is smaller than ϕ_0; see [4] for an example in which this property is obvious. The monotonicity of the functions $y_j(\phi)$, $j=1,\ldots,4$, and the inequalities (4.20) and (4.21) follow by considering the inverse function $\phi(y)$, cf. (4.22). □

At its branch points the function $\zeta(\phi; z_1)$, cf. (4.3), is single valued, so that the values $x_j(\phi) := \zeta(\phi; y_j(\phi))$ are uniquely determined: for $j=1,\ldots,4$,

$$x_j(\phi) = [c_2 + r_1/(b_1 y_j(\phi))]^{-1}[A(\phi) - c_1 y_j(\phi) - (1-r_1)/(b_1 y_j(\phi))]/2. \tag{4.24}$$

LEMMA 4.2. *For $\phi > \phi_0$ the values $x_j(\phi)$, $j=1,\ldots,4$, are monotonous functions of ϕ, which satisfy*

$$-\infty \leqslant x_4(\phi) < x_1(\phi) \leqslant 0 < x_2(\phi) < x_3(\phi) < \infty. \tag{4.25}$$

For all ϕ, $\phi > \phi_0$, $x_1(\phi) < 0$ if $r_1 < 1$, $x_1(\phi) = 0$ if $r_1 = 1$, and $x_4(\phi) > -\infty$ if $c_1 > 0$, $x_4(\phi) = -\infty$ if $c_1 = 0$. For $\phi = \phi_0$,

$$-\infty \leqslant x_4(\phi_0) < x_1(\phi_0) \leqslant 0 < x_2(\phi_0) = x_3(\phi_0) < \infty. \tag{4.26}$$

PROOF. Let $c_1 > 0$, $r_1 < 1$. From the proof of Lemma 4.1 it follows directly that the last term at the right-hand side of (4.24) is negative for $j=1$ and $j=4$, and positive for $j=2$ and $j=3$. Combining (4.7) and (4.24) gives

$$[x_j(\phi)]^2 = [c_2 + r_1/(b_1 y_j(\phi))]^{-1}[1 - r_2 + r_2 y_j(\phi)]/b_2, \quad j=1,\ldots,4. \tag{4.27}$$

Because the function $[1-r_2+r_2 y]/[c_2 + r_1/(b_1 y)]$ is an increasing function of y on the interval $(0, \infty)$, and because $y_j(\phi)$ is monotonous for $\phi > \phi_0$, cf. Proposition 4.2, it follows from (4.27) that $x_j(\phi)$ is monotonous for $\phi > \phi_0$, $j=1,\ldots,4$. The other assertions (also for the cases $c_1=0$ and/or $r_1=1$) follow readily from Proposition 4.2. □

To conclude this section the zeros of the kernel will be studied for ϕ real and for z_1 on the real interval $[y_1(\phi), y_2(\phi)]$, cf. (4.20). Define for $\phi \geqslant \phi_0$,

$$\Lambda(\phi) := \{z_2 \colon K(\phi; z_1, z_2) = 0,\ y_1(\phi) \leqslant z_1 \leqslant y_2(\phi)\}. \tag{4.28}$$

Because the function $D(\phi; z_1)$ is negative for $y_1(\phi) \leqslant z_1 \leqslant y_2(\phi)$, cf. the proof of Lemma 4.1, the set of points $\Lambda(\phi)$ forms a contour in the complex plane, with the real axis as axis of symmetry, and with $x_1(\phi)$ and $x_2(\phi)$ as the only crossing points with the real axis (for $\phi \geqslant \phi_0$).

Lemma 4.3. *For real ϕ, $\phi>\phi_0$, the contour $\Lambda(\phi)$ possesses a tangent at every point. The contour $\Lambda(\phi_0)$ possesses a tangent at every point, except at the point $x_2(\phi_0)$, where it has a corner point.*

Proof. The arguments are similar to those of the proof of Lemma 4 in [4], and are therefore omitted here. □

The interior of the contour $\Lambda(\phi)$ will be denoted by $\Lambda^+(\phi)$, and the exterior by $\Lambda^-(\phi)$, $\phi\geqslant\phi_0$.

Finally, we introduce $\gamma(\phi; w)$ as a conformal mapping of the closed unit disk $|w|\leqslant 1$ onto the region $\Lambda^+(\phi)\cup\Lambda(\phi)$ (see [2, I.6], for relevant references on conformal mappings). Because the real axis is an axis of symmetry of the contour $\Lambda(\phi)$, $\gamma(\phi; w)$ may be chosen such that for $\phi\geqslant\phi_0$, $|w|\leqslant 1$,

$$\gamma(\phi; \bar{w}) = \overline{\gamma(\phi; w)}, \quad \gamma(\phi; -1)=x_1(\phi), \quad \gamma(\phi; 1)=x_2(\phi). \tag{4.29}$$

The conformal mapping $\gamma(\phi; w)$ is determined by (4.29) up to the value of $\gamma(\phi; 0)$, which is restricted to the real interval $(x_1(\phi), x_2(\phi))$. When $r_1<1$ we will choose $\gamma(\phi; 0)=0$, which is possible by Lemma 4.2. The inverse conformal mapping of $z_2=\gamma(\phi; w)$ will be denoted by $w=\gamma_0(\phi; z_2)$. Note that for $\phi\geqslant\phi_0$,

$$\gamma_0(\phi; 0) = 0, \quad \text{if} \quad r_1<1; \quad \gamma_0(\phi; 0)=-1, \quad \text{if} \quad r_1=1. \tag{4.30}$$

5. Formulation and solution of a boundary value problem

Throughout this section ϕ is assumed to be real and $\phi\geqslant\phi_0$. Before we shall formulate a boundary value problem for the function $\Omega_x(\phi; 0, z_2)$, possible zeros of the functions $H_1(\zeta_0(\phi; z_2), z_2)$ and $H_2(\zeta_0(\phi; z_2), z_2)$ on the contour $\Lambda(\phi)$ will be investigated; here $z_1=\zeta_0(\phi; z_2)$ is that branch of the inverse of the function $z_2=\zeta(\phi; z_1)$, cf. (4.1)-(4.3), which maps points of the contour $\Lambda(\phi)$ onto the real interval $[y_1(\phi), y_2(\phi)]$. For this purpose the following quantities are introduced:

$$\chi_1 := [c_2+(r_1/b_1)\sqrt{\rho_1}\,][1-r_2+r_2/\sqrt{\rho_1}\,], \tag{5.1}$$

$$\chi_2 := [c_1+(r_2/b_2)\sqrt{\rho_2}\,][1-r_1+r_1/\sqrt{\rho_2}\,], \tag{5.2}$$

$$\phi_j := -(1-\sqrt{\rho_j}\,)^2/(m_j\beta_j), \quad j=1, 2, \tag{5.3}$$

$$\nu_j(\phi) := [1+m_j\beta_j\phi+\rho_j-\sqrt{(1+m_j\beta_j\phi+\rho_j)^2-4\rho_j}\,]/(2\rho_j), \tag{5.4}$$

for $\phi\geqslant\phi_j$, $j=1, 2$. Notice that $\nu_j(\phi)$ possesses two (real) branch points of which ϕ_j is the larger, $j=1, 2$.

Lemma 5.1. *For all values of the parameters of the network it holds that $\phi_1\geqslant\phi_0$, and $\phi_1=\phi_0$ if and only if $\chi_1=1/b_2$.*

Proof. Taking $y=1/\sqrt{\rho_1}$ in (4.19) gives the following upper bound for ϕ_0:

$$A(\phi_0) \leqslant c_1/\sqrt{\rho_1}+[(1-r_1)/b_1]\sqrt{\rho_1}+2\sqrt{\chi_1/b_2}\,. \tag{5.5}$$

It is readily verified, using (5.3), (5.1) and (4.5), that ϕ_1 satisfies

$$A(\phi_1) = c_1/\sqrt{\rho_1} + [(1-r_1)/b_1]\sqrt{\rho_1} + \chi_1 + 1/b_2. \tag{5.6}$$

Relations (5.5) and (5.6) imply that $\phi_1 \geqslant \phi_0$, and that $\chi_1 = 1/b_2$ is a necessary condition for $\phi_1 = \phi_0$ to hold. On the other hand, it is not difficult to see that $y = 1/\sqrt{\rho_1}$ is a root of (4.9) when $\chi_1 = 1/b_2$, cf. (5.1), (2.9). With (4.8), or with (4.17) when (4.11) holds, this implies that indeed $\phi_1 = \phi_0$ when $\chi_1 = 1/b_2$. □

LEMMA 5.2. *For all values of the parameters of the network it holds that* $\phi_2 \geqslant \phi_0$, *and* $\phi_2 = \phi_0$ *if and only if* $\chi_2 = 1/b_1$.

PROOF. Taking $y = \eta := 1 - r_1 + r_1/\sqrt{\rho_2}$ in (4.19) gives another upper bound for ϕ_0:

$$\begin{aligned} A(\phi_0) \leqslant\ & c_1\eta + (1-r_1)/(b_1\eta) \\ & + 2\sqrt{[c_2 + r_1/(b_1\eta)]/\sqrt{\rho_2}}\,\sqrt{[(1-r_2+r_2\eta)/b_2]\sqrt{\rho_2}}\,. \end{aligned} \tag{5.7}$$

It can be verified by substituting $\eta = 1 - r_1 + r_1/\sqrt{\rho_2}$ and by using (5.3) and (4.5), that ϕ_2 satisfies

$$\begin{aligned} A(\phi_2) =\ & c_1\eta + (1-r_1)/(b_1\eta) \\ & + [c_2 + r_1/(b_1\eta)]/\sqrt{\rho_2} + [(1-r_2+r_2\eta)/b_2]\sqrt{\rho_2}\,. \end{aligned} \tag{5.8}$$

Relations (5.7) and (5.8) imply that $\phi_2 \geqslant \phi_0$, and that a necessary condition for $\phi_2 = \phi_0$ to hold is

$$[c_2 + r_1/(b_1\eta)]/\sqrt{\rho_2} = [(1-r_2+r_2\eta)/b_2]\sqrt{\rho_2}\,.$$

Straightforward calculations show that this condition is equivalent to $\chi_2 = 1/b_1$, cf. (5.2), and that $y = \eta$ is a root of (4.9) when $\chi_2 = 1/b_1$, cf. (5.2), (2.9). Again, this implies with (4.8) or (4.17) that indeed $\phi_2 = \phi_0$ when $\chi_2 = 1/b_1$. □

LEMMA 5.3. *If* $\chi_2 > 1/b_1$ *then the function* $H_1(\zeta_0(\phi; z_2), z_2)$ *has no zeros on the contour* $\Lambda(\phi)$ *for all* ϕ, $\phi \geqslant \phi_0$; *if* $\chi_2 \leqslant 1/b_1$ *then there exists exactly one value of* ϕ *on the interval* $[\phi_0, \infty)$, *say* $\phi = \phi_{H1}$, *for which the function* $H_1(\zeta_0(\phi; z_2), z_2)$ *possesses a zero on the contour* $\Lambda(\phi)$; *this zero is unique and it is located at the point* $z_2 = x_2(\phi_{H1})$; *further,* $\phi_{H1} > \phi_2$ *if* $\chi_2 < 1/b_1$, *and* $\phi_{H1} = \phi_2$ *if* $\chi_2 = 1/b_1$; *finally, when* $\rho_2 < 1$ *then* $\phi_{H1} > 0$ *if and only if* $1/b_2 < c_2 + r_1/b_1$.

PROOF. See Appendix A.

LEMMA 5.4. *If* $\chi_1 > 1/b_2$ *then the function* $H_2(\zeta_0(\phi; z_2), z_2)$ *has no zeros on the contour* $\Lambda(\phi)$ *for all* ϕ, $\phi \geqslant \phi_0$; *if* $\chi_1 \leqslant 1/b_2$ *then there exists exactly one value of* ϕ *on the interval* $[\phi_0, \infty)$, *say* $\phi = \phi_{H2}$, *for which the function* $H_2(\zeta_0(\phi; z_2), z_2)$ *possesses a zero on the contour* $\Lambda(\phi)$; *this zero is unique and it is located at the point* $z_2 = x_2(\phi_{H2})$; *further,* $\phi_{H2} > \phi_1$ *if* $\chi_1 < 1/b_2$, *and* $\phi_{H2} = \phi_1$ *if* $\chi_1 = 1/b_2$; *finally, when* $\rho_1 < 1$ *then* $\phi_{H2} > 0$ *if and only if* $1/b_2 > c_2 + r_1/b_1$.

PROOF. See Appendix B.

REMARK 5.1. When the network is reversible, cf. Remark 4.1, then $\chi_1 = 1/b_2$ is equivalent to $1/\sqrt{\rho_2} = 1 - r_2 + r_2/\sqrt{\rho_1}$, and $\chi_2 = 1/b_1$ to $1/\sqrt{\rho_1} = 1 - r_1 + r_1/\sqrt{\rho_2}$. When the network consists of service stations in series with feedback, i.e. when $r_1 = 1$ and $c_2 = 0$, then $\chi_1 = 1/b_2$ is equivalent to $\sqrt{\rho_1} = \rho_2[1 - r_2 + r_2/\sqrt{\rho_1}]$, and $\chi_2 = 1/b_1$ to $\sqrt{\rho_2} = \rho_1[1 - r_2 + r_2/\sqrt{\rho_2}]$.

In the following theorems a boundary value problem will be formulated and solved for the function $\Omega_x(\phi; 0, z_2)$. See [14, §§ 40, 41], for the definition of and a solution method for Riemann-Hilbert boundary value problems.

THEOREM 5.1. *For real positive ϕ, ϕ so large that $x_2(\phi) < 1$, $\phi > \phi_{H1}$ when $\chi_2 < 1/b_1$, $\phi > \phi_{H2}$ when $\chi_1 < 1/b_2$, the generating function $\Omega_x(\phi; 0, z_2)$ is the solution of a Riemann-Hilbert boundary value problem on the contour $\Lambda(\phi)$, of index $\kappa = -2$, with the following boundary condition: for $z_1 = \zeta_0(\phi; z_2)$, $z_2 \in \Lambda(\phi)$,*

$$\operatorname{Re}[iH_1(z_1, z_2)\Omega_x(\phi; 0, z_2)/H_2(z_1, z_2)] = -\operatorname{Re}[i\alpha z_1^{x_1} z_2^{x_2}/H_2(z_1, z_2)]. \tag{5.9}$$

PROOF. For real ϕ, $\phi \geqslant \phi_0$, and $z_2 \in \Lambda(\phi)$ the points $z_1 = \zeta_0(\phi; z_2)$ lie on the real interval $[y_1(\phi), y_2(\phi)]$; on this interval the generating function $\Omega_x(\phi; z_1, 0)$ is real-valued. Hence, boundary condition (5.9) follows directly from relation (3.6). Further, the generating function $\Omega_x(\phi; 0, z_2)$ is known to be regular for $\operatorname{Re}\phi > 0$, $|z_2| \leqslant 1$. Points $z_2 \in \Lambda(\phi)$ satisfy the relation, cf. (4.2)-(4.4), (4.28),

$$\begin{aligned} |z_2|^2 &= [c_2 + r_1/(b_1 z_1)]^{-1}[1 - r_2 + r_2 z_1]/b_2, \\ z_1 &= \zeta_0(\phi; z_2) \in [y_1(\phi), y_2(\phi)]. \end{aligned} \tag{5.10}$$

The right-hand side of this relation is an increasing function of z_1, and hence it attains its maximum at $z_1 = y_2(\phi)$. This implies that

$$|z_2| < x_2(\phi), \quad \text{for all} \quad z_2 \in \Lambda(\phi). \tag{5.11}$$

Because $x_2(\phi) \downarrow 0$ as $\phi \to \infty$ the contour $\Lambda(\phi)$ lies inside the unit circle for ϕ large enough. Hence, the function $\Omega_x(\phi; 0, z_2)$ is regular in $\Lambda^+(\phi)$ when $x_2(\phi) \leqslant 1$. This implies that a Riemann-Hilbert boundary value problem is defined by relation (5.9). The index κ of this boundary value problem is defined, cf. [14], to be the increment of the function, cf. (3.4), (3.5),

$$(\pi i)^{-1} \log[H_2(\zeta_0(\phi; z_2), z_2)/H_1(\zeta_0(\phi; z_2), z_2)], \tag{5.12}$$

when z_2 traverses the contour $\Lambda(\phi)$ once in positive direction. When $r_1 < 1$ then $y_1(\phi) > 0$ for $\phi \geqslant \phi_0$, cf. Lemma 4.1, and $0 \in \Lambda^+(\phi)$ for $\phi \geqslant \phi_0$, cf. Lemma 4.2. Consequently, the factor $1/z_2$ in (5.12) contributes -2 to the index, and the factor $z_1 = \zeta_0(\phi; z_2)$ contributes 0. Further, it follows that for $z_2 \in \Lambda(\phi)$, as $\phi \to \infty$,

$$[z_2 - 1 - r_2(\zeta_0(\phi; z_2) - 1)]/[\zeta_0(\phi; z_2) - 1 - r_1(z_2 - 1)] \to (r_2 - 1)/(r_1 - 1), \tag{5.13}$$

so that this factor in (5.12) also contributes 0 to the index for ϕ large enough. The index of this factor only changes when a zero of its numerator or its denominator crosses the contour $\Lambda(\phi)$, i.e. at $\phi=\phi_{H1}$ or at $\phi=\phi_{H2}$, cf. Lemmas 5.3 and 5.4. Hence, $\kappa=-2$ when $r_1<1$ and ϕ is large enough. When $r_1=1$ then $y_1(\phi)=0$ for $\phi\geqslant\phi_0$, cf. Lemma 4.1, and $0\in\Lambda(\phi)$ for $\phi\geqslant\phi_0$, cf. Lemma 4.2. For ϕ large enough, the factor $z_2-1-r_2(\zeta_0(\phi;z_2)-1)$ contributes 0 to the index. From (5.10) it follows for $z_2\in\Lambda(\phi)$ that $|z_2|^2\approx(b_1/b_2)(1-r_2)\zeta_0(\phi;z_2)$ as $\phi\to\infty$, so that as $\phi\to\infty$,

$$[\zeta_0(\phi;z_2)/z_2]/[\zeta_0(\phi;z_2)-z_2] \approx [\bar{z}_2/|\bar{z}_2|]^2[\bar{z}_2-(1-r_2)(b_1/b_2)]^{-1}. \tag{5.14}$$

The contribution of this factor of (5.12) to the index is -2. Hence also when $r_1=1$ the index is $\kappa=-2$ for ϕ large enough. □

In order to formulate the solution of the above described Riemann-Hilbert boundary value problem the following functions are introduced (C denotes the unit circle): for real ϕ, $\phi\geqslant\phi_0$, $|w|=1$,

$$F(\phi;w) := iH_1(\zeta_0(\phi;\gamma(\phi;w)),\gamma(\phi;w))/H_2(\zeta_0(\phi;\gamma(\phi;w)),\gamma(\phi;w)), \tag{5.15}$$

$$f_x(\phi;w) := -\mathrm{Re}\,[i\alpha\zeta_0(\phi;\gamma(\phi;w))^{x_1}\gamma(\phi;w)^{x_2}/H_2(\zeta_0(\phi;\gamma(\phi;w)),\gamma(\phi;w))]; \tag{5.16}$$

$$X(\phi;w) := \exp[(2\pi i)^{-1}\int_C \log\{u^2F(\phi;1/u)/F(\phi;u)\}(u-w)^{-1}du],\ |w|<1; \tag{5.17}$$

$$X^+(\phi;u_0) := \lim_{w\to u_0,\,|w|<1} X(\phi;w),\quad |u_0|=1. \tag{5.18}$$

The function $X(\phi;w)$ is only defined for $\phi>\phi_{H1}$ when $\chi_2<1/b_1$, $\phi>\phi_{H2}$ when $\chi_1<1/b_2$; then the limiting values $X^+(\phi;u_0)$, $|u_0|=1$, exist, because the function $F(\phi;w)$ is regular on the unit circle.

THEOREM 5.2. *The solution of the Riemann-Hilbert boundary value problem on the contour $\Lambda(\phi)$, formulated in Theorem 5.1, is given by: for $z_2\in\Lambda^+(\phi)$,*

$$\Omega_x(\phi;0,z_2) = \frac{X(\phi;\gamma_0(\phi;z_2))}{\pi i}\int_C \frac{f_x(\phi;u)}{F(\phi;u)X^+(\phi;u)}\,\frac{du}{u-\gamma_0(\phi;z_2)}. \tag{5.19}$$

PROOF. By inserting $z_2=\gamma(\phi;w)$ in (5.9) the Riemann-Hilbert problem on the contour $\Lambda(\phi)$, formulated in Theorem 5.1, is conformally transformed into a similar problem on the unit circle. The corresponding boundary condition reads, with the notations introduced in (5.15)-(5.18),

$$\mathrm{Re}\,[F(\phi;w)\Omega_x(\phi;0,\gamma(\phi;w))] = f_x(\phi;w),\quad |w|=1. \tag{5.20}$$

For ϕ so large that its index is $\kappa=-2$, this boundary value problem has a solution under the condition that

$$\int_C f_x(\phi;u)/[F(\phi;u)X^+(\phi;u)]du = 0. \tag{5.21}$$

By using the following relations: for $|u|=1$,

$$f_x(\phi; 1/u) = -f_x(\phi; u),$$
$$u^2 F(\phi; 1/u) X^+(\phi; 1/u) = F(\phi; u) X^+(\phi; u), \tag{5.22}$$

which are consequences of (4.29), (5.15)-(5.18), and the Plemelj formulas, cf. [14], it follows that the integral in (5.21) is equal to its opposite, and hence that condition (5.21) is fulfilled. The solution of a Riemann-Hilbert problem with index $\kappa=-2$, given in [14], can be rewritten to give (5.19). □

In Theorem 5.2 the function $\Omega_x(\phi; 0, z_2)$ has been determined for $z_2 \in \Lambda^+(\phi)$ and for ϕ on a real interval. By means of analytic continuation this function is then, in principle, determined at least for $\operatorname{Re}\phi>0$, $|z_2|\leq 1$. Notice that the conformal mapping $\gamma(\phi; w)$ and its inverse, which have been defined only for real values of ϕ in Section 4, are regular functions of ϕ on the interval (ϕ_0, ∞) and can therefore be continued analytically into a part of the complex plane; see also [4]. Further, the function $\Omega_x(\phi; z_1, 0)$ is determined through relation (3.6), by using (4.2), and the function $\Omega_x(\phi; z_1, z_2)$ is then determined through relation (3.2), and the necessary analytic continuations.

The remainder of this section is devoted to a more detailed study of the function $\Omega_x(\phi; 0, 0)$, the Laplace transform of the probability that the network is empty at time t, $t \geq 0$.

COROLLARY 5.1. *For real ϕ, $\phi>\phi_0$, $\phi>\phi_{H1}$ when $\chi_2<1/b_1$, $\phi>\phi_{H2}$ when $\chi_1<1/b_2$, it holds that when $r_1<1$,*

$$\Omega_x(\phi; 0, 0) = X(\phi; 0)/(\pi i) \int_C f_x(\phi; u)/[uF(\phi; u)X^+(\phi; u)]du, \tag{5.23}$$

while in the case that $r_1=1$,

$$\Omega_x(\phi; 0, 0) = X^+(\phi; -1)/(\pi i) \int_C f_x(\phi; u)/[(u+1)F(\phi; u)X^+(\phi; u)]du. \tag{5.24}$$

PROOF. Relation (5.23) follows directly from (5.19) and (4.30); the condition $x_2(\phi)<1$, cf. Theorem 5.1, can be dropped because of analytic continuation. When $r_1=1$ more care is needed, because $\gamma_0(\phi; 0)$ lies on the unit circle in this case, cf. (4.30). From (5.22) it is seen that $f_x(\phi; -1)=0$. Because the function $f_x(\phi; u)$ is regular at $u=-1$, the integral in (5.24) is well defined. □

The integrals in (5.23) and (5.24) become singular at the point $u=1$ for $\phi=\phi_{H2}$ when $\chi_1<1/b_2$ and for $\phi=\phi_{H1}$ when $\chi_2<1/b_1$. When $\chi_1<1/b_2$ and $\phi_0<\phi<\phi_{H2}$, then the factor $z_2-1-r_2(\zeta_0(\phi; z_2)-1)$ in (5.12) is positive at $z_2=x_2(\phi)$ and negative at $z_2=x_1(\phi)$, cf. (5.13), and therefore this factor adds 2 to the index of a Riemann-Hilbert problem for the function $\Omega_x(\phi; 0, z_2)$ with boundary condition (5.9). Similarly, this index increases by 2 when ϕ decreases below ϕ_{H1}, when $\chi_2<1/b_1$, cf. (5.13), (5.14). Noting that the inequalities $\chi_1<1/b_2$ and $\chi_2<1/b_1$ together imply that $\rho_1<1$ and $\rho_2<1$, it is found with the aid of Lemmas 5.3 and 5.4 that the (analytic continuation of the) function

$\Omega_x(\phi; 0, z_2)$ is the solution of a boundary value problem of index $\kappa=0$ in the following cases:

$$\text{for } \phi_0<\phi<\phi_{H1}, \quad \text{when } \chi_1 \geqslant 1/b_2 \text{ and } \chi_2<1/b_1, \tag{5.25a}$$

$$\text{for } \phi_0<\phi<\phi_{H2}, \quad \text{when } \chi_1<1/b_2 \text{ and } \chi_2 \geqslant 1/b_1, \tag{5.25b}$$

$$\text{for } \phi_{H2}<\phi<\phi_{H1}, \quad \text{when } \chi_1<1/b_2, \ \chi_2<1/b_1, \text{ and } 1/b_2<c_2+r_1/b_1, \tag{5.25c}$$

$$\text{for } \phi_{H1}<\phi<\phi_{H2}, \quad \text{when } \chi_1<1/b_2, \ \chi_2<1/b_1, \text{ and } 1/b_2>c_2+r_1/b_1; \tag{5.25d}$$

and of index $\kappa=+2$ in the following case:

$$\text{for } \phi_0<\phi<\min\{\phi_{H1}, \phi_{H2}\}, \quad \text{when } \chi_1<1/b_2, \ \chi_2<1/b_1. \tag{5.26}$$

When the index $\kappa=0$ or $\kappa=+2$, the solution of the Riemann-Hilbert problem with boundary condition (5.9) which is real on the real axis reads, cf. [14]: for $|w|<1$,

$$\begin{aligned}\Omega_x(\phi; 0, \gamma(\phi; w)) = {} & Y(\phi; w)[w^{\kappa/2}\sum_{k=0}^{\kappa/2}(w^k+w^{-k})\Theta_k(\phi) \\ & +(2\pi i)^{-1}\int_C f_x(\phi; u)/[F(\phi; u)Y^+(\phi; u)][1+(w/u)^{\kappa+1}](u-w)^{-1}du].\end{aligned} \tag{5.27}$$

Here $\Theta_k(\phi)$, $k=0,\ldots,\kappa/2$, are real functions of ϕ which are still undetermined, and

$$Y(\phi; w) := \exp\left[\frac{1}{4\pi i}\int_C \log\{-\frac{F(\phi; 1/u)}{u^\kappa F(\phi; u)}\}(1+\frac{w}{u})\frac{du}{u-w}\right], \tag{5.28}$$

for $|w|<1$, while $Y^+(\phi; u_0)$, $|u_0|=1$, is defined analogously to $X^+(\phi; u_0)$ in (5.18). Below it will be indicated how the functions $\Theta_0(\phi)$ and $\Theta_1(\phi)$ can be determined. Suppose first that $\chi_1<1/b_2$ and $\phi_0<\phi<\phi_{H2}$. Then $1-r_2+r_2\nu_1(\phi)<x_2(\phi)$, so that $1-r_2+r_2\nu_1(\phi)\in\Lambda^+(\phi)$. Substituting $z_1=\nu_1(\phi)$ and $z_2=1-r_2+r_2\nu_1(\phi)$ into functional equation (3.2) gives:

$$\begin{aligned}&\Omega_x(\phi; 0, 1-r_2+r_2\nu_1(\phi)) \\ &= m_1\beta_1[1-\nu_1(\phi)]^{-1}[\nu_1(\phi)]^{x_1+1}[1-r_2+r_2\nu_1(\phi)]^{x_2}.\end{aligned} \tag{5.29}$$

In the cases (5.25b) and (5.25d) the only unknown function $\Theta_0(\phi)$ is determined by substituting $w=\gamma_0(\phi; 1-r_2+r_2\nu_1(\phi))$ into (5.27) and by using (5.29). Suppose next that $\chi_2<1/b_1$ and $\phi_0<\phi<\phi_{H1}$. Substituting $z_1=1-r_1+r_1\nu_2(\phi)$ and $z_2=\nu_2(\phi)$ into functional equation (3.2) leads to an expression for $\Omega_x(\phi; 1-r_1+r_1\nu_2(\phi), 0)$ which is analogous to (5.29). It follows from (3.3) that $K(\phi; 1-r_1+r_1\nu_2(\phi), z_2)=0$ for $z_2=\nu_2(\phi)$ and for $z_2=\xi(\phi)$,

$$\xi(\phi) := (b_1/b_2)[r_1r_2+(1-r_1r_2)/\nu_2(\phi)]/[c_2b_1+r_1/\{1-r_1+r_1\nu_2(\phi)\}]. \tag{5.30}$$

With the aid of (A.3) it is seen that $\xi(\phi_{H1})=\nu_2(\phi_{H1})=x_2(\phi_{H1})$. It is readily verified that $\xi(\phi)>x_2(\phi)$ as $\phi\to\infty$, which implies that $\xi(\phi)<x_2(\phi)$, i.e.

$\xi(\phi)\in\Lambda^+(\phi)$, for $\phi_0<\phi<\phi_{H1}$. An expression for $\Omega_x(\phi;0,\xi(\phi))$ can be obtained by substituting $z_1=\mu_2(\phi):=1-r_1+r_1\nu_2(\phi)$ and $z_2=\xi(\phi)$ into relation (3.6):

$$\Omega_x(\phi;0,\xi(\phi)) = (b_1/r_1)\mu_2(\phi)^{x_1+1}\{\alpha[\xi(\phi)^{x_2}-\nu_2(\phi)^{x_2}]/[\xi(\phi)-\nu_2(\phi)] \\ +m_2\beta_2[1-\nu_2(\phi)]^{-1}\nu_2(\phi)^{x_2+1}[c_2+r_1/\{b_1\mu_2(\phi)\}]\}. \tag{5.31}$$

In the cases (5.25a) and (5.25c) the only unknown function $\Theta_0(\phi)$ is determined by substituting $w=\gamma_0(\phi;\xi(\phi))$ into (5.27) and by using (5.31). When the index $\kappa=+2$, cf. (5.26), both (5.29) and (5.31) can be used to determine the two functions $\Theta_0(\phi)$ and $\Theta_1(\phi)$ in (5.27). This procedure, however, does not apply when $\xi(\phi)=1-r_2+r_2\nu_1(\phi)$ for all ϕ, $\phi\geqslant\phi_0$. By comparing both sides of this equation as $\phi\to\infty$ and at $\phi=0$, it is found that it holds if and only if $r_1=1$, $c_2=0$, $b_1=b_2$. By differentiating both sides of equation (3.2) as functions of z_2 and by taking then $z_1=\nu_1(\phi)$ and $z_2=1-r_2+r_2\nu_1(\phi)$, it is obtained that the derivative of the function $\Omega_x(\phi;0,z_2)$ at $z_2=\mu_1(\phi):=1-r_2+r_2\nu_1(\phi)$ is equal to

$$m_1\beta_1[1-\nu_1(\phi)]^{-1}\nu_1(\phi)^{x_1+1}\mu_1(\phi)^{x_2}\Delta(\phi)^{-1}\{[1-\nu_1(\phi)][x_1+r_2x_2\nu_1(\phi)/\mu_1(\phi)] \\ +[1-\rho_1\nu_1^2(\phi)][x_2-\{1-[\nu_1(\phi)/\mu_1(\phi)]^{x_2}\}/\{\mu_1(\phi)/\nu_1(\phi)-1\}]+\rho_1\nu_1^2(\phi)\}, \tag{5.32}$$

$$\Delta(\phi) := [1-r_2+r_2\nu_1(\phi)][1-\rho_1\nu_1^2(\phi)]+r_2\nu_1(\phi)[1-\nu_1(\phi)]. \tag{5.33}$$

This relation can be used, together with (5.29), to determine an expression for the function $\Omega_x(\phi;0,z_2)$ when $\kappa=+2$ in the case $r_1=1$, $c_2=0$, $b_1=b_2$. In [4] such an analysis has been carried out for the case that also $r_2=0$. It is important to note that the function $\Delta(\phi)$ possesses a zero on the interval $[\phi_1,0)$, which will be denoted by ϕ_3. This can be verified as follows. Because $\kappa=+2$, the conditions $\chi_1<1/b_2$ and $\chi_2<1/b_1$ hold and imply that $\rho_1<1$, cf. Remark 5.1. Hence, cf. (5.4), (5.3), $\nu_1(0)=1$ and $\Delta(0)=1-\rho_1>0$, while $\nu_1(\phi_1)=1/\sqrt{\rho_1}$ so that $\Delta(\phi_1)=r_2[1-1/\sqrt{\rho_1}]/\sqrt{\rho_1}\leqslant 0$. Further, $\phi_{H1}=\phi_{H2}=0$. Note that $\phi_3=\phi_1$ when $r_2=0$.

6. The relaxation time

In this section the relaxation times of open queueing networks with two single server stations are determined with the aid of the foregoing analysis. A conjecture is made concerning the relaxation times of open queueing networks with an arbitrary number of single server stations. By the (global) relaxation time of a queueing network we will understand the relaxation time of the distribution of the total number of jobs in the network. This relaxation time will be studied on the basis of an asymptotic analysis of the probability that the network is empty (i.e. that the total number of jobs in the network is 0).

Theorem 6.1. *The relaxation time of an open queueing network with two single server stations, Poisson arrival processes, exponentially distributed service times and random transitions is given by:*

$$T = -1/\phi_0, \quad \textit{if} \quad \chi_1\geqslant 1/b_2 \quad \textit{and} \quad \chi_2\geqslant 1/b_1, \tag{6.1}$$

$$T = -1/\phi_1, \quad \textit{if} \quad \chi_1 \leqslant 1/b_2 \quad \textit{and, when} \quad \chi_2 \leqslant 1/b_1, \quad \textit{if} \quad \phi_1 \geqslant \phi_2, \tag{6.2}$$

$$T = -1/\phi_2, \quad \textit{if} \quad \chi_2 \leqslant 1/b_1 \quad \textit{and, when} \quad \chi_1 \leqslant 1/b_2, \quad \textit{if} \quad \phi_2 \geqslant \phi_1, \tag{6.3}$$

with the understanding that $T=\infty$ *when the network is null-recurrent, i.e. when* $\rho_1=1$ *or* $\rho_2=1$, *cf. (5.3); an exception to (6.2) and (6.3) is formed by the case* $r_1=1$, $c_2=0$, $r_2>0$, $b_1=b_2<1-r_2$, *when* $T=-1/\phi_3$.

PROOF. The relaxation time T of the probability that a queueing network is empty is equal to $-1/\phi_c$, where ϕ_c is the abscissa of convergence of the Laplace transform of this probability as function of time (apart from a pole of this transform at the origin when the network is stable). Hence, the abscissa of convergence of the function $\phi\Omega_x(\phi; 0, 0)$ has to be determined. The abscissa of convergence of the analytic continuation of the conformal mapping $\gamma(\phi; w)$, $|w| \leqslant 1$, is ϕ_0. Because this conformal mapping occurs in every expression for $\Omega_x(\phi; 0, 0)$, cf. (5.23), (5.24), (5.27), (5.15), (5.16), the abscissa of convergence of $\phi\Omega_x(\phi; 0, 0)$ is at least ϕ_0. When $\chi_j \leqslant 1/b_{3-j}$ then ϕ_j is a branch point of $\Omega_x(\phi; 0, 0)$, because it is a branch point of $\nu_j(\phi)$, $j=1,2$, cf. (5.27), (5.29), (5.31); and $\phi_j \geqslant \phi_0$, $j=1,2$, by Lemmas 5.1 and 5.2. Only in the case $r_1=1$, $c_2=0$, $r_2>0$, $b_1=b_2<1-r_2$, the function $\phi\Omega_x(\phi; 0, 0)$ has a pole at ϕ_3, $\phi_3>\phi_1=\phi_2$, cf. (5.32), (5.33). Finally, it should be noted that the function $\phi\Omega_x(\phi; 0, 0)$ is regular at ϕ_{H1} (when $\chi_2<1/b_1$) and at ϕ_{H2} (when $\chi_1<1/b_2$). This is most clearly seen by noting that ϕ_{H1} or ϕ_{H2} may be positive and that $\Omega_x(\phi; 0, 0)$ is regular for positive values of ϕ. In fact, the principal values of the integrals in (5.23), (5.24) and (5.27) exist at ϕ_{H1} and at ϕ_{H2}, except when $\phi_{H1}=\phi_{H2}$. But then $\chi_1<1/b_2$ and $\chi_2<1/b_1$, implying that $\rho_1<1$ and $\rho_2<1$, so that it must hold that $1/b_2=c_2+r_1/b_1$ and $\phi_{H1}=\phi_{H2}=0$, cf. Lemmas 5.3 and 5.4. In this case the function $\Omega_x(\phi; 0, 0)$ has a pole at $\phi=\phi_{H1}=\phi_{H2}=0$. As noted before, $\Omega_x(\phi; 0, 0)$ has a pole at $\phi=0$ whenever $\rho_1<1$, $\rho_2<1$. When $1/b_2>c_2+r_1/b_1$ this pole stems from the factor $[1-\nu_1(\phi)]^{-1}$ in (5.29) and when $1/b_2<c_2+r_1/b_1$ from the factor $[1-\nu_2(\phi)]^{-1}$ in (5.31). □

THEOREM 6.2. *The relaxation time of a stable open queueing network with two single server stations, Poisson arrival processes, exponentially distributed service times and random transitions is given by* $T=-1/\phi_3$ *when* $r_1=1$, $c_2=0$, $r_2>0$, $b_1=b_2$ *and in all other cases (satisfying (2.12)) by:*

$$T = \max_{j=1,2} [m_j\beta_j/(1-\sqrt{\rho_j})^2]; \tag{6.4}$$

the probability that such a network is empty at time t, $t>0$, *has the following asymptotic expansion for* $t\to\infty$:

$$\begin{aligned} &\Pr\{N_1(t)=N_2(t)=0|N_1(0)=x_1,\ N_2(0)=x_2\} \\ &= (1-\rho_1)(1-\rho_2)+C_x(t/T)^M(t/T)^{-3/2}e^{-t/T}[1+O(1/t)]; \end{aligned} \tag{6.5}$$

here C_x *is a constant which depends on the parameters of the network and on the initial state* (x_1, x_2), *and* $M=0$ *except when* $r_1=1$, $c_2=0$, $b_1=b_2$; *then* $M=1$ *when* $r_2=0$ *and* $M=3/2$ *when* $r_2>0$.

Proof. For stable networks ($\rho_1<1$, $\rho_2<1$) we have either $\phi_{H1}>0$, or $\phi_{H2}>0$, or $\phi_{H1}=\phi_{H2}=0$, cf. Lemmas 5.3 and 5.4. Suppose first that $r_1=1$, $c_2=0$, $r_2>0$, $b_1=b_2$ does not hold. Then, the abscissa of convergence of the function $\phi\Omega_x(\phi;0,0)$ is either ϕ_1 or ϕ_2. Suppose that $\phi_1\geqslant\phi_2$. This implies with (5.3) and (2.10) that

$$(c_1+c_2r_2)(1/\sqrt{\rho_1}-1)^2 \leqslant (c_1r_1+c_2)(1/\sqrt{\rho_2}-1)^2. \tag{6.6}$$

This inequality leads to:

$$1/b_2 \geqslant [\sqrt{c_1r_1+c_2}+(1/\sqrt{\rho_1}-1)\sqrt{c_1+c_2r_2}\,]^2/(1-r_1r_2). \tag{6.7}$$

By using the inequalities $\rho_1<1$, $r_1r_2<1$, and

$$\begin{aligned}4(c_1+c_2r_2)(c_1r_1+c_2) > {} & [c_1r_1(1+r_2)+c_2r_2(1+r_1)]^2\\ & =[c_1r_1+c_2r_2+r_1r_2]^2,\end{aligned} \tag{6.8}$$

it follows from (6.7) that

$$\begin{aligned}1/b_2 > {} & [c_1r_1+c_2+(c_1r_1+c_2r_2+r_1r_2)(1/\sqrt{\rho_1}-1)\\ & +r_1r_2(c_1+c_2r_2)(1/\sqrt{\rho_1}-1)^2]/(1-r_1r_2).\end{aligned} \tag{6.9}$$

Straightforward calculation shows that the right-hand side of (6.9) is equal to χ_1, cf. (5.1). In a similar way it can be shown that $\chi_2<1/b_1$ when $\phi_2\geqslant\phi_1$ and $\rho_2<1$. Hence, (6.4) follows from (6.2) and (6.3), cf. (5.3). When $r_1=1$, $c_2=0$, $r_2>0$, $\rho_1=\rho_2<1$, there is a branch point at $\phi_1=\phi_2$, but also a pole at ϕ_3 which is closer to the origin than the branch point and, hence, determines the relaxation time.

The asymptotic expansion (6.5) can be obtained by applying the saddle point method to the inverse transform of $\Omega_x(\phi;0,0)$, cf. [16]. In most cases the transform $\Omega_x(\phi;0,0)$ has just a branch point at ϕ_1 or at ϕ_2, cf. (5.27), (5.29), (5.31). When $r_1=1$, $c_2=0$, $r_2=0$, $b_1=b_2$, then $\Omega_x(\phi;0,0)$ has a branch point as well as a pole at $\phi_1=\phi_2$. This pole stems from the factor $\Delta(\phi)=1-\rho_1\nu_1^2(\phi)$ in (5.32). As a consequence of this pole, an additional factor t/T appears in the asymptotic expansion. See also [4]. When $r_1=1$, $c_2=0$, $r_2>0$, $b_1=b_2$, then $\Omega_x(\phi;\,0,0)$ has a pole at ϕ_3; this leads to an asymptotic expansion with only an exponential function of t/T in the second term. □

Remark 6.1. When $r_1=1$, $c_2=0$, $r_2>0$, $\rho_1=\rho_2<1$, the relaxation time $T=-1/\phi_3$, where ϕ_3 is implicitly determined as the zero on the interval $[\phi_1,0)$ of the function $\Delta(\phi)$, cf. (5.33). To give an impression of the size of this relaxation time we note that $\phi_1/\phi_3\to1+\sqrt{r_2}<2$ as $\rho_1\downarrow0$ and that $\phi_1/\phi_3\to(1+r_2/2)^2/(1+r_2)\leq9/8$ as $\rho_1\uparrow1$, while $\phi_1=\phi_3$ when $r_2=0$ and $\phi_1/\phi_3\to[5+4\rho_1+3\sqrt{1+8\rho_1}\,]/[2(1+\sqrt{\rho_1}\,)]^2<2$ as $r_2\uparrow1$ and ρ_1 is fixed, $\rho_1<1$.

Theorem 6.2 states that the global relaxation time of a two-node open queueing network (which is not a tandem system with equally loaded nodes and with feedback from the last to the first node) is equal to the maximum of the local relaxation times of the nodes; here the local relaxation time of a node is defined as the relaxation time such a node would have when it would operate as an M/M/1 queueing system in isolation, with arrival rate λ_j/m_j and mean service time $m_j\beta_j$, $j=1,2$, cf. (2.5), or, in a slightly different formulation, when it would operate in isolation as an M/M/1 queueing system with feedback, with arrival rate λ_j/m_j, mean service time β_j and feedback probability $(m_j-1)/m_j$, $j=1,2$, cf. [6,§4.2]. Probably, this property is not restricted to networks with two nodes, and this is the motivation for the conjecture formulated below. We notice, however, that this property does not imply in general, that the relaxation time of the marginal queue length distribution at a particular node is equal to the above defined local relaxation time of this node. More likely, the relaxation time of the marginal queue length distribution at a particular node is equal to the maximum of the local relaxation times of the node itself and of those nodes from which this particular node can be reached, either directly or via some intermediary nodes (this depends only on the transition matrix Q); i.e. at most equal to the global relaxation time T and at least equal to the local relaxation time of the node itself.

The local relaxation time of a node depends only on the traffic intensity and the average total service time at the node itself. Both these quantities reflect the structure of the network partly (only nodes from which a particular node can be reached with positive probability play a role). The global relaxation time of a network is only directly dependent of the parameters of the node(s) with the largest local relaxation time. It depends on the parameters of the other nodes only to the extent that it is larger than their local relaxation times. But it will be clear that the time-dependent behavior of a network in which the local relaxation times of all but one node are much smaller than the global relaxation time will be quite different from that of a network in which the local relaxation times of all nodes have (almost) the same value. This difference is reflected in the constant C_x and higher order terms in the expansion (6.5). Yet, important differences exist even among networks in which the average total service times and the traffic intensities (and hence the local relaxation times) are the same at each station. In symmetrical networks ($c_1=c_2$, $b_1=b_2$, $r_1=r_2$) both stations will tend to stochastic equilibrium at the same speed, and the asymptotic behavior of such a network is in essence similar to that of an M/M/1 system. When stations are in series ($c_1=1$, $c_2=0$, $b_1=b_2$, $r_1=1$), however, all traffic for the second station first has to pass through the first station, and therefore the second station will tend somewhat slower to equilibrium than the first station. If there is no feedback in the tandem system ($r_2=0$), this feature is reflected in an additional factor t in the asymptotic expansion (6.5), compared with the expansion of an M/M/1 system. If there exists a positive probability that a job returns to the first station after departure from the last station ($r_2>0$), the above effect is even stronger. Each time a job returns to the first station all stations in the tandem system have to be passed through again. In this specific case, the global relaxation time is larger than the local relaxation times of the stations. This is the motivation for the exceptional case in the conjecture below.

CONJECTURE. *The relaxation time of a stable open queueing network with K ($K \geq 1$) single server stations, Poisson arrival processes, exponentially distributed service times and random transitions is given by:*

$$T = \max_{j=1,\ldots,K} [m_j \beta_j / (1-\sqrt{\rho_j})^2], \tag{6.10}$$

except when the network contains at least one subsystem consisting of two or more equally loaded nodes which can be entered by jobs only at one particular node and which has such a structure that each job which enters the subsystem visits each node of it in a fixed order and may afterwards return to the first node with positive probability; then the right-hand side of (6.10) is a lower bound for the relaxation time of the network.

ACKNOWLEDGEMENT
The author wishes to thank prof. J.W. Cohen for suggesting the topic of this research and for some stimulating discussions.

REFERENCES

1. F. BASKETT, K.M. CHANDY, R.R. MUNTZ, F.G. PALACIOS (1975). Open, closed, and mixed networks of queues with different classes of customers. *J. Assoc. Comput. Mach. 22,* 248-260.
2. J.P.C. BLANC (1982). *Application of the Theory of Boundary Value Problems in the Analysis of a Queueing Model with Paired Services,* Mathematical Centre Tract 153, Amsterdam.
3. J.P.C. BLANC (1984). Asymptotic analysis of a queueing system with a two-dimensional state space. *J. Appl. Probab. 21,* 870-886.
4. J.P.C. BLANC (1985). The relaxation time of two queueing systems in series. *Commun. Statist. - Stochastic Models 1,* 1-16.
5. J.P.C. BLANC (1985). The transient behavior of networks with infinite server nodes. E. GELENBE (ed.). *Performance '84,* North-Holland, Amsterdam, 159-174.
6. J.P.C. BLANC, E.A. VAN DOORN (1986). Relaxation times for queueing systems. J.W. DE BAKKER, M. HAZEWINKEL, J.K. LENSTRA (eds.). *Mathematics and Computer Science,* CWI Monograph 1, North-Holland, Amsterdam, 139-162.
7. J.W. COHEN (1982). *The Single Server Queue,* North-Holland, Amsterdam, 2nd ed.
8. J.W. COHEN, O.J. BOXMA (1983). *Boundary Value Problems in Queueing System Analysis,* North-Holland, Amsterdam.
9. G. FAYOLLE, R. IASNOGORODSKI (1979). Two coupled processors: the reduction to a Riemann-Hilbert problem. *Z. Wahrsch. Verw. Gebiete 47,* 325-351.
10. P.G. HARRISON (1981). Transient behavior of queueing networks. *J. Appl. Probab. 18,* 482-490.
11. J.R. JACKSON (1957). Networks of waiting lines. *Oper. Res. 5,* 518-521.

12. F.P. KELLY (1979). *Reversibility and Stochastic Networks,* Wiley, New York.
13. H. KOBAYASHI (1974). Application of the diffusion approximation to queueing networks II: nonequilibrium distributions and applications to computer modelling. *J. Assoc. Comput. Mach. 21,* 459-469.
14. N.I. MUSCHELISCHWILI (1965). *Singuläre Integralgleichungen,* Akademie-Verlag, Berlin.
15. A.R. ODONI, E. ROTH (1983). An empirical investigation of the transient behavior of stationary queueing systems. *Oper. Res. 31,* 432-455.
16. D.G.M. WISHART (1966). Discussion on Kingman's paper. *J. Roy. Stat. Soc. Ser. B 28,* 442-443.

Appendix A

Proof of Lemma 5.3

For a zero of the function $H_1(\zeta_0(\phi;z_2),z_2)$ on the contour $\Lambda(\phi)$ the following conditions must simultaneously hold:

$$H_1(z_1,z_2) = 0, \quad K(\phi;z_1,z_2) = 0, \quad z_2 \in \Lambda(\phi). \tag{A.1}$$

The first two relations imply, cf. (3.3), (3.4), that $z_1=1-r_1+r_1\nu$, $z_2=\nu$, where ν satisfies the equation

$$\alpha\phi+(c_1r_1+c_2)(1-\nu)+(1-r_1r_2)(\nu-1)/(b_2\nu) = 0. \tag{A.2}$$

For $\phi>\phi_2$ this equation has two positive roots of which $\nu_2(\phi)$ is the smaller, cf. (5.4). This implies that $z_2=x_2(\phi)$ is the only point on the contour $\Lambda(\phi)$ where the function $H_1(\zeta_0(\phi;z_2),z_2)$ may vanish. Substituting $y_2(\phi)=1-r_1+r_1\nu$, $x_2(\phi)=\nu$, into relation (4.27) leads to

$$\nu^2 = (b_1/b_2)[1-r_1+r_1\nu][1-r_1r_2+r_1r_2\nu]/[c_2b_1\{1-r_1+r_1\nu\}+r_1]. \tag{A.3}$$

The three roots of this equation are real and exactly one of them, say $\nu=\nu_{H1}$, is positive. Notice that the function $z_2=\zeta(\phi;z_1)$ is real-valued for z_1 on the real interval $[y_2(\phi),y_3(\phi)]$, and that the graph of this two-valued function on this interval is a contour in the real (z_1,z_2)-plane. For $\phi>\phi_2$, the line $z_1=1-r_1+r_1z_2$ crosses this contour twice. In order that the point ν_{H1} lies on the contour $\Lambda(\phi)$, i.e. $\nu_{H1}=x_2(\phi)$, and not $\nu_{H1}=x_3(\phi)$, it must hold that $\nu_{H1}<\nu_2(\phi_2)=1/\sqrt{\rho_2}$. It follows in a straightforward manner from (A.3) that $\nu_{H1}<1/\sqrt{\rho_2}$ if and only if $\chi_2<1/b_1$, cf. (5.2). In this case we find, cf. (A.2), (2.10),

$$\phi_{H1} = (1-\nu_{H1})(1-\rho_2\nu_{H1})/(m_2\beta_2\nu_{H1}). \tag{A.4}$$

Obviously, $\phi_{H1}>\phi_2$. When $\chi_2=1/b_1$, then $\phi_{H1}=\phi_2=\phi_0$, cf. (A.3) and Lemma 5.2. When $\chi_2>1/b_1$, it has just been shown that $H_1(\zeta_0(\phi;z_2),z_2)$ does not vanish on the contour $\Lambda(\phi)$ for $\phi \geqslant \phi_2$. For $\phi_0 \leqslant \phi < \phi_2$, the function $\nu_2(\phi)$ is either not real, while $\zeta_0(\phi; z_2)$ is real for $z_2 \in \Lambda(\phi)$, or negative, but not equal to $x_1(\phi)$ - this would lead to a contradiction. Hence, $H_1(\zeta_0(\phi; z_2), z_2)$ does not vanish on the contour $\Lambda(\phi)$ for $\phi_0 \leqslant \phi < \phi_2$.

Finally, let $\rho_2 < 1$. Then $\phi_{H1} > 0$ if and only if $\nu_{H1} < 1$, cf. (A.4), and it follows from (A.3) that $\nu_{H1} < 1$ if and only if $1/b_2 < c_2 + r_1/b_1$.

Appendix B

Proof of Lemma 5.4

The main ideas of this proof are similar to those of Lemma 5.3 (see Appendix A). For a zero of the function $H_2(\zeta_0(\phi; z_2), z_2)$ on the contour $\Lambda(\phi)$ the following conditions must simultaneously hold:

$$H_2(z_1, z_2) = 0, \quad K(\phi; z_1, z_2)=0, \quad z_2 \in \Lambda(\phi). \tag{B.1}$$

The first two relations imply, cf. (3.3), (3.5), that $z_1=\nu$, $z_2=1-r_2+r_2\nu$, where ν satisfies the equation

$$\alpha\phi+(c_1+c_2r_2)(1-\nu)+(1-r_1r_2)(\nu-1)/(b_1\nu) = 0. \tag{B.2}$$

For $\phi>\phi_1$ this equation has two positive roots of which $\nu_1(\phi)$ is the smaller, cf. (5.4). This implies that $z_2=x_2(\phi)$ is the only point on the contour $\Lambda(\phi)$ where the function $H_2(\zeta_0(\phi; z_2), z_2)$ may vanish. Substituting $x_2(\phi)=1-r_2+r_2\nu$, $y_2(\phi)=\nu$, into relation (4.27) leads to

$$\nu = (b_2/b_1)[1-r_2+r_2\nu][c_2b_1\nu+r_1]. \tag{B.3}$$

This equation has either two positive roots or none. It is readily seen that this equation has exactly one root smaller than $\nu_1(\phi_1)=1/\sqrt{\rho_1}$ if and only if $\chi_1<1/b_2$. This root is denoted by ν_{H2}, and we find, cf. (B.2), (2.10),

$$\phi_{H2} = (1-\nu_{H2})(1-\rho_1\nu_{H2})/(m_1\beta_1\nu_{H2}). \tag{B.4}$$

Obviously, $\phi_{H2}>\phi_1$. When $\chi_1=1/b_2$, then the two roots of (B.3) are:

$$\nu = 1/\sqrt{\rho_1}, \quad \text{and} \quad \nu=[r_1(1-r_2)/(c_2r_2b_1)]\sqrt{\rho_1}. \tag{B.5}$$

if and only if (2.12) holds then $1/\sqrt{\rho_1}$ is the smaller root of (B.3). Hence, $\phi_{H2}=\phi_1=\phi_0$ when $\chi_1=1/b_2$. Further, this implies that when the nodes of the network have been numbered such that inequality (2.12) holds then

$H_2(\zeta_0(\phi; z_2), z_2)$ does not vanish on the contour $\Lambda(\phi)$ for $\phi \geqslant \phi_1$ if $\chi_1 > 1/b_2$. That this function does not vanish on the contour $\Lambda(\phi)$ for $\phi_0 \leqslant \phi < \phi_1$ follows by a similar argument as in the proof of Lemma 5.3.

Finally, let $\rho_1 < 1$. Then $\phi_{H2} > 0$ if and only if $\nu_{H2} < 1$, cf. (B.4). Equation (B.3) has exactly one root smaller than 1 if and only if $1/b_2 > c_2 + r_1/b_1$. When $1/b_2 = c_2 + r_1/b_1$ then $\nu_{H2} = 1$ is the smaller root of (B.3) if and only (2.12) holds. This implies that equation (B.3) has no root smaller than 1 when $1/b_2 < c_2 + r_1/b_1$.

Two Queues with Alternating Service and Switching Times

O.J. Boxma, W.P. Groenendijk
Centre for Mathematics and Computer Science
P.O. Box 4079, 1009 AB Amsterdam, The Netherlands

This paper is concerned with a system of two queues, attended by a single server who alternately serves one customer of each queue (if not empty). The server experiences switching times in his transition from one queue to the other. It is shown that the joint stationary queue-length distribution, at the instants at which the server becomes available to a queue, can be determined via transformation to a Riemann boundary value problem. The latter problem can be completely solved for general service- and switching-time distributions. The stationary distributions of the waiting times at both queues, and of the cycle times of the server, are also derived. The results obtained, and in particular the extensive numerical data for moments of waiting times and cycle times, yield insight into the behavior of more general cyclic-service models. Such models are frequently used to analyse polling systems.

1. Introduction

Some 15 years ago the analysis of polling systems, employed to multiplex the service requests of several users in computer-terminal communication systems, gave rise to a new class of queueing models: a single server serves a number of queues in some cyclic fashion. Presently, these *cyclic-service* queueing models are finding a new application in local area networks with a ring or bus topology, employing a medium access control protocol based on token passing. Various service disciplines at the queues of the cyclic-service models have been considered, ranging from exhaustive service (when the server visits a queue, he serves its customers until the queue has become empty) to 1-limited service (when the server visits a queue, he serves only one customer, if present). For an extensive discussion of the many results that have recently been obtained for cyclic-service systems, we refer to the book [16] and survey paper [17] of Takagi. Generally speaking, the analysis of cyclic-service systems with exhaustive service is complex but tractable; 1-limited service, on the other hand, gives rise to very intricate mathematical problems, which only have been solved for models with not more than two queues.

The model with two queues, 1-limited service and no switching times has first been tackled in an important study of Eisenberg [8]. In the sequel this model with two queues, one server and 1-limited service discipline will be referred to as the *alternating-service model.* Eisenberg transformed the problem of determining the joint queue-length distribution at the two queues into the problem of solving a singular Fredholm integral equation. Almost simultaneously, FAYOLLE and IASNOGORODSKI in a highly original paper [9] solved another queueing problem with a two-dimensional state space, via transformation of the functional equation for the (bivariate) generating function of the joint queue-length distribution into a Riemann-Hilbert boundary value problem. Of course, the latter type of problem also belongs to the realm of singular integral equations. These two studies strongly stimulated the interest of J.W. Cohen in the analysis of queueing and random walk problems with a two-dimensional state space. He started an extensive research program which has led to a powerful method of transforming functional equations, encountered in such 'two-dimensional problems', into boundary value problems of the Riemann or Riemann-Hilbert type. One of the first fruits of this method has been a detailed analysis [6] of the alternating-service model without switching times. The resulting boundary value problem appeared to be a Dirichlet problem, a special case of a Riemann-Hilbert problem. In [1] the analysis was extended to the case *with* switching times of the server between queues, but the restriction was made that both queues had identical characteristics. This time, the resulting boundary value problem was a Riemann-Hilbert problem.

The goal of the present paper is to give an exact analysis of the alternating-service model with switching times, with arbitrary service-time distributions and switching-time distributions. As in the just mentioned studies, the arrival processes at both queues are independent Poisson processes.

The organization of the paper is as follows. In Section 2 the model is described in detail. Section 3 contains the main analysis. The joint stationary queue-length distribution, at the instants at which the server becomes available to a queue, is determined via transformation into a Riemann boundary value problem. Once the joint queue-length distribution is known, one can easily derive expressions for various important performance measures, like waiting times of customers and cycle times of the server (the time between two successive arrivals of the server at a particular queue). Waiting times are studied in Section 4, with particular attention to the mean waiting times; cycle times are studied in Section 5, with particular attention to second moments of the cycle times (first moments of cycle times are trivially determined). Section 6 is devoted to a numerical evaluation of some important performance measures of the alternating-service model. It is shown that the boundary value problem formulation leads to formulas which can be numerically evaluated in a straightforward manner. The numerical results in this section may also provide additional insight into the behavior of the alternating-service model and, more generally, of cyclic-service models. E.g., it is shown that, while first moments of cycle times do not depend on the number of the queue at which the cycle starts, second moments of cycle times generally differ only slightly. This supports an approximation assumption in [2,3], to the effect that the second moments are equal.

2. Model description

A single server S serves two queues Q_1, Q_2 (with infinite buffer capacities) in cyclic order. The arrival process of customers at Q_i is a Poisson process with rate λ_i, $i=1,2$. The service times at Q_i are independent, identically distributed stochastic variables with distribution $B_i(.)$, with first moment β_i, second moment $\beta_i^{(2)}$ and LST (Laplace-Stieltjes Transform) $\beta_i(\cdot)$. The various arrival and service processes are independent.

The utilization at Q_i, ρ_i, is defined as

$$\rho_i := \lambda_i \beta_i, \quad i=1,2. \tag{2.1}$$

The total utilization of the server, ρ, is defined as

$$\rho := \rho_1 + \rho_2. \tag{2.2}$$

The server serves one customer, if any, from Q_1, and after a switching time he inspects Q_2. He serves one customer, if any, from Q_2, and switches back to Q_1; etc. The successive switching times from Q_i to $Q_{(i+1)\ mod\ 2}$ are independent, identically distributed stochastic variables, also independent of the service times, with distribution $S_i(\cdot)$. Their first moment, second moment and LST are respectively denoted by s_i, $s_i^{(2)}$ and $\sigma_i(.)$.

Let $\mathbf{C}_i$ denote the time between two successive arrivals of S at Q_i, the *cycle time* for Q_i. Clearly each cycle consists of two switches and at most one service at each of the two queues. The first and second moments of the total switching time during one cycle are respectively denoted by

$$s := s_1 + s_2, \tag{2.3}$$

$$s^{(2)} := s_1^{(2)} + 2s_1 s_2 + s_2^{(2)}. \tag{2.4}$$

It is well known, and easily seen (cf. WATSON [19]) that the mean cycle time $E\mathbf{C}_i$ is independent of i, and is given by

$$E\mathbf{C} = \frac{s}{1-\rho}. \tag{2.5}$$

Ergodicity conditions

In the model *without* switching times, $\rho<1$ is easily seen to be a necessary and sufficient condition for ergodicity. KUEHN [14] has shown that, in the model *with* switching times,

$$\rho + \max(\lambda_1 s, \lambda_2 s) < 1, \tag{2.6}$$

is a necessary condition for ergodicity (indeed, the mean number of arrivals at Q_i during a cycle, $\lambda_i s/(1-\rho)$, should be less than one). SZPANKOWSKI and REGO [15] have recently proved that this condition is also sufficient. In the sequel we assume that the system is ergodic, hence necessarily (2.6) holds.

3. Formulation and solution of the boundary value problem

Let $\mathbf{q}_i^{(1)}$, $i=1,2$, denote the number of customers at Q_i at those instants at which S arrives at Q_1; similarly for $\mathbf{q}_i^{(2)}$. Let

$$F_j(z_1,z_2) := E[z_1^{\mathbf{q}_1^{(j)}} z_2^{\mathbf{q}_2^{(j)}}], \quad |z_1| \leqslant 1, \ |z_2| \leqslant 1, \ j=1,2. \tag{3.1}$$

Our goal in this section is to determine $F_1(z_1,z_2)$ and $F_2(z_1,z_2)$. This goal will be accomplished by formulating and solving a so-called Riemann boundary value problem (cf. Gakhov [12]). Once $F_i(z_1,z_2)$ is determined, the LST of the waiting-time distribution at Q_i and of the distribution of the cycle time $\mathbf{C}_i$ can be obtained. In Sections 4 and 5 we will demonstrate this, deriving mean waiting times and second moments of cycle times as by-products.

The vector of queue lengths at Q_1 and Q_2 at successive arrival epochs of server S at a queue forms a vector Markov chain. A study of its transition probabilities yields the following recurrence relations for the generating functions $F_1(z_1,z_2)$ and $F_2(z_1,z_2)$: for $|z_1| \leqslant 1$, $|z_2| \leqslant 1$,

$$F_2(z_1,z_2) = \{[F_1(z_1,z_2) - F_1(0,z_2)]\, z_1^{-1} \beta_1(x) + F_1(0,z_2)\}\, \sigma_1(x), \tag{3.2}$$

$$F_1(z_1,z_2) = \{[F_2(z_1,z_2) - F_2(z_1,0)]\, z_2^{-1} \beta_2(x) + F_2(z_1,0)\}\, \sigma_2(x), \tag{3.3}$$

with for $|z_1| \leqslant 1$, $|z_2| \leqslant 1$:

$$x := \lambda_1(1-z_1) + \lambda_2(1-z_2). \tag{3.4}$$

Substitution of (3.2) into (3.3) yields:

$$\begin{aligned} K(z_1,z_2)F_1(z_1,z_2) = {} & F_1(0,z_2)\{\beta_2(x)\,\sigma_1(x)\,\sigma_2(x)\,(z_1 - \beta_1(x))\} + \\ & F_2(z_1,0)\{z_1\,\sigma_2(x)\,(z_2 - \beta_2(x))\}, \quad |z_1| \leqslant 1, \ |z_2| \leqslant 1, \end{aligned} \tag{3.5}$$

while analogously,

$$\begin{aligned} K(z_1,z_2)F_2(z_1,z_2) = {} & F_2(z_1,0)\{\beta_1(x)\,\sigma_1(x)\,\sigma_2(x)\,(z_2 - \beta_2(x))\} + \\ & F_1(0,z_2)\{z_2\,\sigma_1(x)\,(z_1 - \beta_1(x))\}, \quad |z_1| \leqslant 1, \ |z_2| \leqslant 1; \end{aligned} \tag{3.6}$$

here $K(z_1,z_2)$ is the *kernel of the functional equation,* defined as

$$K(z_1,z_2) := z_1 z_2 - \beta_1(x)\,\beta_2(x)\,\sigma_1(x)\,\sigma_2(x). \tag{3.7}$$

Relation (3.5) is the starting point for our analysis. The main idea is similar to the one in [6,7] for the model without switching times and in [1] for the model with switching times and with identical characteristics of both queues ('the symmetric model'): the determination of $F_1(z_1,z_2)$ from (3.5) will be reduced to the solution of a boundary value problem (BVP) of mathematical physics. In the model without switching times, this BVP was a Dirichlet problem, a special case of a Riemann-Hilbert problem; in the symmetric model with switching times, this BVP was a Riemann-Hilbert problem; and in the present more general model, reduction to a Riemann problem can be accomplished. In fact, in the latter two cases both a Riemann-Hilbert and a Riemann problem formulation are possible; the Riemann formulation seems to be somewhat more natural here. The analysis consists of four steps.

Step 1: the set-up
According to its definition as a generating function, $F_1(z_1,z_2)$ should be regular for $|z_1| < 1$, continuous for $|z_1| \leqslant 1$, for every fixed z_2 with $|z_2| \leqslant 1$; and similarly with z_1 and z_2 interchanged. Hence every zero (z_1,z_2) of the kernel $K(z_1,z_2)$ in (3.5) should be a zero of the right-hand side of (3.5). This condition must lead to the yet unknown functions $F_1(0,z_2)$ and $F_2(z_1,0)$ in the right-hand side of (3.5), and hence to $F_1(z_1,z_2)$.

Step 2: analysis of the kernel
It is not possible to determine, explicitly, exactly one zero z_1 in $|z_1| \leqslant 1$ for each z_2 in $|z_2| \leqslant 1$. Various sets of zero-pairs of the kernel $K(z_1,z_2)$ can be determined; our choice will lead to a Riemann BVP. $K(z_1,z_2)$ is a so-called Poisson kernel (cf. Ch. II.4 of [7]). It has the same structure as the Poisson kernel defined in (2.4) on p. 274 of [7], where the alternating-service model *without* switching times is studied. Therefore we can proceed as in [7] (cf. also [1]). First introduce

$$\beta(x) := \beta_1(x)\, \beta_2(x)\, \sigma_1(x)\, \sigma_2(x),$$

$$\lambda := \lambda_1 + \lambda_2,$$

$$r_1 := \lambda_1/\lambda, \qquad r_2 := \lambda_2/\lambda.$$

Without loss of generality, it will henceforth be assumed that

$$r_1 \geqslant r_2.$$

Finally introducing

$$w_1 := 2r_1z_1, \qquad w_2 := 2r_2z_2,$$

we can rewrite $4r_1r_2K(z_1,z_2)$ as

$$w_1w_2 - 4r_1r_2\beta(\lambda(1-(w_1+w_2)/2)).$$

The symmetry of this expression suggests to look for pairs of zeros of the kernel that are each other's complex conjugates: $(w_1,w_2)=(w,\bar{w})$. These pairs of zeros turn out to supply all the information we need. The following should hold for w:

$$|w|^2 = 4r_1r_2\beta(\lambda(1-\text{Re } w)).$$

Write

$$w = e^{i\phi}2\sqrt{r_1r_2}\sqrt{\beta(\lambda(1-\text{Re } w))}, \qquad 0 \leqslant \phi \leqslant 2\pi.$$

Defining $\delta(\phi)$ to be the unique zero of

$$\delta - 2\sqrt{r_1r_2}\cos(\phi)\sqrt{\beta(\lambda(1-\delta))}, \qquad 0 \leqslant \phi \leqslant 2\pi,\ \text{Re } \delta \leqslant 1, \tag{3.8}$$

it is seen that, when ϕ once traverses the trajectory $[0,2\pi]$,

$$w = w(\phi) = \delta(\phi)(1 + i \tan(\phi))$$

once encircles a simple, smooth contour F that is contained in the unit circle.

F is an egg-shaped contour. Using the notation L^+ (L^-) for the interior (exterior) of a contour L, we have $0 \in F^+$. Every $w \in F$ satisfies the relation $|w|^2 = 4r_1r_2\beta(\lambda(1-\text{Re } w))$; hence $(z_1,z_2) = (w/2r_1,\bar{w}/2r_2)$ forms a pair of zeros of the kernel $K(z_1,z_2)$ for every $w \in F$.

Step 3: formulation of a Riemann boundary value problem

The choice of zero-pairs $(z_1,z_2) = (w/2r_1,\bar{w}/2r_2)$ of the kernel leads, in a natural way, to the formulation of a Riemann BVP. In the formulation and solution of the BVP a few technical difficulties will arise; these are mainly related to the position of the point $2r_2$ with respect to the contour F. Depending on the choice of parameters, this point can be inside, on or outside the contour. For the sake of clarity, we restrict ourselves here to the case $2r_2 \in F^+$; see Remark 3.1 for a short discussion of the case $2r_2 \in F$ (which occurs, e.g., for $r_2 = 1/2$) and the (relatively rare) case $2r_2 \in F^-$.

Basically, the Riemann BVP amounts to finding two functions, one regular inside a certain smooth contour and the other one regular outside that contour, such that a certain linear relation exists between these functions on the contour; see [7] for a short exposition, and see GAKHOV [12] for a detailed discussion. The first part of Step 3 concerns that linear relation between two functions on a contour. The right-hand side of (3.5) should be zero for all those $w \in F$, for which $(w/2r_1,\bar{w}/2r_2)$ forms a pair of zeros of $K(z_1,z_2)$ inside the product of unit circles. Now $|w/2r_1| \leqslant 1$ always holds for $w \in F$, but the possibility that $|\bar{w}/2r_2| > 1$ cannot be excluded. Fortunately, in the latter case analytic continuation can be used (cf. below (3.21)) to show that, for *all* $w \in F$, the right-hand side of (3.5) should be zero. So for *all* $w \in F$, the following linear relation should exist between $F_1(0,\bar{w}/2r_2)$ and $F_2(w/2r_1,0)$:

$$F_1(0,\bar{w}/2r_2)\,[\beta_2(\lambda(1-\text{Re } w))\,\sigma_1(\lambda(1-\text{Re } w))\,\sigma_2(\lambda(1-\text{Re } w)) \times \tag{3.9}$$
$$\{\frac{w}{2r_1} - \beta_1(\lambda(1-\text{Re } w))\}] +$$
$$F_2(w/2r_1,0)\,[\frac{w}{2r_1}\sigma_2(\lambda(1-\text{Re } w))\,\{\frac{\bar{w}}{2r_2} - \beta_2(\lambda(1-\text{Re } w))\}] = 0.$$

Hence

$$F_2(w/2r_1,0) = G(w)\,F_1(0,\bar{w}/2r_2), \quad w \in F, \tag{3.10}$$

with

$$G(w) := -\beta_2(\lambda(1-\text{Re } w))\,\sigma_1(\lambda(1-\text{Re } w))\,\frac{1}{w/2r_1} \times \tag{3.11}$$
$$\frac{w/2r_1 - \beta_1(\lambda(1-\text{Re } w))}{\bar{w}/2r_2 - \beta_2(\lambda(1-\text{Re } w))}.$$

In the standard formulation of the Riemann BVP, the involved smooth contour is the unit circle. A conformal mapping of F^+ onto the interior C^+ of the unit circle C will lead us to a standard Riemann BVP. The second part of

Step 3 concerns this conformal mapping:

$$z = f(w):\ F^+ \to C^+, \tag{3.12}$$

and its inverse, the conformal mapping

$$w = f_0(z):\ C^+ \to F^+. \tag{3.13}$$

One can write (cf. GAIER [11], Section 2.1; see also Section I.4.4 of [7]):

$$f_0(z) = z\ \exp[\frac{1}{2\pi}\int_0^{2\pi} \log\{\frac{\delta(\theta(\omega))}{\cos(\theta(\omega))}\}\ \frac{e^{i\omega}+z}{e^{i\omega}-z}\ d\omega], \quad |z|<1, \tag{3.14}$$

with the angular deformation $\theta(\cdot)$ being uniquely determined as the continuous solution of the *Theodorsen* integral equation

$$\theta(\phi) = \phi - \frac{1}{2\pi}\int_0^{2\pi} \log\{\frac{\delta(\theta(\omega))}{\cos(\theta(\omega))}\}\ \mathrm{cotan}\{\frac{1}{2}(\omega-\phi)\}d\omega, \quad 0\leqslant\phi\leqslant 2\pi; \tag{3.15}$$

$\theta(\phi)$ is a strictly increasing and continuous function of ϕ, and $\theta(\phi)=2\pi-\theta(2\pi-\phi)$. According to the corresponding-boundaries theorem ([7], p. 66), $f_0(z)$ is continuous in $C^+\cup C$. Application of the conformal mapping $f_0(\cdot)$ transforms (3.10) into:

$$F_2(f_0(z)/2r_1,0) = G(f_0(z))\ F_1(0,f_0(1/z)/2r_2), \quad z\in C. \tag{3.16}$$

Introducing the functions

$$\hat{F}_2(z) := F_2(f_0(z)/2r_1,0), \quad z\in C^+\cup C, \tag{3.17}$$

$$\hat{F}_1(z) := F_1(0,f_0(1/z)/2r_2), \quad z\in C\cup C^-, \tag{3.18}$$

$$H(z) := G(f_0(z)) = -\beta_2(\lambda(1-\mathrm{Re}\, f_0(z)))\ \sigma_1(\lambda(1-\mathrm{Re}\, f_0(z)))\times \tag{3.19}$$

$$\frac{1}{f_0(z)/2r_1}\ \frac{f_0(z)/2r_1-\beta_1(\lambda(1-\mathrm{Re}\, f_0(z)))}{f_0(1/z)/2r_2-\beta_2(\lambda(1-\mathrm{Re}\, f_0(z)))}, \quad z\in C,$$

(3.16) can be rewritten as:

$$\hat{F}_2(z) = H(z)\ \hat{F}_1(z), \quad z\in C. \tag{3.20}$$

There are some technical requirements for a Riemann BVP formulation: $H(z)$ should satisfy a so-called Hölder condition on C (this can be easily verified, and will not be further discussed); and $0<|H(z)|<\infty$ for $z\in C$. The third part of Step 3 concerns a proof that $H(z)\neq 0$ for $z\in C$ (the proof that $|H(z)|<\infty$ is left to the reader). We prove the equivalent statement that $G(w)\neq 0$ for $w\in F$. The two points of F on the real axis are the only candidate zeros of $G(w)$, $w\in F$. It is soon clear that we can concentrate on $w=\delta(0)\in F$, and that it remains to show that $\delta(0)/2r_1 - \beta_1(\lambda(1-\delta(0)))\neq 0$. The assumption that $2r_2\in F^+$ implies that $\delta(0)/2r_2 > \beta_2(\lambda(1-\delta(0)))$. The definition of $\delta(0)$, see (3.8), implies that

$$\delta(0) = 2\sqrt{r_1r_2}\,\sqrt{\beta(\lambda(1-\delta(0)))} < 2\sqrt{r_1r_2}\,\sqrt{\beta_1(\lambda(1-\delta(0)))}\,\sqrt{\beta_2(\lambda(1-\delta(0)))}.$$

Hence

$$\delta(0)/2r_1 \geqslant \beta_1(\lambda(1-\delta(0))) \quad \text{and} \quad \delta(0)/2r_2 \geqslant \beta_2(\lambda(1-\delta(0)))$$

are not simultaneously possible. In view of the above,

$$\delta(0)/2r_1 < \beta_1(\lambda(1-\delta(0))). \tag{3.21}$$

We are almost ready to formulate our Riemann BVP. It remains to show that $\hat{F}_1(z)$ is regular for $z \in C^-$, continuous for $z \in C \cup C^-$, and $\hat{F}_2(z)$ is regular for $z \in C^+$, continuous for $z \in C^+ \cup C$. We show, equivalently, that $F_2(w/2r_1,0)$ and $F_1(0,w/2r_2)$ are regular in F^+ and continuous in $F^+ \cup F$. Since by assumption $2r_1 \geqslant 1$, while F is contained in the unit circle, it immediately follows that $F_2(w/2r_1,0)$ is regular in F^+, and continuous in $F^+ \cup F$. It is somewhat more difficult to show that $F_1(0,w/2r_2)$ is also regular in F^+ and continuous in $F^+ \cup F$. First note that $\delta(0) = \max|w|$, $w \in F^+ \cup F$. Subsequently note that $F_1(0,\delta(0)/2r_2)$ is finite, because $F_2(\delta(0)/2r_1,0)/G(\delta(0))$ is finite. These observations, combined with the fact that the coefficients in the series expansion of $F_1(0,w/2r_2)$ are non-negative, lead to the stated regularity and continuity properties of $F_1(0,w/2r_2)$ (cf. [7], p. 277, for a similar reasoning).

We have now arrived at a standard, homogeneous, Riemann BVP on the unit circle:
Determine two functions $\hat{F}_1(z)$ and $\hat{F}_2(z)$, such that
- (3.20) holds, with $H(.)$ satisfying a Hölder condition on C and $0 < |H(z)| < \infty$, $z \in C$;
- $\hat{F}_1(z)$ is regular for $z \in C^-$, continuous for $z \in C \cup C^-$;
- $\hat{F}_2(z)$ is regular for $z \in C^+$, continuous for $z \in C^+ \cup C$;
- $\hat{F}_1(z) \to A$ for $|z| \to \infty$, with A a constant.

Step 4: solution of the Riemann boundary value problem
A crucial role in the solution of the Riemann BVP is played by the *index*, χ, of the function $H(.)$ on C. This index is by definition:

$$\chi := \operatorname{ind}_{z \in C} H(z) = \frac{1}{2\pi} \int_{z \in C} d\{\arg H(z)\} \tag{3.22}$$

$$= \operatorname{ind}_{w \in F} G(w) = \frac{1}{2\pi} \int_{w \in F} d\{\arg G(w)\}.$$

Lemma 3.1. *$\chi = 0$ for $2r_2 \in F^+$.*

Proof. From (3.11),

$$\chi = -\operatorname{ind}_{w \in F} \frac{w}{2r_1} + \operatorname{ind}_{w \in F} [w/2r_1 - \beta_1(\lambda(1-\operatorname{Re} w))] \tag{3.23}$$

$$-\operatorname{ind}_{w \in F} [\overline{w}/2r_2 - \beta_2(\lambda(1-\operatorname{Re} w))] =$$

$$-1 + \text{ind}_{w \in F} [w/2r_1 - \beta_1(\lambda(1-\text{Re } w))]$$

$$+\text{ind}_{w \in F} [w/2r_2 - \beta_2(\lambda(1-\text{Re } w))].$$

The fact that $2r_2 \in F^+$ implies that $w/2r_2 > \beta_2(\lambda(1-\text{Re } w))$ for $w=\delta(0) \in F$. It now readily follows that

$$\text{ind}_{w \in F} [w/2r_2 - \beta_2(\lambda(1-\text{Re } w))] = 1. \tag{3.24}$$

Similarly, from (3.21)

$$\text{ind}_{w \in F} [w/2r_1 - \beta_1(\lambda(1-\text{Re } w))] = 0. \tag{3.25}$$

The lemma follows from (3.23), (3.24) and (3.25). □

The homogeneous Riemann BVP formulated at the end of Step 3, with index 0, has the following solution (cf. [7], Section I.2.3):

$$\hat{F}_1(z) = A \exp[\frac{1}{2\pi i} \int_{t \in C} \frac{\log H(t)}{t-z} dt], \quad z \in C^-, \tag{3.26}$$

$$\hat{F}_2(z) = A \exp[\frac{1}{2\pi i} \int_{t \in C} \frac{\log H(t)}{t-z} dt], \quad z \in C^+, \tag{3.27}$$

with $A = F_1(0,0)$ yet to be determined.
Formulas (3.26) and (3.27) lead, in combination with (3.17), (3.18) and the definition (3.12) of the conformal mapping $f(.)$, to our main result:

THEOREM 3.1.

$$F_1(0,w/2r_2) = A \exp[\frac{1}{2\pi i} \int_{t \in C} \frac{\log H(t)}{t-1/f(w)} dt], \quad w \in F^+, \tag{3.28}$$

$$F_2(w/2r_1,0) = A \exp[\frac{1}{2\pi i} \int_{t \in C} \frac{\log H(t)}{t-f(w)} dt], \quad w \in F^+. \tag{3.29}$$

It remains to determine the constant $A = F_1(0,0)$. Substitution of $w=2r_2$ in (3.28) yields a linear relation between $F_1(0,1)$ and $F_1(0,0)$. $F_1(0,1)$ can be determined in various ways. E.g., substituting $z_2=1$ in (3.5) and subsequently letting $z_1 \to 1$ gives one linear relation between $F_1(0,1)$ and $F_2(1,0)$; applying a similar procedure to (3.6) gives a second linear relation between those quantities. Solution of the two equations yields:

$$F_1(0,1) = 1 - \frac{\lambda_1 s}{1-\rho}, \tag{3.30}$$

$$F_2(1,0) = 1 - \frac{\lambda_2 s}{1-\rho}. \tag{3.31}$$

The following observation also immediately leads to (3.30) (and similarly (3.31)): $1-F_1(0,1)$ is the probability that server S finds Q_1 not empty upon his arrival. Therefore it equals the fraction of times that S serves a customer in Q_1 during his visit. By a balance argument, this fraction also equals the mean number of arrivals at Q_1 during a cycle of the server. According to (2.5), the mean cycle time of the server equals $s/(1-\rho)$; hence the mean number of arrivals at Q_1 during a cycle of the server equals $\lambda_1 s/(1-\rho)$.

The above implies that the constant A is given by:

$$A = F_1(0,0) = (1-\frac{\lambda_1 s}{1-\rho}) \exp[-\frac{1}{2\pi i} \int_{t \in C} \frac{\log H(t)}{t-1/f(2r_2)} dt]. \tag{3.32}$$

REMARK 3.1. Earlier, the restrictive assumption $2r_2 \in F^+$ has been made. However, the cases $2r_2 \in F$ and $2r_2 \in F^-$ can also occur. E.g., if $r_1 = r_2 = 1/2$, then $2r_2 = 1 = \delta(0) \in F$; examples in which $2r_2 \in F^-$ can also be constructed, although one has to be careful not to violate the ergodicity condition (cf. [7], pp. 360-361). We shortly consider Steps 3 and 4 above for these two cases.

(i) $2r_2 \in F$.
The Riemann BVP formulation and the proof that the index $\chi=0$ proceed as before. The special case $r_1 = r_2 = 1/2$ requires some subtlety. Now $\delta(0)=1$; both the numerator and denominator of the right-hand side of (3.11) become zero, but the zeros cancel and again $G(\delta(0)) \neq 0$. Furthermore, (3.23) reduces to $\chi = -1+1/2+1/2=0$. For all cases in which $2r_2 \in F$, the solution of the Riemann BVP proceeds as before, and Theorem 3.1 still holds. A minor difficulty is that $F_1(0,1)$ cannot be obtained from (3.28) by substitution of $w=2r_2$. But application of the so-called *Plemelj-Sokhotski* formula (cf. [7], Formula (I.1.6.4)) to (3.28) leads to an expression for $F_1(0,w/2r_2)$, $w \in F$. In the resulting expression, the substitution $w=2r_2$ can be made.

(ii) $2r_2 \in F^-$.
The formulation of the Riemann BVP proceeds as before. This time verification of the regularity of $F_1(0,w/2r_2)$ in F^+ is trivial. Again the index $\chi=0$, but verification is less straightforward. It is based on the observation that, for $2r_2 \in F^-$ and $w=\delta(0) \in F$, the coefficients of the functions $w/2r_1 - \beta_1(\lambda(1-\mathrm{Re}\, w))$ and $\bar{w}/2r_2 - \beta_2(\lambda(1-\mathrm{Re}\, w))$ in (3.9) are both non-negative, so that these functions have opposite signs. Theorem 3.1 still holds, but a major difficulty now is that $F_1(0,1)$ cannot be obtained by substitution of $w=2r_2$ in (3.28). One might obtain an analytic continuation of $F_1(0,w/2r_2)$ in F^-, but this is numerically impractical. A numerically feasible approach is to obtain numerical values for, say, $F_1(0,1)$ by using a Taylor series expansion of $F_1(0,z)$ around $z=0$. Chapter IV.1 of [7] contains an example of this procedure, for the alternating-service model without switching times.

Expressions for the generating functions $F_1(z_1,z_2)$ and $F_2(z_1,z_2)$ follow from (3.5), (3.6) and Theorem 3.1. In the next two sections we use the results obtained about queue-length generating functions to derive information about the distributions of waiting times and cycle times, and in particular about their moments.

4. WAITING TIMES

In this section we shall derive an expression for $E\mathbf{w}_2$, the mean waiting time at Q_2. $E\mathbf{w}_2$ will be expressed in some given model parameters and in the function $d/dz\ F_1(0,z)$, evaluated at $z=1$. The latter function is obtained from (3.28) after differentiation with respect to w and substitution of $w=2r_2$. $E\mathbf{w}_1$ cannot be obtained from (3.29) in a similar way; the fact that $2r_1 \in F^-$ poses a problem. To reach the point $2r_1$ out of F^+ we might take our refuge to analytic continuation, but this would lead to numerical difficulties. We might also use a Taylor series expansion of $F_2(w/2r_1,0)$ around $w=0$; such an approach would also be useful in case $2r_2 \in F^-$, as discussed at the end of the previous section.

Anyway, once we have calculated $E\mathbf{w}_2$, $E\mathbf{w}_1$ can be obtained from a linear relation between $E\mathbf{w}_1$ and $E\mathbf{w}_2$. One can derive such a relation from (3.1) and (3.2) by repeated differentiation, using (4.2) below. WATSON [19] has done this, more generally, for a single-server, multi-queue system with cyclic service. He has thus derived a linear relation between the mean waiting times at the various queues. For two queues, his result reduces to:

$$\rho_1[1-\frac{\lambda_1 s}{1-\rho}]E\mathbf{w}_1 + \rho_2[1-\frac{\lambda_2 s}{1-\rho}]E\mathbf{w}_2 \tag{4.1}$$

$$= \rho\frac{\lambda_1\beta_1^{(2)}+\lambda_2\beta_2^{(2)}}{2(1-\rho)} + \rho\frac{s^{(2)}}{2s} + \frac{s}{2(1-\rho)}[\rho^2+\rho_1^2+\rho_2^2].$$

See [4,5] for a generalization, with a probabilistic proof, of Watson's result to the case of a single-server, multi-queue system with cyclic service and a mixture of various types of service disciplines at the queues.

$E\mathbf{w}_2$ is obtained in the following way. By a standard M/G/1-type argument (cf. WATSON [19], TAKAGI [16]) we can write:

$$E\{e^{-\lambda_2(1-z)\mathbf{w}_2}\} = \frac{F_2(1,z) - F_2(1,0)}{z(1-F_2(1,0))}. \tag{4.2}$$

Indeed, the customers present in Q_2 at the start of a non-empty service period at that queue, excluding the customer about to be served, are just the customers who had arrived during the waiting time of that customer. Note that (4.2) completely determines the waiting-time distribution at Q_2; a similar relation holds for the transform of the waiting-time distribution at Q_1. From (4.2) and (3.31),

$$E\mathbf{w}_2 = \frac{1-\rho}{\lambda_2^2 s}\{\frac{d}{dz}F_2(1,z)\}_{z=1} - \frac{1}{\lambda_2}. \tag{4.3}$$

The derivative occurring in (4.3) follows from (3.6) after a tedious but straightforward calculation. Denote by β and $\beta^{(2)}$ the first and second moments of the sum of a service time at Q_1, a switching time from Q_1 to Q_2, a service time at Q_2 and a switching time from Q_2 to Q_1. Then

$$\{\frac{d}{dz}F_2(1,z)\}_{z=1} = (1-\frac{\lambda_2 s}{1-\rho})\left[\frac{\lambda_2(\beta_1+s)(1-\lambda_2\beta_2)}{1-\lambda_2\beta}\right. \tag{4.4}$$

$$\left. -\frac{\frac{1}{2}\lambda_2^2\beta_2^{(2)}}{1-\lambda_2\beta} + (1-\lambda_2\beta_2)\frac{\frac{1}{2}\lambda_2^2\beta^{(2)}}{(1-\lambda_2\beta)^2}\right]$$

$$-\{\frac{d}{dz}F_1(0,z)\}_{z=1}\frac{\lambda_2\beta_1}{1-\lambda_2\beta}$$

$$-(1-\frac{\lambda_1 s}{1-\rho})\left[\frac{\lambda_2\beta_1(1+\lambda_2 s_1)+\frac{1}{2}\lambda_2^2\beta_1^{(2)}}{1-\lambda_2\beta}+\frac{\frac{1}{2}\lambda_2^3\beta_1\beta^{(2)}}{(1-\lambda_2\beta)^2}\right].$$

It remains to determine the derivative occurring in the right-hand side of (4.4). From (3.28),

$$\{\frac{d}{dz}F_1(0,z)\}_{z=1} = 2r_2\{\frac{d}{dw}F_1(0,w/2r_2)\}_{w=2r_2} \tag{4.5}$$

$$= 2r_2\ F_1(0,1)\frac{-f^{(1)}(2r_2)}{(f(2r_2))^2}\frac{1}{2\pi i}\int_{t\in C}\frac{\log H(t)}{(t-1/f(2r_2))^2}dt.$$

$E\mathbf{w}_2$ follows from (4.3), (4.4), (4.5) and (3.30). As indicated above, $E\mathbf{w}_1$ subsequently follows from (4.1).

5. Cycle times

In Section 2 the cycle time $E\mathbf{C}_i$ for Q_i has been defined as the time between two consecutive arrivals of S at Q_i. Both from a theoretical and a practical point of view, cycle times are important quantities in cyclic-service systems. Mean cycle times are easily calculated (cf. (2.5)), but in cyclic-service systems with 1-limited service hardly any other exact cycle-time results are known. Only for the special case of two completely symmetric queues, an exact formula for the LST of the cycle-time distribution has been obtained [1]. In the present section we extend this result to the asymmetric case. We are thus able to compare $E\mathbf{C}_1^2$ and $E\mathbf{C}_2^2$, and also to determine

$$E\mathbf{C}_{b,i} := E[\mathbf{C}_i \,|\, A_i], \tag{5.1}$$

with A_i the indicator function of the event 'the cycle contains a service at Q_i'. This quantity plays an important role in several mean waiting-time approximations [2,3,10,14]. Generally speaking, exact cycle-time formulas for the two-queue case give more insight into the accuracy of general approximations for cycle-time distributions, as were proposed by Hashida and Ohara [13] and Kuehn [14].

We now derive an exact expression for the LST of the distribution of $\mathbf{C}_1$;

the analogous result for $\mathbf{C}_2$ is obtained by interchanging all indices. Starting-point of the analysis is the relation

$$E[e^{-\omega \mathbf{C}_1}] = F_1(0,0)\, E[e^{-\omega \mathbf{C}_1} \,|\, \mathbf{q}_1^{(1)}=0,\ \mathbf{q}_2^{(1)}=0] + \tag{5.2}$$
$$[F_1(0,1)-F_1(0,0)]\, E[e^{-\omega \mathbf{C}_1} \,|\, \mathbf{q}_1^{(1)}=0,\ \mathbf{q}_2^{(1)}>0] +$$
$$[F_1(1,0)-F_1(0,0)]\, E[e^{-\omega \mathbf{C}_1} \,|\, \mathbf{q}_1^{(1)}>0,\ \mathbf{q}_2^{(1)}=0] +$$
$$[1-F_1(0,1)-F_1(1,0)+F_1(0,0)]\, E[e^{-\omega \mathbf{C}_1} \,|\, \mathbf{q}_1^{(1)}>0,\ \mathbf{q}_2^{(1)}>0]$$
$$= F_1(0,0)\, \sigma_2(\omega)[\sigma_1(\omega+\lambda_2)+\{\sigma_1(\omega)-\sigma_1(\omega+\lambda_2)\}\beta_2(\omega)] +$$
$$[F_1(0,1)-F_1(0,0)]\, \beta_2(\omega)\sigma_1(\omega)\sigma_2(\omega) +$$
$$[F_1(1,0)-F_1(0,0)]\, \sigma_2(\omega)\times$$
$$[\beta_1(\omega+\lambda_2)\sigma_1(\omega+\lambda_2)+\{\beta_1(\omega)\sigma_1(\omega)-\beta_1(\omega+\lambda_2)\sigma_1(\omega+\lambda_2)\}\beta_2(\omega)] +$$
$$[1-F_1(0,1)-F_1(1,0)+F_1(0,0)]\, \beta_1(\omega)\sigma_1(\omega)\beta_2(\omega)\sigma_2(\omega).$$

$F_1(0,1)$ is given by (3.30). Hence $E[e^{-\omega \mathbf{C}_1}]$ can be expressed in $F_1(0,0)$ and $F_1(1,0)$. Substitution of $z_1=1$, $z_2=0$ into (3.5) leads to a linear relation between those two terms:

$$F_1(1,0) = F_1(0,0)\frac{\beta_1(\lambda_2)-1}{\beta_1(\lambda_2)} + \frac{1-\lambda_2 s/(1-\rho)}{\beta_1(\lambda_2)\sigma_1(\lambda_2)}. \tag{5.3}$$

Differentiation of the expressions in (5.2) w.r.t. ω, and substitution of $F_1(1,0)$ into $F_1(0,0)$ using (5.3), leads to cycle-time moments. A simple calculation yields the mean cycle time given in (2.5); a lengthy calculation yields

$$E\mathbf{C}_1^2 = s^{(2)} + \sum_{i=1}^{2} \lambda_i \beta_i^{(2)} \frac{s}{1-\rho} + 2\beta_1(\beta_2+s)\frac{\lambda_1 s}{1-\rho} + \tag{5.4}$$
$$2\beta_2 s - 2\beta_2 s_2\left(1-\frac{\lambda_2 s}{1-\rho}\right) + 2\beta_2\left(1-\frac{\lambda_2 s}{1-\rho}\right)\left[\frac{\beta_1'(\lambda_2)}{\beta_1(\lambda_2)} + \frac{\sigma_1'(\lambda_2)}{\sigma_1(\lambda_2)}\right] -$$
$$2\beta_2 F_1(0,0)\sigma_1(\lambda_2)\frac{\beta_1'(\lambda_2)}{\beta_1(\lambda_2)}.$$

Note that if the switching time from Q_1 to Q_2 is a constant (s_1), then $E\mathbf{C}_1^2$ only depends on the individual mean switching times via the term involving $F_1(0,0)\sigma_1(\lambda_2)$ - apart from that term, only s and $s^{(2)}$ occur. We'll return to this point in the next section.

We now turn to the cycle-time distribution of $\mathbf{C}_1$ under the condition, A_1, that the cycle contains a service at Q_1. Similarly as (5.2),

$$E[e^{-\omega \mathbf{C}_1} \,|\, A_1] = \frac{F_1(1,0)-F_1(0,0)}{1-F_1(0,1)}\sigma_2(\omega)\times \tag{5.5}$$
$$[\beta_1(\omega+\lambda_2)\sigma_1(\omega+\lambda_2)+\{\beta_1(\omega)\sigma_1(\omega)-\beta_1(\omega+\lambda_2)\sigma_1(\omega+\lambda_2)\}\beta_2(\omega)] +$$

$$\frac{1-F_1(0,1)-F_1(1,0)+F_1(0,0)}{1-F_1(0,1)}\beta_1(\omega)\sigma_1(\omega)\beta_2(\omega)\sigma_2(\omega).$$

A simple calculation leads to

$$\begin{aligned} E\mathbf{C}_{b,1} &= E[\mathbf{C}_1 \mid A_1] \\ &= \beta_1 + s + \beta_2[1-\frac{1-\rho}{\lambda_1 s}\{1-\frac{\lambda_2 s}{1-\rho}-F_1(0,0)\sigma_1(\lambda_2)\}]. \end{aligned} \tag{5.6}$$

A similar expression, with all indices interchanged, holds for $E\mathbf{C}_{b,2}$. Note that the term between curly brackets represents the difference between the probability that S finds Q_2 empty and the probability that S finds first Q_1 and then Q_2 empty. Also observe that $\lambda_1 s/(1-\rho)$ is the probability that S does serve at Q_1. Hence the term between square brackets represents the conditional probability that S does serve at Q_2, under the condition A_1.

6. Numerical analysis

The present section is devoted to a numerical evaluation of some important performance measures of the alternating-service model. Our reasons for including this section are twofold:

(i) We want to show that the BVP formulation leads to formulas which can be numerically evaluated in a straightforward manner;

(ii) We want to contribute to the insight into the behavior of the alternating-service model and, more generally, of cyclic-service models; in particular, the numerical results to be presented may be helpful for devising and testing approximations.

For the sake of (i), we now discuss the numerical evaluation of $F_i(0,0)$ and $E\mathbf{w}_i$; other performance measures are easily evaluated from these quantities. The numerical analysis basically consists of five steps. For details we refer to Ch. IV.1 of [7], in which numerical calculations of this kind have been extensively discussed.

Step 1: Solving Theodorsen's integral equation (cf. (3.15))

Determine $\theta(\phi)$, iteratively, from (cf. [11]):

$$\theta_0(\phi) = \phi, \quad 0 \leqslant \phi \leqslant 2\pi, \tag{6.1}$$

$$\theta_{n+1}(\phi) = \phi - \frac{1}{2\pi}\int_0^{2\pi} \log\{\frac{\delta(\theta_n(\omega))}{\cos(\theta_n(\omega))}\} \operatorname{cotan}\{\tfrac{1}{2}(\omega-\phi)\}d\omega, \quad 0 \leqslant \phi \leqslant 2\pi,$$

where $\delta(\theta_n(\omega))$ is determined from (cf. (3.8)):

$$\delta(\theta_n(\omega)) = 2\sqrt{r_1 r_2}\cos(\theta_n(\omega))\sqrt{\beta(\lambda(1-\delta(\theta_n(\omega))))}, \tag{6.2}$$

using the Newton-Raphson root-finding procedure. In our calculations, the iteration has been continued until the differences between successive iterations of $\theta(\cdot)$ (in the supremum norm) were in absolute value less than 10^{-6}. This required between 6 and 14 iterations.

REMARK 6.1. Due to various symmetry properties we can restrict ourselves in the computations, here and in the sequel, to $\phi \in [0,\pi]$. As various integrands that will have to be computed change more rapidly for ϕ close to 0 than for other values of ϕ, a finer subdivision has been chosen for the interval $[0,\pi/5]$ (20 points) than for the interval $[\pi/5,\pi]$ (40 points). All involved integrals have been evaluated using the repeated trapezoidal rule.

Step 2: Determination of the conformal mapping $f_0(e^{i\phi})$, $0 \leqslant \phi \leqslant 2\pi$
Applying the Plemelj-Sokhotski formula (cf. [7], Formula (I.1.6.4)) to (3.14) yields:

$$f_0(e^{i\phi}) = e^{i\phi} \exp[\log\{\frac{\delta(\theta(\phi))}{\cos(\theta(\phi))}\} + \frac{1}{2\pi i}\int_0^{2\pi} \log\{\frac{\delta(\theta(\omega))}{\cos(\theta(\omega))}\} \operatorname{cotan}\{\tfrac{1}{2}(\omega-\phi)\} d\omega] \tag{6.3}$$

$$= e^{i\theta(\phi)} \frac{\delta(\theta(\phi))}{\cos(\theta(\phi))}$$

$$= \delta(\theta(\phi))[1+i \tan(\theta(\phi))], \quad 0 \leqslant \phi \leqslant 2\pi;$$

this result could also have been derived from the formula below (3.8).

Step 3: Determination of $f(2r_2)$ and $f^{(1)}(2r_2)$
Using (3.14), $f(2r_2)$ is obtained as the solution, on $[0,1]$, of $f_0(z) = 2r_2$. Again we have used the Newton-Raphson root-finding procedure. $f^{(1)}(2r_2)$ can be obtained in two ways:
(i) by numerical differentiation of $f_0(\cdot)$; note that

$$f^{(1)}(2r_2) = \frac{1}{f_0^{(1)}(2r_2)}; \tag{6.4}$$

(ii) by a numerical evaluation of the expression:

$$f_0^{(1)}(2r_2) = \frac{2r_2}{f(2r_2)} + 2r_2 \frac{1}{2\pi} \int_0^{2\pi} \log\{\frac{\delta(\theta(\omega))}{\cos(\theta(\omega))}\} \frac{2e^{i\omega}}{(e^{i\omega}-f(2r_2))^2} d\omega, \tag{6.5}$$

and substitution of the result in (6.4).
For a discussion of (ii) see [7], p. 351. We have used both (i) and (ii), but due to the fact that we have chosen a relatively fine subdivision we have found no significant differences.

Step 4: Calculation of $H(e^{i\phi})$, $0 \leqslant \phi \leqslant 2\pi$
$H(e^{i\phi})$ is obtained from (3.19) by noting that $\operatorname{Re} f_0(e^{i\phi}) = \delta(\theta(\phi))$:

$$H(e^{i\phi}) = -\beta_2(\lambda(1-\delta(\theta(\phi))))\, \sigma_1(\lambda(1-\delta(\theta(\phi)))) \frac{2r_1}{f_0(e^{i\phi})} \times \frac{f_0(e^{i\phi})/2r_1 - \beta_1(\lambda(1-\delta(\theta(\phi))))}{f_0(e^{-i\phi})/2r_2 - \beta_2(\lambda(1-\delta(\theta(\phi))))}, \quad 0 \leqslant \phi \leqslant 2\pi. \tag{6.6}$$

Step 5: Determination of $E\mathbf{w}_i$ *and* $F_i(0,0)$, $i=1,2$

Once we have calculated $\{\frac{d}{dz}F_1(0,z)\}_{z=1}$ from (4.5), we can obtain $E\mathbf{w}_2$ (cf. (4.3), (4.4) and (4.5)), and subsequently $E\mathbf{w}_1$ from (4.1). $F_1(0,0)$ is easily calculated from (3.32); $F_2(0,0)$ is obtained from (cf. (3.29)):

$$F_2(0,0) = F_1(0,0)\exp[\frac{1}{2\pi}\int_0^{2\pi}\log H(e^{i\omega})\,d\omega]. \tag{6.7}$$

With the subdivision we have chosen, each row in the tables below takes about 15 sec. of CPU time on a Cyber 170 model 750, with very small memory requirements. Using fewer $\theta(.)$ iterations and a less fine subdivision of the interval $[0,\pi]$ leads to a considerable reduction of CPU time, without sacrificing too much accuracy. The computer program was written in Pascal.

Numerical results are presented in Tables I and II. The performance measures under consideration are the mean waiting times $E\mathbf{w}_i$, the second moments of cycle times $E\mathbf{C}_i^2$, the conditional first moments of cycle times $E\mathbf{C}_{b,i}$ (cf. (5.1) and (5.6)) and the empty-system probabilities at server-arrival epochs, $F_i(0,0)$. Table I studies the influence of the switching times on these performance measures. Table Ia shows that the choice of switching-time *distributions* has hardly any effect on $E\mathbf{C}_{b,i}$ and $F_i(0,0)$, and only has a considerable effect on $E\mathbf{w}_i$ and $E\mathbf{C}_i^2$ when mean service times are relatively small. Exactly the same statement can be made concerning the choice of s_1 and s_2, for given *total* mean switching time $s=s_1+s_2$. In Table Ib results for $\beta_1=\beta_2=0.8$ are printed, exhibiting almost-insensitivity for the choice of s_1 and s_2. In the (non-printed) case with $\beta_1=\beta_2=0.2$ and all other parameter values as in Table Ib, the largest difference (with respect to $E\mathbf{w}_i$ and $E\mathbf{C}_i^2$) due to changes in s_i is in the order of 25%. For deterministic switching-time distributions, $E\mathbf{w}_i$, $E\mathbf{C}_i^2$ and $E\mathbf{C}_{b,i}$ appear to be completely independent of s_1 and s_2, given their sum s. The structure of (5.4) and (5.6) shows that the same must hold for $F_1(0,0)\,\sigma_1(\lambda_2) = \Pr\{S$ *finds first* Q_1 *and then* Q_2 *empty*$\}$. The robustness of the model for switching times is also being expressed by the pseudo-conservation law for mean waiting times mentioned in Section 4 (cf. (4.1) for the alternating-service model). In the pseudo-conservation law, the expression for a weighted sum of mean waiting times is seen to depend on the switching-time distributions only through the mean s and the second moment $s^{(2)}$ of the *total* switching time - and the influence of the factor involving $s^{(2)}$ is usually small.

Table II presents mean waiting times and cycle-time moments for three different combinations of service-time distributions, viz.:

Case A: both service-time distributions are negative exponential;

Case B: both service-time distributions are hyperexponential distributions with squared coefficient of variation 4 ($H_2(4)$) and balanced means [18];

Case C: $B_1(.)$ is a $H_2(4)$ distribution with balanced means, and $B_2(.)$ is deterministic.

Out of a wide range of distributions and parameter values, we have tried to make a representative choice. The observations from Table I allow us to restrict ourselves to constant switching times, with $s_1 = s_2$. In all cases considered, $s = 0.2$. We discuss all tabulated performance measures in turn.

(i) $E\mathbf{w}_i$
In [3] and [2], a mean waiting-time approximation has been proposed for a cyclic-service model without and with switching times, respectively. It is first argued that

$$E\mathbf{w}_i = \frac{E\mathbf{rc}_i}{1-\lambda_i E\mathbf{C}_{b,i}}, \tag{6.8}$$

with $E\mathbf{rc}_i$ the mean residual cycle time for Q_i. In fact, this is not an exact result. In the alternating-service model it appears to be quite close in most cases, but there are a few exceptions. Taking $E\mathbf{rc}_i = E\mathbf{C}_i^2/2E\mathbf{C}_i$ (acting as if the cycle-time process is a renewal process), formula (6.8) would imply that $E\mathbf{w}_i$ changes linearly with $E\mathbf{C}_i^2$ for fixed first moments of service times and switching times. The table entries for these two quantities suggest that this is indeed more or less the case.

In [2] and [3] two approximation assumptions are introduced to estimate the unknown $E\mathbf{rc}_i$ and $E\mathbf{C}_{b,i}$, viz.:

Assumption 1: $E\mathbf{C}_{b,i} = E\tilde{\mathbf{C}}_{b,i} := (\beta_i + s)/(1-\rho+\rho_i)$
(this approximation is due to KUEHN [14]).

Assumption 2: $E\mathbf{rc}_i$ *is the same for all i.*
Subsequently the pseudo-conservation law (cf. (4.1) for the alternating-service model) is used to estimate the one unknown $E\mathbf{rc}_1$. Below we investigate the accuracy of these assumptions for the alternating-service model.

(ii) $E\mathbf{C}_i^2$
Again taking $E\mathbf{rc}_i = E\mathbf{C}_i^2/2E\mathbf{C}_i$, Assumption 2 above would imply that all $E\mathbf{C}_i^2$ are the same. Indeed, in all considered cases, $E\mathbf{C}_1^2$ and $E\mathbf{C}_2^2$ differ less than 7% (and usually much less). FUHRMANN and WANG [10] suggest another mean waiting-time approximation along similar lines as [2], but they assume that

$$E\mathbf{C}_1^2/E\mathbf{C}_2^2 \approx E\mathbf{C}_{b,2}/E\mathbf{C}_{b,1}; \tag{6.9}$$

our numerical results show that this assumption is not accurate for the alternating-service model. Still, Fuhrmann and Wang improve upon [2] in case of heavy traffic. It is not yet fully clear whether (6.9) becomes more accurate when the number of queues is larger, or whether (6.9) counteracts an inaccuracy in (6.8) or in the approximation for $E\mathbf{C}_{b,i}$.

TABLE I
Mean waiting times and cycle-time moments for the alternating-service model; influence of the switching times

$S_1(.)$	$S_2(.)$	EW_1	EW_2	EC_1^2	EC_2^2	$EC_{b,1}$	$EC_{b,2}$	$F_1(0,0)$	$F_2(0,0)$
det	det	0.236	0.192	0.082	0.082	0.425	0.461	0.796	0.829
det	exp	0.265	0.221	0.093	0.095	0.427	0.462	0.797	0.827
exp	det	0.268	0.216	0.094	0.092	0.426	0.465	0.796	0.830
exp	exp	0.297	0.244	0.104	0.105	0.427	0.466	0.797	0.829
det	det	16.566	2.158	2.100	2.007	1.317	1.745	0.286	0.300
det	exp	16.671	2.172	2.110	2.020	1.318	1.746	0.287	0.299
exp	det	16.674	2.169	2.113	2.017	1.317	1.746	0.286	0.300
exp	exp	16.779	2.183	2.124	2.031	1.318	1.747	0.287	0.300

TABLE Ia. *The influence of the switching-time distributions.* $B_i(.)$ *negative exponential,* $i=1,2$; *in the first four rows* $\beta_1=\beta_2=0.2$, *in the last four* $\beta_1=\beta_2=0.8$. $\lambda=1$, $r_1=0.7$, $s_1=s_2=0.1$.

$S_1(.)$	$S_2(.)$	s_1	s_2	EW_1	EW_2	EC_1^2	EC_2^2	$EC_{b,1}$	$EC_{b,2}$	$F_1(0,0)$	$F_2(0,0)$
det	det	0.05	0.15	16.566	2.158	2.100	2.007	1.317	1.745	0.282	0.310
det	det	0.1	0.1	16.566	2.158	2.100	2.007	1.317	1.745	0.286	0.300
det	det	0.15	0.05	16.566	2.158	2.100	2.007	1.317	1.745	0.290	0.289
det	exp	0.05	0.15	16.802	2.189	2.124	2.037	1.319	1.747	0.283	0.309
det	exp	0.1	0.1	16.671	2.172	2.110	2.020	1.318	1.746	0.287	0.299
det	exp	0.15	0.05	16.592	2.161	2.102	2.010	1.317	1.745	0.290	0.289
exp	det	0.05	0.15	16.593	2.161	2.103	2.010	1.317	1.745	0.282	0.310
exp	det	0.1	0.1	16.674	2.169	2.113	2.017	1.317	1.746	0.286	0.300
exp	det	0.15	0.05	16.808	2.183	2.130	2.031	1.318	1.748	0.290	0.290
exp	exp	0.05	0.15	16.829	2.192	2.127	2.040	1.319	1.747	0.283	0.310
exp	exp	0.1	0.1	16.779	2.183	2.124	2.031	1.318	1.747	0.287	0.300
exp	exp	0.15	0.05	16.834	2.187	2.133	2.034	1.318	1.748	0.291	0.290

TABLE Ib. *The influence of* s_1 *and* s_2 *for given* $s=s_1+s_2$. $B_i(.)$ *negative exponential,* $i=1,2$; $\beta_1=\beta_2=0.8$. $\lambda=1$, $r_1=0.7$.

(iii) $EC_{b,i}$ and $E\tilde{C}_{b,i}$
In all considered cases, the approximation $EC_{b,1} \approx E\tilde{C}_{b,1}$ (see Assumption 1 above) is extremely accurate. The approximation $EC_{b,2} \approx E\tilde{C}_{b,2}$ is much less

accurate: the flow-balancing argument on which the approximation is based, should not be applied to the situation of a rarely occurring cycle $\mathbf{C}_2$ with a - sometimes large - service time at Q_2. The approximation becomes useless in the cases marked with an asterisk, because $E\mathbf{C}_{b,2}$ exceeds the obvious upper bound $\beta_1+\beta_2+s$; in those cases we have printed the latter number.

Finally we observe that $E\mathbf{C}_{b,i}$ is hardly dependent on the choice of the service-time distributions.

(iv) $F_i(0,0)$

$F_i(0,0)$, too, appears to be hardly dependent on the choice of the service-time distributions.

TABLE II

Mean waiting times and cycle-time moments for the alternating-service model

r_1	β_1	β_2	$E\mathbf{W}_1$	$E\mathbf{W}_2$	$E\mathbf{C}_1^2$	$E\mathbf{C}_2^2$	$E\mathbf{C}_{b,1}$	$E\tilde{\mathbf{C}}_{b,1}$	$E\mathbf{C}_{b,2}$	$E\tilde{\mathbf{C}}_{b,2}$	$F_1(0,0)$	$F_2(0,0)$
0.7	0.2	0.2	0.236	0.192	0.082	0.082	0.425	0.426	0.461	0.465	0.796	0.829
0.7	0.2	0.5	0.387	0.330	0.139	0.142	0.478	0.471	0.789	0.814	0.772	0.811
0.7	0.2	0.8	0.726	0.599	0.248	0.256	0.554	0.526	1.114	1.163	0.743	0.786
0.7	0.5	0.2	0.738	0.440	0.237	0.231	0.743	0.745	0.629	0.615	0.734	0.759
0.7	0.5	0.5	1.152	0.671	0.369	0.362	0.824	0.824	0.999	1.077	0.690	0.720
0.7	0.5	0.8	2.131	1.098	0.618	0.612	0.932	0.921	1.361	1.500*	0.630	0.663
0.7	0.8	0.2	2.680	0.977	0.707	0.686	1.062	1.064	0.943	0.909	0.606	0.623
0.7	0.8	0.5	5.088	1.406	1.141	1.096	1.175	1.176	1.351	1.500*	0.494	0.513
0.7	0.8	0.8	16.566	2.158	2.100	2.007	1.317	1.316	1.745	1.800*	0.286	0.300
0.9	0.2	0.2	0.256	0.170	0.081	0.081	0.408	0.408	0.477	0.488	0.767	0.831
0.9	0.2	0.5	0.308	0.208	0.099	0.099	0.422	0.421	0.803	0.854	0.758	0.825
0.9	0.2	0.8	0.400	0.269	0.127	0.129	0.438	0.435	1.121	1.200*	0.749	0.816
0.9	0.5	0.2	1.084	0.416	0.292	0.288	0.714	0.714	0.703	0.727	0.653	0.706
0.9	0.5	0.5	1.279	0.473	0.341	0.336	0.737	0.737	1.050	1.200*	0.633	0.687
0.9	0.5	0.8	1.586	0.557	0.413	0.406	0.762	0.761	1.384	1.500*	0.610	0.664
0.9	0.8	0.2	9.564	0.871	1.197	1.182	1.020	1.020	1.122	1.200*	0.304	0.328
0.9	0.8	0.5	16.161	0.959	1.460	1.420	1.053	1.053	1.459	1.500*	0.215	0.233
0.9	0.8	0.8	43.679	1.077	1.841	1.767	1.087	1.087	1.786	1.800*	0.099	0.108

Case A: $B_i(\cdot)$ *negative exponential,* $i=1,2$; $\lambda=1$, $s_1=s_2=0.1$ *(constant switching times).*

TABLE II (CONT'D)

r_1	β_1	β_2	EW_1	EW_2	EC_1^2	EC_2^2	$EC_{b,1}$	$E\tilde{C}_{b,1}$	$EC_{b,2}$	$E\tilde{C}_{b,2}$	$F_1(0,0)$	$F_2(0,0)$
0.7	0.2	0.2	0.330	0.267	0.114	0.113	0.426	0.426	0.463	0.465	0.796	0.830
0.7	0.2	0.5	0.684	0.547	0.230	0.230	0.481	0.471	0.789	0.814	0.773	0.811
0.7	0.2	0.8	1.546	1.122	0.469	0.470	0.559	0.526	1.112	1.163	0.744	0.785
0.7	0.5	0.2	1.389	0.832	0.435	0.425	0.744	0.745	0.642	0.615	0.735	0.762
0.7	0.5	0.5	2.325	1.273	0.689	0.667	0.828	0.824	1.006	1.077	0.692	0.722
0.7	0.5	0.8	4.660	2.118	1.192	1.154	0.937	0.921	1.363	1.500*	0.632	0.663
0.7	0.8	0.2	5.706	2.073	1.450	1.418	1.063	1.064	0.964	0.909	0.608	0.627
0.7	0.8	0.5	11.124	2.855	2.275	2.188	1.177	1.176	1.361	1.500*	0.496	0.516
0.7	0.8	0.8	37.323	4.258	4.130	3.932	1.319	1.316	1.748	1.800*	0.287	0.301
0.9	0.2	0.2	0.355	0.235	0.112	0.111	0.408	0.408	0.478	0.488	0.767	0.831
0.9	0.2	0.5	0.470	0.309	0.148	0.147	0.422	0.421	0.799	0.854	0.758	0.824
0.9	0.2	0.8	0.689	0.441	0.212	0.211	0.438	0.435	1.115	1.200*	0.749	0.815
0.9	0.5	0.2	2.084	0.795	0.558	0.548	0.714	0.714	0.712	0.727	0.653	0.707
0.9	0.5	0.5	2.509	0.902	0.659	0.637	0.737	0.737	1.050	1.200*	0.633	0.687
0.9	0.5	0.8	3.206	1.069	0.813	0.776	0.762	0.761	1.379	1.500*	0.610	0.664
0.9	0.8	0.2	20.659	1.861	2.564	2.521	1.020	1.020	1.127	1.200*	0.304	0.329
0.9	0.8	0.5	35.133	2.021	3.108	2.988	1.053	1.053	1.460	1.500*	0.215	0.233
0.9	0.8	0.8	95.841	2.248	3.908	3.687	1.087	1.087	1.786	1.800*	0.099	0.107

Case B: *$B_i(\cdot)$ hyperexponential (H_2) with squared coefficient of variation 4 and balanced means; $\lambda=1$, $s_1=s_2=0.1$ (constant switching times).*

r_1	β_1	β_2	EW_1	EW_2	EC_1^2	EC_2^2	$EC_{b,1}$	$E\tilde{C}_{b,1}$	$EC_{b,2}$	$E\tilde{C}_{b,2}$	$F_1(0,0)$	$F_2(0,0)$
0.7	0.2	0.2	0.293	0.235	0.102	0.100	0.425	0.426	0.464	0.465	0.796	0.830
0.7	0.2	0.5	0.400	0.333	0.145	0.143	0.475	0.471	0.794	0.814	0.771	0.813
0.7	0.2	0.8	0.605	0.507	0.220	0.219	0.547	0.526	1.121	1.163	0.741	0.789
0.7	0.5	0.2	1.334	0.795	0.418	0.405	0.743	0.745	0.645	0.615	0.734	0.762
0.7	0.5	0.5	1.862	1.021	0.568	0.540	0.823	0.824	1.015	1.077	0.689	0.725
0.7	0.5	0.8	2.909	1.391	0.815	0.772	0.929	0.921	1.376	1.500*	0.628	0.667
0.7	0.8	0.2	5.602	2.030	1.425	1.388	1.063	1.064	0.966	0.909	0.607	0.628
0.7	0.8	0.5	9.975	2.566	2.067	1.970	1.175	1.176	1.366	1.500*	0.494	0.518
0.7	0.8	0.8	29.029	3.410	3.361	3.157	1.317	1.316	1.752	1.800*	0.286	0.303
0.9	0.2	0.2	0.342	0.226	0.108	0.107	0.408	0.408	0.482	0.488	0.767	0.832
0.9	0.2	0.5	0.382	0.252	0.122	0.121	0.422	0.421	0.812	0.854	0.758	0.826
0.9	0.2	0.8	0.444	0.293	0.143	0.141	0.437	0.435	1.135	1.200*	0.748	0.818
0.9	0.5	0.2	2.061	0.785	0.552	0.541	0.714	0.714	0.718	0.727	0.653	0.708
0.9	0.5	0.5	2.346	0.842	0.619	0.596	0.737	0.737	1.069	1.200*	0.633	0.689
0.9	0.5	0.8	2.728	0.915	0.703	0.666	0.762	0.761	1.407	1.500*	0.610	0.667
0.9	0.8	0.2	20.557	1.851	2.552	2.508	1.020	1.020	1.129	1.200*	0.304	0.329
0.9	0.8	0.5	34.081	1.961	3.021	2.901	1.053	1.053	1.465	1.500*	0.215	0.234
0.9	0.8	0.8	88.887	2.091	3.652	3.431	1.087	1.087	1.789	1.800*	0.099	0.108

Case C: *$B_1(\cdot)$ hyperexponential (H_2) with squared coefficient of variation 4 and balanced means; $B_2(\cdot)$ deterministic; $\lambda=1$, $s_1=s_2=0.1$ (constant switching times).*

ACKNOWLEDGEMENT
The authors are indebted to Dr. J.P.C. Blanc for some useful comments.

REFERENCES

1. O.J. BOXMA (1985). Two symmetric queues with alternating service and switching times. E. GELENBE (ed.). *Performance '84,* North-Holland, Amsterdam, 409-431.
2. O.J. BOXMA, B. MEISTER (1986). Waiting-time approximations for cyclic-service systems with switch-over times. *Performance Evaluation Review 14,* 254-262 (Special issue, Proc. Performance '86).
3. O.J. BOXMA, B. MEISTER (1987). Waiting-time approximations in multi-queue systems with cyclic service. *Performance Evaluation 7,* 59-70.
4. O.J. BOXMA, W.P. GROENENDIJK (1987). Pseudo-conservation laws in cyclic-service systems. *J. Appl. Probab. 24,* 949-964.
5. O.J. BOXMA, W.P. GROENENDIJK (1988). Waiting times in discrete-time cyclic-service systems. *IEEE Trans. Commun. COM-36,* 164-170.
6. J.W. COHEN, O.J. BOXMA (1981). The M/G/1 queue with alternating service formulated as a Riemann-Hilbert problem. F.J. KYLSTRA (ed.). *Performance '81,* North-Holland, Amsterdam, 181-199.
7. J.W. COHEN, O.J. BOXMA (1983). *Boundary Value Problems in Queueing System Analysis,* North-Holland, Amsterdam.
8. M. EISENBERG (1979). Two queues with alternating service. *SIAM J. Appl. Math. 36,* 287-303.
9. G. FAYOLLE, R. IASNOGORODSKI (1979). Two coupled processors: the reduction to a Riemann-Hilbert problem. *Z. Wahrsch. Verw. Gebiete 47,* 325-351.
10. S.W. FUHRMANN, Y.T. WANG (1988). Mean waiting time approximations of cyclic service systems with limited service. P.J. COURTOIS, G. LATOUCHE (eds.). *Performance '87,* North-Holland, Amsterdam, 253-265.
11. D. GAIER (1964). *Konstruktive Methoden der Konformen Abbildung,* Springer Verlag, Berlin.
12. F.D. GAKHOV (1966). *Boundary Value Problems,* Pergamon Press, Oxford.
13. O. HASHIDA, K. OHARA (1972). Line accomodation capacity of a communication control unit. *Review of the Electr. Comm. Laboratories, Nippon Telegraph and Telephone Public Corporation, 20,* 231-239.
14. P.J. KUEHN (1979). Multiqueue systems with nonexhaustive cyclic service. *The Bell System Tech. J. 58,* 671-698.
15. W. SZPANKOWSKI, V. REGO (1987). *Ultimate Stability Conditions for Some Multidimensional Distributed Systems,* Report Dept. of Computer Science, Purdue University.
16. H. TAKAGI (1986). *Analysis of Polling Systems,* MIT Press, Cambridge (MA).
17. H. TAKAGI (1987). A survey of queueing analysis of polling models. *Proceedings of the Third International Conference on Data Communication Systems and their Performance,* Rio de Janeiro.
18. H.C. TIJMS (1986). *Stochastic Modelling and Analysis,* Wiley, New York.

19. K.S. WATSON (1985). Performance evaluation of cyclic service strategies - a survey. E. GELENBE (ed.). *Performance '84,* North-Holland, Amsterdam, 521-533.

On a System with Impatience and Repeated Calls

G. Fayolle, M.A. Brun
INRIA, Domaine de Voluceau, Rocquencourt
B.P. 105 78153 Le Chesnay - Cedex, France

We analyze a mathematical model of a system where customers may become impatient and retry for service after some period of time. A functional equation of two complex variables involving partial derivatives and a boundary term is obtained for the generating function of the joint steady-state distribution of the number in the system. Ergodicity conditions are derived by constructing super-martingales applicable to more general cases. The evaluation of the unknown functions in the expression for the generating function is reduced to the solution of a homogeneous integral equation of the first kind, which, in turn, is shown to be equivalent to an infinite linear system.

1. Introduction and model description

We analyze a mathematical model of a system where customers (or calls, jobs, etc.) may become impatient and retry for service after some period of time. Interest in this kind of queueing models arose out of important applications in telephone switching systems and also in computer and telecommunication networks. As Erlang began the study of telephone systems by using queueing theory early this century, one might believe that most of the basic patterns have been solved by now. It will appear below that this is not the whole story. Among the pioneering works dealing with retrials, we quote Cohen [3], Kosten [14] and Wilkinson [21], some time between three and four decades ago. For an excellent and recent survey on retrial queues, the reader is referred to [22], where it is rightly mentioned that analytic results are generally difficult to obtain. In constrast, there are a great number of numerical and approximation methods.

To our knowledge no solution has been given, even in the one server case, taking into account both impatience and repetition phenomena, when capacities (i.e. number of waiting positions) are unconstrained (unlimited). The purpose of this paper - keeping in mind that it is dedicated to professor J.W. Cohen - is to provide some insight toward a theoretical solution of this tricky problem.

Consider an M/M/1 queue with service discipline which can be either FIFO or Processor Sharing (as far as we are concerned in this study, no further specification is necessary). Congested jobs initially stay, but may become impatient before their service is completed. In the latter case, they leave the queue and eventually, with probability r, each independently re-attempts after a certain delay, as shown in Fig. 1.

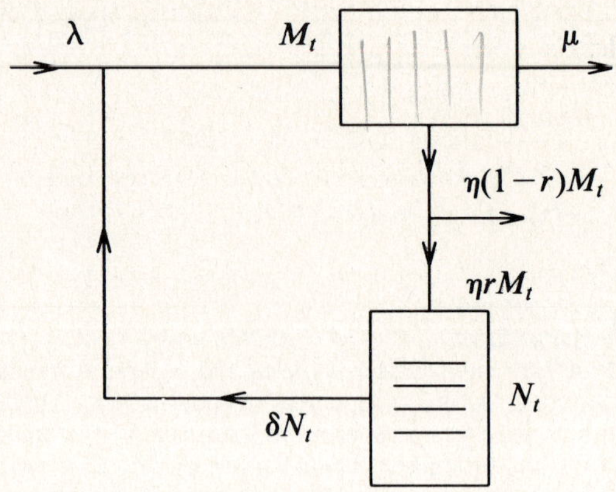

FIGURE 1. *The queueing model with impatience and retrials*

The exogeneous jobs (calls) are initiated as a Poisson process with parameter λ. The service (trunk holding) times are i.i.d. random variables, exponentially distributed with mean $1/\mu$. The calls which will be re-attempted are modeled by an infinite server queue and the time until re-attempt follows a negative exponential distribution with mean $1/\delta$. The impatience times are also exponentially distributed with mean $1/\eta$. All delays, interarrival and service intervals are taken to be mutually independent.

Our primary concern is in the study of the joint steady-state distribution of the number of customers (M_t, N_t) in the M/M/1 queue and 'in orbit' at the infinite server queue, respectively (see Fig. 1), where, according to the terminology proposed in [22], the expression 'in orbit' refers to the customers who will come back and try service later.

Note that in the case of FIFO discipline - in a telephone context - it is legitimate to assume that the customer in service may become impatient: occasionally, a subscriber reneges without knowing that his call started to be processed.

In Section 2, a partial differential equation of two complex variables involving a boundary term is obtained for the generating function of (M_t, N_t) at equilibrium. The ergodicity conditions are derived in Section 3, by using a general approach - applicable to dimension greater than 2 and to non-space-homogeneous random walks - which consists in introducing conic curves as Liapounov functions (supermartingales). Sections 4 and 5 are devoted to the reduction of the functional equation of Section 2 to a singular integral equation of the first kind in one variable. The equivalence to an infinite linear system is also shown.

2. A GENERATING FUNCTION FOR THE NUMBER IN THE SYSTEM

Let M_t, N_t be the numbers of calls at time t in the first and second queue, respectively. The vector process (M_t, N_t), which represents the state of the system at time t, is clearly a Markov process.

Let $p(m, n)$ be the stationary probability of being in the state $(M_t, N_t)=(m, n)$, $\forall(m, n)\in\mathbb{Z}_+^2$, where $\mathbb{Z}_+$ denotes the set of non-negative integers. The $p(m, n)$'s satisfy the following Kolmogorov forward equations:

$$\begin{aligned}(\lambda+\mu I_{(m>0)}+\delta n+\eta m)p(m, n) &= \lambda I_{(m>0)}p(m-1, n)\\ &+\delta(n+1)I_{(m>0)}p(m-1, n+1)+r\eta(m+1)I_{(n>0)}p(m+1, n-1)\\ &+[\mu+\eta(1-r)(m+1)]p(m+1, n),\end{aligned} \tag{2.1}$$

where

$$I_{(B)} = \begin{cases} 1 & \text{if } B \text{ is true,} \\ 0 & \text{otherwise.}\end{cases}$$

Define

$$P(x, y) = \sum_{m\geqslant 0}\sum_{n\geqslant 0} p(m, n)x^m y^n, \quad |x|\leqslant 1,\ |y|\leqslant 1.$$

In the usual way, we multiply (2.1) by $x^m y^n$ and sum over $m\geqslant 0$, $n\geqslant 0$, to obtain the following fundamental functional equation after a routine computation:

$$\begin{aligned}(\lambda x-\mu)(x-1)P(x, y) &= \eta x[(r-1)+x-ry]\frac{\partial P}{\partial x}(x, y)\\ &+\delta x(y-x)\frac{\partial P}{\partial y}(x, y)+\mu(1-x)P(0, y),\end{aligned} \tag{2.2}$$

for all $|x|\leqslant 1$, $|y|\leqslant 1$.

Equation (2.2) is of a non-standard type, although some similar cases have been encountered earlier, see [2]. It will be studied in Sections 4 and 5.

In the next section we present a probabilistic analysis of the conditions for the process (M_t, N_t) to be positive recurrent.

3. ERGODICITY CONDITIONS

The Markov process $X_t=(M_t, N_t)$, $t\geqslant 0$, is a bivariate linear birth and death process. Let $Y_n=(a_n, b_n)$, $n\geqslant 0$, $(a_n, b_n)\in\mathbb{Z}_+^2$, be the imbedded Markov chain of the successive states visited by X_t. Y_n is represented in Fig. 2 as a two-dimensional random walk in $\mathbb{Z}_+^2$, with the following stationary transition probabilities

$$\begin{cases} P\{Y_{k+1} = (m+1, n)|Y_k=(m, n)\}=(\lambda+\mu+\eta m+\delta n)^{-1}\lambda,\\ P\{Y_{k+1} = (m+1, n-1)|Y_k=(m, n)\}=(\lambda+\mu+\eta m+\delta n)^{-1}\delta n,\\ P\{Y_{k+1} = (m-1, n)|Y_k=(m, n)\}=(\lambda+\mu+\eta m+\delta n)^{-1}[\mu+\eta(1-r)m],\\ P\{Y_{k+1} = (m-1, n+1)|Y_k=(m, n)\}=(\lambda+\mu+\eta m+\delta n)^{-1}\eta rm.\end{cases} \tag{3.1}$$

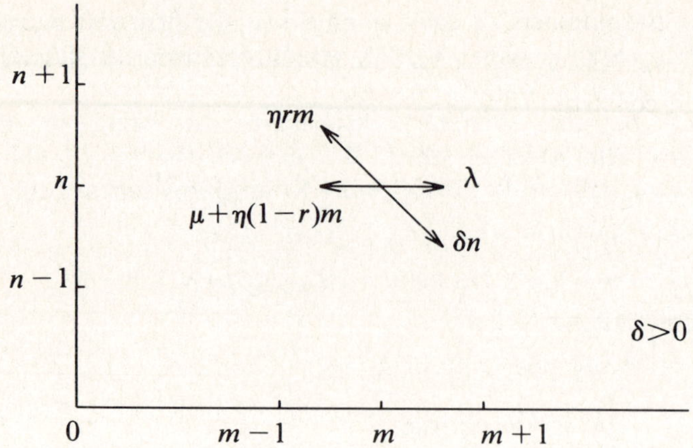

FIGURE 2. *Stationary transition rates for the Y_k-process*

For random walks in $\mathbb{Z}_+^2$ or $\mathbb{Z}_+^3$, satisfying strong homogeneity conditions, general criteria for transience and recurrence in the case of bounded jumps have been given in MALYSHEV ([16,17]), which are closely connected with the well-known Foster criteria [9] (the boundedness conditions have recently been partially relaxed by ROSENKRANTZ [18]). Concerning non-homogeneous n-dimensional random walks in $\mathbb{Z}_+^n$, $n \geq 2$, very few results are available, even when the jumps are of size one.

Our situation fits in the category of linear birth and death processes. To the best of the authors' knowledge, the only slightly general study of ergodicity for such processes appears in KESTEN [13], where some important cases were left out, which could not be handled. In particular, the case $r=1$ of the model presented in Section 1 (delayed feedback) cannot be obtained from [13].

The approach we propose consists in constructing convenient positive supermartingales (i.e. Liapounov functions) outside some compact set. It is worthwhile pointing at this instant at the remarkable work of LAMPERTI [15] who made an extensive use of these tools.

THEOREM 3.1.

1. *If $r \neq 1$, X_t is always positive recurrent, for all λ, μ, δ, η strictly positive numbers.*
2. *If $r=1$, X_t is*
   - *positive recurrent if $\lambda < \mu$;*
   - *transient if $\lambda > \mu$.*

PROOF. Due to the irreducibility assumptions, the Markov process $(X_t)_{t \geq 0}$ is transient if and only if $P\{|X_t| \to \infty | X_0 = (m_0, n_0)\} = 1$, for some (m_0, n_0) (and then for all (m_0, n_0)), where $|X_t| = |M_t + N_t|$.

A similar definition holds for Y_n. First, let us briefly show along the same lines as in [13] that X_t is transient if and only if Y_n is transient. If τ_k stands for the k-th jump instant of X_t, we have $|X_{\tau_k}|\leqslant|X_0|+k$. Then, conditional on the past, $\tau_{k+1}-\tau_k$ has an exponential distribution with mean at least A/k, A being a positive constant. Thus X_t is regular since $\Sigma\frac{1}{k}=\infty$ (see CHUNG [1]), and in particular, it has only finitely many jumps in a finite time interval. The above assertion is proved. From now on we shall deal with Y_n.

Let $\mathfrak{F}_n$ be the increasing family of σ-algebras $\sigma\{Y_k, 0\leqslant k\leqslant n\}$. The components of Y_n, as a rule, will be written (a_n, b_n).

A. The case $r\neq 1$

Introduce the following positive real function f on $\mathbb{Z}_+^2$: $f(a, b)=\frac{\eta a^2+\delta b^2}{2}$, $(a, b)\in\mathbb{Z}_+^2$. For the sake of brevity, we shall write f_n instead of $f(a_n, b_n)$. Define, for any fixed positive constant K, the sets

$$E_K = \{(a, b)\in\mathbb{Z}_+^2 \,|\, \eta a+\delta b\geqslant K\},$$

and

$$\bar{E}_K = \mathbb{Z}_+^2 - E_K,$$

($\bar{E}_K$ is finite).

We prove

LEMMA 3.1. *There exist $\epsilon>0$ and K, $0<K<\infty$, such that*

$$E[f_{n+1}-f_n|\mathfrak{F}_n]<-\epsilon, \quad \textit{for all} \quad (a_n, b_n)\in E_K. \tag{3.2}$$

PROOF. An easy computation yields

$$E[f_{n+1}-f_n|\mathfrak{F}_n] = g_n,$$

where

$$g_n = \frac{Q(a_n, b_n)+R(a_n, b_n)}{D(a_n, b_n)},$$

with

$$\begin{cases} Q(a, b) = -(\eta a)^2-(\delta b)^2+\delta\eta(1+r)ab, \\ R(a, b) = \eta a[\lambda-\mu I_{(a>0)}+\frac{(r\delta+\eta)}{2}]+\frac{\delta b(\eta+\delta)}{2}, \\ D(a, b) = \lambda+\mu I_{(a>0)}+\eta a+\delta b. \end{cases} \tag{3.3}$$

The quadratic form $Q(a, b)$ is negative definite, for all $0\leqslant r<1$. Therefore the following inequalities hold

$$\begin{cases} Q(a, b) \leqslant \dfrac{-(1-r)K^2}{4} \quad \text{for all} \quad (a, b) \in E_K; \\ \dfrac{Q(a, b)}{D(a, b)} \leqslant -A_1 K \quad \text{for some positive constant } A_1. \end{cases} \tag{3.4}$$

On the other hand

$$\frac{R(a, b)}{D(a, b)} < A_2, \quad \forall (a, b) \in \mathbb{Z}_+^2,$$

where A_2 is a fixed finite constant.

Choosing K large enough in (3.4) yields (3.2), and the proof of Lemma 3.1 is concluded. □

We are now in a position to apply Foster's criterion, see FOSTER [9], that we quote hereafter.

THEOREM 3.2. *In order that an irreducible Markov chain with countable state space S and transition matrix $p(i, j)$ be positive recurrent, it is sufficient that there exist a finite set Δ, an $\epsilon > 0$ and a positive function $f(i)$ such that*

$$\sum_j p(i, j) f(j) < f(i) - \epsilon, \quad \forall i \in S - \Delta,$$

$$\sum_j p(i, j) f(j) < \infty, \quad \forall i \in \Delta.$$

The proof of case A now follows from Lemma 3.1.

B. The case $r = 1$

New difficulties arise because, for $r = 1$, the quadratic form $Q(a, b)$ defined above is only negative semi-definite. In fact, $Q(a, b) = 0$ on the singular line $\eta a - \delta b = 0$.

After mature (and somewhat lengthy!) consideration, it appears that an explicit correlation factor between a_n and b_n has to be taken into account (there is none in $f(a, b)$). Define the real function θ on $\mathbb{Z}_+^2$,

$$\theta(a, b) = ua^2 + vb^2 + wab, \quad \text{for all} \quad (a, b) \in \mathbb{Z}_+^2, \tag{3.5}$$

where u, v, w are unspecified positive constants to be properly chosen later. $\theta(a_n, b_n)$ will be rewritten as θ_n. The analogue of Lemma 3.1 reads

LEMMA 3.2. *There exist $\epsilon > 0$, K, $0 < K < \infty$, and positive numbers u, v, w, such that, if $\lambda < \mu$,*

$$E[\theta_{n+1} - \theta_n | \mathcal{F}_n] < -\epsilon, \quad \text{for all} \quad (a_n, b_n) \in E_K,$$

where E_K is as in case A.

Proof. After a straightforward calculation, we get

$$E[\theta_{n+1}-\theta_n|\mathcal{F}_n] = h_n, \tag{3.6}$$

where

$$h_n = \frac{S(a_n, b_n)+T(a_n, b_n)}{D(a_n, b_n)},$$

with

$$\begin{cases} D(a, b) \quad \text{as in (3.3)}, \\ S(a, b) = (\eta a-\delta b)[(2v-w)b+(w-2u)a], \\ T(a, b) = [2ua+wb][\lambda-\mu I_{(a>0)}]+(u+v-w)[\eta a+\delta b]. \end{cases} \tag{3.7}$$

Select now u, v, w according to the following linear system:

$$\begin{cases} w = u+v+c, \\ v-u+c = -\zeta\eta, \\ v-u-c = \zeta\delta, \end{cases} \tag{3.8}$$

ζ being some positive real number. Hence (3.8) and (3.7) imply

$$\begin{cases} c = -\dfrac{\zeta(\eta+\delta)}{2}, \\ v-u = \dfrac{\zeta(\delta-\eta)}{2}, \\ S(a, b) = -\zeta(\eta a-\delta b)^2, \\ T(a, b) = [2ua+(u+v-\dfrac{\zeta(\eta+\delta)}{2})b][\lambda-\mu I_{(a>0)}] \\ \qquad +\dfrac{\zeta(\eta+\delta)(\eta a+\delta b)}{2}. \end{cases} \tag{3.9}$$

We have at our disposal two degrees of freedom, ζ and u. Remarking that $|\dfrac{T(a, b)}{D(a, b)}|$ is bounded for all $(a, b)\in\mathbb{Z}_+^2$, it is possible, according to (3.9), to find u, w, ϵ_1, ζ, positive such that

$$\frac{T(a, b)}{D(a, b)} < -\epsilon_1, \quad \text{for all} \quad (a, b)\in\mathbb{Z}_+^2, \quad a\neq 0. \tag{3.10}$$

It is indeed sufficient to take ζ, $0<\zeta<B(\mu-\lambda)$, where

$$B = \min_{\substack{(a,b)\in\mathbb{Z}_+^2 \\ a\neq 0}} \frac{2[2ua+(u+v)b]}{(\eta+\delta)[\eta a+(\delta+\mu-\lambda)b]}.$$

B being strictly positive, (3.10) is satisfied. Then, choose K large enough (because of the boundary line $a=0$) to ensure

$$\frac{S(a,b)+T(a,b)}{D(a,b)} < -\epsilon \quad \text{for all} \quad (a,b)\in E_K.$$

The proof of Lemma 3.2 is completed. □

Lemma 3.2 allows us to call again upon Foster's criterion. This concludes the proof of positive recurrence if $r=1$ and $\lambda<\mu$. By arguing with the same tools, it is possible to show the transience if $\lambda>\mu$.

Just for amusement, we propose a direct proof of the transience for $\lambda>\mu$. Consider the function $\Psi_n = x^{|Y_n|} = x^{a_n+b_n}$, for some positive x. Then

$$E[\Psi_{n+1}|\mathcal{F}_n] = \Psi_n \frac{[\lambda(x-1)+\mu I_{(a_n>0)}(\frac{1}{x}-1)]}{\lambda+\mu I_{(a_n>0)}+\eta a_n+\delta b_n}.$$

When $\lambda>\mu$, take $x=x_0$ with $\frac{\mu}{\lambda}<x_0<1$. This entails that $(\Psi_n)_{n\geq 0}$ is a positive and uniformly bounded supermartingale. Define the stopping time

$$\tau = \inf\{n>0 | Y_n=(0,0)\}.$$

Assume $P(\tau<\infty)=1$. It is well-known that $(\Psi_{n\wedge\tau}, \mathcal{F}_n)_{n\geq 0}$ also form a positive uniformly bounded supermartingale which converges w.p. 1. From the bounded convergence theorem, it follows that $E(\Psi_\tau)\leq E(\Psi_0)$, for all $(a_0,b_0)\neq(0,0)$. But $\Psi_\tau=1$, while on the other hand $E(\Psi_0)<1$, yielding a contradiction. Thus if $\lambda>\mu$, $P(\tau<\infty)<1$ and the system is transient. The proof of Theorem 3.1 is completed. □

4. An integral equation for $P(0,y)$

In this section, the evaluation of $P(x,y)$ satisfying (2.2) is reduced to the solution of a homogeneous integral equation for the function of one variable $P(0,y)$. We only consider the case that the system is ergodic. To keep this section self-contained as much as possible, rewrite (2.2) as

$$\begin{aligned}(\lambda x-\mu)(x-1)P(x,y) &= \eta x(x-\tau-ry)\frac{\partial P}{\partial x}(x,y) \\ &+\delta x(y-x)\frac{\partial P}{\partial y}(x,y)+\mu(1-x)P(0,y),\end{aligned} \tag{4.1}$$

where we simply have set

$$\tau = 1-r.$$

We proceed to integrate (4.1) by the classical method of characteristic curves - see Smirnov [19]. Define the following functions K, θ, S which will be ubiquitous in the sequel:

$$
\left\{
\begin{aligned}
&K(s, A) = (A-\eta s)^{\frac{\mu}{A}-1}[A-(\delta+\eta)s]^{\theta(A)}(\eta s+\delta-A)e^{\frac{\lambda\eta s}{\delta(\eta+\delta)}}A^{\frac{\lambda A-(\mu+\lambda)(\delta+\eta)}{(\delta+\eta)^2}} \\
&\theta(A) = \frac{-\lambda A^2-\mu(\delta+\eta)^2+A(\delta+\eta)(\lambda+\mu-\delta-\eta)}{A(\delta+\eta)^2}, \\
&e^{-S(y,\, \delta x+\eta y)} = e^{\frac{\lambda\eta y}{\delta(\eta+\delta)}}(\delta x)^{\frac{\mu}{\delta x+\eta y}}[\delta(x-y)]^{\frac{[\lambda(\eta y+\delta x)-\mu(\delta+\eta)]}{(\eta y+\delta x)(\eta+\delta)^2}[\delta+\eta-(\delta x+\eta y)]} \times \\
&\qquad (\delta x+\eta y)^{\frac{[\lambda(\delta x+\eta y)-(\delta+\eta)(\lambda+\mu)]}{(\delta+\eta)^2}}.
\end{aligned}
\right.
\tag{4.2}
$$

The definitions of the above exponential functions are valid in the whole complex plane cut along the negative real axis.

THEOREM 4.1. *For $r=1$, the generating function $P(x, y)$ satisfies the integral equation*

$$P(x, y)e^{-S(y,\, \delta x+\eta y)} = \psi(\delta x+\eta y)+\int_0^y \mu P(0, s)K(s, \delta x+\eta y)ds, \tag{4.3}$$

in the region $|x|\leqslant 1$, $|y|\leqslant 1$, where

$$\psi(\delta x) = P(x, 0). \tag{4.4}$$

PROOF. The partial differential equation (4.1) yields the following characteristic system

$$\frac{dx}{\eta x(x-\tau-ry)} = \frac{dy}{\delta x(y-x)} = \frac{dP}{(x-1)[(\lambda x-\mu)P(x, y)+\mu P(0, y)]}.$$

i) The first family of characteristic curves is given by

$$\delta x(y-x)dx = \eta x(x-\tau-ry)dy,$$

which, after integration, gives

$$x-1 = K\left[\frac{y-1}{x-1}-t_1\right]^{\alpha_1}\left[\frac{y-1}{x-1}-t_2\right]^{\alpha_2},$$

where K is a constant, t_1 and t_2 are the two different real roots of the second degree polynomial $\eta rt^2+(\delta-\eta)t-\delta$, and α_1 and α_2 are given by

$$\frac{\eta(rt-1)}{\delta(1-t)+\eta t(1-rt)} = \frac{\alpha_1}{t-t_1}+\frac{\alpha_2}{t-t_2}.$$

Note that this family has a simple analytic form when $r=1$. Henceforth, until the end of our study, we shall assume

$$r = 1.$$

Thus the first family of characteristic curves corresponds to the linear varieties

$$\delta x + \eta y = A, \tag{4.5}$$

where A is some unspecified constant.

ii) The second family of characteristic curves is obtained by integrating the ordinary differential equation

$$\frac{d\tilde{P}}{dy} = R(y, A)[\tilde{P} + g(y, A)],$$

where P has been renamed $\tilde{P}$ and

$$\begin{cases} R(y, A) = \dfrac{[\lambda(\frac{A-\eta y}{\delta}) - \mu][\frac{A-\eta y}{\delta} - 1]}{(A-\eta y)(y - \frac{A-\eta y}{\delta})}, \\[2ex] g(y, A) = \dfrac{P(0, y)}{\rho(\frac{A-\eta y}{\delta}) - 1}, \\[2ex] \rho = \dfrac{\lambda}{\mu}. \end{cases} \tag{4.6}$$

Its solution is provided by

$$P(y) = e^{S(y, A)}[B + \int_0^y R(s, A) g(s, A) e^{-S(s, A)} ds], \tag{4.7}$$

with

$$S(y, A) = \int_0^y R(s, A) ds.$$

Introducing an arbitrary relationship between A and B, that is

$$B = \psi(A),$$

then substituting into (4.5) and (4.7), we formally get (4.3). But we still have to prove that $\psi(x)$ is analytic in some domain.

Consider, for a fixed x, the limit of the integral

$$\int_0^y \mu P(0, s)[\delta x + \eta y - (\delta+\eta)s]^{\theta(\delta x + \eta y)}[\delta x + \eta y - \eta s]^{\frac{\mu}{\delta x + \eta y} - 1} \times$$

$$(\eta s + \delta - (\delta x + \eta y)) e^{\frac{\lambda \eta s}{\delta(\eta+\delta)}} ds,$$

as y tends to zero along some specified path.

Now choose x and y such that $\text{Re}\, x < 0$ and $\text{Re}\, y < 0$. Then, $\delta x + \eta y - \eta s$ is different from zero for all s in $[0, y]$. On the other hand, $1 + \text{Re}\,(\theta(\delta x + \eta y))$ is strictly positive when $\mathrm{Re}\,(\delta x+\eta y)$ is negative. The function $s \to (\delta x+\eta y-(\delta+\eta)s)^{\theta(\delta x+\eta y)}$ is thus integrable in a neighbourhood of $(\delta x+\eta y)/(\delta+\eta)$.

It follows, by the bounded convergence theorem, that the above integral tends to zero as y tends to zero by negative values. Hence $\lim_{y\to 0} P(x, y)e^{-S(y,\,\delta x+\eta y)} = P(x, 0)$.

Passing to the limit in (4.3), we get

$$\psi(\delta x) = P(x, 0), \quad \text{for} \quad \mathrm{Re}\, x<0 \quad \text{and} \quad |x|<1.$$

Since $P(x, 0)$ is sought to be an analytic function of x in the unit disc, the principle of analytic continuation allows us to write (4.3) and (4.4). The proof of Theorem 4.1 is concluded. □

To obtain more information concerning $P(x, y)$, we remark that the right-hand side member of (4.3) vanishes whenever $e^{-S(y,\,\delta x+\eta y)}=0$, $|x|$, $|y|\leqslant 1$. Define the kernel of (4.3) as the quantity

$$L(x, y) = e^{-S(y,\,\delta x+\eta y)}, \tag{4.8}$$

where $S(\cdot,\cdot)$ was introduced in (4.2).

4.1. The zeros of $L(x, y)$

There are three distinct varieties on which $L(x, y)$ vanishes.

a) $\delta x+\eta y=0,\ (x, y)\neq(0,0)$

Necessarily, $\delta x+\eta y$ must tend to zero by positive values of its real part (for $\delta x=-\eta y$, $\log(1-\frac{y}{x})$ is positive).

This case doesn't provide any new information - see Appendix A. This fact should not surprise the reader as the quantity $\delta x+\eta y$ has already been used to find the characteristic curves.

b) $x=y$

The power of $(x-y)$ in the kernel L can be written:

$$\frac{(\lambda z-\mu)(1-z)}{(\delta+\eta)z} = 1+\theta((\delta+\eta)z), \quad \text{where} \quad z=x=y.$$

Hence any admissible solution z must belong to the domain $\{z\,|\,[1+\mathrm{Re}\,(\theta((\delta+\eta)z))]>0\}$. The algebraic curve, defined by $1+\mathrm{Re}\,(\theta((\delta+\eta)z))=0$ in the entire complex plane, consists of the two contours (L) and (S) shown in Fig. 3.

An easy computation yields

$$1+\mathrm{Re}\,[\theta((\delta+\eta)(u+iv))] = \frac{u(1-u)(\lambda u-\mu)+v^2(\lambda+\mu-\lambda u)}{u^2+v^2},$$

which in turn implies

$$1+\mathrm{Re}\,(\theta((\delta+\eta)iv)) = \lambda+\mu, \quad \text{for all real} \quad v.$$

The region $\mathcal{D}=\{z|[1+\text{Re}\,(\theta((\delta+\eta)z))]>0\}$ is a doubly-connected domain having (S) and (L) as boundary (see Fig. 3). When we traverse (L) clockwise, $\mathcal{D}$ remains on the left.

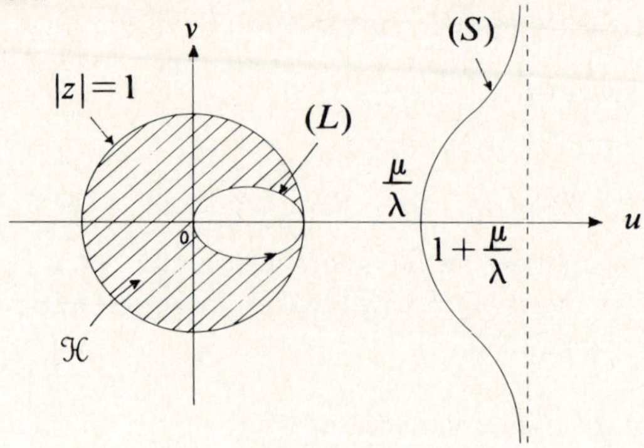

FIGURE 3.

Let $\mathcal{H}$ denote the intersection of $\mathcal{D}$ with the unit disc. For $z\in\mathcal{H}$, the function $K(s,(\delta+\eta)z)$ has only one singularity at the point $s=z$, which is integrable since

$$1+\text{Re}\,[\theta(\eta+\delta)z\,] > 0.$$

Consequently (4.2), (4.3) and (4.8) imply

$$\psi((\delta+\eta)z)+\mu\int_0^z P(0,s)K(s,(\delta+\eta)z)ds \;=\; 0, \quad z\in\mathcal{H}. \tag{4.9}$$

It is important to note that (4.9) provides an analytic continuation for $\psi(\cdot)$.

c) $x=0$

The factor $(\delta x)^{\frac{\mu}{\delta x+\eta z}}$ appearing in $L(x,z)$ vanishes as $x\to 0$, provided that $\text{Re}\,z>0$. Accordingly the singularity $s=z$ in the integral $\int_0^z P(0,s)K(s,\eta z)ds$ is integrable since $1+\text{Re}\,(\frac{\mu}{\eta z}-1)=\text{Re}\,(\frac{\mu}{\eta z})>0$, for $\text{Re}\,z>0$. But $K(s,\eta z)$ has a second algebraic singularity at the point $s=\frac{\eta z}{\delta+\eta}$, with an exponent equal to $\theta(\eta z)$. Introduce the following notation

$$\begin{cases} \alpha = \dfrac{\eta}{\eta+\delta}, \\ (L_\alpha), \text{ the contour obtained from } (L) \text{ under the} \\ \text{mapping } z\to z/\alpha \quad (\text{see Fig. 4}), \\ \mathcal{R}, \text{ the shaded domain (see Fig. 4) bounded by} \\ (L_\alpha), \text{ the imaginary axis and the unit circle.} \end{cases} \tag{4.10}$$

For $z\in\mathcal{R}$, we have $1+\text{Re}\,[\theta(\eta z)]>0$. Hence, arguing as in b) and using (4.2),

(4.3), (4.4), (4.8), (4.10), we get

$$\psi(\eta z)+\mu\int_0^z P(0,s)K(s,\eta z)ds = 0, \quad \text{for all} \quad z\in\mathcal{R}. \tag{4.11}$$

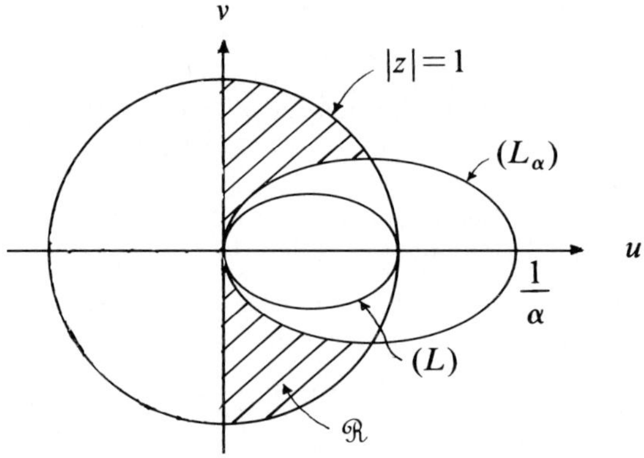

FIGURE 4.

The analysis of the kernel is terminated. For the sake of brevity, introduce the following quantities

$$\begin{cases} f(s) = P(0,s)e^{\frac{\lambda\eta s}{\delta(\eta+\delta)}}, \\ a(z) = \dfrac{\mu}{(\delta+\eta)z}, \\ b(z) = \theta((\delta+\eta)z)+1, \\ c(z) = a(z)+b(z). \end{cases} \tag{4.12}$$

We are in a position to state

THEOREM 4.2. *The three following relations hold:*

$$\begin{cases} \psi((\delta+\eta)z)+\dfrac{\mu}{(\delta+\eta)z}I_1(z) = 0, \quad z\in\mathcal{H}, \\ \psi((\delta+\eta)z)+\dfrac{\mu}{(\delta+\eta)z}I_2(z) = 0, \quad z\in\mathcal{G}, \end{cases} \tag{4.13}$$

$$\int_1^{1/\alpha} f(zt)(1-\alpha t)^{a(z)-1}(1-t)^{b(z)-1}[1-\alpha-z(1-\alpha t)]dt = 0, \quad z\in\mathcal{G}, \tag{4.14}$$

where

$$I_1(z) = \int_0^1 f(zt)(1-\alpha t)^{a(z)-1}(1-t)^{b(z)-1}[1-\alpha-z(1-\alpha t)]dt,$$

$$I_2(z) = \int_0^{1/\alpha} f(zt)(1-\alpha t)^{a(z)-1}(1-t)^{b(z)-1}[1-\alpha-z(1-\alpha t)]dt,$$

and $\mathcal{G}$ is the domain

$$\mathcal{G} = \mathcal{K} \cap \{\text{Re } z > 0\} \cap \{|z| < \alpha\}.$$

PROOF. System (4.13) is an immediate consequence of equations (4.9) and (4.11). Eliminating ψ in (4.13) yields at once (4.14).

It is worth remarking that we have used (4.4) and Subsection 4.1 to continue analytically at one and the same time two functions:

- *Firstly $\psi((\delta+\eta)z)$, over the region $\mathcal{K} \cup \{|z|<1-\alpha\}$;*
- *Secondly $I_2(z)$, the analytic continuation of which is $I_1(z)$.*

The proof of Theorem 4.2 is concluded. □

The next section will be devoted to the (intricate!) analysis of the fundamental operator given by (4.14).

5. Some insight into the inversion of (4.14)

We tackle here a problem of the form $Af=0$ which does not necessarily belong to the class of problems so-called 'correctement posés', according to Hadamard's classification, see FENYÖ and STOLLE [8].

5.1. On the analytic continuation of $f(z)$

As it emerges from the discussion in Theorem 4.2, the difficulty mainly resides in the fact that no explicit expression a priori can be obtained for the analytic continuation of $f(z)$, since this function is given under an integral form in (4.13) and (4.14).

Recall the definition of the following two classical special functions.

i) The hypergeometric function $F(u, v, w; z)$ which, for $\text{Re } w > \text{Re } v > 0$ and $|z|<1$ possesses the integral representation

$$F(u, v, w; z) = \frac{\Gamma(w)}{\Gamma(v)\Gamma(w-v)} \int_0^1 t^{v-1}(1-t)^{w-v-1}(1-tz)^{-u} dt.$$

Thus $F(u, v, w; z)$ gives the analytic continuation of the integral in the right-hand side. See [12].

ii) The 'beta' function

$$B(x, y) = \frac{\Gamma(x)\Gamma(y)}{\Gamma(x+y)}.$$

Making in (4.14) the change of variable $t = 1+(\frac{1}{\alpha}-1)u = 1+\frac{\delta}{\eta}u$, yields

$$\int_0^1 f(z(1+\frac{\delta}{\eta}u))(1-u)^{a(z)-1}u^{b(z)-1}[1-z(1-u)]du = 0, \quad \forall z \in \mathcal{G}. \tag{5.1}$$

Keeping in mind that it is required to find $f(z)$ analytic at least in the unit disc, we have by Cauchy's formula,

$$f(z) = \frac{1}{2\pi i} \int_{\mathcal{C}} \frac{f(t)}{t-z} dt,$$

where $\mathcal{C}$ denotes the unit circle.

THEOREM 5.1. *$f(z)$ satisfies the following integral equation:*

$$f(\frac{z}{\alpha}) = \frac{(2\pi i)^{-1}}{1+\rho(1-z)} \int_{\mathcal{C}} \frac{f(t)}{t-z} F(1, b(z), c(z)+1; \frac{\delta z}{\eta(t-z)})[1-\frac{1-\alpha}{z-\alpha t}]dt, \qquad \forall z \in \mathcal{G}. \tag{5.2}$$

PROOF. In the proof of the theorem we need two lemmas which are formulated and proved below.

LEMMA 5.1.

$$B(b(z), a(z)) \int_{\mathcal{C}} \frac{f(t)}{t-z} [F(1, b(z), c(z); \frac{\delta z}{\eta(t-z)}) - \frac{\mu}{(\delta+\eta)c(z)} F(1, b(z), c(z)+1; \frac{\delta z}{\eta(t-z)})]dt = 0, \quad \forall z \in \mathcal{G}. \tag{5.3}$$

PROOF. Replacing f by its integral representation in (5.1), we obtain:

$$\int_0^1 (1-u)^{a(z)-1} u^{b(z)-1} [1-z(1-u)] (\int_{\mathcal{C}} \frac{f(t)}{t-z(1+\frac{\delta}{\eta}u)} dt) du = 0.$$

For every z in $\mathcal{G}$, the two integrals in this expression are absolutely and uniformly convergent. Changing the order of integration, it is now sufficient to estimate the integral

$$\int_0^1 (1-u)^{a(z)-1} u^{b(z)-1} [1-z(1-u)] [1-\frac{u\delta z}{\eta(t-z)}]^{-1} du,$$

which is indeed equal to

$$B(b(z), a(z))[F(1, b(z), c(z); \frac{\delta z}{\eta(t-z)}) - \frac{\mu}{(\delta+\eta)c(z)} F(1, b(z), c(z)+1; \frac{\delta z}{\eta(t-z)})].$$

Lemma 5.1 is established. □

LEMMA 5.2. *$B(b(z), a(z))$ has no zeros or poles in $\mathcal{G}$.*

PROOF. It is well-known that $\Gamma(z)$ is single-valued and analytic over the entire complex plane, apart from the points $z = 0, -1, -2, \ldots$ where it has simple poles with residue $\frac{(-1)^n}{n!}$. Its reciprocal $\frac{1}{\Gamma(z)}$ is thus an entire function.

The zeros of $B(b(z), a(z))$ occur whenever $c(z)=-n$, for all $n \in \mathbb{Z}_+$. Since $c(z)=\frac{-\lambda}{\delta+\eta}z+\frac{\lambda+\mu}{\delta+\eta}$, we have

$$c(z)+n = \frac{-\lambda}{\delta+\eta}(z-z_n),$$

where

$$z_n = 1+\frac{\mu}{\lambda}+n\left[\frac{\delta+\eta}{\lambda}\right], \quad n=0, 1, 2, \ldots .$$

On the other hand, the poles of $B(b(z), a(z))$ are separated into two families:

i) $a(z)=-n$, $n \in \mathbb{Z}_+$, which corresponds to a set of simple poles on the negative real axis

$$\frac{\mu}{(\delta+\eta)z} = -n, \quad n \geqslant 0. \tag{5.4}$$

ii) $b(z)=-n$, which yields

$$\theta((\delta+\eta)z)+1 = -n,$$

corresponding to the roots of the second degree equation

$$z^2-z_n z+\frac{\mu}{\lambda} = 0, \quad \text{for all } n \geqslant 0. \tag{5.5}$$

Equation (5.5) has two real positive roots α_n and β_n satisfying

$$\alpha_0 = 1, \quad \beta_0 = \frac{\mu}{\lambda};$$

$$\alpha_{n+1}<\alpha_n<1<\beta_n<\beta_{n+1}, \quad \text{for all } n \geqslant 1;$$

$$-1+z_n<\beta_n<z_n, \quad \text{for all } n \geqslant 1.$$

This concludes the proof of Lemma 5.2. □

In general $F(u, v, w; z)$ defines an analytic function having branch points at the points $z=0$, 1 and ∞. If we make a cut in the z-plane along the real axis from $z=1$ to $z=\infty$, the series $F(u, v, w; z)$ yields, for $|z| \geqslant 1$, a single-valued analytic continuation which can be obtained by means of transformation formulas - see [12] (provided $w+1$ is not a natural number and $u-v$ and $w-u-v$ are not integers). On the other hand, after some manipulations on Gauss' recursion formulas (see [12], pp. 1045, 1046), we get

$$(w-v)zF(1, v, w+1; z) = w(z-1)F(1, v, w; z)+w.$$

Instantiating $v=b(z)$, $w=c(z)$ in (5.3) and observing that

$$f\left(\frac{z}{\alpha}\right) = \frac{1}{2\pi i}\int_{\mathcal{C}} \frac{f(t)}{t-\frac{z}{\alpha}}dt,$$

yields equation (5.2). This concludes the proof of Theorem 5.1. □

All the attempts made to explicitly invert the integral equation (4.14) (or the equivalent forms (5.1) and (5.2)) have failed - essentially because of the difficulties which arose concerning the analytic continuation. We just indicate two possible steps, which are the subject of on-going research.

i) First, replace the hypergeometric series by its E.W. Barnes integral representation [20]

$$F(1, v, w; z) = \frac{i\Gamma(w)}{2\Gamma(v)} \int_{\sigma-i\infty}^{\sigma+i\infty} \frac{\Gamma(v+s)(-z)^s}{\Gamma(w+s)\sin \pi s} ds,$$

for any real σ such that $-\min[\mathrm{Re}\, u, \mathrm{Re}\, v] < \sigma < 0$. This should provide analytic continuation to the domain between the curves (L) and (S).

ii) Then, try to set a boundary value problem on a suitable curve [5,6,7]. A good candidate is the circle of radius $\sqrt{\mu/\lambda}$, centered at the origin. The reason for this lies in equation (5.5): the roots z_1, z_2 of $z^2 - 2\xi z + \mu/\lambda = 0$ satisfy $z_1 z_2 = \mu/\lambda$ and, for all real ξ such that $|\xi| < \sqrt{\mu/\lambda}$, they are complex conjugate.

In the second subsection we propose a method which may answer the requirements of the numericians and relies on the fact that the α_n's obtained from (5.5) do have an accumulation point at $z=0$.

5.2. Equivalence to an infinite linear system

The function $f(v) = P(0, v)e^{\frac{\lambda\eta v}{\delta(\eta+\delta)}}$ has a power series expansion

$$f(v) = \sum_{n \geq 0} a_n v^n, \quad |v| \leq 1. \tag{5.6}$$

LEMMA 5.3. *The a_n's defined in (5.6) satisfy the relation*

$$\sum_{n \geq 0} a_n \Big[\sum_{k=0}^{n} \binom{n}{k} \Big(\frac{\delta}{\eta}\Big)^k z^{n-k} \Big(1 + \frac{\mu}{\lambda(z - z_k)}\Big) \prod_{m=0}^{k-1} \frac{(z-\alpha_m)(z-\beta_m)}{(z - z_m)} \Big] = 0, \tag{5.7}$$

for $z \in \mathcal{G}$, where the numbers (α_k), (β_k) and (z_k) have been defined in the proof of Lemma 5.2.

PROOF. Substituting the expansion of f in equation (5.1), we get

$$\int_0^1 \sum_{n \geq 0} a_n \Big(1 + \frac{\delta}{\eta} u\Big)^n z^n (1-u)^{a(z)-1} u^{b(z)-1} [(u-1)z + 1] du = 0, \quad \forall z \in \mathcal{G}. \tag{5.8}$$

Define

$$f_n(u, z) = a_n \Big(1 + \frac{\delta}{\eta} u\Big)^n z^n (1-u)^{a(z)-1} u^{b(z)-1} [(u-1)z+1], \quad \text{for} \quad (u, z) \in]0, 1[\otimes \mathcal{G}.$$

For all $n \geq 0$, f_n is continuous and integrable and (5.8) can be rewritten

$$\int_0^1 \sum_{n \geq 0} f_n(u, z) du = 0, \quad \forall z \in \mathcal{G}. \tag{5.9}$$

Proceeding in stages, we show that the operations $\int$ and Σ in (5.9) can be reversed.

i) $\int_0^1 |f_n(u, z)| du$ has a finite value, because

$$|f_n(u, z)| < |a_n z^n| (1+\frac{\delta}{\eta})^n (1-u)^{\frac{\mu}{(\delta+\eta)}[\frac{\text{Re } z}{|z|^2}-1]} u^{\text{Re } \theta((\delta+\eta)z)} (1+|z|),$$

which is integrable for Re $z>0$ and $1+\text{Re}\,[\theta(\delta+\eta)z\,]>0$.

ii) $\sum_{n\geq 0} |f_n(u, z)|$ is uniformly convergent for $(u, z) \in]0, 1[\otimes \mathcal{G}$, since

$$\forall u \subset [a, b],\ a>0,\ b<1, \quad f_n(u, z) \leq K_{a,b} |a_n| |\frac{z}{\alpha}|^n,$$

which is the general term of a convergent series as $|z| \leq \alpha$.

iii) $\sum_{n\geq 0} \int_0^1 |f_n(u, z)| du < \infty$.

Therefore (5.9) can be integrated term by term. Define

$$g_n(z) = \int_0^1 (1+\frac{\delta t}{\eta})^n (1-t)^{a(z)-1} t^{b(z)-1} [1-(1-t)z\,] dt.$$

Hence we have shown that,

$$\sum_{n\geq 0} a_n z^n g_n(z) = 0, \quad \forall z \in \mathcal{G}.$$

But $g_n(z)$ can be rewritten as

$$g_n(z) = \sum_{k=0}^{n} \binom{n}{k} (\frac{\delta}{\eta})^k [B(a(z), b(z)+k) - zB(a(z)+1, b(z)+k)]$$

$$= \sum_{k=0}^{n} \binom{n}{k} (\frac{\delta}{\eta})^k [1-\frac{\frac{\mu}{\delta+\eta}}{c(z)+k}] B(a(z), b(z)+k).$$

Keeping in mind that

$$b(z)+m = \frac{-\lambda}{(\delta+\eta)z} (z-\alpha_m)(z-\beta_m)$$

and

$$c(z)+m = \frac{-\lambda}{\delta+\eta} (z-z_m),$$

we get (5.7) by a routine computation, with the convention $\Pi_{m=0}^{-1} \frac{b(z)+m}{c(z)+m} = 1$. The proof of Lemma 5.3 is concluded. □

We now examine the equation (5.7) more carefully. For the sake of brevity, we shall write $h_k(n, z)$ instead of $\binom{n}{k} (\frac{\delta}{\eta})^k z^{n-k} [1+\frac{\mu}{\lambda(z-z_k)}]$. Thus (5.7) reads

$$\sum_{n\geqslant 0} a_n \sum_{k=0}^{n} u_k(n, z) = 0, \quad \forall z \in \mathcal{G}, \tag{5.10}$$

with

$$u_k(n, z) = h_k(n, z) \prod_{m=0}^{k-1} \frac{(z-\alpha_m)(z-\beta_m)}{(z-z_m)}.$$

The function of z defined by the left member of equation (5.10) is analytic, by Weierstrass' theorem, at least in the disc of radius α. (The infinite product

$$\prod_{m\geqslant 0} \frac{z-z_m}{z-\beta_m}$$

is uniformly convergent in this region).

By the principle of analytic continuation [11], (5.10) holds for $|z|<\alpha$. Moreover, we know that a function f analytic in a domain U is completely determined by its values on a sequence (s_n) with $s_n \in U$, $s_n \to s$ when $n \to \infty$, $s \in U$. Here, (α_n) tends to zero as n tends to infinity and f, the left member of (5.10), is analytic at the point zero.

Thus the determination of a solution of the integral equation (5.1) is equivalent to the existence of a sequence $(a_n)_{n\in\mathbb{N}} \neq (0)$ satisfying the following infinite linear system

$$\sum_{n\geqslant 0} a_n \sum_{k=0}^{n} u_k(n, \alpha_p) = 0, \quad \forall p \geqslant N_0, \tag{5.11}$$

where $N_0 = \inf\{n \in \mathbb{N} | \alpha_n < \alpha\}$. A priori, N_0 is not necessarily small, depending on the values of the parameters.

REMARK 5.1. By using the monotonicity of the z_k and an exponential majorant for the quantity $\Sigma_{k=0}^{n} u_k(n, z)$, where $u_k(n, z)$ has been defined above, it can be shown that (5.10) indeed holds in the whole unit disc $|z| \leqslant 1$. Hence N_0 can be taken equal to zero.

Noting that

$$\prod_{m=0}^{k-1} \frac{(\alpha_p-\alpha_m)(\alpha_p-\beta_m)}{(\alpha_p-z_m)} = 0$$

as soon as $k \geqslant p+1$, the above system (5.11) is equivalent to

$$\sum_{n\geqslant 0} a_n \sum_{k=0}^{\min(n,p)} u_k(n, \alpha_p) = 0, \quad \forall p \geqslant 0,$$

or, under a more prolix form, to

$$\sum_{n=0}^{p} a_n \sum_{k=0}^{n} u_k(n, \alpha_p) + \sum_{n=p+1}^{\infty} a_n \sum_{k=0}^{p} u_k(n, \alpha_p) = 0, \quad \forall p \geqslant 0. \tag{5.12}$$

At this moment, it is permissible to think that some orthogonalization process could be found w.r.t. some suitable sequence of special polynomials, or in other words, that it should be possible to render the system (5.12) triangular via a proper change of basis. The infinite linear system (5.12) is equivalent to a homogeneous Fredholm-type integral equation of the first kind, when solutions are sought in the class of analytic functions.

6. Final remarks

The example presented in this paper does illustrate strikingly the need for further research to dissect some strange (maybe bizarre...) mathematical objects, such as the one corresponding to equation (4.14), and which are not as rare as one may believe, see [2]. In particular much effort should be put into numerical methods. We must yet confess that (4.14) has been our pleasant bugbear...

Acknowledgement

The authors would like to thank Onno Boxma for his careful reading of the manuscript and suggestions for improving the presentation. G. Fayolle wants to thank Debasis Mitra (AT&T Bell Laboratories) for bringing this old open problem to his attention.

References

1. K.L. Chung (1967). *Markov Chains with Stationary Transition Probabilities,* Springer Verlag, New York.
2. E.G. Coffman, Jr., G. Fayolle, I. Mitrani (1986). Sojourn times in a tandem queue with overtaking: reduction to a boundary value problem. *Stochastic Models 2,* 43-65.
3. J.W. Cohen (1957). Basic problems of telephone traffic theory and the influence of repeated calls. *Philips Telecommunication Review 18,* 49-100.
4. J.W. Cohen (1969). *The Single Server Queue,* North-Holland, Amsterdam.
5. J.W. Cohen, O.J. Boxma (1983). *Boundary Value Problems in Queueing System Analysis,* North-Holland, Amsterdam.
6. G. Fayolle, R. Iasnogorodski (1979). Two coupled processors: reduction to a Riemann-Hilbert problem. *Z. Wahrsch. Verw. Gebiete 47,* 325-351.
7. G. Fayolle, P.J.B. King, I. Mitrani (1982). The solution of certain two-dimensional Markov models. *Adv. in Appl. Probab. 14,* 295-308.
8. S. Fenyö, H.W. Stolle (1984). *Theorie und Praxis der Linearen Integralgleichungen, Vol. 3,* V.E.B. Deutscher Verlag der Wissenschaften, Berlin.
9. F.G. Foster (1953). On the stochastic matrices associated with certain queuing processes. *Ann. Math. Statist. 24,* 355-360.
10. B.A. Fuchs, I. Levin (1961). *Functions of a Complex Variable and Some of their Applications, Vol. 2,* Pergamon Press, New York.
11. B.A. Fuchs, B.V. Shabat (1964). *Functions of a Complex Variable and Some of their Applications, Vol. 1,* Pergamon Press, New York.
12. I.S. Gradshteyn, I.M. Ryshik (1965). *Tables of Integrals, Series and*

Products, Academic Press, New York and London.
13. H. KESTEN (1976). Recurrence criteria for multi-dimensional Markov chains and multi-dimensional linear birth and death processes. *Adv. in Appl. Probab. 8,* 58-87.
14. L. KOSTEN (1947). On the influence of repeated calls in the theory of probabilities of blocking. *De Ingenieur 59,* E123 (in Dutch).
15. J. LAMPERTI (1960). Criteria for the recurrence or transience of stochastic process I. *J. Math. Anal. Appl. 1,* 314-330.
16. V.A. MALYSHEV (1972). Classification of two-dimensional positive random walks and almost linear semimartingales. *Soviet Math. Dokl. 13,* 136-139.
17. V.A. MALYSHEV (1972). An analytical method in the theory of two-dimensional positive random walks. Transl. from *Sibirskii Math. Zhurnal 13,* 1314-1329.
18. W. ROSENKRANTZ (1987). *Ergodicity Conditions for Two-Dimensional Markov Chains with Oblique Reflection,* Report Univ. of Massachusetts.
19. V. SMIRNOV (1984). *Cours de Mathématiques Supérieures,* MIR-Moscou Publishers.
20. E.C. TITCHMARSH (1948). *Introduction to the Theory of Fourier Integrals,* Oxford Univ. Press, Oxford, 2nd ed.
21. R.I. WILKINSON (1956). Theories for toll traffic engineering in the USA. *The Bell System Tech. J. 35,* 421-514.
22. T. YANG, J.G.C. TEMPLETON (1987). A survey on retrial queues. *Queueing Systems 2,* 201-233.

Appendix A

The case $\delta x + \eta y = 0$ (cf. Subsection 4.1)

The problem is to examine the limit of $\int_0^y P(0, t)K(t, \delta x + \eta y)dt$ when $\delta x + \eta y$ tends to 0 by positive values. Remark that $K(0, A) = \frac{\delta}{A^2} - \frac{1}{A}$ for all A different from 0.

Put $A = \delta x + \eta y$. For ϵ sufficiently small, we obtain

$$\int_0^\epsilon P(0, t)K(t, A)dt = \left(\int_0^\epsilon P(0, t)dt\right)K(0, A) + \int_0^\epsilon P(0, t)[K(t, A) - K(0, A)]dt$$
$$= K(0, A)[P(0, 0)\epsilon + \frac{K\epsilon^2}{2} + 0(\epsilon^2)] + I(\epsilon, A),$$

since $t \to P(0, t)$ is analytic around 0. Then

$$P(0, 0) + \mu \int_0^\epsilon P(0, t)K(t, A)dt$$
$$= P(0, 0)[1 + \mu\epsilon \left(\frac{\delta}{A^2} - \frac{1}{A}\right) + \frac{K\epsilon^2}{2}\left(\frac{\delta}{A^2} - \frac{1}{A}\right) + 0(\epsilon^2)] + \tilde{I}(A),$$

with $|I(\epsilon, A)| \leqslant C\epsilon \sup_{t \in (0, \epsilon)} |K(t, A) - K(0, A)|$, where C is a positive constant. Choosing $\epsilon = \frac{-A^2}{\mu(\delta - A)}$, we get

$$P(0, 0) + \mu \int_0^\epsilon P(0, t)K(t, A)dt = \frac{K}{2\mu^2} \frac{A^2}{\delta - A} + 0(A^2) + \tilde{I}(A).$$

Hence

$$\lim_{A\to 0}[P(0,0)+\mu \int_{0}^{\frac{-A^2}{\mu(\delta-A)}} P(0,t)K(t,A)dt] = 0.$$

Keeping $\epsilon = \frac{-A^2}{\mu(\delta-A)}$, we study the limit of

$$\int_{\frac{-A^2}{\mu(\delta-A)}}^{y} P(0,s)K(s,A)ds$$

when A tends to 0 by positive values in the right half-plane, where the path of integration is chosen as the chord $[\frac{-A^2}{\mu(\delta-A)}, y]$. We write

$K(s,A) =$

$$(\eta s+\delta-A)e^{\frac{\lambda\eta s}{\delta(\eta+\delta)}}[\frac{1}{A^2}(\frac{A-\eta s}{A-(\delta+\eta)s})^{\frac{\mu}{A}-1}[1-(\delta+\eta)\frac{s}{A}]^{-2+\frac{\lambda+\mu}{\delta+\eta}-\frac{\lambda A}{(\delta+\eta)^2}}].$$

It can be verified that when Re $A>0$, $A\to 0$,

$$\sup_{s\in[\frac{-A^2}{\mu(\delta-A)},y]} |\frac{1}{A^2}(\frac{A-\eta s}{A-(\delta+\eta)s})^{\frac{\mu}{A}-1}[1-(\delta+\eta)\frac{s}{A}]^{-2+\frac{\lambda+\mu}{\delta+\eta}-\frac{\lambda A}{(\delta+\eta)^2}}| \to 0.$$

REMARK. If y is a positive real number, it seems easier to consider as a path of integration the semi-circle from ϵ to y with radius $\frac{y-\epsilon}{2}$; in this way the uniform convergence to zero is also proved.

As $s\to P(0,s)(\eta s+\delta-A)e^{\frac{\lambda\eta s}{\delta(\eta+\delta)}}$ is a bounded function on the interval $[\frac{-A^2}{\mu(\delta-A)}, y]$, we have the following result:

$$\lim_{A\to 0} \int_{\frac{-A^2}{\mu(\delta-A)}}^{y} P(0,s)K(s,A)ds = 0.$$

Hence, both members of equation (4.3) vanish as was to be proved.

Some Comments on the Work of J.W. Cohen and New Results in the Theory of Queueing Networks

Pierre Le Gall
National Center of Telecommunications,
Paris, France

In this paper, we begin with some recollections about the early work of J.W. Cohen which was more particularly concerned with teletraffic theory and notably with the theory of repeat attempts. We then present an original study of a service system which is basic to the theory of queueing networks: the two-stage tandem queue with Poisson arrivals and a different general service time distribution at each stage. An independent Poisson traffic stream is also offered to the second stage and some customers leave the system after the first stage service. Customers wait at each stage and service is in arrival order (at each stage).

1. Introduction

We are very glad, with the publication of this Liber Amicorum, to have the opportunity of commenting on the already very extensive work of J.W. Cohen in the field of queueing theory. We will, in fact, limit our attention to the early period (before 1960) which was more specifically concerned with teletraffic theory.

His subsequent work, and notably his book on the single server queue [11], makes considerable use of the integral representation in the complex plane introduced by Pollaczek [16]. This is true also of the work of his students De Smit, in his thesis on the *n*-server problem, and Boxma in his work on tandem queues, [3].

We also have been a faithful disciple of Pollaczek [12] and, in this paper, we again use his method to study an important problem in queueing network theory: that of the network nodes.

We generalize the above-mentioned study by Boxma on tandem queues to the case where service time distributions are general and different for the two considered server stages. We further suppose that part of the traffic leaves the system on completing service at the first stage and that additional customers enter directly at the second stage. Waiting is possible at each stage and customers are served in the order at which they arrive at the considered stage. Offered traffic streams are assumed to be Poisson. Except [2] for the first stage, we are unaware of any previous work on the analysis of the tandem queue with general service times despite the fact that this problem has considerable practical importance. It is an excellent opportunity to show, one more time, the effectiveness of Pollaczek's *analytical* method which avoids the need for a tedious detailed analysis of busy-period structures and does not require any subtle, tricky or very detailed probabilistic considerations. For the sake of simplicity, we only intend to consider the case of finite supports for the service time distributions.

2. The contribution of J.W. Cohen to teletraffic theory

We first consider the period 1955-1960. At the time of the first International Teletraffic Congress (ITC) in 1955, the probability theory taught to electrical engineers was very rudimentary: it often consisted simply of the study of random variables and statistics. Stochastic processes were hardly mentioned. Only the more advanced university courses considered this subject which was to gain more and more importance through its technical applications. In particular, in telecommunications, one used to speak of the simple 'statistical equilibrium' theory of the Danish engineer Erlang without recognizing its basis in the theory of Markov chains. This theory only became popular with the publication of Feller's first volume with its simple presentations devoid of the cabalistics signs so dear to the heart of purist mathematicians.

An excellent initiation was to consider the first publications on the problem of a simple trunk group with regenerative arrivals, exponential service time distribution and no waiting. The separate publications of Takács and Cohen [7] awoke the attention of the first teletraffic engineers.

The year 1957, in which [7] appeared, was a very fertile one for Wim:
First, a study on the traffic in the transposed multiples used in the new crossbar exchanges which were just coming out [1]; then, a generalisation of Engset's formula [6] where independent limited sources are considered with the most general assumptions, accessing the trunk group constituting a costly final stage of crossbar switching systems; and last, his important study on repeat attempts [5] which was probably too far in advance of the problems raised by the switching techniques of that time and whose complexity put the results out of the reach of most engineers.

In fact, my first contacts with Wim occurred at the time of his work on derived Markov chains around 1960, [8,9,10]. Its application to queues with grouped arrivals and batch service raised a certain number of probability definition difficulties to establish its conformity with my own work [12].

The developments during this period concerning teletraffic problems in crossbar switching systems have been amply reported in the richly documented book by Syski [19] of which we are glad to note the recently published new edition.

The following period (1960-1970) was devoted to queueing theory and, in particular, to the single server queue. His book [11] has since become the basic reference on this subject, containing all the necessary documentation.

In more recent years, Wim has addressed his attention to the new problems occurring in processors and computer networks and he has produced a large number of papers on these subjects. However, we will remain with the early period coming back, in particular, to his above mentioned study on repeat attempts.

In addition to the rather complex analysis of the correlation between the successive attempts of a given call, his study exposed a simple law in the case where there is a sufficiently long time between attempts: the statistical trunk group occupancy can be considered as being produced by an overall offered traffic equal to the sum of the offered traffic and a certain repeat attempts traffic (which does not however occupy trunks) having the same Poisson nature as the former. The CCITT has since given this overall traffic offered the name 'equivalent' offered traffic, to be distinguished from the traffic offered itself which corresponds to the traffic which would be carried if the number of trunks were sufficiently large and which is the only concept to be used for traffic forecasting. At the time, we were only concerned with the occupation of the trunk group. In the telecommunication networks of the last ten years or so, however, where digital techniques and common channel signalling have greatly reduced call set-up times, it has become necessary to consider also the potentially significant increase in the number of repeat attempts whose interval is shorter and shorter, [13,14].

The work of Wim is carried over into the publication of his students and, in view of its importance in queueing networks, we are going to consider the particular case of tandem queues as studied by one of these students, O.J. BOXMA [3]. We adopt somewhat more general structure and service assumptions.

3. TWO-STAGE TANDEM QUEUES

Fig. 1 shows two successive queues with waiting possible in front of each server and with *service in order of arrival* in each queue.

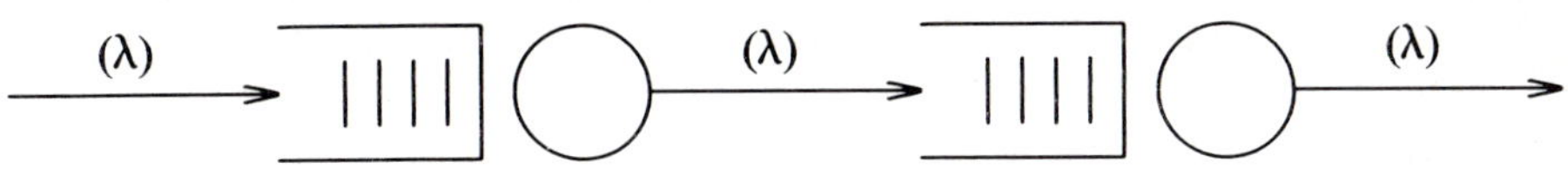

FIGURE 1

3.1. Notation and assumptions

There is a single *Poisson* traffic stream successively crossing the two servers and their respective queues. The arrival intensity is λ. The arrival epoch of the n-th 'customer' is X_n. It has a waiting time τ_n^i $(i=1,2)$ and a service time T_n^i $(i=1,2)$ at stage 1. The sojourn time at stage i is therefore

$$s_n^i = \tau_n^i + T_n^i.$$

The overall sojourn time is

$$S_n = s_n^1 + s_n^2 = A_n^2 + T_n^2,$$

where A_n^2 is the *access time* of the second server:

$$A_n^2 = s_n^1 + \tau_n^2 = \tau_n^1 + T_n^1 + \tau_n^2. \tag{3.1}$$

The arrival intervals $Y_{n-1} = X_n - X_{n-1}$ are assumed to be mutually independent and to have the same exponential distribution:

$$\Pr(Y_{n-1}<t) = 1-e^{-\lambda t}, \quad t\geq 0, \quad n=1,2,\ldots .$$

Similarly, the service times at stage i, T_n^i ($i=1,2$), are assumed to be mutually independent and independent of the arrival process and to have the distribution function $F_i(t)$, $i=1,2$.

We introduce the following characteristic functions, for Re $z\geq 0$:

$$\begin{aligned}\phi_0(z) &= \int_0^\infty e^{-zt}d\Pr\{Y_{n-1}<t\}=\frac{\lambda}{\lambda+z},\\ \phi_i(z) &= \int_0^\infty e^{-zt}dF_i(t), \quad i=1,2,\end{aligned} \tag{3.2}$$

and denoting the *occupancy of server i* by ρ_i we let:

$$\overline{T}_i = \int_0^\infty tdF_i(t), \quad \rho_i = \lambda\overline{T}_i<1, \quad i=1,2. \tag{3.3}$$

To simplify the presentation, *we assume the random variables* T_n^i *are finite* so that $\phi_i(z)$, $i=1,2$, exists for $-\infty<\text{Re } z<+\infty$.

In comparison with [3] the new contribution of the study resides in the fact that the service times T_n^1 and T_n^2 may overlap so that a busy period of the first server can correspond to several busy periods of the second server, thus creating serious complications in analyzing the waiting phenomenon at the second stage.

We aim to determine:

$$A_n(z_1,z_2) = E\{e^{-z_1 s_n^1 - z_2 \tau_n^2}\}, \tag{3.4}$$

giving :

$$\phi_n(z_1,z_2) = E\{e^{-z_1 s_n^1 - z_2 s_n^2}\}=A_n(z_1,z_2)\phi_2(z_2). \tag{3.5}$$

3.2. The equation of the system

The method of Pollaczek requires that we first determine the stochastic relations defining the phenomenon. The integral representation in the complex plane will then allow us to implement a purely *analytical* solution procedure with no necessity to use the properties of the stochastic processes.

The basic stochastic relation characterizing the queue for the i-th server is as follows (using the notation $X^+=\max(0,X)$):

$$\tau_n^i = [s_{n-1}^i - Y_{n-1}^i]^+, \tag{3.6}$$

with for the arrival intervals at stage i ($i=1,2$):

$$Y_{n-1}^1 \overset{def}{=} X_n - X_{n-1} = Y_{n-1},$$

$$
\begin{aligned}
Y_{n-1}^2 &\overset{\text{def}}{=} (X_n+s_n^1)-(X_{n-1}+s_{n-1}^1) \\
&= Y_{n-1}+[T_n^1+(s_{n-1}^1-Y_{n-1})^+]-s_{n-1}^1 \\
&= T_n^1+(Y_{n-1}-s_{n-1}^1)^+.
\end{aligned} \tag{3.7}
$$

The last term in brackets is positive when the n-th customer does not wait at the first stage.

Thus, the second stage arrival interval is T_n^1, except if there is no delay at the first stage. In the latter case it is necessary to add a random duration

$$(Y_{n-1}-s_{n-1}^1)^+ \tag{3.8}$$

corresponding to the time when the first stage is empty. This duration will also complicate calculations.

In the following, we will use the integral representation in the complex plane with the Cauchy integral

$$\int_{C_z} \cdots dz$$

taken over a contour C_z made up of a straight line parallel to the imaginary axis, covered from the bottom to the top and closing at infinity. If this line is against the imaginary axis to the right (left), we use the notation

$$\int_{+0} \quad \Big(\int_{-0}\Big)$$

and we make use of the Pollaczek relation, where Re $z>$Re $\zeta>0$:

$$e^{-z\max(a_1,a_2)}=\frac{1}{2\pi i}\int_{+0} e^{-a_1(z-\zeta)-a_2\zeta}\frac{z\,d\zeta}{(z-\zeta)\zeta}. \tag{3.9}$$

Taking account of (3.6), (3.7) and (3.9), we write for Re $z_2>0$:

$$
\begin{aligned}
e^{-z_1 s_n^1-z_2\tau_n^2} &= e^{-z_1(\tau_n^1+T_n^1)-z_2\tau_n^2}=e^{-z_1(\tau_n^1+T_n^1)-z_2[s_{n-1}^2-T_n^1-(y_{n-1}-s_{n-1}^1)^+]} \\
&= e^{(-z_1+z_2)T_n^1-z_1\tau_n^1-z_2\max[T_n^1,\, s_{n-1}^2-(y_{n-1}-s_{n-1}^1)^+]} \\
&= e^{(-z_1+z_2)T_n^1}\frac{1}{2\pi i}\int_{+0} e^{-(z_2-\zeta)T_n^1-\zeta(T_{n-1}^2+\tau_{n-1}^2)-z_1\tau_n^1+\zeta(y_{n-1}-s_{n-1}^1)^+}\frac{z_2\,d\zeta}{(z_2-\zeta)\zeta}.
\end{aligned}
$$

Then

$$
\begin{aligned}
e^{-z_1\tau_n^1+\zeta(y_{n-1}-s_{n-1}^1)^+} &= e^{-z_1(s_{n-1}^1-y_{n-1})^+ +\zeta(y_{n-1}-s_{n-1}^1)^+} \\
&= \frac{1}{2\pi i}\int_{+0} e^{-\xi(s_{n-1}^1-y_{n-1})}\left[\frac{1}{z_1-\xi}+\frac{1}{\xi-\zeta}\right]d\xi,
\end{aligned} \tag{3.10}
$$

with: Re $z_1>$Re $\xi>$Re $\zeta>0$.

Inserting this expression in the preceding one gives:

$$e^{-z_1 s_n^1-z_2\tau_n^2} = U_0(z_1,z_2)+V_0(z_1,z_2), \tag{3.11}$$

with:

$$U_0(z_1,z_2) = \frac{1}{(2\pi i)^2}\int_{+0}\frac{z_2 d\zeta}{(z_2-\zeta)\zeta}\int_{+0}\frac{d\xi}{z_1-\xi}\times$$
$$e^{(z_1-\zeta)T_n^1-\zeta T_{n-1}^2+\xi y_{n-1}-(\xi s_{n-1}^1+\zeta\tau_{n-1}^2)} \qquad (\text{case } \tau_n^1>0),$$
$$V_0(z_1,z_2) = \frac{1}{(2\pi i)^2}\int_{+0}\frac{z_2 d\zeta}{(z_2-\zeta)\zeta}\int_{+0}\frac{d\xi}{\xi-\zeta}\times \tag{3.12}$$
$$e^{-(z_1-\zeta)T_n^1-\zeta T_{n-1}^2+\xi y_{n-1}-(\xi s_{n-1}^1+\zeta\tau_{n-1}^2)} \qquad (\text{case } \tau_n^1=0).$$

We now apply the expectation operation to this expression with notation (3.4). We obtain the fundamental relation of the system:

$$A_n(z_1,z_2) = B_n(z_1,z_2)+C_n(z_1,z_2), \tag{3.13}$$

with, for Re $z_1>$Re $\xi>$Re $\zeta>0$ and Re $z_2>$Re ζ:

$$B_n(z_1,z_2) = \frac{1}{(2\pi i)^2}\int_{+0}\frac{z_2 d\zeta}{(z_2-\zeta)\zeta}\int_{+0}\frac{d\xi}{z_1-\xi}\times$$
$$[\phi_1(z_1-\zeta)\phi_2(\zeta)]\phi_0(-\xi)A_{n-1}(\xi,\zeta) \qquad (\text{case } \tau_n^1>0),$$
$$C_n(z_1,z_2) = \frac{1}{(2\pi i)^2}\int_{+0}\frac{z_2 d\zeta}{(z_2-\zeta)\zeta}\int_{+0}\frac{d\xi}{\xi-\zeta}\times \tag{3.14}$$
$$[\phi_1(z_1-\zeta)\phi_2(\zeta)]\phi_0(-\xi)A_{n-1}(\xi,\zeta) \qquad (\text{case } \tau_n^1=0).$$

Introduce the generating functions, for $|x|<1$:

$$A(z_1,z_2;x) = \sum_{n=0}^{\infty}A_n(z_1,z_2)x^n,$$
$$A_0(z_1,z_2)\equiv\phi_1(z_1),$$
$$C(z_1,z_2;x) = \sum_{n=1}^{\infty}C_n(z_1,z_2)x^n, \tag{3.15}$$
$$\Phi(z_1,z_2;x) = \sum_{n=1}^{\infty}\Phi_n(z_1,z_2)x^n.$$

We deduce the integral equation of the system:

$$A(z_1,z_2;x)-\frac{x}{(2\pi i)^2}\int_{+0}\frac{z_2 d\zeta}{(z_2-\zeta)\zeta}\int_{+0}\frac{d\xi}{z_1-\xi}[\phi_1(z_1-\zeta)\phi_2(\zeta)]\phi_0(-\xi)A(\xi,\zeta;x)$$
$$= \phi_1(z_1)+C(z_1,z_2;x), \tag{3.16}$$

where we can suppose, for the time being, that the right-hand side is known, $C(z_1,z_2;x)$ corresponding to the case $\tau_n^1=0$.

The above developments are valid for *any regenerative arrival process.* However, the presence of the variable z_1 in the kernel of the integral greatly complicates the solution [12]. We therefore choose to limit our attention to the usual case of Poisson arrivals with expression (3.2) for $\phi_0(z)$. In the domain

Re $\xi > 0$, we then only have two poles $\xi = z_1$ and $\xi = \lambda$ where we can apply the residue theorem. The integral equation (3.16) then has only one variable z_2 (Re $z_2 > 0$), for given z_1 ($0 < \text{Re } z_1 < \lambda$):

$$A(z_1, z_2; x) - \frac{x}{2\pi i} \int_{+0} [\phi_1(z_1 - \zeta)\phi_2(\zeta)]\phi_0(-z_1)A(z_1, \zeta; x)\frac{z_2 d\zeta}{(z_2 - \zeta)\zeta}$$
$$= D(z_1, z_2; x), \quad 0 < \text{Re } \zeta < \text{Re } z_2, \tag{3.17}$$

where we suppose, for the time being, that the right-hand side $D(z_1, z_2; x)$ is known:

$$D(z_1, z_2; x) = \phi_1(z_1)$$
$$+ \frac{x}{2\pi i} \int_{+0} [\phi_1(z_1 - \zeta)\phi_2(\zeta)\phi_0(-\zeta)]\frac{\zeta - z_1}{\lambda} \phi_0(-z_1)A(\lambda, \zeta; x)\frac{z_2 d\zeta}{(z_2 - \zeta)\zeta}, \tag{3.18}$$

where we have used the relation:

$$\phi_0(-\zeta) - \phi_0(-z_1) \equiv \frac{\zeta - z_1}{\lambda} \phi_0(-\zeta)\phi_0(-z_1).$$

3.3. Particular solution

Before we derive the general solution, it is first necessary to consider the following reduced case:

$$A(z_1, z_2; x) - \frac{x}{2\pi i} \int_0 [\phi_1(z_1 - \zeta)\phi_2(\zeta)]\phi_0(-z_1)A(z_1, \zeta; x)\left[\frac{1}{z_2 - \zeta} + \frac{1}{\zeta - u}\right]d\zeta = 1,$$

with Re $z_2 >$ Re $\zeta >$ Re u. We know the solution has the following expression (see Annex in [12]) for $|x| < 1$:

$$\begin{cases} A'(z_1, z_2, u; x) \\ = \exp\left\{\dfrac{-1}{2\pi i} \displaystyle\int_0 \ln[1 - x\phi_1(z_1 - \zeta)\phi_2(\zeta)\phi_0(-z_1)] \left[\dfrac{1}{z_2 - \zeta} + \dfrac{1}{\zeta - u}\right]d\zeta\right\}, \end{cases} \tag{3.19}$$

with Re $u <$ Re $\zeta <$ Re z_2.

It is convenient to express this function in another form. For $0 \leqslant \text{Re } z_1 < \lambda$, Re $z_2 > 0$ and $|x| < 1$, let:

$$H(z_1, z_2; x) = \exp\left\{\frac{-1}{2\pi i} \int_{+\delta} \ln[1 - x\phi_1(z_1 - \zeta)\phi_2(\zeta)\phi_0(-z_1)]\frac{z_2 d\zeta}{(z_2 - \zeta)\zeta}\right\}, \tag{3.20}$$

with Re $z_2 >$ Re ζ and $|\text{Re } \zeta| < \delta$, where δ is in the integral analyticity band. Similarly, let:

$$H_1(z_1, z_2; x) = \exp\left\{\frac{-1}{2\pi i} \int_{+\delta} \ln[1 - x\phi_1(z_1 - \zeta)\phi_2(\zeta)\phi_0(-z_1)]\frac{z_2 d\zeta}{(z_2 - \zeta)\zeta}\right\}, \tag{3.21}$$

with Re $z_2 <$ Re ζ. Taking account of the residue at the pole $\zeta = z_2$, we have the following relation:

$$H_1(z_1, z_2; x) = H(z_1, z_2; x)[1 - x\phi_1(z_1 - z_2)\phi_2(z_2)\phi_0(-z_1)], \quad (3.22)$$

which allows us to extend the validity of H_1 for $-\infty \leqslant \text{Re } z_2 < +\infty$ by the assumption of finite supports for T_n^1 and T_n^2 with the condition: $0 < \text{Re } z_1 < \lambda$. We deduce:

$$A'(z_1, z_2, u; x) = \frac{H(z_1, z_2; x)}{H_1(z_1, u; x)}. \quad (3.23)$$

We can discover a more physical meaning for the function H by developing the logarithm in (3.20). Let:

$$\sum_{r=1}^{n} T_r^i = R_n^i, \quad \sigma_n = \sum_{r=1}^{n} Y_{r-1}. \quad (3.24)$$

Taking account of (3.9) with $a_1 = 0$, expression (3.20) becomes:

$$H(z_1, z_2; x) = \frac{1}{1-x} \exp\left\{ \sum_{n=1}^{\infty} \frac{x^n}{n} \left[E\{e^{-z_1[R_n^1 - \sigma_n] - z_2[R_n^2 - R_n^1]^+}\} - 1 \right] \right\}, \quad (3.25)$$

with $0 \leqslant \text{Re } z_1 < \lambda$, $\text{Re } z_2 \geqslant 0$, $|x| < 1$. Expression (3.22) becomes, with the notation $X^- = \min(0, X)$:

$$H_1(z_1, z_2; x) = \exp\left\{ -\sum_{n=1}^{\infty} \frac{x^n}{n} \left[E\{e^{-z_1[R_n^1 - \sigma_n] - z_2[R_n^2 - R_n^1]^-}\} - E\{e^{-z_1[R_n^1 - \sigma_n]}\} \right] \right\}, \quad (3.26)$$

which makes us think of the formulae of Spitzer. In particular, $H(0, z_2; x)$ corresponds to the case GI/G/1 with arrival intervals T_n^1 and service times T_n^2. Further, we have:

$$H(z, z; x) = \frac{1}{1-x} \exp\left\{ \sum_{n=1}^{\infty} \frac{x^n}{n} \left[E\{e^{-z[\max(R_n^1, R_n^2) - \sigma_n]}\} - 1 \right] \right\}, \quad (3.27)$$

and

$$H_1(z, z; x) = \exp\left\{ -\sum_{n=1}^{\infty} \frac{x^n}{n} \left[E\{e^{-z[\min(R_n^1, R_n^2) - \sigma_n]}\} - E\{e^{-z[R_n^1 - \sigma_n]}\} \right] \right\}. \quad (3.28)$$

3.4. General solution

The general solution of (3.17) has the following expression (see Annex in [12]), for $\text{Re } z_2 > 0$ and $0 \leqslant \text{Re } z_1 < \lambda$:

$$A(z_1, z_2; x) = \frac{1}{2\pi i} \int_{+0} D(z_1, u; x) A'(z_1, z_2, u; x) \frac{z_2 \, du}{(z_2 - u)u}. \quad (3.29)$$

Apply the residue theorem to the pole $\zeta = z_2$ in (3.18). There remains an integral which is holomorphic for $\text{Re } z_2 < 0$ and is zero for $z_2 = 0$. It disappears when this expression is used in (3.29) which becomes:

$$A(z_1, z_2; x) = \frac{1}{2\pi i} \int_{+0} \frac{H(z_1, z_2; x)}{H_1(z_1, u; x)} \times$$

$$[\phi_1(z_1) + x(\phi_1(z_1-u)\phi_2(u)\phi_0(-z_1))\phi_0(-u)\frac{u-z_1}{\lambda} A(\lambda, u; x)] \frac{z_2 du}{(z_2-u)u}, \tag{3.30}$$

with $0 \leq \text{Re } z_1 < \lambda$, $\text{Re } z_2 > 0$, $|x| < 1$.
It will be useful to write (3.30) also in the following form, taking (3.22) into account:

$$[1 - x\phi_1(z_1-z_2)\phi_2(z_2)\phi_0(-z_1)]A(z_1, z_2; x)$$

$$= \frac{1}{2\pi i} \int_{+0} \frac{H(z_1, z_2; x)}{H_1(z_1, u; x)} \{\phi_1(z_1) + x(\phi_1(z_1-u)\phi_2(u)\phi_0(-z_1))\times \tag{3.31}$$

$$\phi_0(-u)\frac{u-z_1}{\lambda} A(\lambda, u; x)\} \frac{z_2 du}{(z_2-u)u}.$$

For the time being, we come back to (3.30) and make use of (3.22) to give:

$$A(z_1, z_2; x) = \frac{1}{2\pi i} \int_{+0} \frac{H(z_1, z_2; x)}{H_1(z_1, u; x)} \times$$

$$\left[\frac{\phi_1(z_1) + x(\phi_1(z_1-u)\phi_2(u)\phi_0(-z_1))\phi_0(-u)\frac{u-z_1}{\lambda} A(\lambda, u; x)}{1 - x\phi_1(z_1-u)\phi_2(u)\phi_0(-z_1)}\right] \frac{z_2 du}{(z_2-u)u}. \tag{3.32}$$

For Re $u > 0$, $H(z_1, u; x)$ has no zeros and the function under the integral sign has the poles $u = z_2$ and $u = \lambda$ together with any occasional zeros of the denominator which, on multiplying by $(\lambda - z_1)$, may be written:

$$(\lambda - z_1) - x\lambda\phi_1(z_1-u)\phi_2(u) = 0. \tag{3.33}$$

Since, by assumption, T_n^1 and T_n^2 have finite supports, the function $\phi_1(z_1-u)\phi_2(u)$ remains finite for finite u and z_1 (the assumption of finite supports has been made to simplify the study of the zeros of equation (3.33)). We can therefore find a value of x of sufficiently small modulus to have the inequality:

$$|x\lambda\phi_1(z_1-u)\phi_2(u)| < |\lambda - z_1|, \tag{3.34}$$

for finite u and z_1. In this domain, since the right-hand side has only one zero, Rouché's theorem can be applied to deduce that, for a given finite u, equation (3.33) has a single zero $z_1(u, x)$. If this zero were to coincide with z_1 in (3.32) it would simply be necessary to choose another value for x to avoid that the denominator disappears for finite u.

It remains that, for any value of x ($|x| < 1$), $A(\lambda, u; x)$ has the same zeros as in equation (3.33), for finite u and consequently also when Re $u \to +\infty$ if $\phi_1(z_1-u)\phi_2(u)$ is not holomorphic in that case. It results that

$$\frac{A(\lambda, u; x)}{H(z_1, u; x)[1 - x\phi_1(z_1-u)\phi_2(u)\phi_0(-z_1)]} = \frac{A(\lambda, u; x)}{H_1(z_1, u; x)} \tag{3.35}$$

remains bounded in any case when Re $u \to +\infty$.

We can now apply the residue theorem at the only poles $u = z_2$ and $u = \lambda$ for Re $u > 0$. Let us use equation (3.31) which has the same poles as (3.32). We deduce the general equation:

$$[1 - x\phi_1(z_1 - z_2)\phi_2(z_2)\phi_0(-z_1)]A(z_1, z_2; x)$$
$$= H_1(z_1, z_2; x)[\phi_1(z_1) + x(1 - \phi_0(-z_2))\left[\frac{\phi_1(z_1 - \lambda)}{H_1(z_1, \lambda; x)}\right]\Phi(\lambda, \lambda; x)] \tag{3.36}$$
$$+ x\frac{z_2 - z_1}{\lambda}(\phi_1(z_1 - z_2)\phi_2(z_2)\phi_0(-z_1))\phi_0(-z_2)A(\lambda, z_2; x),$$

if we note the relation:

$$\frac{1}{2\pi i}\int_{+0} \frac{H_1(z_1, z_2; x)}{H_1(z_1, u; x)}\phi_1(z_1)\frac{z_2 du}{(z_2 - u)u} = H_1(z_1, z_2; x)\phi_1(z_1),$$

due to the pole $u = 0$ where $H_1(z_1, 0; x) = 1$, $H_1(z_1, u; x)$ being defined (without zeros) for Re $u \leqslant 0$.

Recalling (3.5) and (3.15), $\Phi(\lambda, \lambda; x)$ is the generating function of customers who leave the system empty. This is also the generating function of *customers finding the system empty*. Similarly, $A(\lambda, z_2; x)$ corresponds to the *first stage only being empty* while $A(z_1, 0; x)$ corresponds to the first stage considered alone.

Derivation of $\Phi(\lambda, \lambda; x)$

In (3.36), set $z_1 = z_2 = z$. We have, for $0 < \text{Re } z < \lambda$:

$$[1 - x\phi_2(z)\phi_0(-z)]A(z, z; x)$$
$$= H_1(z, z; x)[\phi_1(z) + x(1 - \phi_0(-z))\left[\frac{\phi_1(z - \lambda)}{H_1(z, \lambda; x)}\right]\Phi(\lambda, \lambda; x)]. \tag{3.37}$$

In the domain Re $z_2 > 0$, let $z_2(x)$ be the unique zero of the equation in z_2:

$$1 - x\phi_2(z_2)\phi_0(-z_2) = 1 - x\phi_2(z_2)\frac{\lambda}{\lambda - z_2} = 0. \tag{3.38}$$

For this zero, the left-hand side of (3.37) disappears giving:

$$\Phi(\lambda, \lambda; x) = \frac{\lambda}{z_2(x)}\phi_1(z_2(x))\phi_2(z_2(x))\frac{H_1(z_2(x), \lambda; x)}{\phi_1(z_2(x) - \lambda)}. \tag{3.39}$$

Derivation of $A(\lambda, z_2; x)$

Consider $z_1(z_2, x)$, the zero of (3.33) for $u = z_2$. Let:

$$\zeta(z_2, x) = \frac{\lambda - z_1(z_2(x))}{\lambda}. \tag{3.40}$$

With the same argument as used by BOXMA [3], this expression may also be written

$$\zeta(z_2, x) = E\left\{x^{n_1} e^{-z_2 \sum_{r=1}^{n_1}(T_r^2 - T_r^1)}\right\}, \tag{3.41}$$

where n_1 is the random number of customers served in a busy period of the first server. As we do not always have $T_n^1 \leqslant T_n^2$, this function is only defined for finite $|z_2|$ in the entire z_2 plane (because of the finite supports of T_n^1 and T_n^2). We now insert the value $z_1 = z_1(z_2, x)$ in (3.36). The left-hand side disappears giving:

$$\phi_0(-z_2)A(\lambda, z_2; x) = -\frac{\lambda}{z_2 - \lambda + \lambda\zeta(z_2, x)} H_1(\lambda(1-\zeta(z_2, x)), z_2; x) \times$$
$$\left\{\phi_1(\lambda(1-\zeta(z_2; x))) + x(1-\phi_0(-z_2))\frac{\phi_1(-\lambda\zeta(z_2, x))}{H_1(\lambda(1-\zeta(z_2; x)), \lambda; x)} \Phi(\lambda, \lambda; x)\right\}, \tag{3.42}$$

where $\Phi(\lambda, \lambda; x)$ is given by (3.39). Due to the term $H_1(\lambda(1-\zeta(z_2; x)), z_2; x)$, we may note that the ratio (3.35) remains bounded for $\operatorname{Re} u > 0$. Expression (3.36) then gives $A(z_1, z_2; x)$.

3.5. Stationary limit

We know that the stationary limit exists [17]. First assume $\operatorname{Re} z_i > 0$ and $\rho_i < 1$, $i = 1, 2$, to avoid any problems in passing to the limit, and let x tend to 1. Let us consider the convergence in distribution and let the limits, for $|z_i| < \lambda$, be:

$$\begin{cases}(1-x)A(z_1, z_2; x) & \to A(z_1, z_2), \\ (1-x)A(\lambda, 0; x) & \to A(\lambda, 0) = 1-\rho_1, \\ (1-x)H(z_1, z_2; x) & \to H(z_1, z_2), \\ H_1(z_1, z_2; x) & \to H_1(z_1, z_2), \\ (1-x)\Phi(\lambda, \lambda; x) & \to \Phi(\lambda, \lambda), \\ \zeta(z_2; x) & \to \zeta(z_2).\end{cases}$$

If $z_1 = 0$, corresponding to the case GI/G/1 with arrival intervals T_n^1 and service times T_n^2, $H_1(z_1, z_2)$ may not be bounded. Introduce the function $K_1(z_1, z_2)$ letting:

$$H_1(z_1, z_2) = \frac{K_1(z_1, z_2)}{1-\phi_0(-z_1)\phi_1(z_1)}, \tag{3.43}$$

$K_1(0, z_2)$ is bounded, and we may define $[K_1'(z_1, z_2)]_{z_1=0}$. Expression (3.26) then becomes:

$$K_1(z_1, z_2) = \exp\left\{-\sum_{n=1}^{\infty}\frac{1}{n}[E\{e^{-z_1[R_n^1-\sigma_n]-z_2[R_n^2-R_n^1]^-}\}-1]\right\},$$
$$K_1(0, 0) = 1. \tag{3.44}$$

The derivation is straightforward from (3.36). We let:

$$W(z_1, z_2) = (1-\rho_2)(1-\phi_0(-z_2))[H(z_1, z_2)(1-\phi_0(-z_1)\phi_1(z_1))]$$
$$= (1-\rho_2)\frac{1-\phi_0(-z_2)}{1-\phi_1(z_1-z_2)\phi_2(z_2)\phi_0(-z_1)} K_1(z_1, z_2). \tag{3.45}$$

Let $z_1=z_2=z$ and let $W(z,z)=W(z)$. Recall that the characteristic function of the delay for the second server, when it is assumed to be in isolation, is:

$$W_2(z) = (1-\rho_2)\frac{1-\phi_0(-z)}{1-\phi_0(-z)\phi_2(z)}. \tag{3.46}$$

Expression (3.45) gives:

$$\begin{aligned} W(z) &= W_2(z)K_1(z,z), \\ W(0) &= 1, \end{aligned} \tag{3.47}$$

with:

$$K_1(z,z) = \exp\left\{-\sum_{n=1}^{\infty}\frac{1}{n}[E\{e^{-z[\min(R_n^1,\,R_n^2)-\sigma_n]}\}-1]\right\}. \tag{3.48}$$

The general equation (3.36) becomes (note the condition $A(0,0)=1$ to evaluate $\Phi(\lambda,\lambda)$):

$$\begin{aligned} A(z_1,z_2) &= W(z_1,z_2)\frac{\phi_1(z_1-\lambda)/K_1(z_1,\lambda)}{\phi_1(-\lambda)/K_1(0,\lambda)} \\ &+\frac{z_2-z_1}{\lambda}\,\frac{\phi_1(z_1-z_2)\phi_2(z_2)\phi_0(-z_1)}{1-\phi_1(z_1-z_2)\phi_2(z_2)\phi_0(-z_1)}\,\phi_0(-z_2)A(\lambda,z_2), \end{aligned} \tag{3.49}$$

with, by deduction from (3.42), for the case where *the first stage is empty*:

$$A(\lambda,z_2) = \left[E\{e^{-z_2V_1}\}\right](1-\rho_1)\frac{\phi_1(-\lambda\zeta(z_2))/K_1(\lambda(1-\zeta(z_2)),\lambda)}{\phi_1(-\lambda)/K_1(0,\lambda)}, \tag{3.50}$$

and

$$E\{e^{-z_2V_1}\} = \frac{1-\rho_2}{1-\rho_1}\,\frac{z_2}{z_2-\lambda(1-\zeta(z_2))}\,K_1(\lambda(1-\zeta(z_2)),z_2). \tag{3.51}$$

It can be shown that this is the characteristic function of the waiting time, V_1, at the second server of a customer who started a busy period at the first stage.

From (3.49) we deduce the following theorem:

THEOREM 1. *For the tandem queue $M/G/1\to G/1$, with finite supports for the service time distributions and in the stationary case, the second server access time (from the moment the customer enters the first queue) has the characteristic function*

$$A(z,z) = W(z)\frac{\phi_1(z-\lambda)/K_1(z,\lambda)}{\phi_1(-\lambda)/K_1(0,\lambda)} \tag{3.52}$$

with Re $z\geqslant 0$, *and where $W(z)$ is given by (3.47).*

In the neighbourhood of $z=0$, let: $W(z)=1-z\overline{W}+O(z^2)$, $A(z,z)=1-z\overline{A}+O(z^2)$. We deduce the following expression on taking logarithmic derivatives:

COROLLARY 1. *The mean second server access time (from the moment the customer enters the first queue) is:*

$$\overline{A} = \overline{W} - \frac{\phi_1'(-\lambda)}{\phi_1(-\lambda)} + \frac{[K_1'(z,\lambda)]_{z=0}}{K_1(0,\lambda)}, \tag{3.53}$$

with:

$$\overline{W} = \overline{W}_2 - [K_1'(z,z)]_{z=0}.$$

NOTES

1) If we always have $T_n^1 > T_n^2$, the function $\phi_1(z_1-u)\phi_2(u)$ is holomorphic for $\mathrm{Re}\, u < \delta$ and $H(z_1,z_2;x) = H(z_1,0;x)$. Assume that, in equation (3.30), we have $A(\lambda,u;x) = A(\lambda,0;x)$. The function under the integral sign is then holomorphic for $\mathrm{Re}\, u < \delta$. In this domain the only pole is $u=0$ whose residue is independent of z_2. We therefore have $A(z_1,z_2;x) \equiv A(z_1,0;x)$, which means that there is no waiting at the second stage.
2) If, on the contrary, we always have $T_n^1 < T_n^2$, we have the case already studied by BOXMA [3]. We then have, from (3.26), $H_1(z_1,z_2;x) = 1$ and (3.52) becomes:

$$A(z,z) = W_2(z)\frac{\phi_1(z-\lambda)}{\phi_1(-\lambda)}, \tag{3.54}$$

which is in agreement with Boxma's results. In the particular case of constant service times, [15] and [18] give a detailed analysis with particular attention paid to the relationship between the busy periods at each stage.

3.6. Example

We close this section with a practical example. Consider the expressions:

$$\phi_1(z_1) = e^{-T_1 z_1}, \quad \phi_2(z_2) = q_2 e^{-T_2' z_2} + p_2 e^{-T_2'' z_2}, \tag{3.55}$$

with: $p_2 \geq 0$, $p_2 + q_2 = 1$, $T_2' < T_1$ and $T_2'' > T_1$, where the quantities p_2, q_2, T_1, T_2' and T_2'' are constants. We deduce from (3.44):

$$K_1(z_1,z_2) = \exp\left\{-\sum_{n=1}^{\infty}\frac{1}{n}[e^{-nT_1 z_1}[\phi_0(-z_1)]^n G_n(z_2) - 1]\right\}, \tag{3.56}$$

with

$$G_n(z_2) = \sum_{k=0}^{n}\binom{n}{k} p_2^{n-k} q_2^k e^{[(n-k)(T_1-T_2'')+k(T_1-T_2')]^+ z_2}. \tag{3.57}$$

Expression (3.46) becomes:

$$W_2(z) = (1-\rho_2)\frac{1-\phi_0(-z)}{1-\phi_0(-z)[q_2 e^{-T_2' z} + p_2 e^{-T_2'' z}]}, \tag{3.58}$$

with

$$\rho_2 = \lambda[q_2 T_2' + p_2 T_2''] < 1. \tag{3.59}$$

4. Second stage arrivals of a second Poisson traffic stream

We now introduce an extra Poisson traffic stream, feeding into the second queue. The arrival intensity of this new traffic stream is denoted by λ'. The system is now as illustrated in Fig. 2 and we assume that the calls of both traffic streams are mixed together in one queue for the second server with service in arrival order without regard to the traffic stream concerned.

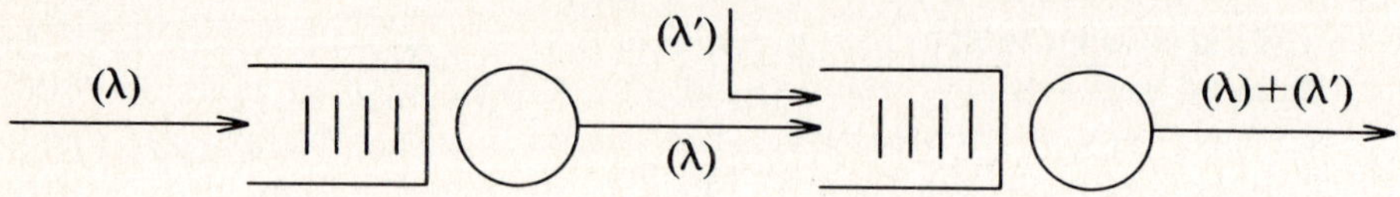

Figure 2

We again aim to evaluate the second server access time for a customer of the first traffic stream of arrival intensity λ. To this end, we evaluate the additional amount of work to be added to T_{n-1}^2 arriving at the second server during an arrival interval of the first traffic stream defined by expression (3.7) for Y_{n-1}^2. To conduct this evaluation, when we apply the expectation operator to

$$e^{\zeta Y_{n-1}^2} = e^{\zeta [T_n^1 + (Y_{n-1} - s_{n-1}^1)^+]},$$

we must make the substitution

$$\zeta \to \zeta + \lambda'(1-s)$$

to get the generating function of s for the number of arrivals in the new traffic stream.

To evaluate the amount of work to be added to T_{n-1}^2, we replace s by $\phi_2(\zeta)$ and we note in (3.12) the grouping $\exp(\zeta Y_{n-1}^2 - \zeta T_{n-1}^2)$. We deduce the final substitution

$$\zeta \to \zeta - \lambda'(1-\phi_2(\zeta)) \tag{4.1}$$

to take account of the term $(-\zeta)$ in $e^{-\zeta T_{n-1}^2}$.

This substitution applies to (3.12) for $e^{\zeta T_n^1}$, but also to (3.10) for

$$e^{\zeta [Y_{n-1} - s_{n-1}^1]^+},$$

and therefore to the denominator $(\zeta - \xi)$ of (3.12). In other words, in (3.14) we make the substitutions:

$$\phi_1(z_1 - \zeta) \to \phi_1(z_1 + \lambda'(1-\phi_2(\zeta)) - \zeta),$$
$$\frac{d\xi}{\xi - \zeta} \to \frac{d\xi}{\xi - \zeta + \lambda'(1-\phi_2(\zeta))}. \tag{4.2}$$

The general integral equation of the system (3.17) becomes, for $\mathrm{Re}\, z_2 > 0$:

$$A(z_1,z_2;x)-\frac{x}{2\pi i}\int\limits_{+0}[\phi_1(z_1+\lambda'(1-\phi_2(\zeta))-\zeta)\phi_2(\zeta)]\times$$
$$\phi_0(-z_1)A(z_1,\zeta;x)\frac{z_2 d\zeta}{(z_2-\zeta)\zeta} = D(z_1,z_2;x), \tag{4.3}$$

with

$$D(z_1,z_2;x) = \phi_1(z_1)$$
$$+\frac{x}{2\pi i}\int\limits_{+0}[\phi_1(z_1+\lambda'(1-\phi_2(\zeta))-\zeta)\phi_2(\zeta)\phi_0(-\zeta+\lambda'(1-\phi_2(\zeta)))]\times \tag{4.4}$$
$$\frac{\zeta-\lambda'(1-\phi_2(\zeta))-z_1}{\lambda}\phi_0(-z_1)A(\lambda,\zeta;x)\frac{z_2 d\zeta}{(z_2-\zeta)\zeta}.$$

The end result is then obtained as in Section 3.

5. Exit from the first stage of part of the first traffic stream of intensity λ

The system is now as illustrated in Fig. 3.

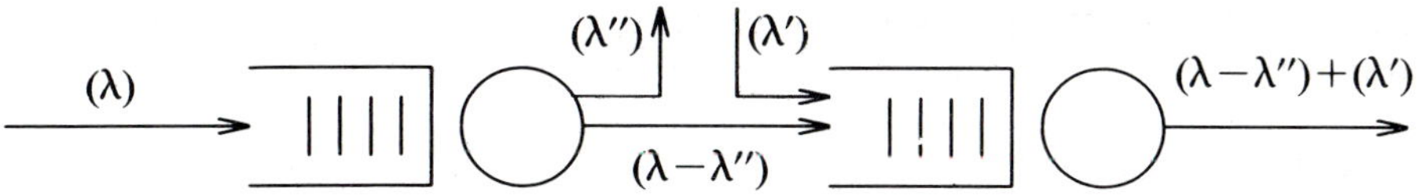

FIGURE 3

The intensity of the total traffic stream carried through the second server is:

$$\rho_2 = \Lambda\bar{T}_2=[(\lambda-\lambda'')+\lambda']\bar{T}_2. \tag{5.1}$$

The second server access time is in no way changed for the traffic $(\lambda-\lambda'')$ if we assume customers of traffic (λ'') also wait to be served by the second server, but with a service time equal to zero.

In the equations (3.36) or (3.45) we may let:

$$\phi_2(z)\rightarrow\frac{\lambda''}{\lambda}+\frac{\lambda-\lambda''}{\lambda}\phi_2(z). \tag{5.2}$$

In the case of Section 4, we do not change $\phi_2(\zeta)$ in the substitution (4.1).

The general equation of the system, deduced from (4.3) and (4.4), becomes:

$$A(z_1,z_2;x)-\frac{x}{2\pi i}\int\limits_{+0}[\phi_1(z_1+\lambda'(1-\phi_2(\zeta))-\zeta)\left[\frac{\lambda''}{\lambda}+\frac{\lambda-\lambda''}{\lambda}\phi_2(\zeta)\right]]\times$$
$$\phi_0(-z_1)A(z_1,\zeta;x)\frac{z_2 d\zeta}{(z_2-\zeta)\zeta} = D(z_1,z_2;x), \tag{5.3}$$

with

$$D(z_1, z_2; x) = \phi_1(z_1) + \frac{x}{2\pi i} \int_{+0} [\phi_1(z_1 + \lambda'(1-\phi_2(\zeta)) - \zeta) \times$$

$$\left[\frac{\lambda''}{\lambda} + \frac{\lambda - \lambda''}{\lambda} \phi_2(\zeta)\right] \phi_0(-\zeta + \lambda'(1-\phi_2(\zeta)))] \times \tag{5.4}$$

$$\frac{\zeta - \lambda'(1-\phi_2(\zeta)) - z_1}{\lambda} \phi_0(-z_1) A(\lambda, \zeta; x) \frac{z_2 d\zeta}{(z_2 - \zeta)\zeta}.$$

The end result is then obtained as in Section 3. But, in (3.32), the pole $u = \lambda$ of $\phi_0(-u)$ becomes the pole $u = \lambda_0$ of

$$\phi_0(-u + \lambda'(1-\phi_2(u))),$$

that is the zero of the equation

$$[(\lambda + \lambda') - z] - \lambda' \phi_2(z) = 0. \tag{5.5}$$

If we write $z_2(x, \lambda)$ for the zero of (3.38), we deduce, for Re $z_2 > 0$:

$$\lambda_0 = z_2 \left[\frac{\lambda'}{\lambda + \lambda'}, \lambda + \lambda'\right]. \tag{5.6}$$

(3.21) becomes, for Re $\zeta >$ Re z_2:

$$H_1(z_1, z_2; x) = \exp\left\{\frac{-1}{2\pi i} \int_{+\delta} \ln(1 - xN(z_1, \zeta)) \frac{z_2 d\zeta}{(z_2 - \zeta)\zeta}\right\}, \tag{5.7}$$

with

$$N(z_1, z_2) = \phi_1(z_1 + \lambda'(1-\phi_2(z_2)) - z_2) \left[\frac{\lambda''}{\lambda} + \frac{\lambda - \lambda''}{\lambda} \phi_2(z_2)\right] \phi_0(-z_1). \tag{5.8}$$

And, similarly to (3.43) for $x = 1$, we let:

$$K_1(z_1, z_2) = (1 - \phi_0(-z_1)\phi_1(z_1)) H_1(z_1, z_2). \tag{5.9}$$

(3.45) becomes, in the stationary case:

$$W(z_1, z_2) = K \frac{z_2}{z_2 - \lambda_0} \frac{K_1(z_1, z_2)}{1 - N(z_1, z_2)}.$$

The constant factor K is deduced from the relation $W(0, 0) = 1$. We get:

$$W(z_1, z_2) = [(1-\rho_2) + \lambda' \overline{T}_2 (1-\rho_1)] \left[\frac{\lambda_0}{\lambda}\right] \frac{z_2}{z_2 - \lambda_0} \frac{K_1(z_1, z_2)}{1 - N(z_1, z_2)}. \tag{5.10}$$

We deduce the general equation of the system *(stationary case)*:

$$A(z_1, z_2) = \frac{1-\rho_2}{1-\rho_2 + \lambda' \overline{T}_2 (1-\rho_1)} W(z_1, z_2) \frac{\phi_1(z_1 - \lambda)/K_1(z_1, \lambda_0)}{\phi_1(-\lambda)/K_1(0, \lambda_0)}$$
$$+ \frac{z_2 - \lambda'(1-\phi_2(z_2)) - z_1}{\lambda} \frac{N(z_1, z_2)}{1 - N(z_1, z_2)} \phi_0(-z_2 + \lambda'(1-\phi_2(z_2))) A(\lambda, z_2). \tag{5.11}$$

Similarly to (3.33) and (3.40) we introduce, for Re $z_2>0$, the zero $\zeta(z_2)$ of the equation in ζ:

$$\zeta = \phi_1(\lambda(1-\zeta)+\lambda'(1-\phi_2(z_2))-z_2)\left[\frac{\lambda''}{\lambda}+\frac{\lambda-\lambda''}{\lambda}\phi_2(z_2)\right], \tag{5.12}$$

with $\zeta(0)=1$. Similarly to (3.50), we get:

$$A(\lambda, z_2) = \psi(z_2)\frac{\phi_1(-\lambda\zeta(z_2))/K_1(\lambda(1-\zeta(z_2)), \lambda_0)}{\phi_1(-\lambda)/K_1(0, \lambda_0)}, \tag{5.13}$$

with

$$\psi(z_2) = (1-\rho_2)\frac{\lambda_0}{\lambda}\frac{z_2}{z_2-\lambda_0}\times$$
$$\frac{\phi_0(-\lambda\zeta(z_2)-z_2+\lambda'(1-\phi_2(z_2)))}{\phi_0(-z_2+\lambda'(1-\phi_2(z_2)))}K_1(\lambda(1-\zeta(z_2)), z_2). \tag{5.14}$$

The general equation (5.11) is well defined. We deduce the following theorem:

Theorem 2. *In the stationary case for Fig. 3, the characteristic function of the second server access time, for the traffic intensity* $(\lambda-\lambda'')$ *and from the moment the customer enters the first queue, is:*

$$A(z, z) = \frac{1-\rho_2}{1-\rho_2+\lambda'\overline{T}_2(1-\rho_1)}W(z, z)\frac{\phi_1(z-\lambda)/K_1(z, \lambda_0)}{\phi_1(-\lambda)/K_1(0, \lambda_0)}$$
$$-\frac{\lambda'}{\lambda}(1-\phi_2(z))\frac{N(z, z)}{1-N(z, z)}\phi_0(-z+\lambda'(1-\phi_2(z)))A(\lambda, z). \tag{5.15}$$

Due to the term $A(\lambda, z)$, this expression is not as simple as the expression (3.52) of Section 3. (5.10) and (5.15) define the Poisson traffic combination in two successive nodes of a queueing network, in the stationary case.

The case considered in [15] corresponds to $\lambda''=0$ with constant service times and $T_2 \geqslant T_1$. But we may note that the pole $u=\lambda_0$ of $\phi_0(-u+\lambda'(1-\phi_2(u))$ has been omitted, corresponding to the supplementary amount of work due to an arrival of the stream (λ') when the first stage is empty. A physical reasoning is not as sure as an analytical method.

References

1. B.J. Beukelman, J.W. Cohen (1957). Call congestion of transposed multiples. *Philips Telecommunication Review 17,* 145-154.
2. J.P.C. Blanc, R. Iasnogorodski, Ph. Nain (1986). *Analysis of the M/GI/1→/M/1 Queueing Model,* Report INRIA, Sophia Antipolis, to appear in *Queueing Systems.*
3. O.J. Boxma (1979). Two queues in series with non-overlapping service times. *Proc. 9th ITC,* Torremolinos.
4. J.W. Cohen (1956). Certain delay problems for a full availability trunk group loaded by two traffic sources. *Proc. 1st ITC (1955) and Communi-*

cation News (Philips, Holland) 16, 105-113.

5. J.W. COHEN (1957). Basic problems of telephone traffic theory and the influence of repeated calls. *Philips Telecommunication Review 18,* 49-100.
6. J.W. COHEN (1957). The generalized Engset formulae. *Philips Telecommunication Review 18,* 158-170.
7. J.W. COHEN (1957). The full availability group of trunks with an arbitrary distribution of the inter-arrival times and a negative exponential holding time distribution. *Wis- en Natuurk. Tijdschr. Simon Stevin 26,* 169-181.
8. J.W. COHEN (1962). Derived Markov chains I, II, III. *Indag. Math. 24,* 55-92.
9. J.W. COHEN (1962). On derived and non-stationary Markov chains. *Theory of Prob. and Applications 7,* 402-423.
10. J.W. COHEN (1963). Applications of derived Markov chains in queueing theory. *Appl. Sci. Res. B 10,* 269-303.
11. J.W. COHEN (1982). *The Single Server Queue,* 2nd ed., North-Holland, Amsterdam.
12. P. LE GALL (1962). *Les Systèmes avec ou sans Attente et les Processus Stochastiques,* Publ. Dunod, Paris.
13. P. LE GALL (1977). Repeated calls and traffic engineering. *Commut. et Electron. 56,* Paris (and 2nd Bulgarian Conference on Telecommunications, Varna, Oct. 1976).
14. P. LE GALL (1984). The repeated call model and the queue with impatience. *3rd ITC Seminar on 'Fundamentals of Teletraffic Theory',* Moscow, paper 36, 278-289.
15. M.A. KAPLAN (1980). A two-fold tandem net with deterministic links and source interference. *Oper. Res. 28,* 512-526.
16. F. POLLACZEK (1957). Problèmes stochastiques posés par le phénomène de formation d'une queue d'attente à un guichet et par des phénomènes apparentés. *Mémorial des Sciences Mathématiques,* Publ. Gauthier Villars 136, Paris.
17. J. SACKS (1960). Ergodicity of queues in series. *Ann. Math. Statist. 31,* 579-588.
18. M. SHALMON, M.A. KAPLAN (1984). A tandem network of queues with deterministic service and intermediate arrivals. *Oper. Res. 32,* 753-773.
19. R. SYSKI (1986). *Introduction to Congestion Theory in Telephone Systems,* Oliver and Boyd, London; 2nd ed. North-Holland, Amsterdam.

PART IV
NETWORKS AND COMPUTERS

A Recursive Aggregation-Disaggregation Method to Approximate Large-Scale Closed Queueing Networks with Multiple Job Types

Jan van Doremalen
Centre for Quantitative Methods, Nederlandse Philipsbedrijven BV
P.O. Box 518, NL-5600MD Eindhoven, The Netherlands

Jaap Wessels
Department of Mathematics and Computing Science,
Eindhoven University of Technology,
P.O. Box 513, NL-5600MB Eindhoven, The Netherlands

This paper deals with a new approximation method for large scale queueing network models with multiple job types. In recent years mean value oriented approximation algorithms have received much attention. Most of the approximation methods are based on decomposition and aggregation arguments and use iteration to obtain a fixed point of an implicitly defined set of non-linear equations for the relevant performance measures, such as mean response times, throughputs and mean queue lengths. In this paper a recursive aggregation-disaggregation method is introduced to bypass the computational problems involved in evaluating the standard multi-dimensional recursive schemes associated with exact mean value analysis in separable queueing networks with multiple job types. As a side result we study the influence of Pollaczek-Khintchine type approximations for the mean response times at first-in first-out single server queues with non-exponential service demand distributions. The power of the method is tested with a closed central server model involving multiple central processors, disk units and job types.

1. Introduction

Networks of queues are used in many areas to model and analyse real-life systems, as for instance computer systems, communication networks, transportation systems, harbour facilities, flexible manufacturing systems and production lines. The use of such models is justified by a common characteristic of such systems: they can be viewed as a collection of interconnected resources providing service to groups of customers with specific job requirements. The resources have finite capacities or service rates and, consequently, waiting lines of jobs will be formed.

Stochastic models and particularly (semi) Markov processes are a widely used tool for the analysis of queueing network systems. Under certain ergodicity conditions the equilibrium distribution of the Markov process may be obtained as the unique strictly positive and normalized solution of a set of linear equations relating the equilibrium probabilities. If a queueing network model apart from these ergodicity conditions satisfies so-called separability conditions, the solution attains an attractive product-form. These models form the class of product-form or separable queueing network models.

The search for separability in queueing networks started in JACKSON [21]. Noteworthy papers in this line are GORDON and NEWELL [18] and BASKETT, CHANDY, MUNTZ and PALACIOS [2]. Important theoretical extensions are covered in, for example, LAM [26], KELLY [22], COHEN [9] and HORDIJK AND VAN DIJK [19, 20].

Apart from the fact that for separable queueing network models analytical expressions have been obtained for the solution of the equilibrium equations, it has been shown that important system characteristics may be computed in a relatively simple and efficient way.

In this paper we shall be concerned with the computation of relevant performance characteristics in so-called closed queueing networks, i.e. for each job type a fixed number of jobs proceeds through the network and the jobs neither enter nor leave the system. For such networks the product-form establishes the equilibrium distribution up to a normalization constant. The evaluation of this constant causes computational problems, as it involves a summation of the unnormalized equilibrium probabilities over the complete state space.

The two main procedures for solving this problem are known as the convolution method (CM) and mean value analysis (MVA), which lead to the CM algorithm and the MVA algorithm respectively. Recently, a third convolution-like procedure has been developed known under the name of recursion by chain algorithm (RECAL). This approach almost immediately found its mean value sister in the mean value analysis by chain algorithm (MVAC).

The CM algorithm, introduced for the simplest situation in BUZEN [7], is a recursive procedure for the computation of the normalization constant. It appears that the system characteristics can be expressed in the recursively computed values and thus may be evaluated efficiently. In REISER and KOBAYASHI [34] the procedure has been extended to cover a large class of separable queueing network models.

The MVA algorithm is based on a set of recursive relations between important system characteristics, such as mean response times, mean queue lengths and throughputs. It has been introduced in REISER [32] and extensions can be found in REISER and LAVENBERG [35], ZAHORJAN and WONG [43], KRZESINSKI, TEUNISSEN and KRITZINGER [25] and BRUELL, BALBO and AFSHARI [6].

Whereas CM and MVA are essentially based on a recursion in the number of job types, RECAL and MVAC are based on a recursion in the number of workstations. RECAL is described in CONWAY and GEORGANAS [10] and MVAC in CONWAY, DE SOUZA E SILVA and LAVENBERG [11].

For notes on the implementation of the CM and MVA algorithms we refer to BRUELL and BALBO [5], REISER [33] and ZAHORJAN and WONG [43].

Though separable queueing network models are an important tool to study queueing network systems, the application area is restricted. Most realistic models for practical problems do not satisfy the separability conditions, whereas separable models tend to be very large and, therefore, cannot be evaluated using standard methods. For both types of problems the use of approximation methods seems a natural way out.

The importance of the MVA relations in the development of simple and efficient approximation methods lies in the fact that they have an attractive intuitive interpretation in terms of Little's formula (cf. LITTLE [28] and STIDHAM [40]) and an arrival theorem, that couples the equilibrium distributions at the departure and arrival moments of individual jobs with the equilibrium distribution of the network model with one job less (cf. LAVENBERG and REISER [27] and SEVCIK and MITRANI [37]). An extra property of MVA based approximation methods is that they, in general, can be implemented very easily in existing MVA codes.

Since the presentation of MVA a wave of papers has appeared on the subject, see e.g. DE SOUZA E SILVA, LAVENBERG and MUNTZ [39] and VAN DOREMALEN and WESSELS [13] for an overview and appraisal.

In this paper we shall be concerned with a combination of the modelling and computational problems. The line of argument is as follows. We start with an introduction of the MVA algorithm for a class of closed queueing networks with multiple job types in Section 2. In Section 3 a heuristic extension of the basic MVA relation for the mean response times at first-in first-out single servers is introduced. Our main contribution to the literature on queueing network models is the recursive aggregation-disaggregation method introduced in Section 4. The method is developed for the approximation of large scale closed queueing network models with multiple job types. The basic idea is to aggregate groups of job types. This leads to a substantial reduction of the number of dimensions in the multidimensional MVA scheme and, consequently, to a reduction in the computational complexity. In order to maintain the influences of the original job type structure, an elementary disaggregation step is performed in every recursion step of the reduced scheme. In Section 5 we investigate the merits of the method by means of a numerical example stemming from computer performance evaluation: a closed central server model with multiple central processors, disk units and job types.

Section 6 concludes with some final remarks, related topics and points of future research.

2. THE MVA ALGORITHM FOR CLOSED QUEUEING NETWORK MODELS WITH MULTIPLE JOB TYPES

2.1. The queueing network model

Consider a closed queueing network with N workstations and R job types. The K_r jobs of type r proceed through (part of) the network in accordance with a Markov routing defined by an irreducible stochastic matrix P_r, $r = 1, 2, \ldots, R$. The relative visiting frequencies $f_{n,r}$ of jobs of type r at the workstations $n = 1, 2, \ldots, N$ are defined as the limiting distribution of the discrete Markov chain generated by the matrix P_r. The service demands of type r jobs at workstation n are stochastically independent random variables with finite mean $w_{n,r}$ and variance $\sigma_{n,r}^2$. We allow for three service disciplines: first-come first-served (FCFS), processor sharing (PS) and infinite server (IS). It is assumed that the service rates at each of the workstations are fixed and normalized to unity and that the service demands are exponentially distributed with the same mean for each job type at FCFS workstations.

The relevant steady state system characteristics are:

$S_{nr}(\mathbf{K})$: expected sojourn time of a type r job at workstation n,
$\Lambda_{nr}(\mathbf{K})$: throughput of type r jobs at workstation n, and
$L_{nr}(\mathbf{K})$: expected number of type r jobs at workstation n,
where the argument $\mathbf{K}$ denotes the dependence on the population vector that is introduced as $\mathbf{K}=(K_1,\ldots,K_R)$.

2.2. The MVA algorithm

In Reiser and Lavenberg [35] it is proved that the system characteristics are related through the following so-called mean value relations. These relations constitute the kernel of the MVA algorithm.

The first relation couples the expected sojourn times at population vector $\mathbf{K}$ with the mean number of jobs at the population vectors $\mathbf{K}-\mathbf{e}_1$, $\mathbf{K}-\mathbf{e}_2,\ldots,\mathbf{K}-\mathbf{e}_R$, where $\mathbf{e}_r$ denotes the r-th unit vector,

$$S_{n,r}(\mathbf{K}) = \begin{cases} \left(\sum_{l=1}^{R} L_{n,l}(\mathbf{K}-\mathbf{e}_r)+1\right) w_{n,r}, & \text{if } n \text{ is FCFS or PS} \\ w_{n,r}, & \text{if } n \text{ is IS.} \end{cases} \tag{2.1}$$

The relation (2.1) is based on the arrival theorem mentioned in Section 1 (cf. [27,37]).

The second relation is a version of Little's formula and couples throughputs and expected sojourn times,

$$\Lambda_{n,r}(\mathbf{K})=\frac{f_{n,r}K_r}{\sum_{m=1}^{N} f_{m,r}S_{m,r}(\mathbf{K})}. \tag{2.2}$$

The last relation is a direct application of Little's formula and couples mean numbers of jobs, mean sojourn times and throughputs,

$$L_{n,r}(\mathbf{K}) = \Lambda_{n,r}(\mathbf{K})S_{n,r}(\mathbf{K}). \tag{2.3}$$

For more details on the interpretation of the MVA relations we refer to Reiser [32], Reiser and Lavenberg [35] and Van Doremalen [15]. Bard [1] uses these interpretations for constructing approximations in non-separable systems.

2.3 Notes on the complexity of the MVA algorithm

The MVA relations (2.1), (2.2) and (2.3) constitute a recursive algorithm for the evaluation of the system characteristics over a range of population vectors. The initialization is by $L_{n,r}(\mathbf{0})=0$ for each couple n and r. The recursion runs through all the vectors in the range of $(0, 0, \ldots, 0)$ up till $(K_1, K_2 \ldots, K_R)$. Hence, the computational complexity is largely determined by the fact that $\Pi_{r=1}^{R}(K_r+1)$ recursion steps are necessary. For larger values of R, $K_1, \ldots, K_R$ this will prohibit an exact evaluation of the scheme within a reasonable amount of time. Furthermore, the storage requirements also grow exponentially with R, as the larger part of all computed information must be stored.

3. Non-identical and non-exponential service demands at FCFS single server queues

Consider the network introduced in Section 2. However, now the service demands of type r jobs at a FCFS workstation n are distributed with mean $w_{n,r}$ and variance $\sigma_{n,r}^2$ without extra conditions. This relaxation of the model assumptions leads to a violation of the separability conditions and, consequently, the MVA relations are no longer valid.

In this section we shall adopt an approximation method advocated in Reiser [31] that lends itself to a direct implementation in any existing MVA algorithm. First we shall describe the method for lucidity of presentation. Then we will make some comments.

A first approximation for the evaluation of expected sojourn times in the non-separable queueing network at FCFS single server queues would be the following. Using an adjustment of (2.1)

$$S_{n,r}(\mathbf{K}) = \sum_{l=1}^{R} L_{n,l}(\mathbf{K}-\mathbf{e}_r)w_{n,l}+w_{n,r}, \tag{3.1}$$

as if the arrival theorem holds (which is not the case) and, consequently, a type r job would see, upon its arrival at workstation n, an expected number of $L_{n,l}(\mathbf{K}-\mathbf{e}_r)$ jobs of type l in front of it. The mean service demand of a type l job is $w_{n,l}$. So, summing these mean service demands and adding the mean service demand of the job itself, yields relation (3.1).

The approximation may be improved upon. If a type l job is being served upon the arrival moment of a type r job, its expected residual service demand, in general, will not equal $w_{n,l}$. It seems a better guess to apply the mean residual life time formula at an arbitrary time instant. So, if a job of type l is in service, we use

$$\frac{\sigma_{n,l}^2+w_{n,l}^2}{2w_{n,l}} \tag{3.2}$$

to approximate the expected residual service demand at the arrival moment of a type r job.

Note that $\Lambda_{n,l}(\mathbf{K}-\mathbf{e}_r)\,w_{n,l}$ is an obvious approximation for the probability that a type r job finds upon its arrival at workstation n a type l job in service.

This yields the following improvement of relation (3.1).

$$S_{n,r}(\mathbf{K}) = \sum_{l=1}^{R}(L_{n,l}(\mathbf{K}-\mathbf{e}_r)-\Lambda_{n,l}(\mathbf{K}-\mathbf{e}_r)w_{n,l})w_{n,l} + \sum_{l=1}^{R}\Lambda_{n,l}(\mathbf{K}-\mathbf{e}_r)w_{n,l}\frac{\sigma_{n,l}^2+w_{n,l}^2}{2w_{n,l}}+w_{n,r}. \tag{3.3}$$

The similarity between this formula and the Pollaczek-Khintchine formula for M/G/1 queues should be observed, cf. OLIVER [30] and KLEINROCK [23,24] for similar reasonings to obtain the expected sojourn times in M/G/1 queues.

Observe that (3.3) with (2.2) and (2.3) forms a strictly recursive scheme for the evaluation of approximations of the system characteristics. It is easily verified that, in a straightforward way, the scheme can be implemented in any existing MVA algorithm.

In REISER [31] it has been reported that the approximation performs quite well, especially for values of the coefficient of variation $v_{n,r}=\sigma_{n,r}/w_{n,r}$ in the range of $0 \leqslant v_{n,r} \leqslant 1$. Recently, in BONDI and WHITT [3], the influence of the service demand variability in closed queueing networks with FCFS single server queues has been analysed in more detail. Bondi and Whitt compare four approximation methods, namely

1. Reiser's MVA-based approximation (REISER [31]),
2. the extended product form method (SHUM and BUZEN [38]),
3. the generalized product form method (TRIPATHI [41]) and
4. Marie's device complement method (MARIE [29]).

The authors conclude that Marie's method performs best. However, Reiser's method, especially for smaller values of the coefficient of variation, performs reasonably well also. If we consider the question of implementation, especially for larger closed networks with many job types, the MVA adjustment proposed by Reiser has to be advocated. Our numerical results support the findings in REISER [31] and BONDI and WHITT [3].

4. A RECURSIVE AGGREGATION-DISAGGREGATION METHOD

4.1. Introduction

The computational complexity and storage requirements of the exact MVA algorithm introduced in Section 2 prohibit the evaluation of closed queueing network models with a large number of job types. The complexity is largely caused by the fact that the number of recursion steps grows exponentially with the number of job types.

In the literature much attention has been paid to this complexity problem and a number of approximation methods have been proposed and analysed. Three lines of research can be discerned.

The first approach is to avoid the recursion and to concentrate on the evaluation of the performance measures at the desired population vectors. A typical representative of this approach is the widely appraised Schweitzer method, introduced in SCHWEITZER [36], and its refinements suggested in

Chandy and Neuse [8].

The second approach is to decompose the queueing network model by analysing adjusted queueing network models per job type. The mutual influence of the job types is approximated e.g. by adjusting the operation of the service centres by slowing down the service rates as in Reiser [31] and Reiser and Lavenberg [35] or by adjusting the MVA scheme as in Van Doremalen [15].

The third approach is to aggregate the groups of job types in some way or another. The standard way to do so is to define a smaller number of new job types and to assign each original job type to one of the newly defined types. The problem is then to assign job type characteristics to the newly formed job types. In most instances this is done in an iterative way.

Typical of the methods presented in the literature is that they are iterative in nature. In fact, most methods can be viewed as successive approximation methods to determine a fixed point of an implicitly defined set of non-linear equations.

We shall present a recursive aggregation-disaggregation method that is based on the aggregation of job types, but that tries to preserve the typical features of the distinct job types in the recursive disaggregation. We start with a description of the aggregation method in its most extreme form: a total aggregation. Then a refinement, the partial aggregation, will be discussed. Eventually, a so-called depth-improvement will be introduced.

4.2. A total aggregation method

We shall introduce an approximation method based on a recursive aggregation-disaggregation of the job types. The aggregation concentrates on the sojourn time relation (2.1). The disaggregation concentrates on the throughput relation (2.2). Starting point of the analysis is the exact MVA algorithm of Subsection 2.2. In this we assume exponentially distributed service times, but in a similar way as in Section 3 the method can be extended to the non-exponential case.

The complexity of the MVA algorithm is caused by the multidimensional recursion in the population vector. We suggest to aggregate the set of job types into a single type of job. As a consequence the adjusted MVA algorithm will run through the integers $0, \ldots, T$, where

$$T = \sum_{r=1}^{R} K_r. \tag{4.1}$$

A recursion step starts with a sojourn time relation. The expected sojourn time with k jobs in the system, $k = 1, 2, \ldots, T$, is given by

$$S_{n,r}(k) = \begin{cases} \left(\sum_{l=1}^{R} L_{n,l}(k-1)+1\right)w_{n,r}, & n \text{ is FCFS or PS} \\ \\ w_{n,r}, & n \text{ is IS}, \end{cases} \tag{4.2}$$

for a type r job at workstation n.

The basic idea of the disaggregation is that the jobs which are lumped together in a single type in the aggregation step, may be distributed over the original types in a disaggregation step. Let, therefore, $\alpha_r, r=1,\ldots,R$ be the fraction of the total number of jobs being of type r, i.e.

$$\alpha_r = \frac{K_r}{T}. \tag{4.3}$$

Then we approximate the throughput of type r jobs at workstation n by

$$\Lambda_{n,r}(k) = \frac{\alpha_r f_{n,r} k}{\sum_{m=1}^{N} f_{m,r} S_{m,r}(k)}. \tag{4.4}$$

The last relation in the MVA-algorithm remains unchanged, i.e.

$$L_{n,r}(k) = \Lambda_{n,r}(k) S_{n,r}(k). \tag{4.5}$$

The algorithm is easy to implement and has the considerable advantage over most existing methods that it is non-iterative.

4.3. Partial aggregation

In this subsection we discuss a refinement of the total aggregation method. It seems a good idea to use a partitioning of the set of job types. Each of the created subsets forms the basis for an aggregate closed job type.

A partitioning of C, where C is defined as $\{1,\ldots,R\}$, in I subsets, $C_1,\ldots,C_I$ say, is the basis of the partial aggregation method. The sets C_i are disjoint and their union is C. With the subset C_i we associate a population size B_i, where

$$B_i = \sum_{r\in C_i} K_r. \tag{4.6}$$

This defines a population vector $\mathbf{B}=(B_1,\ldots,B_I)$ for the aggregated model. We introduce for all $r\in C_i$ the quantity α_r as the fraction of the number of jobs in the aggregate C_i which is of type r, i.e.

$$\alpha_r = \frac{K_r}{B_i}. \tag{4.7}$$

The approximating MVA-algorithm again is an adjusted mean value analysis scheme involving an I-dimensional recursion on the population vector $\mathbf{B}$.

The recursion is defined by the following four relations at an arbitrary population vector $\mathbf{b}=(b_1,\ldots,b_I)$ in the range of $(0,\ldots,0)$ through $(B_1,\ldots,B_I)$. For all $r\in C_i$, $i=1,\ldots,I$ and $n=1,\ldots,N$ evaluate

$$S_{n,r}(\mathbf{b}) = \begin{cases} \left(\sum_{l=1}^{R} L_{n,l}(\mathbf{b}-\mathbf{e}_i)+1\right) w_{n,r}, & n \text{ is FCFS or PS} \\ \\ w_{n,r}, & n \text{ is IS,} \end{cases} \tag{4.8}$$

$$\Lambda_{n,r}(\mathbf{b}) = \frac{\alpha_r b_i f_{n,r}}{\sum_{m=1}^{N} f_{m,r} S_{m,r}(\mathbf{b})}, \tag{4.9}$$

$$L_{n,r}(\mathbf{b}) = \Lambda_{n,r}(\mathbf{b}) S_{n,r}(\mathbf{b}). \tag{4.10}$$

Note that the fractions α_r remain constant during the recursion. In fact, it would be difficult to define a local value for these fractions.

The recursion runs through all the vectors in the range of $(0, \ldots, 0)$ through $(B_1, \ldots, B_I)$ and, consequently,

$$\prod_{i=1}^{I} (B_i + 1) \tag{4.11}$$

recursion steps are to be evaluated. The amount of work per recursion step is very similar to that in the original MVA-algorithm. As a consequence the total amount of work to be done will be much less than for the original MVA-algorithm if the value of I is not too large.

We have introduced the partial aggregation method. The problem that remains, is a good choice of a partitioning. The accuracy of the method will tend to be better for a more detailed partitioning, but a refined partitioning will cost more computation time and larger storage facilities than a simple one. One has to look for a partitioning in the range of $I = R$ corresponding with the exact MVA-algorithm, and $I = 1$ corresponding with the total aggregation, which yields the right balance between desired accuracy and acceptable computational costs.

The number of recursion steps gives a good indication of the computational complexity and may be used to give a prior estimate of the efficiency of a given partitioning.

The accuracy of the method for a given partitioning is hard to measure as no bounds are presently known for the resulting procedures (compare the final remark of the next subsection). Here, numerical experiments and intuition have to be combined to come up with guidelines. For more details we refer to Section 5.

4.4. Depth improvement

The total aggregation method is highly efficient and fairly accurate, even for quite substantial differences in the workloads of the respective types (see Section 5). However, if a higher accuracy is required and one is willing to spend extra computation time and storage facilities, a straightforward depth improvement may be considered.

The basic idea is to start with the computation of approximations via the total aggregation method for the behavioural characteristics at the population vectors $\mathbf{K} - \mathbf{e}_r$, $r = 1, \ldots, R$, and then to evaluate the last step in the recursive MVA-algorithm exactly. We shall refer to this improvement as a first order depth improvement. It is a special case of a range of improvements which we have called depth improvements.

The MVA-algorithm runs through all vectors in the range of $(0,\ldots,0)$ up to $(K_1,\ldots,K_R)$. With V we denote the set of all R-dimensional integer valued vectors in this range, i.e.

$$V = \left\{ \mathbf{k}=(k_1,\ldots,k_R) \,\middle|\, k_r \in \{0,\ldots,K_r\},\ r=1,\ldots,R \right\}. \tag{4.12}$$

The MVA-algorithm is recursive in the number of jobs of the distinct chains. So, it makes sense to introduce the sets $V(t)$, $t = 0, 1,\ldots, T$, as

$$V(t) = \left\{ \mathbf{k} \in V \,\middle|\, \sum_{r=1}^{R} k_r = t \right\} \text{ with } T = \sum_{r=1}^{R} K_r. \tag{4.13}$$

Observe that the sets $V(t)$ are disjoint and their union is the set V. These sets form the basis for an efficient implementation of the MVA-algorithm as well, cf. ZAHORJAN and WONG [43]. To evaluate the characteristics at the vectors in the set $V(t)$ the behavioural characteristics at the population vectors in the set $V(t-1)$ are needed only.

We now introduce the t-th order depth improvement. First, approximations via total aggregation for the behavioural characteristics at the population vectors in the set $V(T-t)$ are evaluated. Afterwards, an evaluation of the original MVA-algorithm is performed for the population vectors in the sets $V(T-t+1)$ through $V(T)$.

The depth improvement is a generally applicable method and it yields very satisfactory results.

For most purposes a first order depth improvement, with $t=1$, will suffice. An interesting use of higher order depth improvements is the Performance Bound Hierarchy method introduced in EAGER and SEVCIK [16,17]. The method generates upper and lower bounds on certain performance characteristics by an appropriate choice of initial sets of characteristics.

4.5. *Non-identical and non-exponential service demands*

So far, the computational methods have been developed under the service demand restrictions for separable networks. However, Section 3 shows how an MVA-algorithm can be adapted in the case of non-identical and non-exponential service demands at single server FCFS-stations. These adaptations do not lead to exact solutions, but the approximations appear to be reasonable (see Section 3). If necessary the methods described in this section can and will be analogously amended. In the numerical examples of Section 5 this is executed for the disks in the relevant cases.

5. A NUMERICAL EXAMPLE

5.1. *The test environment*

The power of the introduced class of approximation methods has been tested with numerical experiments. The basic model is a closed central server model with multiple central processor units (CPUs), a shared set of background storage devices (DISKs) and multiple job classes. Such models can be used to evaluate optimal configurations, job allocations and multi-programming levels.

In our examples the computer system comprises three CPUs and nine DISKs. The CPUs operate in a processor sharing (PS) fashion and the service discipline at the DISKs is first-come first-served (FCFS).

We discern three types of jobs. The jobs are characterized by their routing probabilities and their workloads at the DISKs. For job type i, $i=1,2,3$, the probability to visit DISK j, $j=1,\ldots,9$, after a visit to one of the CPUs is $P_{i,j}$. The probabilities are pictured in Table 1.

$P_{i,j}$	DISK UNIT								
	1	2	3	4	5	6	7	8	9
TYPE 1	.25	.25	.25	.05	.05	.05	.06	.04	.00
TYPE 2	.15	.15	.15	.10	.10	.10	.05	.05	.15
TYPE 3	.00	.00	.05	.05	.15	.15	.20	.20	.20

TABLE 1. *The routing probabilities*

Each job is associated with a single CPU. So, we can discern nine job classes. It is assumed that to each CPU a fixed number of jobs from each type has been allocated. $K_{i,j}$ is the number of jobs of type i, $i=1,2,3$, associated with CPU j, $j=1,2,3$, in the system. In Table 2 the three population vectors that we have used in our experiments have been shown.

Note that our definitions lead to a closed queueing network model with nine job chains.

$K_{i,j}$	case a	CPU			case b	CPU			case c	CPU		
	$T=16$	1	2	3	$T=20$	1	2	3	$T=26$	1	2	3
TYPE 1		3	1	1		4	2	1		4	2	2
TYPE 2		2	2	1		2	3	2		3	4	2
TYPE 3		1	2	3		2	2	2		3	3	3

TABLE 2. *The population vectors*

The service demands at the CPUs are deterministic. This assumption is relatively arbitrary. In the separable case it is unimportant, but in the other cases it might have some influence on the results. For the service demands at the DISKs we have tested deterministic demands (coefficient of variation 0), exponential demands (coefficient of variation 1) and hyperexponential demands (coefficient of variation 2). Those demand distributions have been tested with the four sets of mean service demands as shown in Table 3. Note that only those models are separable that have job type independent exponential service demands at the DISKs.

CASE A	CPU			DISKS	DISKS	DISKS	DISKS
	1	2	3	1,2,3	4,5	6,7	8,9
TYPE 1	15	15	15	30	50	60	80
TYPE 2	25	25	25	30	50	60	80
TYPE 3	40	40	40	30	50	60	80

CASE B	CPU			DISKS	DISKS	DISKS	DISKS
	1	2	3	1,2,3	4,5	6,7	8,9
TYPE 1	15	15	15	36	60	72	96
TYPE 2	25	25	25	36	60	72	96
TYPE 3	40	40	40	36	60	72	96

CASE C	CPU			DISKS	DISKS	DISKS	DISKS
	1	2	3	1,2,3	4,5	6,7	8,9
TYPE 1	10	15	15	30	50	60	80
TYPE 2	20	25	30	30	50	60	80
TYPE 3	30	40	50	30	50	60	80

CASE D	CPU			DISKS	DISKS	DISKS	DISKS
	1	2	3	1,2,3	4,5	6,7	8,9
TYPE 1	15	15	15	25	45	55	75
TYPE 2	25	25	25	30	50	60	80
TYPE 3	40	40	40	35	55	65	85

TABLE 3. *Mean service demands*

5.2. The numerical experiments

The following evaluation techniques have been used:

SIM Via a discrete event simulation the performance characteristics have been approximated. We have used a sub-run argument to obtain 90% confidence intervals.

MVA A direct implementation of the MVA algorithm with the Pollaczek-Khintchine adjustments at FCFS workstations. No aggregation involved.

TA Aggregation of all job types into a single type.

TA-FODI The first order depth improvement of TA.

PA-CPU A partial aggregation where all job types associated with a specific CPU have been aggregated. The number of job types has been reduced to three.

PA-TYPE A partial aggregation where all job types associated with a given disk-routing are aggregated. The number of job types has been reduced to three.

TABLE 4. *Mean and maximum relative error in the evaluation of the utilizations at the CPUs in percentages.*

COEFFICIENT OF VARIATION	$VC=0$		$VC=1$		$VC=2$	
	MEAN	MAX	MEAN	MAX	MEAN	MAX
PA-CPU	1.87	2.94	1.66	6.73	3.50	7.41
PA-TYPE	8.53	11.17	5.99	8.62	5.11	8.18
TA	9.69	13.14	7.36	10.33	6.52	9.72
TA-FODI	2.71	4.32	1.22	2.96	1.72	4.87

TABLE 4a.

NUMBER OF JOBS	$T=16$		$T=20$		$T=26$	
	MEAN	MAX	MEAN	MAX	MEAN	MAX
PA-CPU	2.50	7.38	2.48	7.41	2.05	6.61
PA-TYPE	4.75	11.17	6.71	10.98	6.24	10.62
TA	8.32	13.14	7.97	12.21	7.29	11.36
TA-FODI	1.78	4.87	2.02	4.25	1.85	4.32
MVA	1.38	5.70	--	--	--	--

TABLE 4b.

DEMAND AT DISKS	CASE A		CASE B		CASE C		CASE D	
	MEAN	MAX	MEAN	MAX	MEAN	MAX	MEAN	MAX
PA-CPU	2.23	7.41	3.38	7.38	2.58	6.73	2.70	6.80
PA-TYPE	7.37	10.54	4.75	8.15	6.77	11.17	8.28	10.98
TA	8.51	12.23	6.45	9.44	6.08	13.14	8.39	11.56
TA-FODI	2.04	4.10	1.83	4.87	1.66	3.94	1.99	4.32

TABLE 4c.

TABLE 5. *Mean and maximum relative error in the evaluation of the utilizations at the DISKs in percentages.*

COEFFICIENT OF VARIATION	$VC=0$		$VC=1$		$VC=2$	
	MEAN	MAX	MEAN	MAX	MEAN	MAX
PA-CPU	2.72	4.78	2.50	5.35	3.51	7.07
PA-TYPE	9.23	12.02	6.79	10.76	5.07	9.30
TA	10.34	12.98	8.11	11.74	6.49	10.42
TA-FODI	3.24	5.94	1.90	4.77	2.03	5.63

TABLE 5a.

NUMBER OF JOBS	$K=16$		$K=20$		$K=26$	
	MEAN	MAX	MEAN	MAX	MEAN	MAX
PA-CPU	3.05	7.07	2.89	5.88	2.79	7.04
PA-TYPE	7.05	11.76	7.19	12.02	6.86	11.05
TA	8.65	12.98	8.38	12.81	7.90	11.74
TA-FODI	2.29	5.27	2.36	5.40	2.51	5.94
MVA	1.78	5.63	--	--	--	--

TABLE 5b.

DEMAND AT DISKS	CASE A		CASE B		CASE C		CASE D	
	MEAN	MAX	MEAN	MAX	MEAN	MAX	MEAN	MAX
PA-CPU	2.73	7.04	3.22	7.07	3.06	6.05	2.63	5.26
PA-TYPE	7.82	11.23	5.63	9.01	6.78	12.02	7.90	11.85
TA	8.93	12.60	7.28	10.02	8.09	12.81	8.94	12.98
TA-FODI	2.55	5.63	1.90	4.91	2.27	4.76	2.83	5.40

TABLE 5c.

The MVA-algorithm has only been used for the cases with $T = 16$ in order to keep computation time within reasonable bounds.

The Tables 4 and 5 show the results of our numerical experiments. The performance measure at stake is the utilization of the CPUs. Pictured are the means and maxima of the relative differences in the results of the approximation methods and the simulation results in percentages. We have grouped the numerical results in accordance with three model criteria: 1. the coefficient of variation of the service demands at the DISKs (Table 4a), 2. the total number of jobs in the system (Table 4b) and 3. the mean service demand characteristics (Table 4c). For the DISKs the same results have been shown in the Tables 5a, 5b and 5c.

5.3. Conclusions

The results show that the refinement of the total aggregation method with a first order depth improvement (TA-FODI) and the partial aggregation with respect to CPU-routing (PA-CPU) both have a very good overall performance. Quite remarkable is the relative insensitivity with respect to the choice of system characteristics. Particularly, the coefficient of variation has no severe influence on the accuracy of the proposed methods (cf. Tables 4a and 5a). So, the deviations in the model from the exponential distribution ($VC = 1$) are well captured by the Pollaczek-Khintchine type approximations.

The TA-FODI approximation has to be recommended, as it is easier to implement and in general will require much less computation time.

Noteworthy is that the partial aggregation with respect to CPU-routing (PA-CPU) has a very good performance, whereas the results of the partial aggregation with respect to DISK-routing (PA-TYPE) are poor and can be compared with the total aggregation method (TA).

This leads to an important observation: a partial aggregation should be done in such a way that job types visiting approximately the same set of workstations are lumped together. This observation has far reaching consequences for the analysis of large queueing networks with many job types which, for example, are typical in the analysis of communication networks. In such networks partial aggregation has to be dissuaded, as the routing patterns of the job types will in general be highly disjoint.

In another context we have studied the performance of the aggregation method against other approximation methods (cf. VAN DOREMALEN [12], VAN DOREMALEN, WESSELS and WIJBRANDS [14] and WIJBRANDS [42]). It has appeared that the Schweitzer method (cf. SCHWEITZER [36]) with a first order depth improvement has the same performance characteristics as the total aggregation method with a first order depth improvement. It should be observed that straightforward Schweitzer gives relatively poor results. The more so, when the method is integrated in further adjustments of the MVA scheme, as for example the Pollaczek-Khintchine adjustments. Linearizer, the refinement of the Schweitzer method proposed in CHANDY and NEUSE [8], is slightly better than our TA-FODI method.

6. CONCLUDING REMARKS

The total aggregation method with a first order depth improvement offers an interesting alternative to well-known existing methods, such as the Schweitzer method (with a first order depth improvement) and the Chandy and Neuse refinements (cf. SCHWEITZER [36] and CHANDY and NEUSE [8]). Apart from the fact that the performance of the aggregation method resembles that of related methods, it has the advantage that no iteration is involved.

Furthermore, the method can be efficiently implemented on a personal computer. In fact, our tests have been run on a Philips P-3200, an IBM-AT compatible PC, in a TURBO-PASCAL environment.

Finally, we would like to note that the method can be easily extended to cover approximation ideas in different situations. For example, in VAN

Doremalen, Wessels and Wijbrands [14] and Wijbrands [42] approximations for priority queues have been developed and successfully tested. At the moment we are testing M/G/c adjustments in the MVA scheme based on the M/G/c approximations advocated in Boxma, Cohen and Huffels [4]. The first results are promising.

Acknowledgement
We thank our student Jeroen Kuipers for his help in executing and evaluating the numerical experiments and our colleague Rudo Wijbrands for his advice and helpful discussions.

References

1. Y. Bard (1979). Some extensions to multiclass queueing network analysis. M. Arato, A. Butrimenko, E. Gelenbe (eds.). *Fourth Int. Symp. on Modelling and Performance Evaluation of Computer Systems,* North-Holland, Amsterdam, 51-62.
2. F. Baskett, K.M. Chandy, R. Muntz, F. Palacios (1975). Open, closed and mixed networks of queues with different classes of customers. *J. Assoc. Comput. Mach. 22,* 248-260.
3. A.B. Bondi, W. Whitt (1986). The influence of service-time variability in closed networks of queues. *Performance Evaluation 6,* 219-234.
4. O.J. Boxma, J.W. Cohen, N. Huffels (1979). Approximations of the mean waiting time in an M/G/s queueing system. *Oper. Res. 27,* 1115-1127.
5. S.C. Bruell, G. Balbo (1980). *Computational Algorithms for Closed Queueing Networks,* North-Holland, Amsterdam.
6. S.C. Bruell, G. Balbo, P.V. Afshari (1984). Mean Value Analysis of mixed multiple class BCMP-networks with load dependent service stations. *Performance Evaluation 4,* 241-260.
7. J.P. Buzen (1973). Computational algorithms for closed queueing networks with exponential servers. *Commun. ACM 16,* 527-531.
8. K.M. Chandy, D. Neuse (1982). Linearizer: A heuristic algorithm for queueing network models of computing systems. *Commun. ACM 25,* 126-134.
9. J.W. Cohen (1979). The multiple phase service network with generalized processor sharing. *Acta Informatica 12,* 245-284.
10. A.E. Conway, N.D. Georganas (1986). A new method for computing the normalization constant of multiple-chain queueing networks. *INFOR 24,* 184-198.
11. A.E. Conway, E. De Souza e Silva, S.S. Lavenberg (1988). Mean value analysis by chain of product form queueing networks. To appear in *IEEE Trans. Comput.*
12. J.B.M. van Doremalen (1984). Mean Value Analysis in multichain queueing networks: an iterative approximation. H. Steckhan, W. Buhler, K.E. Jager, Ch. Schneeweiss, J. Schwarze (eds.). *Operations Research Proceedings 1983,* Springer Verlag, Berlin, 441-448.

13. J.B.M. VAN DOREMALEN, J. WESSELS (1985). Iterative approximations for networks of queues. F. ARCHETTI, G. DI PILO, M. LUCERTINI (eds.). *Stochastic Programming*, Springer Verlag, Berlin, 95-106.
14. J.B.M. VAN DOREMALEN, J. WESSELS, R. WIJBRANDS (1986). Approximate analysis of priority queueing networks. O.J. BOXMA, J.W. COHEN, H.C. TIJMS (eds.). *Teletraffic Analysis and Computer Performance Evaluation*, North-Holland, Amsterdam, 117-131.
15. J.B.M. VAN DOREMALEN (1986). *Approximate Analysis of Queueing Network Models*, Ph.D. Thesis, Department of Mathematics and Computing Science, Eindhoven University of Technology, Eindhoven.
16. D.L. EAGER, K.C. SEVCIK (1983). Performance bound hierarchies for queueing networks. *ACM Trans. on Comp. Syst. 1*, 99-115.
17. D.L. EAGER, K.C. SEVCIK (1986). Bound hierarchies for multiple class queueing networks. *J. Assoc. Comput. Mach. 33*, 179-206.
18. W.J. GORDON, G.F. NEWELL (1967). Closed queueing systems with exponential servers. *Oper. Res. 15*, 254-265.
19. A. HORDIJK, N.M. VAN DIJK (1982). Stationary probabilities for networks of queues. R.L. DISNEY, T.J. OTT (eds.). *Applied Probability - Computer Science: The Interface*, Birkhäuser, Boston, 423-451.
20. A. HORDIJK, N.M. VAN DIJK (1984). Network of queues: Parts I and II. F. BACCELLI, G. FAYOLLE (eds.). *Modelling and Performance Evaluation Methodology*. Springer Verlag, Berlin, 151-205.
21. J.R. JACKSON (1957). Networks of waiting lines. *Oper. Res. 5*, 518-521.
22. F.P. KELLY (1979). *Reversibility and Stochastic Networks*, Wiley, New York.
23. L. KLEINROCK (1975). *Queueing Systems, Volume 1: Theory*, Wiley, New York.
24. L. KLEINROCK (1976). *Queueing Systems, Volume 2: Computer Applications*, Wiley, New York.
25. A. KRZESINSKI, P. TEUNISSEN, P. KRITZINGER (1982). *Mean Value Analysis for Load Dependent Servers in Mixed Multiclass Queueing Networks*, ITR 82-01-00 University of Stellenbosch, Inst. for Appl. Comp. Sci., Stellenbosch, South Africa.
26. S.S. LAM (1977). Queuing networks with population size constraints. *IBM J. Res. Develop. 21*, 370-378.
27. S.S. LAVENBERG, M. REISER (1980). Stationary state probabilities at arrival instants for closed queueing networks with multiple types of customers. *J. App. Probab. 17*, 1048-1061.
28. J.D.C. LITTLE (1961). A proof of the queueing formula $L = \lambda W$. *Oper. Res. 9*, 383-387.
29. R. MARIE (1979). An approximate analytical method for general queueing networks. *IEEE Trans. Software Eng. SE-5*, 530-538.
30. R.M. OLIVER (1964). An alternative derivation of the Pollaczek-Khintchine formula. *Oper. Res. 12*, 158-159.
31. M. REISER (1979). A queueing network analysis of computer communication networks with window flow control. *IEEE Trans. Commun. COM-27*, 1199-1209.

32. M. REISER (1979). Mean Value Analysis of queueing networks: a new look at an old problem. M. ARATO, A. BUTRIMENKO, E. GELENBE (eds.). *Fourth Int. Symp. on Modelling and Performance Evaluation of Computer Systems,* North-Holland, Amsterdam, 63-77.
33. M. REISER (1981). Mean Value Analysis and convolution method for queue-dependent servers in closed queueing networks. *Performance Evaluation 1,* 7-18.
34. M. REISER, H. KOBAYASHI (1975). Queuing networks with multiple closed chains: theory and computational algorithms. *IBM J. Res. Develop. 19,* 283-294.
35. M. REISER, S.S. LAVENBERG (1980). Mean Value Analysis of closed multichain queueing networks. *J. Assoc. Comput. Mach. 27,* 313-322.
36. P.J. SCHWEITZER (1979). Approximate analysis of multiclass closed networks of queues. *International Conference on Stochastic Control and Optimization, Amsterdam.*
37. K.C. SEVCIK, I. MITRANI (1981). The distribution of queueing network states at input and output instants. *J. Assoc. Comput. Mach. 28,* 358-371.
38. A.W. SHUM , J.P. BUZEN (1977). The EPF technique: A method for obtaining approximate solutions to closed queueing networks with general service times. H. BEILNER, E. GELENBE (eds.). *Measuring, Modeling and Evaluating Computer Systems,* North-Holland, Amsterdam, 201-220.
39. E. DE SOUZA E SILVA, S.S. LAVENBERG, R. MUNTZ (1984). A perspective on iterative methods for the approximate analysis of closed queueing networks. G. IAZEOLLA, P.J. COURTOIS, A. HORDIJK (eds.). *Mathematical Computer Performance and Reliability,* North-Holland, Amsterdam, 225-244.
40. S. STIDHAM (1974). A last word on $L = \lambda W$. *Oper. Res. 22,* 417-421.
41. S.K. TRIPATHI (1981). On approximate solution techniques for queueing network models of computer systems. *Ph.D. Thesis, Tech. Rep. CSRG-106, CSRG,* University of Toronto.
42. R. WIJBRANDS (1985). *On an Approximation Method for Priority Queueing in CPU-Disk Models,* Memorandum COSOR 85-18, Eindhoven University of Technology, Dept. of Math. and Comp. Sci., Eindhoven.
43. J. ZAHORJAN, E. WONG (1981). The solution of separable queueing networks using Mean Value Analysis. *ACM Sigm. Perf. Eval. Rev. 3,* 80-85.

Reduced State-Space Heuristics for Symmetric Markovian Queueing Systems

Micha Hofri
Department of Computer Science
The Technion, Haifa 32000 - Israel

Several techniques that help in treating Markovian models with inherently large state-space are presented. The basic idea is expressed by a Representation Theorem, that states the equivalence of results obtained either through a detailed solution or a properly estimated, reduced transition matrix. The techniques either use 'combinatorial placement models' that are very successful in spite of their arbitrary nature, or rely on the Maximum-Entropy Principle. A symmetric message passing system is used to illustrate the techniques, and some numerical examples are presented. These often display surprising indifference to the quality or type of the approximation used.

מעוות לא-יוכל לתקן, וחיסרון לא-יוכל להמנות (קהלת א , 15).

1. Introduction

As our computing and communications facilities increase in number and complexity, modeling them becomes at once more of a necessity and of a challenge. The difficulty stems mainly from their sheer size. Note that complexity as such is not necessarily a problem, especially if we are reconciled to make do with approximations: complex systems often display the phenomenon that simplified models that do not take into account many details of the systems operations are quite adequate, in the sense that the important performance measures turn out to be insensitive to the fine details of the systems structure, or of their operations. Also, it is well-known that in many such cases higher moments of the distributions of the random variables that govern the system evolution, hardly impact performance measures, such as mean queue lengths and blocking probability.

The purpose of this note is to present a particular situation where a detailed specification of system behavior is computationally unthinkable. Consequently, the description of its dynamics has to be reduced by various means. In particular, we discuss in detail an approach where the Markovian description has been replaced by simple-minded arbitrary combinatorial models. We have limited the treatment to completely symmetrical systems, a situation in which the above trade-off between size and complexity tilts in the direction that eases the modeler's task. The results have proved to be very good approximations to the steady-state behavior of the system (or rather, of the exact model).

The original impetus came from a series of papers of Chlamtac and Ganz, see [5] and [6] for recent examples. We place the computational method used there in a wider context and show some extensions.

Section 2 presents the specific model - which is very close to the one used in [5] - that will serve for the various computational approaches. Section 3 discusses the basis for the approximations, and the following sections show several different methods, with some numerical examples, collected for ease of comparison in a single table. Section 7 brings the table together with a discussion of the results.

We are gratified for the opportunity to associate this paper with prof. Cohen, whose work over a long period of time was directed at the investigation of congestion phenomena in communications systems.

2. Model specification

The model to be presented is idealized in several senses, mainly because our interest in this note is not to portray accurately any particular network, but to investigate the computational process. A justification for several of the model assumptions and discussions of their limitations may be found in [2,5,6] and references therein.

Consider a 'communications system' containing N *stations.* All the stations can communicate over a common *bus* that can carry one message at a time. The approaches discussed below can be used to handle more than one bus, with no essential difference, but at some cost in requiring more involved transition probabilities (as was done in [5]). The stations view time as a sequence of equal-duration intervals called *slots.* They are all temporally synchronized with each other; communication is performed by transferring messages between the stations; a message transmission only starts at a slot beginning and requires one slot. A station can store up to L messages. All the stations receive messages from the outside according to independent and identically distributed Bernoulli processes: every time slot there is a fixed probability r for each station to generate such an arrival. A message generated at a station with a full buffer is lost.

REMARK 2.1. The loss of a message occurs even if the station is transmitting during that slot, since the buffer space of the sent message is not freed until the slot is over. The contrary assumption, implying the use of an auxiliary buffer when transmitting, results in a poorer utilization of storage - i.e. higher blocking probability for the same amount of storage allocated.

TRANSMISSION PROTOCOL. We shall assume that transmission errors do not occur. We also assume *perfect scheduling*: every slot one of the stations with non-empty buffers transmits one message. It is of no consequence in the present discussion how this selection is performed, and we shall assume the choice among such stations is done with uniform distribution. The papers

[2,5,6,12] propose other transmission regimes as well, and ways to implement them.

Performance measures that need to be considered are message delays, buffer occupancies and particularly the probability that a message is blocked and lost. This is used indirectly to determine the size of the buffers - L - that should be used, to achieve a prescribed blocking level. To estimate these quantities we require the probabilities of the possible occupancies of the stations.

The first concession we make is that we only enquire about the steady-state behavior of the system. If the station identities played a role, a complete Markovian description would consist of $(L+1)^N$ states. Taking advantage of the complete symmetry of the stations we can make do with a more parsimonious description, using occupancy vectors that record the number of stations with any number of buffered messages: $\boldsymbol{n}=(n_0,n_1,\ldots,n_L)$, where n_j is the number of stations that harbor j messages. These vectors specify a first order Markov chain with $\binom{N+L}{L}$ states. Since normally N would be much larger than L, this number is *very* much smaller than the previous state count, but for any realistic purpose it is still impossibly large, and in order to arrive at manageable computations we must resort to approximations. These will only give us an estimate of the steady-state behavior of the collection of stations.

3. Reducing the state space

Using fewer states to describe the system behavior implies that we forgo, at least in part, a specification of the system dynamics through the reduced model we can afford. The following theorem provides the basis and a rationale for the approach commonly used, even if implicitly:

THEOREM 3.1. *Let $S=\{s_i,\ 1\leqslant i\leqslant n\}$ be the state space of a first-order Markov chain governed by the transition probability matrix P, and having the stationary probability vector (spv) $\boldsymbol{\pi}$. Let $\{A_k\}$ be an m-partition of S (i.e. - the A_k are all disjoint and their union equals S). The partition defines m aggregate states for the chain. Denote by q_k (k-th component of $\boldsymbol{q}$) the limiting probability of the chain being in A_k:*

$$q_k = \sum_{i \mid s_i \in A_k} \pi_i, \quad 1\leqslant k\leqslant m. \tag{3.1}$$

Define the $m\times m$ matrix Q with elements Q_{rv} such that

$$Q_{rv} = \sum_{l \mid s_l \in A_r} \sum_{j \mid s_j \in A_v} \pi_l P_{lj}/q_r, \quad 1\leqslant r,\ v\leqslant m. \tag{3.2}$$

Then the matrix Q is stochastic, and its spv equals $\boldsymbol{q}$, as defined in equation (3.1).

The proof is a simple computation that we relegate to Appendix A.

COMMENTS. Many works on Markov chains (or processes) deal with the processes that can be derived from the original ones by transformations of the state space. Representative examples are in [3,8,11]. These treatments differ from our thrust here in two respects:

a) the transformations there are not restricted to transformations that reduce the size of the state space, and

b) the interest is centered on the conditions under which the derived process is Markovian (or at least, that its transition probability functions satisfy the Chapman-Kolmogorov equations - this is immediate in our case). This interest holds also when the transformation does result in reducing the state-space size. Common terms for such a transformation are *aggregation, collapsing, grouping* and *lumping.* Our interest is in the costs and benefits that may result from such a reduction even if it produces a non-Markovian process. We compare the accuracy vs. computation cost trade-off, that such a reduction entails.

This 'Representation Theorem' is surely part of the arsenal and folklore of anyone doing modeling via Markov chains; still, as we have seen it nowhere explicitly stated...

A stationary distribution over a reduced state space is useful if the required performance measures can be computed - or estimated - in terms of this distribution. That this is the case here will be shown below, using a particular specification of aggregate states. The first issue to tackle is the matrix Q. Equation (3.2) shows clearly that we have no hope of computing its exact value. Hence, the objective is to obtain an estimate, denoting it by $\hat{Q}$. It appears reasonable, if vague, that the better is our value for $\hat{Q}$, the closer we should hope is the associated spv $\hat{\boldsymbol{q}}$ to the vector $\boldsymbol{q}$, which should imply better estimates.

Hence, it is intuitively appealing to select a reduced description that affords a close relation to the complete dynamics of the system. Several approximations that try this have been published; we describe one which reports on the quality of the approximation. More works that provide such analyses are listed in Appendix C.

The following is an outline of the approach that was used in [2] and [12]. The authors propose the reduced state description $\{S=(n,j)\}$, where j is the number of messages in an arbitrary but fixed station, designated the 'reference station', and n is the number of stations with buffered messages in the rest of the system. The idea is that since the system is symmetric, describing one station in detail yields an adequate estimate of station-specific measures, and the number of other non-empty stations provides for its main interaction with the rest of the system: the probability of the reference station to be selected to transmit when $j>0$.

Since $0 \leqslant n \leqslant N-1$ and $0 \leqslant j \leqslant L$, this description generates $N(L+1)$ aggregate states. Developing an estimate $\hat{Q}$ for this description is straightforward enough; the only component which cannot be readily obtained from the specification (n,j) is the probability that a transmission from any non-empty station other than the reference station creates an empty station. To estimate this we need to know the probability that the transmission is from a station with a single message. The authors of [2,12] chose to approximate this, when the state is $(n, \cdot)$, by the probability of the event 'the reference station is chosen to transmit at the state $(n-1, 1)$', conditioned on the reference station being non-empty. This conditioning introduces elements of $\hat{\boldsymbol{q}}$ into $\hat{Q}$. Hence the set of equations $\hat{\boldsymbol{q}} = \hat{\boldsymbol{q}}\hat{Q}$ becomes nonlinear in $\hat{\boldsymbol{q}}$. The authors report that direct iterative solution of a linearized form of these equations (with some effort to take advantage of the band-matrix form of $\hat{Q}$), converged very nicely. Results were compared with simulation and were within 95% confidence intervals, with mean deviations of the order of one percent and usually less than that.

4. The Bose-Einstein approximation

We consider now in more detail a different, and to us a more intriguing reduction. It was first proposed by Chlamtac and Ganz. They use a drastically reduced representation of the state space, specifying only the *total number of messages in the stations.* The number of states is only slightly less than the one produced by the above method. It appears that this choice offers an excellent trade-off between the reduction in state space - which largely determines the computational difficulties, since their approach only requires the solution of *linear* equations - and the quality of approximation. This holds for many values of the parameters (N, L and r in the model discussed here).

Using the Theorem in the present context requires $m = NL$ aggregate states. (The remark in Section 2 implies that the state $n = NL$ cannot be reached.) Looking at a moderately small network, with $L=3$, $N=10$, we find

$$(L+1)^N = 1048576, \quad \binom{N+L}{L} = 286, \quad N(L+1) = 40, \quad NL = 30.$$

The savings are much larger when interesting networks are considered.

Estimating Q. Let n denote the total number of buffered messages, $0 \leqslant n < NL$. This will specify the aggregate states. During a slot there is one transmission whenever $n > 0$. The number of newly arrived messages has the binomial distribution $\mathrm{B}(N - n_L, r)$, where n_L is the number of fully occupied stations, also called blocking stations. In the present transmission protocol this is the only quantity that has to be estimated.

The estimate proposed in [5,6] to compute the distribution of n_L uses occupancy vectors $\boldsymbol{c} = (c_1, c_2, \ldots, c_N)$, that specify the number of messages at each of the N stations. Further, it assumes the following:

At the beginning of every time slot the buffered messages are rearranged, independently of the past, in such a manner, that for a given n, all the feasible distinct occupancy vectors $\boldsymbol{c} = (c_1, c_2, \ldots, c_N \mid \Sigma c_i = n)$ are equally likely.

This is precisely the distributional assumption called 'Bose-Einstein (BE) statistics' in statistical mechanics. The choice of this assumption here seems to have been inspired by an oblique remark in [9, p.158], in a rather different context.

Note that this approach considers the stations as similar in every respect - but *distinguishable!*

As a result of the buffer upper limit L, the number of the occupancy vectors is not the customary one $\binom{n+N-1}{n}$ (see e.g. [4] or [7]). We present a few ways for its efficient evaluation.

Denote the required number by $B_L(n, N)$. To compute it we can condition on the number of empty buffers, to obtain the recursion

$$B_L(n, N) = \sum_{j \geq 0} \binom{N}{j} B_{L-1}(n-(N-j), N-j), \quad L \geq 2, \tag{4.1}$$

with the initial value $B_1(n, N)=\binom{N}{n}$.

Two other recursions are obtained if we observe that the generating function (in n) of $B_L(n, N)$ is given by $(1+x+\cdots+x^L)^N$. Hence*

$$B_L(n, N) = [x^n](1+x+\cdots+x^L)^N. \tag{4.2}$$

The generating function can be handled in two ways: first, use the binomial expansion, with x^L as one term and the rest lumped together to derive

$$B_L(n, N) = [x^n]\sum_{j=0}^{N} \binom{N}{j} x^{Lj}(1+x+\cdots+x^{L-1})^{N-j} \tag{4.3}$$

$$= \sum_{j=0}^{N} \binom{N}{j} B_{L-1}(n-jL, N-j), \quad L \geq 2.$$

This has the initial value as above. Another way is to split the power N in the generating function to 1 and $N-1$, whence the recursion

$$B_L(n, N) = \sum_{j=0}^{\min(n, L)} B_L(n-j, N-1), \quad N \geq 2, \tag{4.4}$$

with the initial value $B_L(n, 1)=1$ for $0 \leq n \leq L$. Each of these recursions can be reduced to an explicit solution (also available, naturally, by directly extracting the coefficient from the generating function):

$$B_L(n, N) = \sum_{j=0}^{\lfloor n/(L+1) \rfloor} (-1)^j \binom{N}{j} \binom{n+N-j(L+1)-1}{N-1}. \tag{4.5}$$

Computationally, this solution is quite poor; note that for quite modest networks, the right-hand side contains binomial coefficients that are outside the range which is directly representable by most machines, and in any case, much larger than the left-hand side. We found that recursion (4.4) is the most convenient, and it was used in the numerical experiments.

To obtain a similar count, conditioned on the number of buffers having a prescribed occupancy - say t - we can use the inclusion-exclusion formula to obtain

* The symbol $[x^n]f(x)$ is the coefficient of x^n in the power development of $f(x)$ around $x=0$.

$$B_L(n, N; n_t = i) = \begin{bmatrix} N \\ i \end{bmatrix} \sum_{j \geqslant 0} (-1)^j \begin{bmatrix} N-i \\ j \end{bmatrix} B_L(n - t(i+j), N - i - j). \tag{4.6}$$

The same expression results from subtracting x^t from the polynomial in the generating function, and evaluating the coefficient directly. But a more convenient procedure for numerical work results from the direct conditioning on the required occupancy. This is particularly convenient for $t = L$, which is the value actually required for our computations:

$$\begin{aligned} B_L(n, N; n_L = i) &= \begin{bmatrix} N \\ i \end{bmatrix} [x^{n-Li}](1 + x + \cdots + x^{L-1})^{N-i} \\ &= \begin{bmatrix} N \\ i \end{bmatrix} B_{L-1}(n - Li, N - i), \quad L \geqslant 2. \end{aligned} \tag{4.7}$$

For single-message buffers ($L = 1$) clearly $B_1(n, N; n_1 = i) = \delta_{ni}\binom{N}{i}$.

The essence of the present approximation is that the values of the conditional counts in equation (4.7) provide the estimate of the distribution of the number of blocking stations:

$$p(n_L = i \mid n) \equiv B_L(n, N; n_L = i)/B_L(n, N). \tag{4.8}$$

Now the desired estimate of the transition probabilities is immediate:

$$\hat{Q}_{0m} = \binom{N}{m} r^m (1-r)^{N-m}, \tag{4.9}$$

$$\hat{Q}_{nm} = \sum_{i \geqslant 0} p(n_L = i \mid n) \binom{N-i}{m-n+1} r^{m-n+1} (1-r)^{N-i-m+n-1},$$

$$m \geqslant n - 1 \geqslant 0.$$

REMARK 4.1. Even for moderate networks, the numbers generated by equations (4.4) grow extremely rapidly. So much so that they easily exceed the range of numerical representation on most machines. Since they are not needed in themselves, but only their ratios are required to estimate the conditional probabilities, there is a simple solution to the difficulty: generate them separately, using a package of multiprecision routines, or preferably within a system of symbolic computation such as Maple or Macsyma, and evaluate $\hat{Q}$ there. These estimates are then used for the computation of the vector $\hat{\boldsymbol{q}}$ and the performance measures, which can all be done in normal representation quite satisfactorily, even for fairly large configurations (such as $N = 50$, $L = 8$).

PERFORMANCE MEASURES. The following performance measures were computed for all the approximation methods described here and below.

The 'throughput' of the network, denoted by ρ, is defined as the mean number of messages transmitted per slot. Due to the 'perfect scheduling' our estimate is clearly

$$\rho = 1 - \hat{q}(0). \tag{4.10}$$

The expected combined buffer occupancy, denoted by γ, is estimated by

$$\gamma = \sum_{n>0} n\hat{q}(n), \tag{4.11}$$

and its variance is similarly given by $\Sigma n^2\hat{q}(n)-\gamma^2$. The expected message delay is immediately available as γ/ρ.

Finally, the probability of a message being blocked, denoted by β, is given by

$$\beta = \sum_n \hat{q}(n)\sum_i p(n_L=i\,|\,n)\frac{i}{N}. \tag{4.12}$$

From the requirement that all non-blocked messages be ultimately transmitted, with probability one, we get the simpler expression

$$\beta = 1-\frac{\rho}{Nr}. \tag{4.13}$$

Results derived from the BE approximation are displayed in Section 7.

5. The Maxwell-Boltzmann approximation

The results obtained via the expressions computed in equations (4.4) to (4.9) are good, often very good, but one can try and improve on them. Clearly the assignment of equal likelihood to each occupancy vector **c**, for a given n, is not exact. When one considers the actual dynamics and the way these assignments may differ from the correct ones, it becomes apparent that vectors with highly skew occupancies - particularly those with some empty stations and some full ones - should be reached less frequently, under the actual mechanism, than configurations in which the occupancies do not differ as much. The reader may find it instructive to observe very small scale configurations - such as $N=2$ or 3, and $L=2$, where the (correct) transition probabilities are very easy to compute. One then observes that states that have no full stations are entered with higher frequency than states with skew message distribution. The Maxwell-Boltzmann (MB) statistics, also familiar from statistical mechanics (see [4] and [7]) does produce precisely this bias. So it is a natural candidate for the estimation of the matrix Q here. Combinatorially, the MB statistics correspond to a redistribution of the messages at every slot beginning, as above; but now, not only the stations but the messages themselves are viewed as distinguishable. Again, all the possible distinct arrangements that can thus arise are considered equiprobable. This introduces an assumption no less arbitrary; but intuitively it appears better suited for the mechanism at hand, and we shall consider its effects.

Let us use $M_L(n,N)$ and $M_L(n, N;n_L=i)$ under the MB statistics instead of $B_L(n, N)$ and $B_L(n, N;n_L=i)$ that were used for the BE assumption. There do not seem to be any simple explicit forms now either; the enumerator (over n) of $M_L(n, N)$ is clearly given by

$$M_L(z, N) = (1+z+\frac{z^2}{2}+\cdots+\frac{z^L}{L!})^N, \tag{5.1}$$

and again we can obtain two recursions, using the same manipulations. First,

$$M_L(n, N) = n! \sum_{k=0}^{\lfloor n/L \rfloor} \binom{N}{k} \frac{M_{L-1}(n-kL, N-k)}{L!^k (n-kL)!}, \quad L \geq 2, \tag{5.2}$$

and then similarly

$$M_L(n, N) = \sum_{i=0}^{\min(n, L)} \binom{n}{i} M_L(n-i, N-1), \quad N \geq 2. \tag{5.3}$$

These recursions have the corresponding initial values $M_1(n, N) = N!/(N-n)!$, $0 \leq n \leq N$, and $M_L(n, 1) = 1$, $0 \leq n \leq L$. The second recursion is somewhat healthier and was used in the numerical work.

Similarly, for the count of configurations with a specified number of stations at a given occupancy level t, the inclusion-exclusion formula can be used to produce an expression analogous to equation (4.6); we only need the value $t = L$ though, which is easier to represent via the enumerator:

$$M_L(n, N; n_L = i) = \binom{N}{i} \frac{n!}{L!^i} [z^{n-iL}] \left(1 + z + \frac{z^2}{2} + \cdots + \frac{z^{L-1}}{(L-1)!}\right)^{N-i} \tag{5.4}$$

$$= \binom{N}{i} \frac{(iL)!}{L!^i} \binom{n}{iL} M_{L-1}(n - iL, N - i), \quad L \geq 2.$$

The single-message buffer case provides an initial value for the recursion, similarly as in equation (4.7), at $M_1(n, N; n_1 = i) = \delta_{ni} N!/(N-i)!$.

The estimate for the matrix Q is completely analogous to that given by equations (4.8) and (4.9).

Note that the numbers generated in this approximation are much larger than those produced in Section 4. The remark at the end of that section applies here as well, and even for smaller networks (when NL exceeds 50).

The corresponding numerical valuations are presented in Section 7.

6. Using the maximum-entropy principle

The above sections used 'combinatorial estimates' for the matrix $\hat{Q}$ needed in the Representation Theorem. No matter how successful we were in evaluating the main performance measures, there remains a doubt as to the range of applicability, and the very arbitrariness of these estimates rankles. Another approach, replacing these estimates by the supposedly more disciplined Maximum-Entropy Principle (MEP) will be presented below. First we show how the combinatorial estimates can be used *within* this formalism. The idea is not to use the Theorem, but rather employ the only external constraint we have on the system operation: all the messages accepted into the buffers are ultimately transmitted, with probability one. The input rate of messages depends on the number of blocking stations, and thus the input-output constraint we shall use, within the reduced representation, is given by

$$1 - q(0) = \sum_{n=0}^{NL-1} q(n) \sum_i p(n_L = i \mid n)(N-i)r. \tag{6.1}$$

The probabilities $q(n)$ will be estimated by the MEP. The conditional probabilities $p(n_L=i\,|\,n)$ will be taken directly from equation (4.8) - using either the BE estimates or the MB ones. Thus we have not rid ourselves of the arbitrariness of these. Introducing the MEP provides us however with computations that are much simpler: the constraint (6.1) is in the so-called 'Gibbs form', with the associated variable being the change in the number of buffered messages (see Appendix B for an outline). In terms of this formalism, we have a single constraint, $m=1$, and as a result, there is only *one* equation to solve.

Comparing equations (6.1) and (B.2) we find the constraint function:

$$f(n) = \sum_{i=0}^{N} p(n_L=i\,|\,n)(N-i)r, \quad 1 \leqslant n < NL, \quad f(0)=1+Nr, \tag{6.2}$$

and the constraint constant F has the value 1. The probabilities then are given by

$$p(n) = \frac{1}{Z(\lambda)}\exp(-\lambda f(n)), \tag{6.3}$$

where λ is determined as the root of the equation

$$\sum_{n=0}^{NL-1} (1-f(n))e^{-\lambda f(n)} = 0. \tag{6.4}$$

This equation turns out to be easy to solve, using the Newton-Raphson procedure, with some attention to selecting the initial value (since the function in equation (6.4) is quite volatile for negative λ). The results obtained by this procedure are also collected in the next section.

Rather out of methodological interest, than from hoping to improve on the estimates obtained in Sections 4-5 above, we proceeded to use the MEP so as to avoid the arbitrary combinatorial crutches. It turned out that while feasible in principle, success was rather limited: to get a reasonable accuracy the computation times are comparable with those required for the exact analysis (this is not an accurate statement - the balance depends on the considered configuration).

The MEP can provide, as explained in Appendix B, estimates of state probabilities subject to 'external constraints'. The only external knowledge we have about the system behavior devolves from requirements of statistical equilibrium. These can be stated for any choice of aggregate states, and it is natural we choose again the set $\{n\}$. The system states, over which these constraints are used will be $\{(n,i)\}$ - aggregate states as well - with n and i denoting the total message population size and the number of full stations, respectively. Note that we estimate directly here the joint probabilities of these aggregate states, rather than the conditional ones used above.

$$q(n,i) = q(n)p(n_L=i\,|\,n), \quad 0 \leqslant n \leqslant NL-1, \quad 0 \leqslant i \leqslant N-1. \tag{6.5}$$

The constraints are the equations that specify statistical balance for the aggregate states $\{n\}$ - essentially equations (4.9). We rewrite them in the form:

$$\begin{aligned} q(n) = \sum_i q(n,i) &= q(0)\begin{bmatrix} N \\ n \end{bmatrix} r^n(1-r)^{N-n} \\ &+ \sum_{i,j} q(j,i)\begin{bmatrix} N-i \\ n-j+1 \end{bmatrix} r^{n-j+1}(1-r)^{N-i-n+j-1}. \end{aligned} \tag{6.6}$$

Note that these NL equations are dependent, hence only $NL-1$ of them were used. Also, instead of the constraint with $n=0$ in (6.6) we used equation (6.1), to obtain a slightly better result (at the fourth significant digit and below, in the examples we did). The constraints were all written so as to be homogeneous in $q(n,i)$, hence the constraint constants all vanish: $F_k=0$, for $0 \leqslant k \leqslant NL-2$. The constraint functions for $n>0$ are given by:

$$f_n(j,i) = \delta_{nj} - \delta_{j0}\delta_{i0}\begin{bmatrix} N \\ n \end{bmatrix} r^n(1-r)^{N-n} - \begin{bmatrix} N-i \\ n-j+1 \end{bmatrix} r^{n-j+1}(1-r)^{N-i-n+j-1}, \tag{6.7}$$

and from the constraint equation 0 (i.e. equation (6.1)) we have

$$f_0(j,i) = \begin{cases} (N-i)r-1, & j>0, \quad 0 \leqslant i \leqslant N-1, \\ Nr, & j=0, \quad 0 \leqslant i \leqslant N-1. \end{cases}$$

The equations for the Lagrange multipliers are

$$g(n) = \sum_{j,i} f_n(j,i)\exp\{-\sum_{r=0}^{NL-2} \lambda_r f_r(j,i)\}=0, \quad 0 \leqslant n \leqslant NL-2, \tag{6.8}$$

and finally

$$q(j,i) = \frac{1}{Z(\boldsymbol{\lambda})}\exp\{-\sum_{r=0}^{NL-2} \lambda_r f_r(j,i)\}. \tag{6.9}$$

Again the Newton-Raphson procedure is invoked, to solve equations (6.8). Here the computational difficulty is much more serious (imagine searching even in a 29-dimensional space - as was required for the modest numerical examples below!). The choice of the initial point for the Newton-Raphson procedure is not critical here, in principle, since J, the Jacobian of the $g(n)$ functions, is a real symmetric matrix; hence it is the Hessian of a convex function - one that has a unique minimum. The partition function $Z(\boldsymbol{\lambda})$ itself can serve as this so-called primitive function during the iterations. The computational difficulty remains however, as we run against the well-known numerical instabilities associated with fitting data (here - the constraint requirements) to exponential functions: the resultant equations are extremely ill-conditioned: during the experiments described below condition numbers of J smaller than 10^{10} were observed, but infrequently. The procedure we used incorporated an idea borrowed from AGMON ET AL. [1]: they suggest that at every iteration the length of the step taken in the direction of steepest descent - as given by $-J^{-1}g(\boldsymbol{\lambda})$ - is determined by finding the point in that direction where the primitive function reaches a local minimum. The actual computation is done by fitting a parabola to the values it takes along this direction. Numerous experiments were performed, and the overall conclusion is that nearly always, within the numerical accuracy provided by 64-bit number representation, the values obtained for the system-wide performance criteria listed in the table were inferior to those produced by the combinatorial estimates. The values for the conditional distribution of the messages in the stations were however often more accurate. Selected results are given in the next section.

7. Discussion

The methods described in the last three sections have been used on a variety of system configurations. The enclosed table summarizes the gross system characteristics - throughput, mean and variance of the total buffer occupancy and message loss probability, in this order. The system studied has ten stations and buffers that can hold three messages, and we considered several new-message generation probabilities (r). The modest size is a result of our wanting to also obtain the exact results in manageable time. The vacant cells attest to serious numerical difficulties in using the MEP. Several features are evident in the table; some of our conclusions were also derived from comparisons of more detailed information - the vectors $\hat{q}$ that were obtained in the process, and estimates of the conditional probabilities $p(n_L=i \mid n)$; these would be too space-consuming to present.

Computational comment: all the spv evaluations, both the approximate and the exact, were done using an iteration (the Gauss-Seidel form), rather than elimination. This is particularly attractive here in terms of the possible precision, since the matrices P and Q contain only non-negative terms.

The most striking fact is the agreement of the results derived from the combinatorial estimates with the exact solution. The agreement extends to the distributions of the total buffer contents as well. However, the conditional distributions (or those of the occupancy vectors) display substantial differences, as could be expected. The MB-derived estimates of these were uniformly the best. Since the detailed distributions are combined in equations (4.9) to satisfy the balance equations, the errors there balance out as well. Comparing in the table the BE and the MB approximations produces a mixed picture: the MB appears to be somewhat more accurate, especially with the loss probabilities, which is, again, as expected (recall the rationale behind its introduction in the first place). Interestingly, this relative advantage is most evident precisely in the range of 'realistic' service grades, 0.01 - 0.0001. Note that the approximations for γ usually bracket the correct values.

Encouraged and intrigued by this combination of arbitrary estimates and respectable results, we also checked what happens under the 'nearly most arbitrary' assumption of them all: assigning, for each n, the same likelihood to all the feasible $p(n_L=i \mid n)$. The results are in the table, in the column headed by UNIFORM. They are not quite as good, but embarrassingly close. One may quote him that was 'wiser than all men' (I Kings IV, 31), in the (very arbitrary) translation of the Authorized Version*, but it appears that here the crooked *may be assumed* straight, when the state space is too large to be numbered.

The MEP with a single constraint - column MEP-I in the table - performs

* The Hebrew original appears at the head of the paper. The authorized translation is 'That which is crooked cannot be made straight, and that which is wanting cannot be numbered'. The context is moral values.

r	BE	MB	UNIFORM	MEP-I	MEP-II	EXACT
0.001	0.00999 0.010045 0.0100357 $\sim 5.6 \times 10^{-10}$	0.00999 0.010045 0.0100357 $\sim 1.2 \times 10^{-10}$	0.00999 0.010045 0.0100357 $\sim 6.14 \times 10^{-9}$	0.00997 0.14793 2.8660 0.002615		0.01000 0.010045 0.0100357 $< 1.0 \times 10^{-11}$
0.01	0.09999 0.10499 0.10429 $\sim 7.2 \times 10^{-7}$	0.09999 0.10499 0.10429 $\sim 1.6 \times 10^{-7}$	0.09999 0.10500 0.10428 $\sim 7.5 \times 10^{-6}$	0.09774 1.35796 23.517 0.022587		0.10000 0.105000 0.10429 4.0×10^{-9}
0.05	0.49983 0.72371 0.80528 0.0003360	0.49995 0.72461 0.80888 0.0000941	0.49892 0.71788 0.78615 0.002161	0.45990 5.56092 62.824 0.08021	0.49914 0.71938 0.76169 0.001727	0.499968 0.724718 0.809267 0.0000635
0.07	0.69836 1.40723 2.04219 0.0023558	0.69939 1.42328 2.12436 0.0008735	0.69364 1.35279 1.83830 0.009082	0.62638 7.43844 68.3036 0.10517	0.69566 1.38191 1.98496 0.006165	0.69929 1.42116 2.11014 0.0010151
0.1	0.95350 5.58965 13.34896 0.046496	0.96159 6.58228 17.56017 0.038415	0.93923 4.62911 10.75313 0.060766	0.84401 10.49433 70.6323 0.15599	0.96325 8.26945 34.95575 0.036747	0.9571375 5.984746 14.97095 0.042863
0.15	0.999999 17.68551 8.75424 0.333333	0.999999 19.35092 7.62603 0.333333	1.0 18.83839 10.33188 0.33333	0.98696 17.01257 75.3066 0.34203	0.99998 17.62883 17.70646 0.33333	0.999999 18.16034 9.15183 0.333333
0.2	0.99999 22.05112 5.00654 0.50000	1.0 23.05856 4.19406 0.50000	1.0 22.67365 3.99204 0.50000	0.99683 21.24694 60.8482 0.50158	1.0 20.89246 10.24883 0.49995	1.0 22.387766 5.21688 0.4999999
0.3	1.0 25.49211 2.45372 0.66667	1.0 25.92707 2.06572 0.66667	1.0 25.25486 2.25535 0.66667	0.99931 25.07565 29.87942 0.66689	1.0 24.00864 5.27691 0.66667	1.0 25.692575 2.48358 0.66667
0.5	1.0 27.67453 0.96636 0.8	1.0 27.79446 0.85097 0.8	1.0 27.23759 1.14726 0.8	0.99996 27.62689 5.65411 0.80001	1.0 26.49897 2.21039 0.8	1.0 27.762072 0.92814 0.8

TABLE 1. *Results for a system with ten stations and buffers of three messages ($N=10$, $L=3$). In each cell, from top to bottom: Throughput (ρ), mean number of buffered messages (γ), their variance and the blocking probability (β).*

very poorly, except at high arrival rates, and even then, the variance estimate is entirely wrong and useless. Considering the drastically simpler calculations this approach requires, we still rate it attractive, even if only for a limited range of the input parameters. The form of the probabilities produced by equation (6.3) displays the technical reason well: the single λ there is not large enough for the probabilities of high occupancies to decay sufficiently fast. The displayed results were obtained with the conditional probabilities estimated via MB statistics; the BE one gave similar values.

The results obtained from equations (6.6)-(6.9) were rather disappointing (see the column headed by MEP-II in the table). The computation times were as long - sometimes longer - than those required for the exact solution. Moreover, though we seem to be using there no less information than provided by the estimate (4.9), the results are poorer than those produced by either the BE or the MB statistics that use (4.9); they are better - small consolation that this is - than those in the UNIFORM column. Pending better understanding of the problem we tend to ascribe this to numerical difficulties. It is a sobering reflection on the many situations where the MEP is used for the lack of any other tool.

At the end of Section 3 we described the method of computation used in [2,12]. We call that approach 'direct analysis', whereas the others we have considered are 'indirect', inasmuch as they rely on arbitrary assumptions, that do not result from the specification of the dynamical behavior of the system. In Appendix C we collected some references to other works using such 'direct analyses'. Our adoption of the MB statistics shows that one could use the system dynamics to choose between indirect methods. Thus it is possible to avoid an exact or direct attack of such a problem, and still obtain useful results, as good - or better - as those obtainable by direct methods, and at vastly reduced computational effort. Furthermore, the analyst can be satisfied with the knowledge that his results are not sensitive to the 'reasonableness' of the underlying assumptions.

ACKNOWLEDGEMENT

Discussions with David Margulis were very helpful in clarifying the relevance of the combinatorial estimates. We are grateful to him for validating a few of the computations. Prof. Syski showed us that processes derived from Markov chains by state-space transformations are ubiquitous in the literature and supplied numerous examples.

REFERENCES

1. N. AGMON, Y. ALHASSID, R.D. LEVINE (1979). An algorithm for finding the distribution of maximal entropy. *J. Comput. Physics 30,* 250-258.
2. T.K. APOSTOPOULOS, E.N. PROTONOTARIOS (1986). Queueing analysis of buffered CSMA/CD protocols. *IEEE Trans. Commun. COM-34,* 898-905.
3. E.B. DYNKIN (1965). *Markov Processes,* Vol. I, Springer-Verlag, Berlin, Section 10.6.
4. W. FELLER (1968). *An Introduction to Probability Theory and its*

Applications, Vol. I, 3rd Edition, John Wiley & Sons, New York, 40-41.
5. A. GANZ, I. CHLAMTAC (1988). Finite buffer queueing models for single and multiple transmissions channel access protocols. G. IAZEOLLA, P.J. COURTOIS, O.J. BOXMA (eds.). *Computer Performance and Reliability,* North-Holland, Amsterdam, 321-336.
6. A. GANZ, I. CHLAMTAC (1988). Finite buffer queueing models for p-persistent CSMA/CD protocols. P.J. COURTOIS, G. LATOUCHE (eds.). *Performance '87,* North-Holland, Amsterdam, 481-496.
7. M. HOFRI (1987). *Probabilistic Analysis of Algorithms,* Springer-Verlag, New York, 59.
8. J.G. KEMENY, J.L. SNELL (1960). *Finite Markov Chains,* Van Nostrand, New York, Section 6.3.
9. L. KLEINROCK (1975). *Queueing Systems,* Vol. I, Wiley, New York.
10. R.D. LEVINE, M. TRIBUS (eds.) (1978). The maximum entropy formalism. *Conf. Proceedings, MIT 2-4 March 1978,* The MIT Press.
11. M. ROSENBLATT (1970). *Markov Processes. Structure and Asymptotic Behavior,* Springer-Verlag, Berlin, Chapter III.
12. E.D. SYKAS, D.E. KARVELAS, E.N. PROTONOTARIOS (1986). Queueing analysis of some buffered random multiple access schemes. *IEEE Trans. Commun. COM-34,* 790-798.

Appendix A

Proof of the Representation Theorem

We have to show that $q_k = \sum_{r=1}^{m} q_r Q_{rk}$. Let us compute directly the right-hand side. Substitute from equation (3.2):

$$\sum_{r=1}^{m} q_r Q_{rk} = \sum_{r=1}^{m} q_r \sum_{i \mid s_i \in A_r} \pi_i \sum_{j \mid s_j \in A_k} P_{ij}/q_r = \sum_{j \mid s_j \in A_k} \Big(\sum_{r=1}^{m} \sum_{i \mid s_i \in A_r} \pi_i P_{ij} \Big).$$

The summations over r and i combine to cover all of S, and since $\boldsymbol{\pi}$ is the stationary probability vector of P we get

$$\sum_{r=1}^{m} q_r Q_{rk} = \sum_{j \mid s_j \in A_k} \pi_j = q_k,$$

as required.

Appendix B

An Outline of the Maximum-Entropy Formalism

Let the possible states of a discrete system be denoted by $\{s_i\}$, $1 \leqslant i \leqslant n$. The $\{s_i\}$ may be associated with variables $\boldsymbol{X} = \{X_j\}$, so that at the state s_i they have the values $\boldsymbol{x}_i$. These values need not be distinct. The formalism assumes that the phrase 'the state s_i is realized with probability p_i' is meaningful; and that if there are any testable or measurable data on the system they can be expressed in terms of the $\{p_i\}$ and the associated variables.

Then, it states, under constant conditions the system state probabilities assume values that maximize the entropy function H,

$$H = -\sum_{i=1}^{n} p_i \log p_i, \tag{B.1}$$

in such a way that all the constraints are satisfied.

The 'Gibbs form' obtains when the constraints can all be expressed in terms of expected values of the associated variables (rather than, say, tail probabilities or inequalities):

$$\sum_{i=1}^{n} p_i f_k(\boldsymbol{x}_i) = F_k, \quad 1 \leqslant k \leqslant m. \tag{B.2}$$

Now that the maximization problem has only linear constraints, the Lagrangian can be explicitly handled. The solution at the maximum is:

$$p_i = \frac{1}{Z(\boldsymbol{\lambda})} \exp\{-\sum_{k=1}^{m} \lambda_k f_k(\boldsymbol{x}_i)\}, \quad 1 \leqslant i \leqslant n, \tag{B.3}$$

where $Z(\boldsymbol{\lambda})$ - commonly called the 'partition function' of the system - is a normalization constant; it is then given by

$$Z(\boldsymbol{\lambda}) = \sum_{i=1}^{n} \exp\{-\sum_{k=1}^{m} \lambda_k f_k(\boldsymbol{x}_i)\}. \tag{B.4}$$

The $\boldsymbol{\lambda}$ are the m Lagrange multipliers, determined by the following set of simultaneous equations:

$$F_k = -\frac{\partial}{\partial \lambda_k} \log Z(\boldsymbol{\lambda}) = \frac{1}{Z(\boldsymbol{\lambda})} \sum_{i=1}^{n} f_k(\mathbf{x}_i) \exp\{-\sum_{k=1}^{m} \lambda_k f_k(\mathbf{x}_i)\}. \qquad \text{(B.5)}$$

When the constraints have the Gibbs form the solution is guaranteed to be unique (see Section 6). The claim of the MEP is that these $\{p_i\}$ constitute the 'best prediction of the system behavior, granted the evidence of equation (B.2)'.

A useful source of information about the MEP is [10].

Appendix C

A Sample of 'Direct Analyses' of Similar Models

The following are papers which propose what we called in Section 7 'direct analysis', at various levels of approximation (and success - but rarely with supporting evidence for the quality of the approximation).

S.S. KAMAL, S.A. MAHMOUD (1979). A study of users' buffer variations in random access satellite channels. *IEEE Trans. Commun. COM-27,* 857-868.
Corrigendum: same volume, 1745.
Comments on the paper:
T. SAADAWI, A. EPHREMIDES (1980). *IEEE Trans. Commun. COM-28,* 1436-1437 (with reply).
K.K. MITAL, A.N. VENETSANOPOULOS (1981). *IEEE Trans. Commun. COM-29,* 919. Reply and rejoinder in *IEEE Trans. Commun. COM-30,* 1801-1802 (1982).

L. KLEINROCK, S.S. LAM (1975). Packet switching in a multiaccess broadcast channel: performance evaluation. *IEEE Trans. Commun. COM-23,* 410-423.

D.V. RAMANA, V.V.S. SARMA (1982). Performance of slotted ALOHA satellite channels with finite buffer. *IRE Proc. 129 Pt. F,* 19-24.

T. SAADAWI, A. EPHREMIDES (1981). Analysis, stability and optimization of slotted ALOHA with a finite number of buffered users. *IEEE Trans. Aut. Cont. AC-26,* 680-689.
Comments on the paper:
D.V. RAMANA, V.V.S. SARMA (1985). *IEEE Trans. Aut. Cont. AC-30,* 316-319 (with reply).

M. SIDI, A. SEGALL (1983). Two interfering queues in packet-radio networks. *IEEE Trans. Commun. COM-31,* 123-129.

Insensitivity, Sensitivity and Partial Insensitivity for Some Queueing Models

Arie Hordijk
Dept. of Mathematics and Computer Science, University of Leiden
P.O. Box 9512, 2300 RA Leiden, The Netherlands

The insensitivity property for some queueing models of Erlang and Engset type is analyzed. Partial insensitivity and Beneš' formula are clarified. An introduction to jobmark processes, job local balance, adjoint transition rates and detailed product form is given.

1. Introduction

One of the themes of the scientific work of J.W. Cohen is the insensitivity property. His first research paper COHEN [4] on the Engset formulae already appeared in the fifties. It belongs certainly to the pioneering scientific work in this area. Much more fundamental research by Cohen on this theme followed during the thirty years since then. The insensitivity property was also the subject of our discussion when he invited me some ten years ago for the first time in his study at his home in the Hague. The insensitivity theme appeared many times in our discussions since that time. Therefore, it was the natural topic for my paper in J.W. Cohen's Liber Amicorum. Since an introduction to and a historic overview of the insensitivity property can be found in Cohen's papers, I have chosen to analyze some specific queueing models. In Section 2 Erlang and Engset type models are focussed on. An extension of the known results is the restricted availability of free lines. It seems that the insensitivity of models in which when congested, one of the calls chosen at random is interrupted, is a new result.

In Section 3 a grading called 'transposed multiple' is analyzed. This grading was the topic of the paper BEUKELMAN and COHEN [3]. Recently, Cohen raised the question whether this grading has the insensitivity property. It is shown that the answer is no. Beukelman and Cohen give an approximation of the call-congestion. The relation with the bounding method recently developed by Hordijk, Ridder and Van Dijk is briefly discussed.

In Section 4 partial insensitivity is illustrated through the M/G/1 model. It is well-known that the stationary distribution of the number of customers is sensitive. However, the probability of zero customers is insensitive. This phenomenon is clarified and a general conjecture on partial insensitivity is posed. Section 4 continues with an alternative derivation and simple explanation of the Beneš formula for the stationary distribution of the virtual waiting time in the M/G/1-model.

The analysis in this paper is based on earlier work of Hordijk and Van Dijk on jobmark processes. We continue this section with an introduction to jobmark processes, job local balance and adjoint transition rates. A precise definition of (detailed) product form is given. It gives the relation to the earlier mentioned bounding method.

In HORDIJK and VAN DIJK [11] we have analyzed queues and queueing networks through a process which we called the jobmark process. The jobmarks contain all the relevant information of the jobs present in the system. This information can be very detailed: for example the node in the network where the job is located together with its place in the queue of that node, its priority or class number etc. The jobmarks can also be used to count only the number of jobs present. So they can model processes with micro states (detailed information) as well as processes with macro states. A transition in the jobmark process occurs when a job enters or leaves the system or if one of the present jobs changes its jobmark. It is assumed that never more than one of these events occur at the same time. Clearly, the jobmark process can model a large variety of queueing processes.

In the last ten years there has been a strong interest in queueing networks which have the celebrated product form stationary distribution (cf. COHEN [5]). What this product form precisely means, is not so clear, it is as far as I know nowhere defined. It is commonly agreed that the open and closed Jackson networks (cf. WHITTLE [19]) do have the product form property. Do the Erlang and Engset loss systems have it? Well, the first model can be seen as an open Jackson network with one node and the second model is a closed Jackson network with two nodes and both networks have the standard form of the stationary distribution.

Let us denote the states with the generic notation E, $\mathcal{E}$ is the set of all states, and let $q(E_i, E_j)$ be the transition rate from state E_i to state E_j with $P(E)$ the stationary probability of state E. If the process is reversible then, as is well-known (cf. KELLY [16]), the following representation of the stationary distribution holds:

$$P(E) = P(E_0)\frac{q(E_0, E_1)}{q(E_1, E_0)}\frac{q(E_1, E_2)}{q(E_2, E_1)} \cdots \frac{q(E_k, E)}{q(E, E_k)} \tag{1.1}$$

for any path of states $E_0, E_1, \ldots, E_k, E$ such that the denominators are positive. A Jackson network with routing or branching matrix (p_{ij}) (the probability that a job who leaves node i will require the next service at node j) is a reversible process if and only if the matrix (p_{ij}) is a reversible matrix. But, also for a nonreversible routing matrix the stationary distribution is of the form (1.1). Of course not with the original transition rates but with transition rates where the routing probability p_{ij} is replaced by λ_j, the solution of the routing equations (see Section 3 of HORDIJK [13] and Chapter 1 of RIDDER [17]). Therefore, it is tempting to call a stationary distribution of detailed product form if it satisfies the relation

$$P(E) = P(E_0)\frac{\overline{q}(E_0, E_1)}{\overline{q}(E_1, E_0)}\frac{\overline{q}(E_1, E_2)}{\overline{q}(E_2, E_1)}\cdots\frac{\overline{q}(E_k, E)}{\overline{q}(E, E_k)}, \tag{1.2}$$

where the $\overline{q}$ rates are properly chosen. In case and only in case that the process is reversible the $\overline{q}$ rates can be chosen equal to the original rates. In this general setting it turns out that all jobmark processes have a detailed product form representation of the stationary distribution. Therefore, we shall make a natural restriction for the choice of the $\overline{q}$ rates. First some more notation is needed. In the jobmark process the state $E=\{l_1, \ldots, l_k\}$ denotes that k jobs are present in the system having jobmarks l_1 to l_k. The transition rate from state $E+l=\{l_1, \ldots, l_k, l\}$ to state $E+l'=\{l_1, \ldots, l_k, l'\}$ is denoted by $q(E+l, E+l')$. For any E we focus on a local chain consisting of all these transitions from $E+l$ to $E+l'$ for all $l, l' \in L$, with L the set of jobmarks. So a local chain means that all jobs in E are frozen, they have fixed jobmarks, and one extra job is followed in its transitions. In the next sections we shall illustrate these notions with specific models. The (global) balance equations for the stationary probabilities $P^*(E+l)$ of the local chain at E are:

$$\begin{cases} P^*(E+l)\sum_{l'} q(E+l, E+l') = \sum_{l'} P^*(E+l')q(E+l', E+l), \\ \sum_{l'} P^*(E+l') = 1. \end{cases} \tag{1.3}$$

If the conditional stationary probabilities of the original chain $P(E+l)(\Sigma_{l'}P(E+l'))^{-1}$ satisfy the relation (1.3) the process is said to have the *job local balance property* on $V(E):=\{E+l, l \in L\}$. In HORDIJK and VAN DIJK [11] sufficient and necessary conditions are derived for job local balance (i.e. job local balance on any $V(E)$, $E \in \mathfrak{S}$). For that purpose the notion of adjoint process was introduced. On the transitions between states on $V(E)$ the *adjoint* transition rates $q_A(E+l, E+l')$ are chosen such that for all $l, l' \in L$,

$$X(E+l)q_A(E+l, E+l') = X(E+l')q_A(E+l', E+l), \tag{1.4}$$

with $X(E+l)$, $l \in L$ a solution of (1.3) which is positive on all the recurrent states of the local chain on $V(E)$. Clearly, the adjoint rates are not uniquely determined by the relation (1.4). But any choice with $q_A(E+l, E+l')>0$ if $E+l$ and $E+l'$ are recurrent in the local chain suffices for our purposes. In the next section we will see that in natural models it is not difficult to find adjoint rates. Often they are equal to the original transition rates. In this paper they are equal in all cases, in networks with more than two nodes they generally are different. Let us return to the detailed product form representation of the stationary distribution of the jobmark process and let us define the precise form of this representation by requiring that the $\overline{q}$ rates have to be adjoint transition rates q_A. Then the following theorem holds (cf. HORDIJK and VAN DIJK [11]):

(a) The stationary distribution has the detailed product form representation if and only if the process has the job local balance property.

(b) The job local balance property holds if (and only if) an (any) adjoint chain is reversible.

With this criterion of the reversibility of adjoint chains we shall analyze specific models in the next section. If an adjoint chain turns out to be reversible then we have an explicit expression for the stationary distribution through the representation (1.2). If an adjoint chain is not reversible then neither the detailed product form nor the job local balance property holds.

An other important property for jobmark processes is the insensitivity property. For queueing networks with service times it says: 'The stationary distribution depends on the service time distributions only through their means'. For jobmark processes the sojourn time of the local chain at E in state $E+l$ is exponentially distributed with parameter $\sum_{l'} q(E+l, E+l')$. If the nominal lifetime of jobmark l (the equivalent of the service time) is exponentially distributed with mean μ_l^{-1} then we have to assume that l ages through a rate $\mu_l^{-1}\sum_{l'} q(E+l, E+l')$ if E is the set of the other jobmarks. This ageing rate is the equivalent of the service capacity given to the job with jobmark l. Insensitivity for the jobmark process then says that the stationary distribution is the same for all distributions of the nominal lifetime of job l with the same mean μ_l^{-1}. Mostly insensitivity with respect to the distributions of all jobmarks is considered. For this insensitivity the job local balance property is a sufficient and necessary condition (cf. HORDIJK and VAN DIJK [12]). So the two interesting properties detailed product form and insensitivity always go together.

2. ERLANG AND ENGSET TYPE MODELS

Let us analyze Erlang and/or Engset type models. Let m be the number of lines or trunks and in the Engset model let n be the number of sources. There may be a restricted availability of free lines, e.g., a call attempt may hunt k lines at random. In this case the probability that the attempt finds a free line when there are already i lines occupied is $a_i := 1 - \binom{i}{k} / \binom{m}{k}$. We note that in the analysis we do not use these specific values of the a_i's, the results are more general. If we are interested in the insensitivity with respect to the service time distribution of calls from one particular source we can use as states in the jobmark process i $(i+l)$, $i \geq 0$ meaning there are i lines (i lines plus one line for the designated source) occupied. So for the nondesignated sources we only count the number of busy lines. Also for Poisson input we will analyze the insensitivity for the service time distribution for one job only. The reason to do this is mainly the representation. The more detailed state description necessary for the insensitivity of all call length distributions makes the model notationally much more complicated. However, the proof of the insensitivity of more or all holding times goes similarly. A secondary reason for the choice of models is their relative novelty and their illustration of the versatility of our approach. The transition rates of the models we focus on are depicted below.

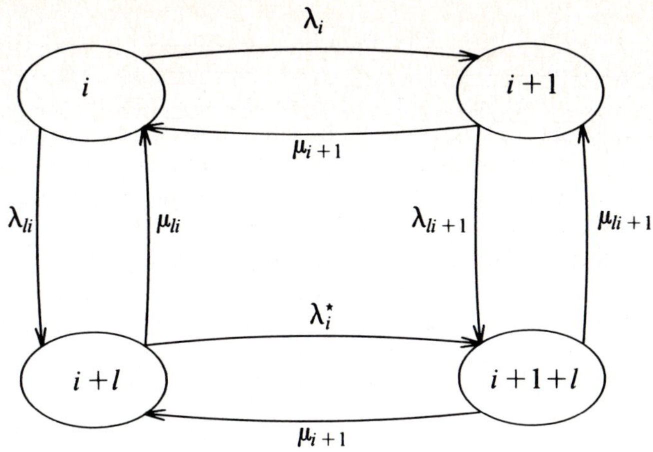

For the Erlang model we take

$$\lambda_i = a_i(1-p)\lambda, \quad \lambda_i^* = a_{i+1}(1-p)\lambda, \quad \mu_i = i\mu, \quad \lambda_{li} = pa_i\lambda \quad \text{and} \quad \mu_{li} = \mu.$$

For the Engset model we take

$$\lambda_i = (n-1-i)a_i\lambda, \quad \lambda_i^* = (n-1-i)a_{i+1}\lambda, \quad \mu_i = i\mu, \quad \lambda_{li} = a_i\lambda \quad \text{and} \quad \mu_{li} = \mu.$$

For the Erlang model p is the probability that an arriving job will be a new labelled job if the last designated one has departed. For the Engset model we have λ as the intensity of a node. We will analyze the insensitivity property for the labelled job. According to Section 1 we have to compute adjoint transition rates. To identify the local chains we need to clarify the connection between the states in the model above and those of the jobmark process. Therefore, we assume three types of jobmarks, say $L=\{0, 1, l\}$. For the set of jobmarks E we count the number of ones, say i and if E does not (does) contain jobmark l then we identify E with i $(i+l)$. For E is i $(i+l)$ the set $V(E)$ becomes $\{i, i+1, i+l\}$ $(\{i+l, i+1+l\})$. It is easily seen that the local chains on these sets are reversible. Therefore, the adjoint rates can be chosen to be equal to the original rates. As a consequence of the theorem of Section 1 we conclude that the insensitivity property holds if and only if the original process is reversible. We use Kolmogorov's criterion (in the sequel K-criterion) to check this. Some reflection gives that the only cycles to check are $(i, i+1, i+1+l, i+l)$ for any $i \leq m-2$. The cycle for i in clockwise direction gives for the product of transition rates,

$$\lambda_i \cdot \lambda_{li+1} \cdot \mu_{i+1} \cdot \mu_{li}.$$

In the opposite direction we get,

$$\lambda_{li} \cdot \lambda_i^* \cdot \mu_{li+1} \cdot \mu_{i+1}.$$

It is easily checked that these expressions are equal for both the Erlang and Engset case. We conclude that the insensitivity for the holding time distribution is guaranteed also for the restricted availability case. The insensitivity for the latter model was studied in IVERSEN [15]. However, the proof given there is in our opinion not correct (cf. BEMELMANS [1]). In COHEN [6] it is shown that the following variant of the usual model has the same stationary distribution. In case a call attempt finds all lines busy the call with the shortest residual holding time is interrupted and the incoming call occupies the freed line. The idea is that the service capacity given to the jobs is made infinitely large when more than m jobs are present.

A question which was raised in a discussion with prof. Cohen is: what happens when the call which will be interrupted is chosen at random? Let us try to analyze this. Again in order to avoid too complex notation we consider at most one identified call. The transition rates for $i \leq m-2$ are similar as before. For $i = m-1$ we have for the Erlang model the following transition rates.

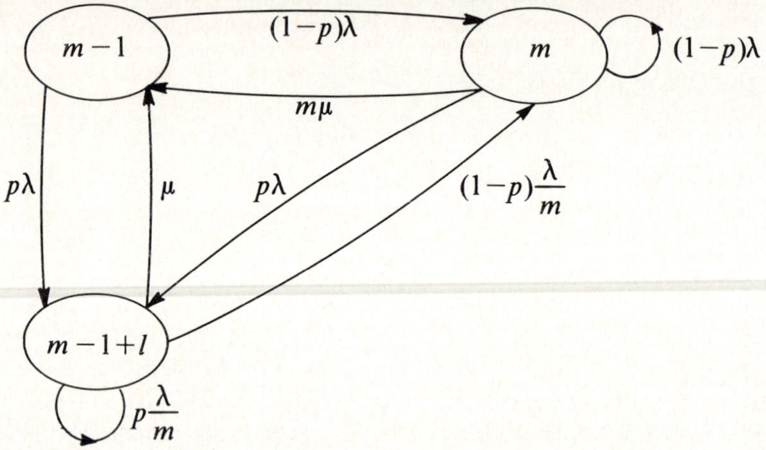

Indeed, in state $m-1+l$ with probability $1/m$ the identified call is interrupted. And with probability p $(1-p)$ the new call is (is not) identified. To find the adjoint rates we remark that the local chain on $\{m-1, m, m-1+l\}$ satisfies the K-criterion. Hence the local chain is reversible and the adjoint rates can again be taken equal to the original rates. The adjoint and original rates satisfy again the K-criterion on the whole state space and both processes are reversible and have the insensitivity criterion. Similar arguments are true for the Engset model. This leads to the proposition that the Erlang and Engset loss formulae remain valid when in a saturated state at random an occupied line is freed.

3. A MORE COMPLEX GRADING

In the Engset system each of the n sources is connected with all m trunks. In teletraffic exchanges more efficient gradings have been implemented. In BEUKELMAN and COHEN [3] the following grading called a transposed multiple is analyzed. In their example there are hundred subscribers, numbered from 00 up to and including 99, which are arranged in a square matrix. The subscribers in a row (column) have complete access to a common group of n_1 (n_2) trunks. There are $10(n_1+n_2)$ trunks available to handle the calls of the 100 subscribers. When a subscriber initiates a call, the n_1 row outlets and the n_2

column outlets will be tested in succession for a free line. The switching equipment is so designed that a call will have no preference for testing the row or the column outlets first.

In BEUKELMAN and COHEN [3] an approximative method is developed to compute the call-congestion. Recently, Cohen raised the question whether the transposed multiple gradings have an insensitive stationary distribution. In case they have, the stationary distribution has the detailed product form and therewith a closed form expression. Then the call-congestion is fairly easy to compute. Unfortunately, the detailed product form and consequently also the insensitivity property do not hold. We shall explain this for a mini grading, but the arguments for more realistic dimensions are strictly similar. Suppose there are 4 subscribers arranged in a square matrix of two by two. The number of trunks per row and column is one.

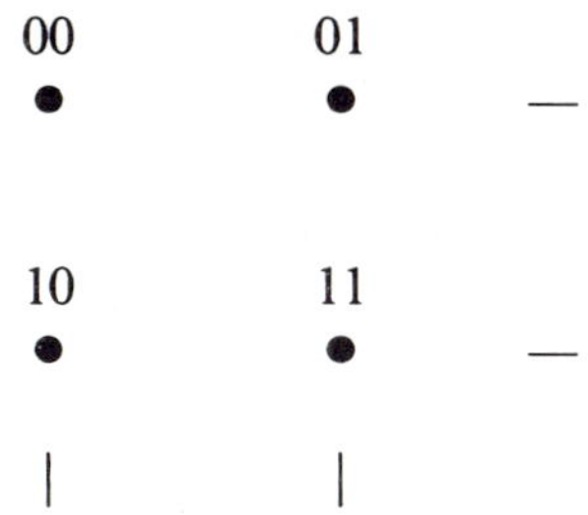

Each of the subscribers has its own jobmark. The jobmark l_i of subscriber i is equal to 0, r, c if the subscriber does not occupy a trunk, occupies a row, occupies a column trunk respectively. Let E be a set of three jobmarks e.g. $E=\{l_{00}, l_{01}, l_{10}\}$ then the local chain at E has as its states $E+0$, $E+r$, $E+c$. Since a transition between the states $E+r$ and $E+c$ has rate zero, we conclude that only simple cycles of two states are possible. Therefore, the local chains are reversible and again the adjoint rates can be chosen equal to the original rates. Hence we have the detailed product form if and only if the K-criterion holds for the original chain.

This is not the case. Denote the state with the values of $(l_{00}, l_{01}, l_{10}, l_{11})$ and take the cycle ((0,0,0,0) $(r,0,0,0)$, $(r,0,c,0)$, $(r,0,c,r)$, $(r,0,0,r)$, $(r,0,0,0)$). In clockwise (counter clockwise) direction we get as the product of rates,

$$\frac{\lambda}{2}\cdot\frac{\lambda}{2}\cdot\frac{\lambda}{2}\cdot\mu\cdot\mu\cdot\mu \quad \left(\frac{\lambda}{2}\cdot\frac{\lambda}{2}\cdot\lambda\cdot\mu\cdot\mu\cdot\mu\right).$$

Hence the K-criterion fails.

Indeed, as is remarked in BEUKELMAN and COHEN [3] the exact calculation of the call-congestion seems practically impossible. Possibly, there is a fast recursive scheme to compute it, we have not thought about that. Beukelman and Cohen give an approximation method. In an ingenious way they compute an approximation of the total load offered to the row (column) outlets including the overflow from the traffic first offered to the column (row) outlets.

As approximation of the call-congestion or blocking probability they use $p_1 p_2$ with p_1 (p_2) the approximation of the row (column) blocking probability.

In recent work by HORDIJK and RIDDER [14], RIDDER [17] and VAN DIJK [18] a methodology is developed to approximate and bound nonproduct form queueing networks through networks which have the product form and are insensitive. It is interesting to note that the approximation of Beukelman and Cohen is of this type. They assume that the offered traffic to the row and column outlets are independent of each other and they approximate the offered traffic correcting it for the overflow.

In HORDIJK and RIDDER [14] there are two bounding methods given which can be applied to this model. In the first one an upper (lower) bound for the call-congestion is found by taking the maximum (minimum) of the expression (1.1) over all paths. In the case of this grading it means that for the upper [lower] bound the offered traffic to any row (column) is always taken as if all lines of the column (row) are [are not] occupied. The approximation of Beukelman and Cohen is a combination of these bounds.

In the second bounding method of HORDIJK and RIDDER [14] the maximum (minimum) is taken over all shortest paths. Clearly this gives better bounds. However, to evaluate them the normalization constant has to be computed. Probably, there is a recursive scheme to do this.

4. PARTIAL INSENSITIVITY

The well-known insensitive queueing models have the property that the complete stationary distribution is insensitive. One can ask for models in which some stationary probabilities are insensitive and other stationary probabilities are not. Probably, the most famous example is the stationary probability P_0 of the empty system in the M/G/1 model. The probability $1-P_0$ is equal to ρ, the traffic intensity, and therefore independent of the service time distribution. Unfortunately for users and probably fortunately for analysts, the other stationary probabilities of the M/G/1 model are not equal to those of the M/M/1 model. We start with a jobmark process and lump states together say to aggregated states. Then it is obvious that the insensitivity with respect to some or all nominal lifetimes is preserved in the aggregated process. The aggregated process is in general not a jobmark process. However, if it is then the job local balance property necessarily holds (cf. HORDIJK and VAN DIJK [11]).

The service discipline in the M/G/1 model is a 'first come first serve' discipline. In HORDIJK and VAN DIJK [11] it is shown in the more general case of a network of queues, that symmetric service disciplines are necessary and sufficient for the job local balance property. The 'first come first serve' discipline is not symmetric. (cf. KELLY [16]) and therefore the stationary distributions of the M/M/1 and M/G/1 model are not the same. The 'last come first serve pre-emptive' discipline is symmetric. It is clear that the jobmark processes for these two service disciplines are different (cf. HORDIJK [13]). Let us consider their aggregated processes in which we only count the number of jobs present in the system. Note that these processes are no longer jobmark processes since the detailed information about the service capacity given to each job is lost. If the service times are exponentially distributed then both processes are Markov chains. Moreover, as a consequence of the Markov property of the exponential distribution the transition rates of both chains are the same. This explains that for both models the stationary distribution has the same detailed product form,

$$P(N=j) = (1-\rho)\rho^j, \quad j \geqslant 0,$$

with N the number of jobs in the system and ρ the traffic intensity. Let us focus now on nonexponential service time distributions. The standard method in jobmark processes is to include the residual service requirement and/or the already received service time measured in a number of phases as components of the jobmarks (cf. HORDIJK and VAN DIJK [12], HORDIJK [13]). Clearly the supplemented jobmark processes, as they are called, are different. However, the probability of the empty system, $P_0 = P(N=0)$, is the same for both service disciplines. The 'last come first serve pre-emptive' discipline has an insensitive stationary distribution. This gives an explanation for the insensitivity of P_0 in the 'first come first serve' discipline system i.e. the M/G/1 model. For the systems with partial insensitivity we have seen so far, similar explanations were possible. This leads us to the conjecture: for any system with partial insensitivity and any insensitive stationary probability P of this system, there is a jobmark process with the insensitivity property such that for some event the stationary probability is equal to P for all service time distributions.

Let us return to the M/G/1 model. In BENEŠ [2] the following formula for the stationary distribution of virtual waiting time, say F_W, is derived,

$$F_W(t) = (1-\rho)\sum_{j=0}^{\infty} \rho^j F_R^{j*}(t), \tag{4.1}$$

with F_R^{j*} the j-fold convolution of the stationary distribution of the residual service time. In COOPER [8] this formula is commented on page 174 as follows (a similar comment can be found in the second edition): 'Although not introduced into the analysis directly, we see that the distribution function of remaining service time at arrival points appears, but in a rather mysterious way. Despite its simple form, the equation (4.1) does not seem to yield to an intuitive term-by-term interpretation except in the special case of exponential service times. Equation (4.1) provides a dramatic counterexample to the oft-stated (and oft-true) sentiment that simple results have simple explanations'.

Well, this result has a simple explanation and a term-by-term interpretation. The virtual waiting time in the M/G/1 model with 'first come first serve' discipline is equal to the total residual workload process for the 'last come first serve pre-emptive' discipline. As already mentioned above the latter service discipline has the insensitivity property. The stationary probability of j jobs is equal to $(1-\rho)\rho^j$. It is well-known that for insensitive models the residual service requirements have identical and independent distributions equal to F_R (cf. COHEN [5], HORDIJK and VAN DIJK [10]). Hence the term with index j in (4.1) is precisely the stationary probability for the 'last come first serve pre-emptive' discipline that there are j jobs in the system and the residual workload is not larger than t.

ACKNOWLEDGEMENT
I want to thank Onno Boxma and an anonymous referee for their helpful comments. They pointed out to me that the interpretation of the Beneš formula has been made before. It can be found in KELLY [16] (p. 81 and p. 213) and it is explained without using insensitivity in COOPER and NIU [9].

REFERENCES

1. R. BEMELMANS (1981). *Analysis of Iversen's Formula* (in Dutch), Master-thesis, University of Leiden.
2. V.E. BENEŠ (1957). On queues with Poisson arrivals. *Ann. Math. Statist. 28,* 670-677.
3. B.J. BEUKELMAN, J.W. COHEN (1957). Call congestion of transposed multiples. *Philips Telecommunication Review 17,* 145-154.
4. J.W. COHEN (1957). The generalized Engset formulae. *Philips Telecommunication Review 18,* 158-170.
5. J.W. COHEN (1979). The multiple phase service network with generalized processor sharing. *Acta Informatica 12,* 245-284.
6. J.W. COHEN (1979). On networks with generalized processor sharing (and a new property of the Erlang B formula). *Proceedings of the 9th International Teletraffic Congress.*
7. J.W. COHEN (1980). Sensitivity and insensitivity. *Delft Progress Report 5, no.3* (papers presented to the symposium 'Traffic and Control' dedicated to the scientific work of prof.dr.ir. L. Kosten), 159-173.
8. R.B. COOPER (1972). *Introduction to Queueing Theory,* The Macmillan Company, New York. Second edition (1981), North-Holland, New York.
9. R.B. COOPER, S.-C. NIU (1986). Beneš's formula for M/G/1-FIFO 'explained' by preemptive LIFO. *J. Appl. Probab. 23,* 550-554.
10. A. HORDIJK, N.M. VAN DIJK (1982). Stationary probabilities for networks of queues. R.L. DISNEY, T.J. OTT (eds.). *Applied Probability-Computer Science: The Interface, Vol. II,* Birkhäuser, Basel, 423-451.
11. A. HORDIJK, N.M. VAN DIJK (1983). Networks of queues. Part I: Job local balance and the adjoint processes, Part II: General routing and service characteristics. F. BACCELLI, G. FAYOLLE (eds.). *Modelling and Performance Evaluation Methodology,* Lecture Notes in Control and Information Sciences 60, Springer, Berlin, 151-205.
12. A. HORDIJK, N.M. VAN DIJK (1983). Adjoint processes, job local balance and insensitivity for stochastic networks. *Proceedings of the 44th Session ISI,* 776-788.
13. A. HORDIJK (1984). Insensitivity for stochastic networks. G. IAZEOLLA, P.J. COURTOIS, A. HORDIJK (eds.). *Mathematical Computer Performance and Reliability,* North-Holland, Amsterdam, 77-94.
14. A. HORDIJK, A. RIDDER (1988). Insensitive bounds for the stationary distribution of nonreversible Markov chains. To appear in *J. Appl. Probab. 25.*

15. V.B. IVERSEN (1980). The A-formula. *Teleteknik* (English edition) *2*, 64-79.
16. F.P. KELLY (1979). *Reversibility and Stochastic Networks,* Wiley, New York.
17. A. RIDDER (1987). *Stochastic Inequalities for Queues,* Ph.D. Thesis, University of Leiden.
18. N.M. VAN DIJK (1987). Simple and insensitive bounds for a grading and an overflow model. *Oper. Res. Letters 6,* 73-76.
19. P. WHITTLE (1986). *Systems in Stochastic Equilibrium,* Wiley, New York.

The Optimization of Queueing and Loss Networks

F.P. Kelly[1]
University of Cambridge
Statistical Laboratory, 16 Mill Lane
Cambridge CB2 1SB, England

How should demands be routed or capacity allocated in a network so as to improve the performance of the network? Here we consider the question, paying particular attention to the common features of queueing and loss networks. We describe how product-form solutions and related approximation procedures can be used to provide important structural insights into optimization issues. In particular, we show that there exist implicit shadow prices associated with each route and with each component of the network, and that the equations defining these prices have a local character. There are a number of implications concerning the extent to which control and planning can be decentralized. At one level the results suggest adaptive routing schemes, capable of responding automatically to local overloads or failures. At another level the shadow prices can be used as a basis for pricing policy or the apportionment of revenue between different sections of a network operation.

1. Introduction

Advances in technology are leading to computer and telecommunications networks which have the potential to behave as coherent, almost intelligent, organisms. At the most basic level networks are able to respond to failures and overloads by rapidly rerouting traffic and reallocating resources. One consequence is that the classical concerns of teletraffic theory, the understanding of loss probabilities and queueing delays, now have impact over a much wider range of design and management issues. Where once capacity was matched to forecast traffic levels every year or so, it is now possible to measure traffic in real time and to reallocate capacity in seconds. Where once a single link could be dimensioned in isolation it is now necessary to treat the network as a whole, since local overloads may have repercussions throughout the system.

A challenge for the theorist is to develop models which can shed light on these complex and interrelated issues. The models must be sufficiently sophisticated to capture the predominant effects present; at the same time they must be sufficiently tractable that meaningful conclusions can be drawn. In this paper an attempt is made to tackle the interrelated issues of routing and capacity allocation, by pursuing some of the optimization questions which naturally arise in connection with product-form networks.

1. Work supported by the Nuffield Foundation.

Many of us have cause to recall the kindness and wisdom of Wim Cohen, and I know that the direction of my work has been influenced by his perceptive remarks, so often delivered with his memorable dry humour. Professor Cohen has long stressed the close relationship between teletraffic analysis and computer performance evaluation, and I hope he will feel this paper to be a contribution to both.

2. Overview

In a queueing network an arriving demand makes use of different resources sequentially in time, and congestion causes delay. In a loss network an arriving demand requests simultaneous use of different resources, and congestion causes blocking. Despite these differences there is a remarkable similarity between the models used to analyze queueing and loss networks. In both areas product-form solutions arise as exact equilibrium distributions for idealized models, and independence assumptions have led to widely useful approximation procedures. In this paper we aim to display the close parallels that exist in connection with optimization issues.

Throughout we especially emphasize the structural features of derivatives rather than the consequential properties of optima. This leads to a more unified mathematical treatment, and there are other advantages. Usually the mathematical formulation of an optimization problem does not capture all of the considerations relevant in the practical problem: derivative information can then be useful in assessing trade-offs between the criteria captured in the mathematical formulation and those not. Also the form of derivatives often gives important structural insights. We shall see that derivatives can usually be expressed in terms of shadow prices, implied costs and surplus values - quantities with tangible interpretations. Often we shall find that the equations defining these quantities have a local character. This has implications concerning the extent to which control and planning can be decentralized. At one level the results suggest adaptive routing schemes, capable of responding automatically to local overloads or failures. At another level the costs and prices can be used as a basis for pricing policy or the apportionment of revenue between different sections of a network operation.

In Section 3 we discuss open queueing networks. These are well understood, and optimization questions were to the fore in much of the early work ([4,17]). See Gerla and Kleinrock [9] for a survey of mathematical programming and network flow approaches to the design of distributed computer communication networks. The underlying analytical model of an open network has a number of simplifying features. In particular, an increase in the capacity of a queue will decrease the mean delay at that queue, but will not affect the mean delay at other queues. Similarly an increase in traffic along a particular route will increase delay only at queues along that route. A consequence is the existence of a shadow price for a queue which can be estimated by the queue from its traffic statistics alone, and Gallager [8] has shown that this can provide the basis for a decentralized routing algorithm. Recently Whittle ([25,26]) has presented a Lagrangian analysis of routing in open queueing networks, and related optima to marginal costs associated with individual queues. The results of Section 3 are, then, well known. Our presentation is less standard, but provides a straightforward introduction to concepts we shall find useful in the discussion of other forms of networks.

Section 4 considers loss networks. Here an increase in the offered traffic along a particular route will increase the blocking at links along that route; this in turn will affect the traffic carried along other routes through these links, and knock-on effects may propagate throughout the network. It remains possible to define a shadow price for a link, but this now depends not just upon the traffic statistics for the link but also upon the shadow prices of adjacent other links within the network. These shadow prices can be recast as an implied cost for a link, measuring the expected net effect upon the performance of the network of using one circuit from that link, and a surplus value for a route, measuring the expected net effect upon the performance of the network of accepting a call on that route. By averaging shadow prices over time or over differing traffic conditions, we can assess the effect on the network of adding additional capacity to the various links of the network. Or we can use a real time distributed computation to estimate implied costs and surplus values, and to drive a decentralized routing algorithm. Important stability questions arise: the effect of a local disturbance to capacity or routing may diminish rapidly with distance from the point of disturbance, or it may not. This indicates the importance of treating a network as a whole: the apparent local benefits of a capacity or routing change may be completely negated by adverse consequences elsewhere in the network. The results of Section 4 are taken from KELLY [14].

In Section 5 we consider closed queueing networks. In many respects these networks are intermediate between those of Sections 3 and 4. An increase in the population size of a particular cycle within a closed queueing network will increase delays at the queues on that cycle; this in turn will affect other cycles through these queues, and again knock-on effects may extend throughout the network. Nevertheless we are able, under a simplified analytical model, to find shadow prices associated with each cycle and with each queue of the network. As in the case of open queueing networks or loss networks, these prices can be used to guide the allocation of additional capacity or as the basis for decentralized routing algorithms.

3. OPEN QUEUEING NETWORKS

We begin with a brief description of an open multi-class network of $\cdot/M/1$ queues (for a more leisurely introduction see [11]). Let the network have J queues, numbered $1, 2, \ldots, J$, and suppose that a customer entering the system is labelled with the route he will follow through the network. More specifically, suppose that customers labelled $r=(r(1), r(2), \ldots, r(M_r))$ arrive at the system in a Poisson stream of rate ν_r and pass through the sequence of queues

$$r(1),\ r(2), \ldots, r(M_r)$$

before leaving the system. Thus the queue which a customer labelled r visits at stage $m(=1, 2, \ldots, M_r)$ of his passage through the network is queue $r(m)$. Write R for the set of possible routes, and assume that as r varies over R it indexes independent Poisson streams. Let each queue have a single server operating a first come first served discipline, and suppose that a customer visiting queue j has a service time there which is exponentially distributed with parameter ϕ_j and independent of all other service times and of the arrival processes at the network.

Write $j \in r$ if route r passes through queue j at some stage, and, for simplicity of notation, suppose no route passes through the same queue more than once. Let

$$\rho_j = \sum_{r:\, j\in r} \nu_r$$

and suppose $\rho_j < \phi_j$, $j=1, 2, \ldots, J$. Thus ρ_j measures the throughput of queue j.

For the network described above, the stationary probability that queue j contains n_j customers, for $j=1, 2, \ldots, J$, is

$$\pi(n_1, n_2, \ldots, n_J) = \prod_{j=1}^{J} \pi_j(n_j) \tag{3.1}$$

where

$$\pi_j(n_j) = \left(1 - \frac{\rho_j}{\phi_j}\right)\left(\frac{\rho_j}{\phi_j}\right)^{n_j}, \tag{3.2}$$

the geometric distribution familiar as the stationary distribution of an $\cdot/M/1$ queue with arrival rate ρ_j. From the distribution (3.2) and Little's formula it follows that the mean sojourn time of a customer in queue j is $(\phi_j - \rho_j)^{-1}$.

In many applications it is possible to vary the parameters ν_r or ϕ_j. For example if the model represents a packet-switched computer communication network there may be a variety of routes r through the network capable of linking two nodes, or there may be the possibility of changing the transmission rate ϕ_j of a channel. How should we judge variations of this sort? To provide some overall measure of the performance of the network suppose that unit delay of a customer on route r incurs a cost of w_r. The mean cost per unit time will then be

$$W(\nu;\phi) = \sum_r w_r \sum_{j\in r} \frac{\nu_r}{\phi_j - \rho_j} \tag{3.3}$$

where $\nu = (\nu_r, r\in R)$ and $\phi = (\phi_j, j=1, 2, \ldots, J)$.

THEOREM 3.1.

$$\frac{d}{d\nu_r} W(\nu;\phi) = \sum_{j\in r}\left(c_j + \frac{w_r}{\phi_j - \rho_j}\right) \tag{3.4}$$

and

$$\frac{d}{d\phi_j} W(v;\phi) = -c_j, \tag{3.5}$$

where

$$c_j = (\phi_j - \rho_j)^{-2} \sum_{r:\, j\in r} w_r v_r. \tag{3.6}$$

PROOF. These results follow directly from the definition (3.3). □

Equation (3.4) shows that the effect of increasing traffic on route r can be assessed by the following rule of thumb: an additional customer on route r will himself suffer a mean delay $(\phi_j - \rho_j)^{-1}$, and thus incur a direct cost of $w_r(\phi_j - \rho_j)^{-1}$, at queue j; in addition he will produce an *implied cost* of c_j to other customers following through queue j. The right-hand side of equation (3.4) adds these direct and indirect costs over each queue j on the route r. Observe that the implied cost c_j of an additional customer at queue j does not depend upon the route r of that customer. Equation (3.5) shows that c_j also has an interpretation as the *shadow price* for any increase in the capacity ϕ_j of queue j.

Finally observe that the implied cost c_j can be estimated, through equation (3.6), from parameters and measurements which are available locally at queue j: the queue's service rate ϕ_j, its throughput ρ_j, and the throughput v_r and worth w_r of each route r passing through queue j.

The above results follow directly from the essential simplicity of the marginal distribution (3.2). If the queues of the network are such that a product-form distribution fails to hold then approximations may be necessary to obtain similar results. A plausible general approximation may be that the mean sojourn time at queue j is a function $D_j(\rho_j, \phi_j)$ of just the throughput ρ_j and capacity ϕ_j. For example, in a network of $\cdot$/G/1 queues D_j might be calculated from the Pollaczek-Khinchin formula for an M/G/1 queue. The mean cost per unit time under the general approximation is

$$W(v;\phi) = \sum_r \sum_{j\in r} w_r v_r D_j(\rho_j, \phi_j). \tag{3.7}$$

Differentiation of this form yields

$$\frac{d}{dv_r} W(v;\phi) = \sum_{j\in r} (c_j + w_r D_j(\rho_j, \phi_j)) \tag{3.8}$$

where

$$c_j = \left[\sum_{r:\, j\in r} w_r v_r\right] \frac{d}{d\rho_j} D_j(\rho_j, \phi_j). \tag{3.9}$$

Clearly

$$\frac{d}{d\phi_j} W(v;\phi) = \left[\sum_{r:\, j\in r} w_r v_r\right] \frac{d}{d\phi_j} D_j(\rho_j, \phi_j), \tag{3.10}$$

and, if it is desired, the relationship (3.5) can be forced to hold by choosing the parametrization of capacity, ϕ_j, so that

$$\frac{d}{d\phi_j} D_j(\rho_j, \phi_j) = -\frac{d}{d\rho_j} D_j(\rho_j, \phi_j). \tag{3.11}$$

For some purposes a more natural parametrization is to use ϕ_j to represent total cost; then $D_j(\rho_j, \phi_j)$ is the approximate mean sojourn time at queue j when an amount ϕ_j has been invested in service capacity at this queue. We might, for example, want to minimize $W(v;\phi)$ over capacity allocations ϕ satisfying the linear cost constraint

$$\sum_j \phi_j \leqslant F.$$

If $D_j(\rho_j, \phi_j)$ is convex in ϕ_j, for each j, then the function $W(v;\phi)$ will be a convex function of ϕ. In this case the derivatives (3.10) can be used to locate the global minimum of $W(v;\phi)$ over a convex region for ϕ, by Lagrangian methods [24]. For example a linear constraint on ϕ leads to Kleinrock's square root channel capacity assignment ([17,18]) in the case of a network of $\cdot/\mathrm{M}/1$ queues.

The derivatives (3.8) are useful when traffic between a pair of nodes can be partitioned over a number of routes. Suppose that two routes serve the same pair of nodes and they have different derivatives (3.8). Then $W(v;\phi)$ can be decreased by moving some traffic from the route with the higher derivative to the route with the lower derivative. If $w_r = 1$, $r \in R$, then (3.7) becomes

$$W(v;\phi) = \sum_j \rho_j D_j(\rho_j, \phi_j).$$

Hence W will be a convex function of v if $\rho_j D_j(\rho_j, \phi_j)$ is convex in ρ_j, for each j. This case is considered in detail by GALLAGER [8], who describes a decentralized routing algorithm based on locally estimated implied costs which is able to converge to the global optimum. If the worths w_r vary from route to route then W as a function of v may have multiple local minima, even in the $\cdot/\mathrm{M}/1$ case covered by Theorem 3.1.

The minimization of $W(v;\phi)$ over *both* v and ϕ is generally a difficult problem [9]: economies of scale usually produce functions $D_j(\rho_j, \phi_j)$ which encourage concentration of flow through a small number of high capacity queues. A mathematical problem is then multiple local minima. A practical problem is that the extreme form of the minima may conflict with criteria not explicitly represented in the function $W(v;\phi)$; for example, robustness considerations often militate against networks with a small number of high capacity queues. One approach is to fix on a particular network form and then use the derivatives (3.8) and (3.10) to find a local minimum, perhaps subject to constraints representing robustness considerations. A possible variant on this approach is to exhibit the various implied costs and shadow prices, and use these to assess trade-offs between network efficiency, as captured in the function $W(v;\phi)$, and the various criteria not explicitly represented in the mathematical formulation.

4. Loss networks

The classical example of a loss network is a telephone network, and we shall phrase our discussion in terms of calls, circuits and routes. However, the reader will observe that our model applies more widely to systems in which before a request (which may be a call, or a task, or a customer) is accepted it is first checked that sufficient resources are available to deal with the request.

Consider then a network with J links, labelled $1, 2, \ldots, J$, and suppose that link j comprises C_j circuits. A subset $r \subset \{1, 2, \ldots, J\}$ identifies a route. Calls requesting route r arrive as a Poisson stream of rate ν_r, and as r varies it indexes independent Poisson streams. A call requesting route r is blocked and lost if on any link $j \in r$ there are no free circuits. Otherwise the call is connected and simultaneously holds one circuit on each link $j \in r$ for the holding period of the call. The call holding period is randomly distributed with unit mean and independent of earlier arrival and holding times. Write R for the set of possible routes. Let n_r be the number of calls in progress using route r. Define the vectors $n = (n_r, r \in R)$, $\nu = (\nu_r, r \in R)$ and $C = (C_1, C_2, \ldots, C_J)$. Then n has a unique stationary distribution given by

$$\pi(n_1, n_2, \ldots, n_J) = G(C)^{-1} \prod_r \frac{\nu_r^{n_r}}{n_r!}, \quad n \in S(C), \tag{4.1}$$

where

$$S(C) = \{n: \sum_{r: j \in r} n_r \leqslant C_j, \ j = 1, 2, \ldots, J\}$$

and $G(C)$ is the normalizing constant (or partition function)

$$G(C) = \sum_{n \in S(C)} \prod_r \frac{\nu_r^{n_r}}{n_r!}. \tag{4.2}$$

The stationary probability that a call requesting route r is accepted is

$$\sum_{n \in S(C - e_r)} \pi(n) = G(C)^{-1} G(C - e_r),$$

where the vector $e_r = (I\{j \in r\}, j = 1, 2, \ldots, J)$ lists the circuits needed by a call on route r. Let

$$1 - L_r^* = G(C)^{-1} G(C - e_r); \tag{4.3}$$

L_r^* is then the loss probability for route r.

Suppose that a call carried on route r generates an expected revenue w_r (or, equivalently, interpret w_r as the cost of losing a call on route r). Then the rate of return from the network will be

$$W^*(\nu; C) = E_\pi(\sum_r w_r n_r),$$

where the expectation is taken with respect to the stationary distribution (4.1).

Throughout we shall use stars to mark quantities calculated from this exact distribution.

THEOREM 4.1.

$$\frac{d}{dv_r} W^*(v;C) = (1-L_r^*)(w_r-c_r^*), \tag{4.4}$$

where

$$c_r^* = W^*(v;C) - W^*(v;C-e_r). \tag{4.5}$$

PROOF. $W^*(v;C) = G(C)^{-1} \sum_{n\in S(C)} \left[\prod_r \frac{v_r^{n_r}}{n_r!}\right] \sum_{r'} n_{r'} w_{r'}$.

Thus

$$\frac{d}{dv_r} W^*(v;\, C) = G(C)^{-1} \sum_{n\in S(C-e_r)} \left[\prod_r \frac{v_r^{n_r}}{n_r!}\right] \sum_{r'} (n_{r'} + I\{r=r'\}) w_{r'}$$

$$- G(C)^{-2} G(C-e_r) \sum_{n\in S(C)} \left[\prod_r \frac{v_r^{n_r}}{n_r!}\right] \sum_{r'} n_{r'} w_{r'}$$

$$= G(C)^{-1} G(C-e_r)[(W^*(v;C-e_r) + w_r - W^*(v;C)].$$

The result now follows from the definitions (4.3) and (4.5). □

Equation (4.4) shows that the effect of increasing traffic on route r can be assessed by the following rule of thumb: an additional call offered to route r will be accepted with probability $1-L_r^*$; if accepted it will earn w_r directly, but at an implied cost of c_r^* to other calls arriving at the network later. The relation (4.5) shows that the implied cost c_r^* also has an interpretation as the shadow price for a decrease in the capacity of the network from C to $C-e_r$.

The exact forms (4.1), (4.3) and (4.4) are awkward to use in practice, mainly owing to the difficulty in handling the partition function (4.2). Fortunately an approximation is to hand. Let B_j, $j=1, 2, \ldots, J$, solve the nonlinear equations

$$B_j = E(\rho_j, C_j), \tag{4.6}$$

where

$$\rho_j = \sum_{r:j\in r} v_r \prod_{k\in r-\{j\}} (1-B_k) \tag{4.7}$$

and for scalar v and C

$$E(v, C) = \frac{v^C}{C!}\left[\sum_{n=0}^{C} \frac{v^n}{n!}\right]^{-1}. \tag{4.8}$$

Thus $E(v, C)$ is just Erlang's formula for the proportion of calls lost on a single link of capacity C offered Poisson traffic at rate v. Then an approximation for L_r^*, the proportion of calls requesting route r that are lost, is

$$L_r = 1-\prod_{j\in r}(1-B_j). \tag{4.9}$$

This approximation is a very natural one to consider and has appeared frequently in the telecommunications literature. The idea behind the approximation is that the stream of rate v_r is thinned by a factor $1-B_k$ at each link $k\in r-\{j\}$ before being offered to link j. If these thinnings could be assumed independent both from link to link and over all routes containing link j (they clearly are not), then the traffic offered to link j would be Poisson with rate ρ_j as given by (4.7). By the Brouwer fixed point theorem the equations (4.6) and (4.7) have a solution $B=(B_1, B_2, \ldots, B_J)$, and it is known that the solution is unique [12]. To emphasize its dependence on the parameters $v=(v_r, r\in R)$ and $C=(C_1, C_2, \ldots, C_J)$ write $B=B(v;C)$. For recent discussions of the approximation (4.9) see [13,23].

Under the above approximation the rate of return from the network is given by

$$W(v;C) = \sum_r w_r\lambda_r$$

where

$$\lambda_r = v_r\prod_{j\in r}(1-B_j)$$

corresponds to the carried traffic on route r and $B=B(v;C)$. Let

$$\delta_j = \rho_j(E(\rho_j, C_j-1)-E(\rho_j, C_j)). \tag{4.10}$$

Extend the definition (4.8) to non-integral values of scalar C by linear interpolation, and at integer values of C_j define the derivative of $W(v;C)$ with respect to C_j to be the left derivative.

THEOREM 4.2.

$$\frac{d}{dv_r}W(v;C) = (1-L_r)s_r \tag{4.11}$$

and

$$\frac{d}{dC_j}W(v;C) = c_j \tag{4.12}$$

where $s=(s_r, r\in R)$, $c=(c_1, c_2, \ldots, c_J)$ *solve the linear equations*

$$s_r = w_r-\sum_{j\in r}c_j, \tag{4.13}$$

$$c_j = \delta_j\frac{\sum_{r:j\in r}\lambda_r(s_r+c_j)}{\sum_{r:j\in r}\lambda_r}. \tag{4.14}$$

PROOF. For a proof of this result see [14]. □

We can interpret s_r as the *surplus value* of a call on route r: if such a call is accepted it will earn w_r directly, but at an implied cost of c_j for each link $j \in r$. The implied costs c measure the knock-on effects of accepting a call upon the other routes in the network. From (4.12) it follows that the implied costs have a dual role, with c_j measuring the sensitivity of the rate of return to the capacity C_j of link j.

Comparing Theorems 4.1 and 4.2 we see that the approximation has allowed the exact quantity c_r^* to be replaced by the additive form $\sum_{j \in r} c_j$.

Expression (4.10) for δ_j is called Erlang's improvement formula ([2,7]), and its use in capacity expansion decisions is known as Moe's principle [21]. Observe that δ_j is simply the increase in the rate at which calls are blocked if a single link offered Poisson traffic at rate ρ_j has its capacity reduced by one circuit, and that δ_j increases from zero to one as ρ_j increases from zero to infinity. Equation (4.14) defines c_j in terms of δ_j and an average, weighted over all routes r through link j, of $s_r + c_j$. Comparing equations (3.6) and (4.14) we see that the implied cost c_j now depends not just upon the traffic statistics for link j but also upon the surplus values of routes through link j, and hence, through equation (4.13), upon the implied costs of other links within the network.

Next we describe how equations (4.13) and (4.14) can be used as the basis for decentralized control of routing through the network. As motivation it is helpful to think in terms of the following model of a distributed computation. Suppose there is limited intelligence in the form of arithmetical processing ability available for each link j and for each route r. This intelligence may be located centrally or it may be distributed over the nodes of the network; for example the processing for route r might be carried out by the source node for calls on route r. Suppose also that there is the possibility of limited communication between the intelligence of link j and route r provided $j \in r$. Consider now equations (4.13) and (4.14). One method for attempting a solution to these equations is repeated substitution. Choose a vector c; substitute it in equation (4.13) to obtain a vector s; substitute these into equation (4.14) to obtain a revised vector c, and repeat. But this computation can be distributed over the intelligences of links and routes, since equation (4.13) for s_r involves implied costs c_j only for links j on the route r, while equation (4.14) for c_j involves only surplus values s_r for routes r passing through link j. In [14] it is shown that if

$$\sum_{j \in r - \{k\}} \delta_j < 1, \quad k \in r, \quad r \in R, \tag{4.15}$$

then repeated substitution converges to the unique solution of the equations. When the condition (4.15) is not satisfied it may be necessary to damp the repeated substitution to obtain convergence. This can still be implemented by a distributed computation, but the individual intelligences may require some knowledge of the network beyond that locally available, in order sufficiently to damp the repeated substitution. In fact it is enough for the intelligences to know just one item of global information, namely J, the total number of links in the network.

The quantities δ_j and λ_r appearing in equations (4.13) and (4.14) are not fixed and known. However they can be estimated by intelligences of links and routes from, for example, local measurements of carried loads. The estimates can then be used in a distributed computation of the vector s. Finally the derivatives (4.11) can be used to implement a decentralized hill-climbing search procedure able to vary routing patterns in response to changes in the demands on the network (see [14] for a fuller discussion). Thus the structural form of the derivatives exposed by Theorem 4.2 directly suggests a decentralized routing algorithm.

The inequality (4.15) limits the size of knock-on effects and is a form of light traffic condition. If it is satisfied then equations (4.13) and (4.14) have a certain stability property: any perturbation of the parameter $w_{r'}$ may cause a change in the entire vector $(s_r, r \in R)$ of surplus values, but the change in the component s_r diminishes rapidly with the extent of the separation between routes r' and r. Without condition (4.15) shadow prices may influence each other over arbitrarily great distances.

It is known that phenomena akin to phase transitions are possible in circuit-switched networks, and examples have been deduced from the exact stationary distribution (4.1): see [14] for references. The possibility emphasizes the importance of treating a network as a whole: the local benefits of a capacity or routing change may be completely negated by adverse consequences elsewhere in the network.

5. Closed queueing networks

In the open network considered in Section 3, customers labelled r entered the system in a Poisson stream of rate ν_r and passed through the sequence of queues

$$r(1), r(2), \ldots, r(M_r) \tag{5.1}$$

before leaving the system. Suppose now that customers labelled r return to queue $r(1)$ after leaving queue $r(M_r)$ and repeat the cycle (5.1) through the system. The network will become closed with customers neither entering nor leaving the system. The number of customers labelled r, N_r say, will remain fixed for $r \in R$. As in Section 3 suppose that each queue has a single server, operating a first come first served discipline, and that a customer visiting queue j has a service time there which is exponentially distributed with parameter ϕ_j and independent of other service times. For ease of exposition suppose the cycle of queues (5.1) contains no repeats and write $j \in r$ if queue j appears in this cycle. Let n_j be the number of customers in queue j, and let x_{ij} be the label r of the customer in position i in queue j, for $i = 1, 2, \ldots, n_j$. Write $n = (n_j, j = 1, 2, \ldots, J)$, $x = (x_{ij}, i = 1, 2, \ldots, n_j, j = 1, 2, \ldots, J)$, and let $\phi = (\phi_j, j = 1, 2, \ldots, J)$ and $N = (N_r, r \in R)$. Observe that n is a function of x.

For the network described above x has a unique stationary distribution given by [11]

$$\pi_N(x) = G(N)^{-1}\prod_j \phi_j^{-n_j}, \quad x \in S(N), \tag{5.2}$$

where

$$S(N) = \{x: \sum_i \sum_j I\{x_{ij}=r\}=N_r, r\in R\}$$

and $G(N)$ is the normalizing constant

$$G(N) = \sum_{x\in S(N)} \prod_j \phi_j^{-n_j}.$$

Define the throughput of cycle r to be

$$\theta_r^* = G(N)^{-1}G(N-e_r) \tag{5.3}$$

where e_r is the unit vector $e_r = (I\{r=r'\}, r'\in R)$. From the distribution (5.2) it follows that

$$\theta_r^* = \sum_{x\in S(N):\, x_{1j}=r} \pi_N(x)\phi_j$$

for each $j\in r$. Thus θ_r^* is just the rate at which customers labelled r leave queue j, and this must be equal at each queue j around the cycle r. Suppose now that a customer labelled r generates an expected revenue w_r per cycle. Then the rate of return from the network will be

$$W^*(\phi; N) = \sum_r w_r \theta_r^*. \tag{5.4}$$

The expected number of customers in queue j is

$$E(n_j) = \sum_{x\in S(N)} \pi_N(x) n_j.$$

Let

$$E_r(n_j) = \sum_{x\in S(N-e_r)} \pi_{N-e_r}(x) n_j,$$

the expected number of customers in queue j in a system with one less customer labelled r. The following theorem is a special case of a result due to LOUTH [19].

THEOREM 5.1.

$$\frac{d}{d\phi_j} W^*(\phi; N) = \phi_j^{-1}\sum_r w_r\theta_r^*[E(n_j)-E_r(n_j)]. \tag{5.5}$$

PROOF. From (5.3) and (5.4)

$$W^*(\phi; N) = G(N)^{-1}\sum_r w_r G(N-e_r).$$

The result follows from differentiation of this form, using the identities

$$\frac{d}{d\phi_j}\, G(N) = -\phi_j^{-1} E(n_j) G(N)$$

and

$$\frac{d}{d\phi_j}\, G(N-e_r) = -\phi_j^{-1} E_r(n_j) G(N-e_r). \quad \square$$

The quantity $E_r(n_j)$, the expected number of customers in queue j in a system with one less customer labelled r, has a number of other interpretations. For example, it is the expected number of customers in queue j at the points in time when a customer labelled r arrives at any queue on the cycle r ([11], Theorem 3.12). Thus the quantities involved on the right-hand side of equation (5.5) are potentially measurable. Unfortunately the equation is hardly local: the summation in (5.5) is over all cycles $r \in R$, and not just those which pass through queue j.

Theorem 5.1 is capable of generalization to any closed network of quasi-reversible queues, with ϕ_j a scale factor affecting the service effort at queue j. However for many applications, particularly in the area of large scale distributed systems, the lack of localization in equation (5.5) is a significant drawback.

We attempt to circumvent this drawback by replacing the exact distribution (5.2) with a more tractable approximation. A number of approximations have been considered in the literature, motivated by the difficulty in handling the normalizing constant in the exact distribution (5.2). For reviews see [6,22]. We describe here what is perhaps the simplest. For fixed ϕ and N let θ_r, $r \in R$, solve the equations

$$\sum_{j \in r} \frac{\theta_r}{\phi_j - \rho_j} = N_r \tag{5.6}$$

where

$$\rho_j = \sum_{r:\, j \in r} \theta_r. \tag{5.7}$$

The idea behind this approximation is straightforward: if the various arrival streams at queue j could be assumed independent and Poisson, with customers labelled r arriving at rate θ_r, then $\theta_r(\phi_j - \rho_j)^{-1}$ would be the expected number of customers labelled r in queue j. Equation (5.6) then fixes the expected number of customers labelled r in the network. By the Brouwer fixed point theorem, equations (5.6) and (5.7) have a solution $\theta = (\theta_r, r \in R)$, and Chow [5] has shown it to be unique. To emphasize its dependence on the parameter vectors ϕ and N, write $\theta = \theta(\phi; N)$.

The above approximation has a natural extension beyond exponential single server queues, similar to that discussed for open networks in Section 3. Suppose that we are prepared to represent the mean sojourn time at queue j as a function $D_j(\rho_j, \phi_j)$ of just the throughput ρ_j and capacity ϕ_j. Then $\theta_r D_j(\rho_j, \phi_j)$ is the expected number of customers labelled r in queue j, and the analogue of equation (5.6) becomes

$$\sum_{j\in r}\theta_r D_j(\rho_j, \phi_j) = N_r. \tag{5.8}$$

Impose the following mild regularity condition on the functions D_j, $j=1, 2, \ldots, J$: suppose that as a function of its first argument D_j is continuous and non-decreasing.

THEOREM 5.2. *Equations (5.7) and (5.8) have a unique solution for* $\theta=\theta(\phi; N)$.

PROOF. Consider the function

$$-\sum_j \int_0^{\rho_j} D_j(\rho, \phi_j)d\rho+\sum_r N_r \log\theta_r, \tag{5.9}$$

where ρ_j is given by equation (5.7). The derivative of this function with respect to θ_r is

$$-\sum_{j\in r}D_j(\rho_j, \phi_j)+N_r\theta_r^{-1}.$$

Thus equations (5.8) locate a stationary point of the continuously differentiable function (5.9) with respect to variations in θ. Now the function (5.9) is also a strictly concave function of θ, with an interior maximum. Hence it has a unique stationary point. □

Let

$$W(\phi; N) = \sum_r w_r\theta_r$$

where $\theta=\theta(\phi; N)$ solves (5.7) and (5.8). Observe that $W(\phi; N)$ is defined for real, and not just integer, values of N_r, $r\in R$. Suppose that D_j is differentiable in both its arguments, and let

$$\alpha_j = -\left[\frac{d}{d\rho_j} D_j(\rho_j, \phi_j)\right]^{-1} \frac{d}{d\phi_j} D_j(\rho_j, \phi_j). \tag{5.10}$$

THEOREM 5.3.

$$\frac{d}{d\phi_j} W(\phi; N) = c_j \tag{5.11}$$

and

$$\frac{d}{dN_r} W(\phi; N) = s_r, \tag{5.12}$$

where

$$c_j = \left[\frac{d}{d\rho_j} D_j(\rho_j, \phi_j)\right] \sum_{r:j\in r} \theta_r s_r \tag{5.13}$$

and

$$s_r = \left[\sum_{j\in r} D_j(\rho_j, \phi_j)\right]^{-1} (w_r - \sum_{j\in r} \alpha_j c_j). \tag{5.14}$$

PROOF. We establish this result by differentiating W on various specially constructed manifolds around (ϕ, N). Alter ϕ_j, but adjust N_r for r such that $j\in r$ so that θ_r, $r\in R$, are left constant. The compensating change in N_r is given by

$$dN_r = \theta_r \left[\frac{d}{d\phi_j} D_j(\rho_j, \phi_j)\right] d\phi_j$$

for r such that $j\in r$. Since θ_r, $r\in R$, are left constant we have the following explicit evaluation of a differential form:

$$\left\{\frac{d}{d\phi_j} - \sum_{r:j\in r} \theta_r \left[\frac{d}{d\phi_j} D_j(\rho_j, \phi_j)\right] \frac{d}{dN_r}\right\} W(\phi; N) = 0. \tag{5.15}$$

Next alter N_r, but adjust ϕ_j, $j\in r$, so that $\theta_{r'}$, $r'\neq r$, are left constant. The change dN_r induces a change

$$d\rho_j = d\theta_r, \quad j\in r,$$

where, from (5.8),

$$\left[\sum_{j\in r} D_j(\rho_j, \phi_j)\right] d\theta_r = dN_r.$$

The compensating change in ϕ_j is given by

$$d\phi_j = -\alpha_j^{-1} d\rho_j,$$

from (5.10). We thus have the following explicit evaluation of a differential form:

$$\left\{\frac{d}{dN_r} + \left[\sum_{j\in r} D_j(\rho_j, \phi_j)\right]^{-1} \sum_{j\in r} \alpha_j \frac{d}{d\phi_j}\right\} W(\phi; N) = \left[\sum_{j\in r} D_j(\rho_j, \phi_j)\right]^{-1} w_r. \tag{5.16}$$

Equations (5.13) and (5.14) follow from (5.15) and (5.16), under the identification (5.11) and (5.12). □

If (3.11) holds then $\alpha_j = 1$. For example if $D_j(\rho_j, \phi_j) = (\phi_j - \rho_j)^{-1}$, corresponding to the $\cdot$/M/1 approximation procedure (5.6) and (5.7), then

$$c_j = (\phi_j - \rho_j)^{-1} \sum_{r:j\in r} \theta_r s_r \tag{5.17}$$

and

$$s_r = \left[\sum_{j\in r} (\phi_j - \rho_j)^{-1}\right]^{-1} (w_r - \sum_{j\in r} c_j). \tag{5.18}$$

The similarities with earlier sections are striking. Comparing (5.13) and (5.17) with (3.9) and (3.6) we see that the implied costs c_j are of the same form as in an open queueing network, but with the worth w_r replaced by the surplus value s_r, and the open arrival rate ν_r replaced by the throughput θ_r. Surplus values did not arise in open queueing networks, but they did in loss networks. Comparing equations (5.14) and (5.18) with equation (4.11) we see that the major difference is the additional factor $[\sum_{j\in r} D_j(\rho_j, \phi_j)]^{-1}$, which corresponds to the reciprocal of the mean round-trip time for a customer on cycle r.

6. Concluding remarks

We have attempted to provide a unified treatment of optimization issues concerning routing and capacity allocation in both queueing and loss networks. To this end we have concentrated on the simplest possible models which illustrate our themes. However, the basic approach described is capable of considerable generalization. For example, in the open queueing networks of Section 3 an obvious extension is to allow priority disciplines, based on the worths w_r of differing customers. The approach of Section 3 follows through; we simply represent the mean sojourn time of a customer on route r at queue j by a function of the throughputs ν_r of routes r passing through queue j. This extension is more interesting for the closed queueing networks of Section 5. In closed networks it is not the case that the parameters w_r should determine the priorities at a queue, irrespective of what is happening elsewhere in the network. Rather priorities at a queue should depend upon the surplus values of the customers passing through it, with these values capturing knock-on effects elsewhere in the network.

A similar extension is available for the model of Section 4. In a loss network priority can be given to routes through a link by, for example, trunk reservation: non-priority calls are not allowed through the link when the link is nearly full. Another important extension for loss networks is alternative routing, where a call blocked on a route may be allowed to try again on another route. In [15] it is shown how shadow prices can be defined for loss networks with both trunk reservation and alternative routing,

Many of the results we have described are based on approximate analytical models, and an important issue is the accuracy of these models. The reader interested in this area should see [3,12,22,23,27]. There is a growing body of theory which suggests that the models are most appropriate for large, complex networks, with many overlapping routes, large link capacities or large population sizes.

Another important issue we shall do no more than mention concerns time scales. Any adaptive routing scheme must make implicit, if not explicit, comparisons between the time periods over which offered loads vary, estimates respond, and optimization procedures converge. To illustrate this point suppose that the rate of a Poisson arrival stream is stable for a period of order T. Then an estimate of this rate will have standard error of order at least $T^{-1/2}$. Thus the convergence of an optimization procedure may often be constrained by the slow rate at which errors in load estimates can be reduced. This estimation problem emphasizes the importance of routing schemes which are themselves robust to uncertainties in load parameters, such as alternative routing schemes with either thresholds or shortest paths in queueing networks, or with trunk reservation in loss networks. For example, the scheme described for loss networks in ([10,16]) can react quickly and robustly to sudden changes in load parameters, with estimation and fine tuning of trunk reservation parameters taking place over a longer period.

We have emphasized the parallels that exist between queueing and loss networks: the differences are perhaps more obvious. One difference concerns the controls which are available. In packet-switched computer communication networks an acknowledgement can easily convey back to a source node the time it took to traverse a route: this allows shortest path routing algorithms [1] to be implemented. In contrast, in circuit-switched telephone networks trunk reservation and alternative routing are often implemented. A second difference concerns the various time scales involved, even when offered loads are stable. In a queueing network information that it has just taken a certain period to traverse a route remains relevant for a large number of such periods, especially in heavy traffic [20]. In a loss network information that a link has a certain number of free circuits remains relevant for only a short period, especially in heavy traffic [12].

References

1. D.P. Bertsekas, R.G. Gallager (1987). *Data Networks,* Prentice-Hall, Englewood Cliffs (N.J.).
2. E. Brockmeyer, H.L. Halstrom, A. Jensen (1948). *The Life and Works of A.K. Erlang,* Academy of Technical Sciences, Copenhagen.
3. T.C. Brown, P.K. Pollett (1982). Some distributional approximations in Markovian queueing networks. *Adv. in Appl. Probab. 14,* 654-671.
4. D.G. Cantor, M. Gerla (1974). Optimal routing in a packet switched computer network. *IEEE Trans. Comput. 23,* 1062-1069.
5. W.-M. Chow (1983). Approximations for large scale closed queueing networks. *Performance Evaluation 3,* 1-12.
6. E. de Souza e Silva, S.S. Lavenberg, R.R. Muntz (1984). A perspective on iterative methods for the approximate analysis of closed queueing networks. G. Iazeolla, P.J. Courtois, A. Hordijk (eds.). *Mathematical Computer Performance and Reliability,* North-Holland, Amsterdam, 225-244.
7. A.K. Erlang (1924). On the rational determination of the number of circuits. In [2], 216-221.
8. R.G. Gallager (1977). A minimum delay routing algorithm using distributed computation. *IEEE Trans. Commun. 25,* 73-85.
9. M. Gerla, L. Kleinrock (1977). On the topological design of distributed computer networks. *IEEE Trans. Commun. 25,* 48-60.

10. R.J. GIBBENS, F.P. KELLY, P.B. KEY (1988). Dynamic alternative routing - modelling and behaviour. *Proc. International Teletraffic Congresses 12,* Turin.
11. F.P. KELLY (1979). *Reversibility and Stochastic Networks,* Wiley, Chichester.
12. F.P. KELLY (1986). Blocking probabilities in large circuit-switched networks. *Adv. in Appl. Probab. 18,* 473-505.
13. F.P. KELLY (1986). Blocking and routing in circuit-switched networks. O.J. BOXMA, J.W. COHEN, H.C. TIJMS (eds.). *Teletraffic Analysis and Computer Performance Evaluation,* North-Holland, Amsterdam, 37-45.
14. F.P. KELLY (1988). Routing in circuit-switched networks: optimization, shadow prices and decentralization. *Adv. in Appl. Probab. 20,* 112-144.
15. F.P. KELLY (1989). Routing and capacity allocation in networks with trunk reservation. *Math. Oper. Res. 14* (to appear).
16. P.B. KEY, M.J. WHITEHEAD (1988). Cost-effective use of networks employing dynamic alternative routing. *Proc. International Teletraffic Congresses 12,* Turin.
17. L. KLEINROCK (1964). *Communication Nets: Stochastic Message Flow and Delay,* McGraw-Hill, New York.
18. L. KLEINROCK (1976). *Queueing Systems, Vol. 2: Computer Applications,* Wiley, New York.
19. G.M. LOUTH (1986). Personal communication.
20. M.I. REIMAN (1984). Open queueing networks in heavy traffic. *Math. Oper. Res. 9,* 441-458.
21. R. SYSKI (1986). *Introduction to Congestion Theory in Telephone Systems,* 2nd edition, North-Holland, Amsterdam.
22. W. WHITT (1984). Open and closed models for networks of queues. *AT&T Bell Lab. Tech. J. 63,* 1911-1979.
23. W. WHITT (1985). Blocking when service is required from several facilities simultaneously. *AT&T Bell Lab. Tech. J. 64,* 1807-1856.
24. P. WHITTLE (1971). *Optimization under Constraints,* Wiley, New York.
25. P. WHITTLE (1985). Scheduling and characterization problems for stochastic networks (with discussion). *J. Roy. Statist. Soc. B47,* 407-428.
26. P. WHITTLE (1986). *Systems in Stochastic Equilibrium,* Wiley, Chichester.
27. I. ZIEDINS (1987). *Stochastic Models of Traffic in Star and Line Networks,* Thesis, University of Cambridge.

PART V
QUEUEING MODELS AND RELATED TOPICS

A Note on the Random Walk for the Transport and Deposition of Particles

J. Gani

Statistics and Applied Probability Program
University of California, Santa Barbara, CA 93106, USA

This note considers a bivariate random walk model $\{X(t), Y(t); t \geqslant 0\}$ for the transport and deposition of a particle in a fluid flowing in a tank, or travelling in a porous medium. It is assumed that the number of steps $N(t)$ in the x direction is a Poisson process, the sizes of the steps $\xi_j > 0$ being i.i.d. random variables, also independent of $N(t)$. The number of steps $M(t)$ in the y direction is also taken to be a Poisson process, the sizes of the steps $\eta_k > 0$ being i.i.d. random variables, independent of $M(t)$. Both the cases of $N(t)$, $M(t)$ independent of each other, as well as $N(t)$, $M(t)$ correlated are considered. Some expressions are obtained for the Laplace and Laplace-Stieltjes transforms of the probabilities of suspension, deposition and exit of the particle in the tank or porous medium.

1. Introduction

Many engineering and environmental problems involve the transport of particles in a fluid flow or a porous medium, and their permanent deposition after they have travelled a certain distance. Particle deposition may take the form of sedimentation in a tank or a river, or the trapping of material in a layer of porous medium. In such cases, a useful model for the motion of a particle is the bivariate random walk.

Consider a particle whose position at time $t \geqslant 0$ is specified by $\{X(t), Y(t)\}$, and whose initial position at $t=0$ is $\{X(0)=0, Y(0)=0\}$, as in Fig. 1. $X(t)$ may correspond to the horizontal movement of a particle along the length of a tank or river, while $Y(t)$ is the vertical distance travelled, up to a maximum of H, the depth of the tank or river. Alternatively, $X(t)$ may be the longitudinal

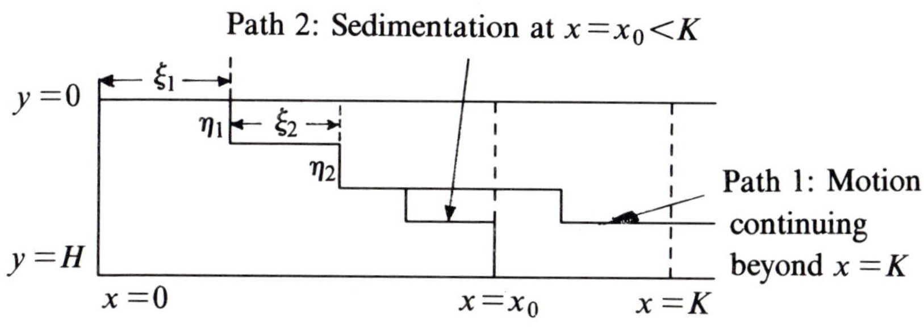

FIGURE 1. *Motion and sedimentation of a particle*

distance travelled by a particle dispersing in a porous medium, and $Y(t)$ its movement orthogonal to the x-axis.

The two processes $\{X(t); t\geq 0\}$ and $\{Y(t); t\geq 0\}$ are taken in the first instance to be mutually independent, and are defined in general by

$$X(t) = \sum_{j=0}^{N(t)} \xi_j \ , \quad 0\leq X(t)<\infty,$$
$$Y(t) = \sum_{k=0}^{M(t)} \eta_k \ , \quad 0\leq Y(t)<\infty. \tag{1.1}$$

For simplicity, let $N(t)$ be a homogeneous Poisson process with parameter λ, independent of the $\xi_j>0$, these being themselves independently and identically distributed (i.i.d.) random variables with the distribution function (d.f.) $A(x)$, while $M(t)$ is another homogeneous Poisson process with parameter μ, independent of the $\eta_k>0$, these being i.i.d. random variables with d.f. $B(y)$, also independent of the ξ_j.

For a tank of depth H, we shall assume that the process $Y(t)$ has an absorbing barrier at $y=H$; this case may correspond in practical terms to the sedimentation of a particle injected into the fluid in the tank at its surface, when it has travelled the vertical distance H and reached the bottom of the tank. It may also correspond to the random distance travelled by a particle in a porous medium before it becomes trapped; in this case H may be random with d.f. $L(h)$.

To provide a readily understandable practical context, we develop the model as one for the sedimentation of a particle in a fluid, following the work of GANI and TODOROVIC [1,2,3]. For the equivalent interpretation of a particle subject to dispersion in a porous medium, the reader is referred to TODOROVIC [6,7], and TODOROVIC and GANI [8]. A review of solute transport in porous media is given by SPOSITO, GUPTA and BHATTACHARYA [5]. Three cases will be considered:

(a) the infinite tank: $0\leq X(t)<\infty$, $0\leq Y(t)\leq H$, corresponding to sedimentation for $Y(t)=H$, and unrestricted non-negative motion along the x-axis otherwise;

(b) the finite tank with unfiltered exit at $x=K$: $0\leq X(t)\leq K$, $0\leq Y(t)\leq H$, where particles sediment for $Y(t)=H$, and are otherwise considered to exit unfiltered from the system with probability 1 when $X(t)>K$;

(c) the finite tank with filtered exit at $x=K$: $0\leq X(t)\leq K$, $0\leq Y(t)\leq H$, where particles sediment when $Y(t)=H$, and exit from the system through a filter with probability less than 1 when $X(t)>K$; there is thus an accumulation of filtered sedimentary particles at $x=K$.

2. THE INFINITE TANK

Let us begin by considering the time T_H to sedimentation (absorption) of a particle in an infinite tank; in Fig. 1, this is the time at which the particle in Path 2 hits the barrier $y=H$. It is readily seen that

$$\Pr\{T_H > t\} = \Pr\{Y(t) < H\},$$

so that the d.f. $G(t) = \Pr\{T_H \leqslant t\} = 1 - \Pr\{Y(t) < H\}$. For the process $Y(t)$ defined in (1.1), we have directly that

$$\Pr\{Y(t) < H\} = e^{-\mu t}\{\sum_{j=0}^{\infty} \frac{(\mu t)^j}{j!} B_j(H-)\}$$

where $B_j(y)$ is the j-th convolution of the d.f. $B(y)$, and $B_0(y) = 1$ by definition.

It follows from this that if, for simplicity, $B(y)$ is continuous to the left,

$$G(t) = 1 - e^{-\mu t}\{\sum_{j=0}^{\infty} \frac{(\mu t)^j}{j!} B_j(H)\}, \tag{2.1}$$

whose Laplace transform with respect to time t is given by

$$\int_0^{\infty} e^{-\theta t} G(t) dt = G^*(\theta) = \frac{1}{\theta} - \sum_{j=0}^{\infty} \frac{\mu^j}{(\mu+\theta)^{j+1}} B_j(H). \tag{2.2}$$

Note that since $\Sigma_{j=0}^{\infty} B_j(H) < \infty$ for any finite H, the d.f. $G(t)$ is proper. The probability density function (p.d.f.) of the time to sedimentation is obtained directly from (2.1) as

$$G'(t) = \mu e^{-\mu t}\{\sum_{j=0}^{\infty} \frac{(\mu t)^j}{j!} [B_j(H) - B_{j+1}(H)]\}. \tag{2.3}$$

If H is a random variable with d.f. $L(h)$, as it may be for particles in a porous medium, then we need to evaluate

$$\int_0^{\infty} G(t) dL(h) = 1 - e^{-\mu t}\{\sum_{j=0}^{\infty} \frac{(\mu t)^j}{j!} \int_0^{\infty} B_j(h) dL(h)\}, \tag{2.4}$$

where $B_j(H)$ in (2.1) is now replaced by $B_j = \int_0^{\infty} B_j(h) dL(h)$. This presents no additional difficulty, and we shall therefore not refer to it further.

Let us now derive the d.f. $F(x,t) = \Pr\{X(t) \leqslant x, Y(t) < H\}$ of a particle in suspension. This is easily seen to be

$$\begin{aligned} F(x,t) &= \Pr\{X(t) \leqslant x | T_H > t\} \Pr\{T_H > t\} \\ &= e^{-(\lambda+\mu)t} \sum_{k=0}^{\infty} \sum_{j=0}^{\infty} \frac{\lambda^k \mu^j t^{k+j}}{k! j!} A_k(x) B_j(H), \end{aligned} \tag{2.5}$$

whose Laplace transform with respect to time is given by

$$\int_0^{\infty} e^{-\theta t} F(x,t) dt = F^*(x,\theta) =$$

$$\sum_{k=0}^{\infty} \sum_{j=0}^{\infty} \frac{1}{(\lambda+\mu+\theta)} (\frac{\lambda}{\lambda+\mu+\theta})^k (\frac{\mu}{\lambda+\mu+\theta})^j \binom{k+j}{j} A_k(x) B_j(H). \tag{2.6}$$

By taking the Laplace-Stieltjes transform of $F^*(x,\theta)$ with respect to x, one obtains the simpler expression

$$f^*(s,\theta) = \int_0^\infty e^{-sx} d_x F^*(x,\theta) =$$

$$= \sum_{j=0}^{\infty} \frac{1}{(\lambda+\mu+\theta)} \left(\frac{\mu}{\lambda+\mu+\theta}\right)^j \left(1-\frac{\lambda A^*(s)}{\lambda+\mu+\theta}\right)^{-(j+1)} B_j(H)$$

$$= \sum_{j=0}^{\infty} \frac{\mu^j B_j(H)}{\{\mu+\theta+\lambda(1-A^*(s))\}^{j+1}}$$

where $A^*(s)=\int_0^\infty e^{-sx} d_x A(x)$.

We must now account for the probability of sedimentation of a particle in the time interval $0<\tau\leqslant t$, at or below the position x; this is

$$P(x,t) = \Pr\{X(T_H)\leqslant x,\ T_H\leqslant t\} = \Pr\{X(T_H)\leqslant x,\ Y(s)<H, 0\leqslant s<T_H\leqslant t\}$$

$$= \int_0^t \Pr\{X(\tau)\leqslant x \mid Y(s)<H, 0\leqslant s\leqslant \tau,\ T_H>\tau\} \Pr\{\tau<T_H<\tau+d\tau\}$$

$$= \int_0^t \frac{F(x,\tau)}{1-G(\tau)} G'(\tau) d\tau$$

$$= \int_0^t \mu e^{-(\lambda+\mu)\tau} \sum_{k=0}^{\infty} \sum_{j=0}^{\infty} \frac{\lambda^k \mu^j \tau^{k+j}}{k!j!} A_k(x)\{B_j(H)-B_{j+1}(H)\} d\tau, \tag{2.7}$$

with the Laplace transform with respect to time

$$P^*(x,\theta) = \sum_{k=0}^{\infty} \sum_{j=0}^{\infty} \frac{\mu}{\theta(\lambda+\mu+\theta)} \left(\frac{\lambda}{\lambda+\mu+\theta}\right)^k \left(\frac{\mu}{\lambda+\mu+\theta}\right)^j \times$$

$$\binom{k+j}{j} A_k(x)\{B_j(H)-B_{j+1}(H)\}. \tag{2.8}$$

It is readily seen that $P(\infty,t)=\Pr\{T_H\leqslant t\}$ and $P(x,\infty)=\Pr\{X(T_H)\leqslant x, T_H<\infty\}=\Pr\{X(T_H)\leqslant x\}$. As $x\to\infty$, we have that $F(\infty,t)+P(\infty,t)=1$; for since $\Pr\{X(t)<\infty \mid T_H>t\}=1$,

$$\Pr\{T_H>t\}+\int_0^t \Pr\{\tau<T_H<\tau+d\tau\} = 1-G(t)+G(t) = 1.$$

The Laplace-Stieltjes transform of $P^*(x,\theta)$ with respect to x is clearly

$$p^*(s,\theta) = \int_0^\infty e^{-sx} d_x P^*(x,\theta) = \sum_{j=0}^{\infty} \frac{\mu^{j+1}\{B_j(H)-B_{j+1}(H)\}}{\{\mu+\theta+\lambda(1-A^*(s))\}^{j+1}}.$$

Finally, it will be obvious since sedimentation is an absorbing state, that $\lim_{t\to\infty} F(x,t)=0$ for all values of x, while $\lim_{t\to\infty} P(x,t)=P_x$ is the stationary probability of sedimentation at or below x, such that $\lim_{x\to\infty} P_x=1$. We see directly from (2.6) that

$$\lim_{t\to\infty} F(x, t) = \lim_{\theta\to 0} \theta F^*(x, \theta) = 0,$$

while from (2.8), we derive

$$P_x = \lim_{\theta\to 0} \theta P^*(x, \theta)$$

$$= \sum_{k=0}^{\infty} \sum_{j=0}^{\infty} \frac{\mu}{\lambda+\mu} \left(\frac{\lambda}{\lambda+\mu}\right)^k \left(\frac{\mu}{\lambda+\mu}\right)^j \binom{k+j}{j} A_k(x)\{B_j(H)-B_{j+1}(H)\} \tag{2.9}$$

which provides us with the stationary distribution of the sedimentation of a particle at or below the position x. Its Laplace-Stieltjes transform with respect to x is given by

$$\lim_{\theta\to 0} \theta p^*(s, \theta) = \int_0^\infty e^{-sx} d_x P_x = \sum_{j=0}^{\infty} \frac{\mu^{j+1}\{B_j(H)-B_{j+1}(H)\}}{\{\mu+\lambda(1-A^*(s))\}^{j+1}}.$$

3. The finite tank case with unfiltered and filtered exit at $x = K$

For the finite tank with unfiltered exit at $x = K$ of 1(b), we assume that for $0 \leqslant x < K$ in Fig. 1, the distribution of particles in suspension is $F(x, t)$, while that of sedimentary particles is $P(x,t)$. The probability of exit of a particle at $x = K$ will be given by

$$Q(t) = 1 - F(K, t) - P(K, t)$$

$$= 1 - e^{-(\lambda+\mu)t} \sum_{k=0}^{\infty} \sum_{j=0}^{\infty} \frac{\lambda^k \mu^j}{k!j!} t^{k+j} A_k(K) B_j(H)$$

$$- \int_0^t \mu e^{-(\lambda+\mu)\tau} \sum_{k=0}^{\infty} \sum_{j=0}^{\infty} \frac{\lambda^k \mu^j}{k!j!} \tau^{k+j} A_k(K)\{B_j(H)-B_{j+1}(H)\} d\tau. \tag{3.1}$$

We see that the limiting probability of exit will be

$$\lim_{t\to\infty} Q(t) = 1 - \lim_{t\to\infty} P(K,t) \tag{3.2}$$

$$= 1 - \sum_{k=0}^{\infty} \sum_{j=0}^{\infty} \frac{\mu}{\lambda+\mu} \left(\frac{\lambda}{\lambda+\mu}\right)^k \left(\frac{\mu}{\lambda+\mu}\right)^j \binom{k+j}{j} A_k(K)\{B_j(H)-B_{j+1}(H)\}.$$

This corresponds, for example, to the unfiltered exit of particles in a porous medium, after they have travelled a distance K in suspension, without being trapped.

The finite tank with filtered exit at $x = K$ of 1(c) is more interesting; we now assume that there is a filtering compartment C for $x \geqslant K$ from which particles may either sediment or exit. This corresponds more closely, for example, to the motion of particles at the end of a tank of length K. In $0 \leqslant x < K$, the distribution of particles in suspension remains $F(x,t)$, while that for sedimentary particles is $P(x,t)$. Within C, however, particles may float, sediment or exit with probabilities $F_C(t)$, $P_C(t)$ and $Q(t)$ respectively.

The simplest model for the compartment C is one for which, during the interval $(t, t+\delta t)$ of time, a particle may sediment with probability $\beta\delta t+o(\delta t)$, exit with probability $\alpha\delta t+o(\delta t)$, and continue to float in suspension with probability $1-(\alpha+\beta)\delta t-o(\delta t)$. It is readily seen that in this case, the Kolmogorov equations for $F_C(t)$, $P_C(t)$ and $Q(t)$ are

$$\begin{aligned} F'_C(t) &= -(\alpha+\beta)F_C-F'(K, t)-P'(K, t), \\ P'_C(t) &= \beta F_C, \\ Q'(t) &= \alpha F_C, \end{aligned} \tag{3.3}$$

where at any time $t\geqslant 0$, $F_C(t)+P_C(t)+Q(t)=1-F(K,t)-P(K,t)$.

Laplace transforms with respect to time yield the equations

$$\begin{aligned} \theta F^*_C(\theta) &= -(\alpha+\beta)F^*_C(\theta)-\{\theta F^*(K, \theta)-F(K,0)\}-\theta P^*(K,\theta), \\ \theta P^*_C(\theta) &= \beta F^*_C(\theta), \\ \theta Q^*(\theta) &= \alpha F^*_C(\theta), \end{aligned} \tag{3.4}$$

where $F(K, 0)=1$, and $P(K, 0)=F_C(0)=P_C(0)=0$. From these, we obtain

$$F^*_C(\theta) = \frac{1}{\alpha+\beta+\theta}\{1-\theta F^*(K,\theta)-\theta P^*(K,\theta)\}, \tag{3.5}$$

and we can see directly that as $t\to\infty$,

$$\lim_{t\to\infty} F_C(t) = \lim_{\theta\to 0}\theta F^*_C(\theta)=0,$$

as expected.

We now have, however, that

$$\begin{aligned} \lim_{t\to\infty} P_C(t) &= \lim_{\theta\to 0}\theta P^*_C(\theta) = \lim_{\theta\to 0}\beta F^*_C(\theta) \\ &= \frac{\beta}{\alpha+\beta}\{1-\sum_{k=0}^{\infty}\sum_{j=0}^{\infty}\frac{\mu}{\lambda+\mu}(\frac{\lambda}{\lambda+\mu})^k(\frac{\mu}{\lambda+\mu})^j \binom{k+j}{j} A_k(K)\{B_j(H)-B_{j+1}(H)\} \end{aligned} \tag{3.6}$$

from (2.8), where the sedimentation may be regarded as a deposit accumulated at $x=K$. The limiting probability of exit is given by

$$\begin{aligned} \lim_{t\to\infty} Q(t) &= \lim_{\theta\to 0}\theta Q^*(\theta) = \lim_{\theta\to 0}\alpha F^*_C(\theta) \\ &= \frac{\alpha}{\alpha+\beta}\{1-\sum_{k=0}^{\infty}\sum_{j=0}^{\infty}\frac{\mu}{\lambda+\mu}(\frac{\lambda}{\lambda+\mu})^k(\frac{\mu}{\lambda+\mu})^j \binom{k+j}{j} A_k(K)\{B_j(H)-B_{j+1}(H)\} \end{aligned} \tag{3.7}$$

from which we readily verify that $\lim_{t\to\infty}\{P(K, t)+P_C(t)+Q(t)\}=1$.

4. A SIMPLE EXAMPLE

By analogy with the discrete state space models considered by GANI and TODOROVIC [1], we illustrate the previous model for the very simple case where all the $\eta_k\geqslant H$. It follows then that $\Pr\{T_H>t\} = \Pr\{Y(t)<H\}=e^{-\mu t}$, and thus in (2.1)

$$G(t) = 1-e^{-\mu t}, \tag{4.1}$$

with the Laplace transform with respect to time

$$G^*(\theta) = \frac{1}{\theta} - \frac{1}{\mu+\theta}.$$

From (2.5), we derive the probability $F(x,t)$ that a non-sedimentary particle is located at or below x at time t as

$$F(x,t) = e^{-(\lambda+\mu)t} \sum_{k=0}^{\infty} \frac{(\lambda t)^k}{k!} A_k(x), \tag{4.2}$$

with the Laplace transform with respect to time

$$F^*(x, \theta) = \sum_{k=0}^{\infty} \frac{\lambda^k}{(\lambda+\mu+\theta)^{k+1}} A_k(x)$$

and Laplace-Stieltjes transform with respect to x

$$f^*(s, \theta) = \int_0^{\infty} e^{-sx} d_x F^*(x, \theta) = \{\mu+\theta+\lambda(1-A^*(s))\}^{-1}.$$

The probability of sedimentation (2.7) is now

$$P(x,t) = \int_0^{t} \mu e^{-(\lambda+\mu)\tau} \sum_{k=0}^{\infty} \frac{(\lambda \tau)^k}{k!} A_k(x) d\tau, \tag{4.3}$$

with the Laplace transform with respect to time

$$P^*(x, \theta) = \sum_{k=0}^{\infty} \frac{\mu}{\theta(\lambda+\mu+\theta)} \left(\frac{\lambda}{\lambda+\mu+\theta}\right)^k A_k(x).$$

Its Laplace-Stieltjes transform with respect to x is given by

$$p^*(s,\theta) = \int_0^{\infty} e^{-sx} d_x P^*(x,\theta) = \frac{\mu}{\theta\{\mu+\theta+\lambda(1-A^*(s))\}}.$$

It is readily seen that for all $x \geqslant 0$,

$$\lim_{t\to\infty} F(x, t) = \lim_{\theta\to 0} \theta F^*(x, \theta) = 0,$$

and

$$P_x = \lim_{t\to\infty} P(x, t) = \lim_{\theta\to 0} \theta P^*(x, \theta) = \sum_{k=0}^{\infty} \frac{\mu}{\lambda+\mu} \left(\frac{\lambda}{\lambda+\mu}\right)^k A_k(x), \tag{4.4}$$

where $\lim_{x\to\infty} P_x = 1$. This has the Laplace-Stieltjes transform with respect to x

$$\lim_{\theta\to 0} \theta p^*(s,\theta) = \int_0^{\infty} e^{-sx} d_x P_x = \frac{\mu}{\{\mu+\lambda(1-A^*(s))\}}, \tag{4.5}$$

from which, by differentiating with respect to s, we see that

$$\int_0^{\infty} x d_x P_x = -\frac{\lambda}{\mu} \frac{dA^*(s)}{ds}\Big|_{s=0} = \frac{\lambda}{\mu} \int_0^{\infty} x d_x A(x).$$

For the unfiltered exit case, the relevant distributions are $F(x, t)$, $P(x, t)$ and $Q(t)=1-F(K, t)-P(K, t)$. These can be written directly from the equations (4.2) and (4.3) above. The limiting distribution of exit of the particle will be

$$\lim_{t\to\infty} Q(t) = 1-P_K=1-\sum_{k=0}^{\infty}\frac{\mu}{\lambda+\mu}\left(\frac{\lambda}{\lambda+\mu}\right)^k A_k(K). \tag{4.6}$$

For the filtered exit case, the distributions required are $F(x, t)$, $P(x, t)$ which are known, as well as $F_C(t)$, $P_C(t)$ and $Q(t)$, whose Laplace transforms with respect to time are

$$F_C^*(\theta) = \frac{1}{\theta+\alpha+\beta}\{1-\theta F^*(K,\theta)-\theta P^*(K,\theta)\}$$

$$= \frac{1}{\theta+\alpha+\beta}\{1-\theta\sum_{k=0}^{\infty}\frac{\lambda^k A_k(K)}{(\lambda+\mu+\theta)^{k+1}}-\sum_{k=0}^{\infty}\frac{\mu\lambda^k A_k(K)}{(\lambda+\mu+\theta)^{k+1}}\},$$

$$P_C^*(\theta) = \beta\theta^{-1}F_C^*(\theta), \quad Q^*(\theta)=\alpha\theta^{-1}F_C^*(\theta).$$

The limiting values of the probabilities of suspension, sedimentation and exit are

$$\lim_{t\to\infty} F_C(t) = \lim_{\theta\to 0}\theta F_C^*(\theta)=0,$$

$$\lim_{t\to\infty} P_C(t) = \lim_{\theta\to 0}\beta F_C^*(\theta) = \frac{\beta}{\alpha+\beta}\{1-\sum_{k=0}^{\infty}\frac{\mu\lambda^k A_k(K)}{(\lambda+\mu)^{k+1}}\}, \tag{4.7}$$

$$\lim_{t\to\infty} Q(t) = \lim_{\theta\to 0}\alpha F_C^*(\theta) = \frac{\alpha}{\alpha+\beta}\{1-\sum_{k=0}^{\infty}\frac{\mu\lambda^k A_k(K)}{(\lambda+\mu)^{k+1}}\}.$$

Some further simplification will ensue if the ξ_j follow the negative exponential distribution so that

$$A_k(x) = \int_0^x e^{-au}a^k\frac{u^{k-1}}{(k-1)!}du, \quad k\geq 1.$$

In this case, we find that

$$F(x,t) = e^{-(\lambda+\mu)t}\{1+\sum_{k=1}^{\infty}\frac{(\lambda t)^k}{k!}\int_0^x e^{-au}a^k\frac{u^{k-1}}{(k-1)!}du\}$$

$$= e^{-(\lambda+\mu)t}\{1+\sqrt{a\lambda t}\int_0^x e^{-av}v^{-\frac{1}{2}}I_1(2\sqrt{a\lambda tv})dv\}, \tag{4.8}$$

$$P(x,t) = \int_0^t \mu e^{-(\lambda+\mu)\tau-ax}\sum_{k=0}^{\infty}\sum_{j=k}^{\infty}\frac{(\lambda\tau)^k}{k!}\frac{(ax)^j}{j!}d\tau,$$

and the limiting probability of sedimentation takes the very simple form

$$P_x = \lim_{t\to\infty}P(x,t) = \sum_{k=0}^{\infty}\sum_{j=k}^{\infty}\frac{\mu}{\lambda+\mu}\left(\frac{\lambda}{\lambda+\mu}\right)^k e^{-ax}\frac{(ax)^j}{j!}$$

$$= 1-(\frac{\lambda}{\lambda+\mu})e^{-(a\mu x/(\lambda+\mu))}. \tag{4.9}$$

Thus, for the unfiltered exit case, we have

$$P_k = 1-(\frac{\lambda}{\lambda+\mu})e^{-\frac{a\mu K}{\lambda+\mu}},$$

$$\lim_{t\to\infty} Q(t) = 1-P_K=(\frac{\lambda}{\lambda+\mu})e^{-\frac{a\mu K}{\lambda+\mu}}, \tag{4.10}$$

while for the filtered exit case, we find that

$$\lim_{t\to\infty} P_C(t) = \frac{\beta}{\alpha+\beta}\{(\frac{\lambda}{\lambda+\mu})e^{-\frac{a\mu K}{\lambda+\mu}}\},$$

$$\lim_{t\to\infty} Q(t) = \frac{\alpha}{\alpha+\beta}\{(\frac{\lambda}{\lambda+\mu})e^{-\frac{a\mu K}{\lambda+\mu}}\}. \tag{4.11}$$

5. The correlated random walk case

Suppose now that the two processes $\{X(t); t\geqslant 0\}$, $\{Y(t); t\geqslant 0\}$ are no longer independent, in the sense that $N(t)$ and $M(t)$ in (1.1) are correlated. For example, let

$$\begin{aligned} N(t) &= U(t)+V(t), \\ M(t) &= U(t)+W(t), \end{aligned} \tag{5.1}$$

where $U(t)$, $V(t)$, $W(t)$ are independent homogeneous Poisson processes with parameters ν, λ, μ respectively. The joint p.g.f. of these processes is known to be

$$\phi(u,v,t) = E(u^{N(t)}v^{M(t)})=e^{-\nu t(1-uv)-\lambda t(1-u)-\mu t(1-v)}.$$

We note that $N(t)$ is a Poisson process with parameter $\nu+\lambda$, $M(t)$ is Poisson with parameter $\mu+\nu$, but $N(t)$ and $M(t)$ are no longer independent. This is the stochastic process version of the bivariate Poisson distribution described by Johnson and Kotz ([4]; Ch. 11). The referee has pointed out that other forms of dependence are also possible.

We can now obtain results similar to those of Section 2 for this more realistic case, where the number of jumps in the x direction is related to that in the y direction. Once again

$$\begin{aligned} \Pr\{T_H>t\} &= \Pr\{Y(t)<H\} \\ &= e^{-(\mu+\nu)t}\sum_{j=0}^{\infty}\frac{((\mu+\nu)t)^j}{j!}B_j(H-), \end{aligned} \tag{5.2}$$

from which it follows that

$$G(t) = 1-e^{-(\mu+\nu)t}\{\sum_{j=0}^{\infty}\frac{((\mu+\nu)t)^j}{j!}B_j(H)\}, \tag{5.3}$$

with Laplace transform with respect to time t of the form

$$G^*(\theta) = \frac{1}{\theta} - \sum_{j=0}^{\infty} \frac{(\mu+\nu)^j}{(\mu+\nu+\theta)^{j+1}} B_j(H). \tag{5.4}$$

The p.d.f. of time to sedimentation will now be

$$G'(t) = (\mu+\nu)e^{-(\mu+\nu)t}\{\sum_{j=0}^{\infty} \frac{((\mu+\nu)t)^j}{j!}[B_j(H)-B_{j+1}(H)]\}. \tag{5.5}$$

We see that the d.f. $F(x,t)=\Pr\{X(t)\leq x,\ Y(t)<H\}$ will now be given by

$$F(x,t) = \sum_{i=0}^{\infty} e^{-\nu t}\frac{(\nu t)^i}{i!}\sum_{j=0}^{\infty} e^{-\lambda t}\frac{(\lambda t)^j}{j!}A_{i+j}(x)\sum_{k=0}^{\infty} e^{-\mu t}\frac{(\mu t)^k}{k!}B_{i+k}(H) \tag{5.6}$$

$$= \sum_{i=0}^{\infty}\sum_{j=0}^{\infty}\sum_{k=0}^{\infty} e^{-(\lambda+\mu+\nu)t}\frac{\nu^i}{i!}\frac{\lambda^j}{j!}\frac{\mu^k}{k!}t^{i+j+k}A_{i+j}(x)B_{i+k}(H),$$

with the Laplace transform with respect to time

$$F^*(x,\theta) = \sum_{i=0}^{\infty}\sum_{j=0}^{\infty}\sum_{k=0}^{\infty}\left(\frac{\nu}{\lambda+\mu+\nu+\theta}\right)^i\left(\frac{\lambda}{\lambda+\mu+\nu+\theta}\right)^j\left(\frac{\mu}{\lambda+\mu+\nu+\theta}\right)^k \times \tag{5.7}$$

$$\frac{(i+j+k)!}{i!j!k!}\frac{A_{i+j}(x)B_{i+k}(H)}{(\lambda+\mu+\nu+\theta)}.$$

If one takes, once again, the Laplace-Stieltjes transform with respect to x, then

$$f^*(s,\theta) = \int_0^{\infty} e^{-sx}d_xF^*(x,\theta) = \sum_{i=0}^{\infty}\sum_{k=0}^{\infty}\frac{(\nu A^*(s))^i\mu^k\binom{i+k}{i}B_{i+k}(H)}{\{\mu+\nu+\theta+\lambda(1-A^*(s))\}^{i+k+1}}. \tag{5.8}$$

The probability of sedimentation at or below x at time t, following the results of (2.7), will be

$$P(x,t) = \int_{\tau=0}^{t} \Pr\{X(\tau)\leq x|T_H>\tau\}\Pr\{\tau<T_H<\tau+d\tau\} \tag{5.9}$$

$$= \int_0^t \frac{F(x,\tau)}{\Pr\{T_H>\tau\}}G'(\tau)d\tau = \int_0^t \frac{F(x,\tau)}{1-G(\tau)}G'(\tau)d\tau$$

$$= \int_0^t \frac{\sum_{i,j,k=0}^{\infty} e^{-(\lambda+\mu+\nu)\tau}\frac{\nu^i}{i!}\frac{\lambda^j}{j!}\frac{\mu^k}{k!}\tau^{i+j+k}A_{i+j}(x)B_{i+k}(H)}{e^{-(\mu+\nu)\tau}\sum_{l=0}^{\infty}((\mu+\nu)\tau)^l B_l(H)/l!} \times$$

$$(\mu+\nu)e^{-(\mu+\nu)\tau}\sum_{m=0}^{\infty}\frac{((\mu+\nu)\tau)^m}{m!}\{B_m(H)-B_{m+1}(H)\}d\tau$$

for which the Laplace transform is difficult to calculate. We can, however, express the limiting probability of sedimentation at or below x as

$$P_x = \int_0^\infty \frac{\sum_{i,j,k=0}^{\infty} (\mu+\nu)e^{-(\lambda+\mu+\nu)\tau} \frac{\nu^i}{i!}\frac{\lambda^j}{j!}\frac{\mu^k}{k!}\tau^{i+j+k} A_{i+j}(x)B_{i+k}(H)}{\sum_{l=0}^{\infty} \frac{((\mu+\nu)\tau)^l}{l!} B_l(H)} \times$$

$$\sum_{m=0}^{\infty} \frac{((\mu+\nu)\tau)^m}{m!}\{B_m(H)-B_{m+1}(H)\}d\tau. \tag{5.10}$$

Illustrating once again by the example of Section 4, where all the $\eta_k \geqslant H$, then from (5.2),

$$G(t) = 1-e^{-(\mu+\nu)t}$$

and from (5.7)

$$F^*(x, \theta) = \sum_{j=0}^{\infty} \frac{\lambda^j}{(\lambda+\mu+\nu+\theta)^{j+1}} A_j(x). \tag{5.11}$$

Taking the Laplace-Stieltjes transform of (5.11) with respect to x, we find

$$f^*(s,\theta) = \int_0^\infty e^{-sx} d_x F^*(x, \theta) = \frac{1}{\mu+\nu+\theta+\lambda(1-A^*(s))}. \tag{5.12}$$

Note that in this case $G'(t)=(\mu+\nu)e^{-(\mu+\nu)t}=(\mu+\nu)\Pr\{T_H>t\}$ so that from (5.9)

$$P(x,t) = \int_0^t F(x,\tau)(\mu+\nu)d\tau$$

and

$$p^*(s,\theta) = \int_0^\infty e^{-sx} d_x \int_0^\infty e^{-\theta t} dt \int_0^t F(x,\tau)(\mu+\nu)d\tau \tag{5.13}$$

$$= \frac{\mu+\nu}{\theta\{\mu+\nu+\theta+\lambda(1-A^*(s))\}}$$

whence

$$\lim_{\theta\to 0} \theta p^*(s, \theta) = \int_0^\infty e^{-sx} d_x P_x = \frac{\mu+\nu}{\mu+\nu+\lambda(1-A^*(s))}, \tag{5.14}$$

which has the same form as (4.5) with $\mu+\nu$ replacing μ We note intuitively that when there is no jump in the $M(t)$, whose parameter is $(\nu+\mu)$, this implies that $U(t)=0$, so that $N(t)$ then has the parameter λ only.

In principle, results for both the general and particular cases we have considered, when applied to the finite tank (unfiltered and filtered) exit models, will follow the same lines as when $X(t)$ and $Y(t)$ are independent, with some obvious modifications. It would be of practical interest to analyze data on particles in a fluid flow or a porous medium, with a view to fitting the most appropriate correlated bivariate random walk model.

ACKNOWLEDGEMENT

It is a pleasure to be able to contribute to Professor J.W. Cohen's *Liber Amicorum*, and to wish so distinguished a scholar the happiest 65th birthday. Thanks are due to the referee for his comments, which greatly improved the first draft of the paper. The author also gratefully acknowledges ONR's support for this research through Contract N00014-86-K-0019.

REFERENCES

1. J. GANI, P. TODOROVIC (1983). A model for the transport of solid particles in a fluid flow. *Stoch. Proc. Appl. 14*, 1-17.
2. J. GANI, P. TODOROVIC (1987). On the persistent release of particles in a fluid flow. A.E. GELFAND (ed.). *Contributions to the Theory and Application of Statistics: A Volume in Honor of Herbert Solomon*, Academic Press, Boston, 43-74.
3. J. GANI, P. TODOROVIC (1987). Transport and sedimentation of particles in a fluid flow. *Stoch. Hydrol. Hydraul. 1*, 209-216.
4. N.L. JOHNSON, S. KOTZ (1969). *Distributions in Statistics: Discrete Distributions*, Houghton Mifflin Company, Boston.
5. G. SPOSITO, V.K. GUPTA, R.N. BHATTACHARYA (1979). Foundational theories of solute transport in porous media: a critical review. *Advances in Water Resources 2*, 59-68.
6. P. TODOROVIC (1975). A stochastic model of dispersion in a porous medium. *Water Resources Res. 11*, 348-354.
7. P. TODOROVIC (1980). Stochastic modeling of longitudinal dispersion in a porous medium. *Math. Scientist 5*, 45-55.
8. P. TODOROVIC, J. GANI (1983). A model of longitudinal dispersion in non-homogeneous porous media. *Stoch. Anal. Appl. 1*, 327-339.

Asymptotic Efficiency Results for the Method of Moments with Application to Estimation for Queueing Processes

C.C. Heyde
Department of Statistics, Institute of Advanced Studies,
Australian National University,
GPO Box 4, Canberra, ACT 2601, Australia

This paper is concerned with estimation of the mean value in a stationary process. It is shown under conditions which are relevant to a broad range of queueing applications that the method of moments estimator (which assigns equal weight to each observation) has the same asymptotic variance as the minimum variance linear unbiased estimator.

1. Introduction

There are many system identification or parameter estimation problems for stochastic processes in which favoured standard estimation methods having desirable properties over a broad range of conditions are not tractable or even available. For example, the use of the method of maximum likelihood requires a complete knowledge of the true distribution of the process under consideration. Much less is required for least squares type estimators, namely a knowledge of the covariance structure of the process, but even this may not be feasible to use in practice. A last resort, from a theoretical point of view, is the method of moments which, for point estimation, relies on nothing more than an ergodic theorem and involves no significant structural assumptions. Nevertheless, for stationary processes it turns out that the method of moments (which estimates $EX_i = \mu$ by $n^{-1}\Sigma_{i=1}^{n} X_i$) produces an estimator which has the same asymptotic variance as the minimum variance unbiased estimator under broad conditions on the spectral density or covariance function. This is of considerable practical significance.

The results given here provide a substantial extension of those of Halfin [7] who dealt with processes having a covariance function which is a finite sum of exponential terms (such as are common for Markovian queueing models with limited queue size).

2. Details

Let $\{X_t, t = \ldots, -1, 0, 1, \ldots\}$ be a stationary ergodic process with $EX_t = \mu$, which is to be estimated, and $EX_t^2 < \infty$. Write

$$\gamma(t-s) = \text{cov}(X_s-\mu,\ X_t-\mu) = \int_{-\pi}^{\pi} e^{i(t-s)\lambda} f(\lambda)d\lambda$$

for the covariance function, where

$$f(\lambda) = (2\pi)^{-1}\left[\gamma(0)+2\sum_{j=1}^{\infty}\gamma(j)\cos j\lambda\right], \quad \lambda\neq 0,\ |\lambda|\leqslant\pi$$

is the spectral density, the spectral function being assumed to be absolutely continuous. It should be noted that all standard queueing processes are purely nondeterministic (i.e. contain no component which is exactly predictable) and hence have an absolutely continuous spectrum (e.g. HANNAN [9, Theorem 3′ p. 154]).

We confine our attention to the discrete time case for convenience; analogous results related by deterministic time-transformations usually hold for continuous time.

We shall be concerned with estimating μ on the basis of observations $(X_1, \ldots, X_n)$ and we shall restrict consideration to linear functions of the observations $\Sigma_{i=1}^{n}\alpha_{i,n}X_i$ where the (constant) weights $\alpha_{i,n}$, $i=1,2,\ldots,n$ satisfy $\Sigma_{i=1}^{n}\alpha_{i,n}=1$ so that the estimators are unbiased.

Amongst this class of unbiased linear estimators, the (unique) *minimum variance* one, $\mu_{MV}=\Sigma_{i=1}^{n}\alpha_{i,n}^{0}X_i$, has weights $\alpha_{i,n}^{0}$ which satisfy the equations

$$\sum_{i=1}^{n}\gamma(j-i)\alpha_{i,n}^{0} = c_n \quad \text{(constant)}, \quad j=1,2,\ldots,n. \tag{2.1}$$

To see this, we first note that

$$\text{cov}\left[\sum_{i=1}^{n}\alpha_{i,n}X_i,\ \sum_{i=1}^{n}\alpha_{i,n}^{0}X_i\right] = \sum_{i=1}^{n}\sum_{j=1}^{n}\alpha_{i,n}^{0}\alpha_{j,n}\gamma(j-i)$$

$$= c_n = \text{var}\left[\sum_{i=1}^{n}\alpha_{i,n}^{0}X_i\right]$$

in view of (2.1). Thus,

$$\text{var}\left[\sum_{i=1}^{n}\alpha_{i,n}X_i\right] - \text{var}\left[\sum_{i=1}^{n}\alpha_{i,n}^{0}X_i\right] = \text{var}\left[\sum_{i=1}^{n}(\alpha_{i,n}-\alpha_{i,n}^{0})X_i\right] \geqslant 0.$$

The minimum variance unbiased estimator is also optimal in the sense of GODAMBE [4]. This follows directly from (2.1) using (12) of [4]. For explanation of the significance of this concept of optimality and its quasi-likelihood connotations see GODAMBE and HEYDE [5].

To obtain an expression for the minimum variance c_n, we write $\Gamma(n)$ for the $n\times n$ matrix $(\gamma(i-j))$ and let $\alpha^0(n)$ and $e(n)$ denote the n-vectors $(\alpha_{1,n}^{0}, \ldots, \alpha_{n,n}^{0})'$ and $(1,1,\ldots,1)'$ respectively, the prime denoting transpose. Then, (2.1) gives

$$\Gamma(n)\alpha^0(n) = c_n e(n),$$

while

$$(e(n))'\alpha^0(n) = 1,$$

and hence

$$c_n = [(e(n))'(\Gamma(n))^{-1}e(n)]^{-1}. \tag{2.2}$$

The explicit expressions for $\alpha^0(n)$ and c_n are, however, of limited practical value as, even when the covariances $\gamma(n)$ are known, they involve calculation of the inverse of the Toeplitz matrix $\Gamma(n)$ which will usually be of high order.

On the other hand, the *method of moments* estimator

$$\mu_{MM} = n^{-1}\sum_{i=1}^{n} X_i$$

uses equally weighted observations and requires no direct knowledge of the underlying distribution. We shall proceed to show that, under a range of conditions which are of quite broad applicability,

$$\operatorname{var} \mu_{MV} = [(e(n))'(\Gamma(n))^{-1}e(n)]^{-1} \tag{2.3}$$

and

$$\operatorname{var} \mu_{MM} = n^{-2}(e(n))'\Gamma(n)e(n) \tag{2.4}$$

have the same asymptotic behavior. That is, the estimator μ_{MM} is asymptotically efficient relative to μ_{MV}.

THEOREM 1. *Suppose that the spectral density $f(\lambda)$ of $\{X_t\}$ is continuous and positive at $\lambda=0$. Then*

$$n \operatorname{var} \mu_{MM} \to 2\pi f(0) \quad as \quad n\to\infty. \tag{2.5}$$

If, in addition, $f(\lambda)$ is everywhere positive and continuous, then

$$n \operatorname{var} \mu_{MV} \to 2\pi f(0) \quad as \quad n\to\infty, \tag{2.6}$$

and

$$n^{\frac{1}{2}}(\mu_{MV}-\mu_{MM}) \xrightarrow{P} 0 \quad as \quad n\to\infty, \tag{2.7}$$

P denoting convergence in probability.

PROOF. The result (2.5) is a well-known consequence of the representation

$$\operatorname{var} \mu_{MM} = \frac{1}{n^2}\int_{-\pi}^{\pi}\left[\frac{\sin(n\lambda/2)}{\sin(\lambda/2)}\right]^2 f(\lambda)d\lambda;$$

see for example IBRAGIMOV and LINNIK [11, p. 322]. The result (2.6) follows from a theorem of GRENANDER and SZEGÖ [6, p. 211].

To obtain (2.7), we note that

$$nE(\mu_{MV}-\mu_{MM})^2 = nE\left[\sum_{i=1}^{n}(\alpha_{i,n}^0-n^{-1})X_i\right]^2$$

$$= n\sum_{i=1}^{n}(\alpha_{i,n}^0-n^{-1})\sum_{j=1}^{n}(\alpha_{j,n}^0-n^{-1})\gamma(i-j)$$

$$= n\sum_{i=1}^{n}(\alpha_{i,n}^0-n^{-1})\left[c_n-n^{-1}\sum_{j=1}^{n}\gamma(i-j)\right]$$

$$= -\sum_{i=1}^{n}(\alpha_{i,n}^0-n^{-1})\sum_{j=1}^{n}\gamma(i-j)$$

$$= n(\operatorname{var}\mu_{MM}-\operatorname{var}\mu_{MV})\to 0$$

as $n\to\infty$ using (2.5) and (2.6) and the result follows. This completes the proof. □

It should be noted from (2.7) that if a central limit result holds for $n^{\frac{1}{2}}(\mu_{MM}-\mu)$, then the corresponding one also holds for $n^{\frac{1}{2}}(\mu_{MV}-\mu)$. Various conditions of asymptotic independence lead to the result

$$n^{\frac{1}{2}}(\mu_{MM}-\mu) = n^{-\frac{1}{2}}\sum_{i=1}^{n}(X_i-\mu)\xrightarrow{d} N(0, 2\pi f(0)),$$

namely convergence in distribution to the normal law with zero mean and variance $2\pi f(0)$. For example, this holds if there is a σ-field $\mathcal{M}_0$ such that X_0 is $\mathcal{M}_0$-measurable and

$$\sum_{k=0}^{\infty} E|E(X_k|\mathcal{M}_0)-\mu|<\infty$$

(HALL and HEYDE [8, Theorem 5.4, p. 136]). This is a very general result from which many special cases, such as for mixing sequences, follow.

For the results of Theorem 1 to be useful it is essential that $f(0)>0$. An example where this may not be the case is furnished by the moving average model

$$X_t = \mu+\epsilon_t-r\epsilon_{t-1}, \quad |r|\leqslant 1,$$

where the ϵ_i are stationary and orthogonal with mean 0 and variance σ^2. Here

$$f(\lambda) = \sigma^2(1+r^2-2r\cos\lambda)/2\pi,$$

so that $f(0)>0$ requires $r<1$. If $r=1$ it turns out that ([6, p. 211]),

$$\operatorname{var}\mu_{MM} \sim 2\sigma^2 n^{-2}, \quad \operatorname{var}\mu_{MV} \sim 12\sigma^2 n^{-3}$$

as $n\to\infty$, so that the method of moments estimator is far from efficient in this case. In fact, if $f(0)=0$, it holds rather generally that $\operatorname{var}\mu_{MV}=o(\operatorname{var}\mu_{MM})$ as $n\to\infty$ (ADENSTADT [1], VITALE [13]).

Now in order to use Theorem 1 to show that μ_{MV} and μ_{MM} have the same asymptotic variance it is necessary to know that the spectral density $f(\lambda)$ is everywhere positive and continuous. Such information is ordinarily not directly available but a useful sufficient condition in terms of covariances is given in the following theorem.

THEOREM 2. *If the covariance $\gamma(n)$ decreases monotonically to zero as $n \to \infty$, then $\{X_t\}$ has a spectral function which is absolutely continuous. The spectral density $f(\lambda)$ is non-negative and continuous, except possibly at zero, being continuous at zero when $\Sigma_{n=1}^{\infty}\gamma(n)<\infty$. If $\{\gamma(n)\}$ is convex and $\gamma(0)+\gamma(2)>2\gamma(1)$, then the spectral density is strictly positive.*

PROOF. The first two parts of the theorem are given in Theorem 3 of DALEY [3]. For the last part it should be noted that when $\{\gamma(n)\}$ is convex, namely

$$\Delta^2\gamma(n) = \gamma(n)+\gamma(n+2)-2\gamma(n+1)\geqslant 0, \quad n\geqslant 0,$$

the spectral density $f(\lambda)$ can be expressed in the form

$$f(\lambda) = (2\pi)^{-1}\sum_{n=0}^{\infty}(n+1)\Delta^2\gamma(n)K_n(\lambda), \quad |\lambda|\leqslant\pi,$$

where

$$K_n(\lambda) = \frac{2}{n+1}\left\{\frac{\sin((n+1)\lambda/2)}{2\sin(\lambda/2)}\right\}^2 \geqslant 0$$

is the Fejér kernel (ZYGMUND [14, p. 183]). Then, since $K_0(\lambda)=\frac{1}{2}$ and $\Delta^2\gamma(n)K_n(\lambda)\geqslant 0$ for $n\geqslant 1$, $f(\lambda)>0$ if $\Delta^2\gamma(0)>0$ as required. □

COROLLARY 1. *If the covariance $\gamma(n)$ decreases monotonically to zero as $n\to\infty$, $\Sigma_{j=1}^{\infty}\gamma(j)<\infty$ and $\{\gamma(n)\}$ is convex with $\gamma(0)+\gamma(2)>2\gamma(1)$, then*

$$\operatorname{var}\mu_{MV} \sim \operatorname{var}\mu_{MM} \sim n^{-1}\left\{\gamma(0)+2\sum_{j=1}^{\infty}\gamma(j)\right\}$$

as $n\to\infty$.

This result follows immediately from Theorems 1 and 2.

The particular case of the result of Corollary 1 for which the covariance function $\gamma(n)$ is of the form

$$\gamma(n) = \sum_{i=1}^{K} a_i e^{-r_i|n|} \tag{2.8}$$

where $a_i>0$, $i=1,2,\ldots,K$, $0<r_1<r_2<\cdots<r_K$, has been obtained by HALFIN [7]. Covariance functions of the type (2.8) are common for Markovian queueing models with limited queue space and various applications are given in [7].

Corollary 1, however, is very much more widely applicable and many examples can be noted from the review article of REYNOLDS [12] on the covariance structure of queues and related processes. As one example, we mention the queue length process in an M/G/∞ queue ([12, p. 386]) where

$$\rho(n) = \gamma(n)/\gamma(0)=\beta \int_n^{\infty} G(x)dx$$

with $\beta^{-1}=ES$, the mean service time and

$$G(x) = P(S \geqslant x).$$

Then, it is easily checked that $\Sigma_{n=0}^{\infty}\gamma(n)<\infty$ is equivalent to $ES^2<\infty$, while convexity of $\{\gamma(n)\}$ is automatically satisfied and $\gamma(0)+\gamma(2)>2\gamma(1)$ requires

$$\int_0^1 G(x)dx > \int_1^2 G(x)dx.$$

Another example involves the waiting time $\{W_n\}$ in an M/G/1 queue where [12, p. 401])

$$\rho(n) = 1-\left[n\mu(1-\rho)-\left[\sum_{r=1}^{n} C_r\right]^2\right]/\alpha\sigma^2, \quad n\geqslant 1,$$

for certain constants $C_r \uparrow$, with $\mu=EW_n$, $\sigma^2=\text{var}\, W_n$, ρ the traffic intensity and α the mean arrival rate. Here it is again easily checked that $\{\rho(n)\}$ is monotone and convex with

$$\rho(0)+\rho(2) > 2\rho(1)$$

while $\Sigma_{n=0}^{\infty}\rho(n)<\infty$ if and only if $ES^4<\infty$.

Now if the spectral density has zeros the above theory breaks down, but even then μ_{MM} may have the same asymptotic variance as μ_{MV}. This is the case, for example, for the queue length process in an M/D/∞ queue where the constant service time $T=2$.

For the M/D/∞ queue with constant service time T, $\rho(j)=\gamma(j)/\gamma(0)$ has the particularly simple form

$$\begin{aligned}\rho(j) &= 1-|j|/T\ , \quad |j|\leqslant T,\\ &= 0 \qquad\qquad , \quad \text{otherwise},\end{aligned}$$

(REYNOLDS [12, p. 387] and $\gamma(j)=\alpha T^2\rho(j)$ where α denotes the mean arrival rate. When $T=2$ the spectral density is given by

$$2\pi f(\lambda) = 4\alpha(1+\cos\lambda)$$

and we find that

$$n\,\text{var}\,\mu_{MV} \rightarrow 8\alpha, \quad n\,\text{var}\,\mu_{MM} \rightarrow 8\alpha \tag{2.9}$$

as $n\rightarrow\infty$ despite the fact that $f(-\pi)=f(\pi)=0$.

The result for $\text{var}\,\mu_{MM}$ in (2.9) follows from Theorem 1 and to obtain the result for $\text{var}\,\mu_{MV}$ we use(2.1)together with $\Sigma_{i=1}^{n}\alpha_{i,n}^{0}=1$. We find, after some algebra, that

$$\begin{aligned} \alpha^0_{2j+1,n} &= c_n(n-2j)/(n+1) \\ \alpha^0_{2j,n} &= 2c_n(j+1)/(n+1) \end{aligned} \qquad j=0,1,2,\ldots$$

and that

$$\alpha^0_{j,n} = \alpha^0_{n-j,n}, \quad j=0,1,2,\ldots,n,$$

whence,

$$c_n = \operatorname{var}\mu_{MV} \sim 8\alpha n^{-1} \quad \text{as} \quad n\to\infty.$$

To obtain the asymptotic behavior of $\operatorname{var}\mu_{MV}$ more generally than is available through Theorem 1 it is necessary, as in the above example, to obtain the asymptotic behavior of the inverse of the Toeplitz matrix $\Gamma(n)$ or, more precisely, the sum of the elements of this inverse (see (2.3)). However, the studies of the asymptotics of inverses of Toeplitz matrices which appear in the literature focus on the case where the spectral density does not vanish (e.g. HARTWIG and FISHER [10], BLEHER [2]). These considerations do not assist in augmenting Theorem 1.

REFERENCES

1. R.K. ADENSTADT (1974). On large-sample estimation for the mean of a stationary random sequence. *Ann. Statist. 2,* 1095-1107.
2. P.M. BLEHER (1981). Inversion of Toeplitz matrices. *Trans. Moscow Math. Soc. 40,* 201-229.
3. D.J. DALEY (1969). Integral representations of transition probabilities and serial covariances of certain Markov chains. *J. Appl. Probab. 6,* 648-659.
4. V.P. GODAMBE (1985). The foundations of finite sample estimation in stochastic processes. *Biometrika 72,* 419-428.
5. V.P. GODAMBE, C.C. HEYDE (1987). Quasi-likelihood and optimal estimation. *Int. Statist. Rev. 55,* 231-244.
6. U. GRENANDER, G. SZEGÖ (1958). *Toeplitz Forms and their Applications,* Univ. Calif. Press, Berkeley and Los Angeles.
7. S. HALFIN (1982). Linear estimators for a class of stationary queueing processes. *Oper. Res. 30,* 515-529.
8. P. HALL, C.C. HEYDE (1980). *Martingale Limit Theory and its Application,* Academic Press, New York.
9. E.J. HANNAN (1970). *Multiple Time Series,* Wiley, New York.
10. R.E. HARTWIG, M. FISHER (1969). Asymptotic behaviour of Toeplitz matrices and determinants. *Arch. Rat. Mech. Anal. 32,* 190-225.
11. I.A. IBRAGIMOV, YU. V. LINNIK (1971). *Independent and Stationary Sequences of Random Variables,* Wolters-Noordhoff, Groningen.
12. J.F. REYNOLDS (1975). The covariance structure of queues and related processes — a survey of recent work. *Adv. in Appl. Probab. 7,* 383-415.
13. R.A. VITALE (1973). An asymptotically efficient estimate in time series

analysis. *Quart. Appl. Math. 30,* 421-440.
14. A. ZYGMUND (1959). *Trigonometric Series Vol. 1,* 2nd ed., Cambridge Univ. Press, Cambridge.

Regenerative Sets and their Applications to Markov Storage Systems

Haya Kaspi[1]
School of Operations Research and MSI
Cornell University
Ithaca, New York 14853, USA

Michael Rubinovitch
Faculty of Industrial Engineering and Management
Technion-Israel Institute of Technology
Haifa, Israel

A general class of storage processes is considered where the content of the storage is a Hunt process on $\mathbf{R}_+$. This allows for downwards jumps and for state dependent jump measures. The theory of regenerative sets (that are not necessarily discrete) and the concept of level crossings is used to obtain the main result, which is a representation for the invariant distribution of the content process. The regeneration set is the zero set of the content process which is assumed here to be positive recurrent.

1. Introduction

The theory of regenerative processes has been one of the most important tools in analyzing queueing and storage systems. Recall that a stochastic process $(Z_t)_{t \geqslant 0}$ is regenerative if there exists a sequence $\{S_n\}_{n \geqslant 0}$ of stopping times of Z so that: (i) the sequence $\{S_n\}$ is a renewal sequence, and (ii) if ϕ_t represents the future of Z after t, and $\mathcal{F}_t$ is the past of Z up to time t, then

$$P_{\phi_{S_n} | \mathcal{F}_{S_n}} = P_0,$$

where P_0 is a specified probability measure. In other words the past and future of Z at each S_n are independent and the law that governs the future is P_0.

Many queueing and storage processes are regenerative in the above sense. The sequence $\{S_n\}$ could be the times when busy periods end in queueing processes, and the times of downcrossings of a specified level, of the content process, in dam processes. The 'regenerative method' for analyzing such

1. Permanent address: Faculty of Industrial Engineering and Management, Technion-Israel Institute of Technology, Haifa, Israel.

Partially supported by the Mathematical Science Institute of Cornell University.

processes is based on the application of classical results from renewal theory to the appropriately chosen sequence $\{S_n\}$. *The Single Server Queue* [7] by J.W. Cohen, his beautiful monograph *On Regenerative Processes in Queueing Theory* [8], as well as his many papers on level crossings in queues and dams, provide a most convincing evidence for the richness of this theory and for its wide scope of applications.

During the last twenty years much work has been done on the theory of regeneration beyond the scope covered by classical renewal theory. This led to the now highly developed theory of regenerative sets and, more generally, of regenerative systems.

A process $(Z_t)_{t\in\mathbf{R}}$ possesses a regenerative set, if there exists a closed set $M\subset\mathbf{R}$, optional with respect to the history generated by Z, for which

$$P_{\phi_T|\mathcal{F}_T} = P_0$$

for any stopping time T that falls in M^d. Here M^d is the minimal right closed set with closure M and ϕ_t and $\mathcal{F}_t$ are as defined above. The set M is no longer required to be discrete and optionality replaces the requirement that $\{S_n\}$ consists of $(\mathcal{F}_t)$ stopping times. Counting the number of regeneration points may no longer be possible, but still, a 'renewal property' is in force. Many of the questions addressed by renewal theory are still interesting and, as it turns out, many of the results can be extended in a natural way to this situation as well. The theory of regenerative systems is an extension of regenerative sets in the same sense that Markov renewal theory [4] extends renewal theory. We do not use here regenerative systems and the reader is referred to the work of MAISONNEUVE [15,17,18], to whom many of the most important contributions to this theory are due.

Consider now the classical dam process with input according to a process with stationary independent increments and linear output. Let $(V_t)_{t\geq 0}$ be the content process. For such processes COHEN and RUBINOVITCH [9] (see also [21]) used the set $\{t\colon V_t=a\}$, with $a>0$, to characterize the stationary distribution of V, making use of the fact that this set is a discrete regenerative set, and appealing to known results from renewal theory. However, if one considers more general dam processes this set need not be discrete and renewal theory does not apply.

In this paper we study a class of general storage systems for which the content process, $(V_t)_{t\geq 0}$, is strong Markov. They will be called Markov Storage Systems. We shall use the theory of regenerative sets to obtain a representation for the stationary distribution of the content process and to extend the level crossing identities obtained in [3, 9, 10, 14, 21].

The paper is organized as follows. The next section is an introduction to regenerative sets. Most of the results cited are well-known, and are brought here to make the paper self-contained. We discuss the notions of regenerative sets, stationary regenerative sets, and the representation of their stationary distributions. The approach we take, as well as the results cited, are due to Maisonneuve. We refer to [19, 20] for further details and for the proofs. We end the section with a discussion of regenerative sets of the form $\{t\colon X_t=b\}$

where (X_t) is a strong Markov process and b is a point in its state space. We also discuss the notion of Harris recurrence which is strongly connected to the persistence of M. Section 3 is devoted to an application of the above results to Markov Storage Systems for which the set $M=\{t\colon V_t=0\}$ is positive recurrent. We study downcrossings and derive a representation for the stationary distribution of the content process.

The paper is devoted to Professor J.W. Cohen on the occasion of his 65-th birthday. The many references, above, to his work are the best proof of our debt to him.

2. Regenerative sets

2.1. Preliminaries and examples

Let $(\Omega,\mathfrak{F},\mathfrak{F}_t,t\in\mathbb{R},P)$ be a filtered probability space satisfying the usual hypotheses of [11] (i.e. right continuity and completeness of $(\mathfrak{F}_t)$) with

$$\mathfrak{F} = \bigvee_{t\in\mathbb{R}}\mathfrak{F}_t.$$

Let M be a closed random set of $\mathbb{R}$, $D_t=\inf\{s>t\colon s\in M\}$ and $R_t=D_t-t$ with $R=R_0$. Assume that D_t is an $(\mathfrak{F}_s)$ stopping time for each $t\in\mathbb{R}$, and note that M is the closure of the set $\{D_r\colon r \text{ rational}\}$ so it is $(\mathfrak{F}_t)$ optional.

Definition 2.1. Let $(\phi_t)_{t\in\mathbb{R}}$ be a measurable process taking values in $(\Omega,\mathfrak{F})$, and assume that (M,ϕ) satisfy the following homogeneity and regeneration properties. For $t\in\mathbb{R}$ and $s\geqslant 0$

$$R_{t+s} = R_s\circ\phi_t, \tag{2.1}$$

$$\phi_{t+s} = \phi_t\circ\phi_s \quad (t\in\mathbb{R},\ s\geqslant 0);$$

and for all $A\in\mathfrak{F}$

$$P(\phi_{D_t}\in A\,|\,\mathfrak{F}_{D_t}) = P_0(A) \text{ on } \{D_t<\infty\}, \tag{2.2}$$

where P_0 is a fixed probability law. Then $(M,\phi)=(\Omega,\mathfrak{F},\mathfrak{F}_t,M,\phi_t,P,P_0)$ is called a *homogeneous regenerative set, M* is called the *regeneration set,* and the process (ϕ_s) is called the *future process.*

Example. Let $(\Omega,\mathfrak{F},\mathfrak{F}_t,P)$ be a realization of a right continuous strong Markov process. That is, Ω is the set of all right continuous functions from $\mathbb{R}$ into a state space $E\cup\Delta$, where E is a nice space (Lusin or locally compact with countable base) and Δ is a pre-birth, post-death point. Take a fixed point $b\in E$ and let

$$M = \overline{\{t\colon \omega(t) = b\}}$$

(here and in the sequel $\overline{C}$ is the closure in $\mathbb{R}$ of the set C), $D_t=\inf\{s>t\colon \omega(s)=b\}$ and $R_t=D_t-t$. Letting ϕ_t be the future process of the underlying Markov process (i.e. $\phi_t=\{\omega(t+s),\ s\geqslant 0;\ \Delta \text{ for } s<0\}$) we see that (M,ϕ) is a homogeneous regenerative set. The filtration $(\mathfrak{F}_t)$ is constructed here as follows. Set $\mathcal{F}_t^0 = \sigma\{\omega(s);\ s \leq t\}$ and

$$\mathcal{F}^0 = \bigvee_{t \in \mathbf{R}} \mathcal{F}_t^0.$$

Let $\mathcal{F}$ be the P completion of $\mathcal{F}^0$ and

$$\mathcal{F}_t = \bigcap_{s>t} (\mathcal{F}_s^0 \vee \mathcal{N}),$$

with $\mathcal{N}$ being the P-null sets of $\mathcal{F}$. The regeneration property follows from the strong Markov property with $P_0 = P_b$ - the law of $(\omega(t))_{t \geq 0}$ given that $\omega(0) = b$.

Note that if we let

$$M_t = \overline{(M-t) \cap (0,\infty)},$$

then the regeneration property implies, in particular, that

$$P(M_{D_t} \in \cdot | \mathcal{F}_{D_t}) = P_0(M_0 \in \cdot) \text{ on } \{D_t < \infty\}. \tag{2.3}$$

2.2. Local time and exit system

Let (M,ϕ) be a homogeneous regenerative set. A *local time*, L, of M is a random measure on $\mathbb{R}$, with support M, which is $(\mathcal{F}_{D_t})$ predictable, $(\mathcal{F}_t)$ adapted, and satisfies

$$L(t,t+s] = L(0,s] \circ \phi_t \quad (t \in \mathbb{R},\ s > 0). \tag{2.4}$$

The existence of L, and its uniqueness up to a multiplicative constant is proved in [18]. It is also shown there that every regenerative set is either a.s. discrete or a.s. perfect. In the discrete case L may be defined by

$$L(du) = P_0(1-e^{-R}) \cdot \sum_{s \in M} \epsilon_s(du) \tag{2.5}$$

where ϵ_s is the Dirac measure with mass at s, so that L is a counting measure multiplied by the constant $P_0(1-e^{-R})$. (Here, and in the sequel, $Q(f)$ is the expected value of f with respect to the measure Q.) When M is perfect, then L is diffuse (i.e. it doesn't charge points) and $1_M(t)dt = m \cdot L(dt)$, where m is a constant in [0,1] (see [18]). In both the discrete and continuous cases it is customary to normalize L so that

$$P_0 \int_0^\infty e^{-s} L(ds) = 1. \tag{2.6}$$

We see that when M is discrete L is a process that counts the points in M (multiplied by a constant). When M is perfect (counting is then not possible) L measures the 'time spent in M'.

Let $G = \{t \in \mathbb{R}:\ R_{t-} = 0,\ R_t > 0\}$ be the set of left end points of intervals contiguous to M. The following theorem deals with exits from M and is due to MAISONNEUVE [20].

THEOREM 2.1. *Let (M,ϕ) be a homogeneous regenerative set, Z a process which is $(\mathcal{F}_{D_t})$ predictable, and H an $\mathcal{F}$-measurable positive function. Then there exists a* *measure* $^*P$ *on* $(\Omega, \mathcal{F})$ *such that*

$$P(\sum_{s \in G} Z_s H \circ \phi_s) = {}^*P(H) P \int Z_s L(ds) \tag{2.7}$$

where

$${}^*P(H) = P_0 \sum_{\substack{s \in G \\ s>0}} e^{-s} H \circ \phi_s. \tag{2.8}$$

An interpretation of this result could proceed as follows. Let $A \in \mathcal{F}$ and take $H = 1_A$. Then $^*P(H)$ is the rate (with respect to the clock L) of exits from M to a future that lies in A. If, for example, $A = \{R > t\}$, then $^*P(A)$ is the rate of excursions from M of durations longer than t.

Since $^*P$ is a rate it is not a probability measure. In the discrete case it is a finite measure, while if M is perfect it is an infinite σ-finite measure satisfying $^*P(1-e^{-R}) = 1$. Since this measure will play an important role in what follows, we shall consider here a few special cases:

(i) In classical renewal theory let $M = \{S_n : n \geq 0\}$ be the set of renewal points. Then M is discrete and therefore $G = M$. For $H = 1_{\{R(\omega)>t\}}$ we obtain

$$P_0 \sum_{\substack{s \in G \\ s>0}} e^{-s} 1_{\{R>t\}} = P_0 \sum_n e^{-S_n} P_0(R>t).$$

Here $P_0(R>t) = 1 - F(t)$ where F is the distribution of the inter-renewal epochs. Thus

$${}^*P(R>t) = \frac{(1-F(t))F^*(1)}{1-F^*(1)}$$

where F^* is the Laplace transform of F.

(ii) In the example of Section 2.1, suppose that $E = \mathbb{R}$ and that $\{t : \omega(t) = b\}$ is a union of (half closed) intervals, i.e. that b is a holding point for the process. Then the lengths of these intervals are independent exponential random variables all with the same parameter λ_b, say. Let $K(b, dy)$ be the jump measure out of b. Let $A = \{\omega : \sup_{t < R(\omega)} \omega_t > b + a\}$ and $H = 1_A$. Here $^*P$ is the rate of excursions out of b that hit the interval $(b+a, \infty)$ and G consists of the times when the process leaves b. Since G consists of stopping times it follows that

$$\begin{aligned} P_0 \sum_{s \in G} e^{-s} H \circ \phi_s &= P_0 \sum_{s \in G} e^{-s} \int K(b, dy) P_y(A) \\ &= \int K(b, dy) P_y(A) \frac{\lambda_b}{\lambda_b + 1 - g^*(1)\lambda_b}, \end{aligned}$$

where $g^*(\theta) = \int K(b, dy) E_y(e^{-\theta R})$. In the same manner one can show that for any A in $\mathcal{F}_R$ (where $\mathcal{F}_R = \{A \in \mathcal{F} : A \cap \{R \leq t\} \in \mathcal{F}_t \text{ for all } t\}$) we have

$${}^*P(A) = \text{constant} \cdot \int \lambda_b K(b, dy) P_y(A).$$

(iii) Similar results can be shown to hold for points that are not necessarily holding points, but from which the process exits by jumps.

When M is perfect let

$$\tau_t = \inf\{s>0:\ L(0,s]>t\}, \tag{2.9}$$

for $t>0$. Then, under P_0, $(\tau_t)_{t\geq 0}$ is a process with stationary independent increments. Its Lévy measure is $\lambda(dt)={}^*P(R\in dt)$ and its drift is equal to the constant m that is defined in the discussion following (2.5).

2.3. Stationary regenerative sets

A homogeneous regenerative set (M,ϕ) is said to be *stationary* if the law of ϕ_t under P is independent of t. If one takes Ω to be the sample paths of (R_t), and ϕ_t to be the future of R after t, then stationarity of (M,ϕ) is equivalent to the stationarity of the Markov process (R_t). This is the case in classical renewal theory when one is interested only in the set M.

For stationary regenerative sets we have the following result (see [19, 20]).

THEOREM 2.2. *Suppose (M,ϕ) is stationary and non-empty. Let $g_t=\sup\{s\leq t:\ s\in M\}$ and $a_t=t-g_t$. Let h be a measurable function on $(\mathbb{R}_+,\mathcal{R}_+)$ and H a measurable function on $(\Omega,\mathcal{F})$. Then*

$$P(H(\phi_t)h(a_t)) = \frac{1}{\gamma}\Big(mP_0(H)h(0)+\int_{\mathbb{R}_+} da\,{}^*P(R>a;\,H(\phi_a))h(a)\Big)$$

where P_0 is defined in (2.2) and

$$\gamma = m+\int_0^\infty a\,{}^*P(R\in da).$$

In particular,

$$P(h(a_t)) = P(h(R_t)) = \frac{1}{\gamma}\Big(mh(0)+\int_{(0,\infty)} da\,{}^*P(R>a)h(a)\Big).$$

We note that when M is discrete then ${}^*P(R>t)=c(1-F(t))$, where c is a constant and $F(t)$ is the distribution of the times between successive regeneration points. In this case the last equation in Theorem 2.2 is the familiar result on the age and forward recurrence time of stationary renewal processes, namely,

$$P(R>t) = \frac{1}{\mu}\int_t^\infty(1-F(u))du \quad \Big(\mu = \int x\,dF(x)\Big).$$

In the non-discrete case things are very much the same since then (see [19])

$$\lim_{t\to\infty} P_0(R_t>a) = \frac{1}{\gamma}\int_a^\infty {}^*P(R>u)du. \tag{2.10}$$

REMARK. Most of the results cited here have natural extensions to regenerative systems. We refer the reader to [15] for details.

2.4. The Markovian case

We next restrict our attention to the situation treated in the example of Section 2.1, where X is a strong Markov process, b is a point in its state space and

$$M = \overline{\{t: X_t = b\}}.$$

The local time at M is the local time at b and $R = \inf\{t>0: X_t = b\}$. The following result is proven in SILVERSTEIN [22], GETOOR [13], and in a slightly disguised form in GEMAN and HOROWITZ [12].

THEOREM 2.3. *Suppose that* $^*P(R=\infty)=0$. *Then the measure* ν *on* E *defined by*

$$\nu(A) = m 1_A(b) + {}^*P\Big(\int_0^R 1_A(X_u)du\Big) \tag{2.11}$$

is an invariant measure for X. *It is* σ*-finite and is concentrated on* $\{x: P_x(R<\infty)=1\}$.

We note that when M is a discrete, a-periodic and persistent random set then, with some constant c, we have

$$\nu(A) = cE_b \int_0^R 1_A(X_u)du.$$

Furthermore, if, in addition, $E_b(R)<\infty$, then the process X is positive recurrent and its unique stationary distribution is

$$\mu(A) = E_b \int_0^R 1_A(X_u)du / E_b(R). \tag{2.12}$$

The last equation is a well-known result that follows from the key renewal theorem, and (2.11) is its extension to general regenerative sets. In Section 3 we shall use (2.11) to study level crossings of general storage processes.

It follows from (2.10) that ν is a finite measure if, and only if, $^*P(R)<\infty$. What we need now is a condition under which this measure is unique. Here the concept of Harris-recurrence will play an important role since, as it is well-known, Harris-recurrent processes always have a unique invariant measure.

DEFINITION 2.2. A Markov process X is Harris-recurrent if there exists a σ-finite measure ν on its state space E so that $\nu(A)>0$ implies that

$$P_x\Big[\int_0^\infty 1_A(X_s)ds = \infty\Big] = 1$$

for all $x \in E$.

Theorem 2.4. *Let X be a strong Markov process, b a point in its state space,*

$$M = \overline{\{t: X_t = b\}},$$

$R=\inf\{t>0: X_t=b\}$ and assume that M is a perfect set. Then $P_x(R<\infty)=1$ for all $x\in E$ implies that X is Harris-recurrent.

Proof. Let τ_s be as in (2.9) and A a Borel set in E. Let

$$\nu(A) = E_b \int_0^{\tau_1} 1_A(X_s)ds.$$

This measure is σ-finite ([16]) and we wish to show that it satisfies the condition of Definition 2.2. For this we recall that (τ_s) is a process with stationary independent increments and hence $\tau_s<\infty$ for $s<\infty$, a.s., if, and only if, $\lambda\{\infty\}=0$. But from the strong Markov property of X we may write

$$\lambda\{\infty\} = {}^*P\{R=\infty\} = {}^*P(R>t;\, P_{X_t}(R=\infty)) = 0,$$

since for all $x\in E$, $P_x(R=\infty)=0$. Hence for $s<\infty$, $\tau_s<\infty$ a.s. and $\tau_s\uparrow\infty$. It follows that

$$\int_0^{\infty} 1_A(X_u)du = \lim_{s\to\infty}\int_0^{\tau_s} 1_A(X_u)du.$$

For $s\geq 0$ let

$$B_s = \int_0^{\tau_s} 1_A(X_u)du. \tag{2.13}$$

An easy computation reveals that

$$B_{s+t} = B_s + B_t\circ\phi_{\tau_s}.$$

Since $X_{\tau_s}=b$ for all $s<\infty$, it follows that

$$E_b(B_{s+t}) = E_b(B_s)+E_b(B_t),$$

from which one obtains that

$$\nu(A) = \frac{1}{s}E_b(B_s), \text{ for all } s>0.$$

To complete the proof we need to show that $\nu(A)>0$ implies that $B_\infty=\infty$ P_x a.s. for all x. For this let n be an integer. Then

$$B_n = B_1 + B_1\circ\phi_{\tau_1} + \cdots + B_1\circ\phi_{\tau_{n-1}}.$$

Since $X_{\tau_s}=b$ for any $s<\infty$, it follows from the strong Markov property at τ_s that under P_b $\{B_1\circ\phi_{\tau_l}\}_{l=1}^{\infty}$ are i.i.d. r.v.'s and by the strong law of large numbers if $E_b(B_1)>0$ then $\lim_{n\to\infty}B_n=\infty$. But $E_b(B_1)=\nu(A)$ which is positive by our assumption. Since $t\to B_t$ is non-decreasing $B_\infty=\lim_{t\to\infty}B_t=\infty$ P_b a.s. If $x\in E$ then

$$P_x(B_\infty=\infty) \geqslant P_x(R<\infty;\, P_{X_R}(B_\infty=\infty)) = 1$$

since $X_R=b$ P_x a.s., and $R<\infty$ P_x a.s. by our assumptions. This completes the proof. □

REMARK. An almost identical argument with

$$\nu(A) = E_b\int_0^R 1_A(X_s)ds$$

proves the result for the case when M is discrete.

3. MARKOV STORAGE SYSTEMS AND LEVEL CROSSING IDENTITY

In this section we treat a general class of storage processes, which we call Markov storage processes. The objective is to use the concept of regeneration for extending results on level crossings and on the stationary distribution of the storage process.

In the classical theory of dam processes the content of the dam is defined in terms of an input, which is usually a process with stationary independent increments, and an output which is either at constant rate or at a rate that is state dependent. The resulting content process is a special Markov process. Here we define a more general content process directly as follows.

A process $V=(V_t)_{t\geqslant 0}$ will be called a *Markov storage process* if it satisfies the following assumptions.

(i) V is a strong Markov process with sample paths that take values in $\mathbb{R}_+$, that are right continuous and have left limits. Jump times of V are totally inaccessible (i.e. if T is a jump time of V, then there is never a sequence of stopping times $\{T_n\}$ so that $T_n<T$ and $T_n\uparrow T$).

(ii) V is a process of bounded variation.

A process that satisfies (i) is a Hunt process on $\mathbb{R}_+$. With (ii) it is also a semi-martingale. We make here assumption (ii) since we are modeling storage systems. The general structure of Hunt semi-martingales is given in [6] from where we cite the following important result adapted to the present situation.

Recall that a process (M_t) is a local martingale if there exists a sequence of stopping times $\{T_n\}, T_n\uparrow\infty$ a.s. so that for every n, $(M_{t\wedge T_n})$ is a martingale.

THEOREM 3.1. *Let V be a Hunt semi-martingale that satisfies (i)-(ii). Then there exist continuous additive functionals (A_t), (B_t) of (V_t), and a kernel $N(x,\cdot)$ from $\mathbb{R}_+$ into $\mathbb{R}_+$, so that for every $f\in C^1(\mathbb{R})$ (continuously differentiable function),*

$$f(V_t) + \int_0^t f'(V_s)dA_s - \int_0^t dB_s \int_{\mathbf{R}} N(V_s,dy)[f(y)-f(V_s)] \tag{3.1}$$

is a local martingale.

Theorem 3.1 extends the well-known Dynkin formula. Indeed, if f belongs to the domain of the infinitesimal generator G of V, then according to Dynkin's formula

$$f(V_t) - \int_0^t Gf(V_s)ds$$

is a martingale. Further if $A_t = \int_0^t a(V_s)ds$, $B_t = \int_0^t b(V_s)ds$ where a and b are sufficiently smooth functions, and $x \rightarrow N(x,f)$ is continuous for continuous f, then $f \in C^1$ belongs to the domain of G and

$$Gf(x) = -a(x)f'(x) + \int_0^t b(x)N(x,dy)\,[f(y)-f(x)]$$

so that (3.1) is actually Dynkin's formula. The processes A_t and B_t and the kernel N are called the local characteristics of V. (For general semi-martingales the local characteristics also include a positive definite matrix as in [6]).

It follows from (3.1) and the Optional Sampling Theorem that if T is a stopping time satisfying $E_x(T) < \infty$, then for $f \in C^1$

$$E_x f(V_T) = f(V_0) - E_x \int_0^T f'(V_s)dA_s + E_x \int_0^T dB_s \int_{\mathbf{R}_+} N(V_s,dy)\,[f(y)-f(V_s)].$$

All classical storage processes with input according to a process with stationary independent increments and continuous output (like the models in [3, 5, 9, 10, 14, 21]) satisfy the above assumptions and follow the representation of (3.1). For these processes

$$A_t = \int_0^t (r(V_s) - \rho)ds,$$

where $\rho \geqslant 0$ is the drift of the input process, r is the output rate, $B_t = t$, and $N(x,dy) = \lambda(dy - x)$, with λ being the Lévy measure of the input process. In the present (more general) setup the V process may have downward jumps and all jumps (up and down) may have state dependent probability laws.

In order to study level crossings we shall make the following additional assumptions.

(iii) The additive functionals A and B of Theorem 3.1 are absolutely continuous with respect to the Lebesgue measure, i.e. there exist functions a and b such that for all $s > 0$

$$A_s = \int_{[0,s)} a(V_t)dt, \quad B_s = \int_{[0,s)} b(V_t)dt. \tag{3.2}$$

(iv) Let $K(x,dy) = N(x, dy + x)$. Then

$$\sup_x \int_{\mathbf{R}} |y| K(x,dy) < \infty. \tag{3.3}$$

(v) In the notation of Section 2, with $X = V$, $b = 0$ and $R = \inf\{t>0: V_t = 0\}$, for any x that can be reached from 0 we have $P_x(R < \infty) = 1$, and the measure ν defined by (2.11) is finite.

Regarding assumption (iii) we note that for any A and B as in Theorem 3.1 we can define

$$C_t = t+A_t+B_t,$$
$$S_t = \inf\{u: C_u>t\} = (C^{-1})_t,$$
$$\hat{V}_t = V_{S_t}.$$

Then $\hat{V}_t$ satisfies assumption (iii) (see [6]) and, moreover, the relationship between the stationary distributions of V_t and $\hat{V}_t$ is well established (see [1]). Therefore this assumption is not restrictive here since our interest is mainly in the representation of the stationary distribution of the storage model. Without loss of generality we shall also assume that $B_t=t$ so that $b(x)$ defined in (3.2) gets absorbed in $N(x,\cdot)$. Assumption (iv) is also not very restrictive but assumption (v) is restrictive. The collection of functions, a, and kernels $N(x,\cdot)$, leading to such processes is large and is not easily characterized even in the cases described in [21]. There are some cases where a and N satisfy some smoothness assumptions (normally less restrictive than those needed to ensure that (3.1) has a unique solution) where a positive solution to (3.5) below with $\pi(0)>0$ is also sufficient for (v) to hold. We refer the reader to [3] for examples, and hope to treat this problem in greater detail in the future.

Let V be a Markov storage process that satisfies all the above assumptions. Let $M=\{t: V_t=0\}$ and adopt the notation of Section 2 with V playing the role of X and $b=0$. Then (M,ϕ) is a homogeneous regenerative set which is Harris-recurrent (Theorem 2.4) and hence has a unique invariant measure. The following is the main result of this section.

THEOREM 3.2. *Let π be the unique stationary distribution defined through (2.11). Then the measure*

$$\mu(A) = \int_A \pi(dy)a(y) \quad \text{on } (0,\infty)$$

is absolutely continuous with respect to the Lebesgue measure and its density is equal to

$$\int_0^y \pi(du)K(u,(y-u,\infty)) - \int_y^\infty \pi(du)K(u,(-\infty,y-u)). \tag{3.4}$$

In particular if $a(y)<0$ for $y>0$ and $a(0)=0$, then on $(0,\infty)$ π is absolutely continuous with respect to the Lebesgue measure and its density g satisfies

$$g(y) = -\frac{1}{a(y)}[\pi\{0\}K(0,(y,\infty)) + \int_{(0,y)} g(u)K(u,(y-u,\infty))du$$
$$- \int_{(y,\infty)} g(u)K(u,(-\infty,y-u))du]. \tag{3.5}$$

PROOF. It was shown in [17] that $^*P$ is a strong Markov law, i.e. for any stopping time T, and $\mathcal{F}$-measurable Z, $^*P(Z\circ\phi_T)={}^*P(P_{V_T}(Z))$. It therefore follows from (3.1) that for $f\in C^1$

$$^*P(f(V_R)-f(V_{0-})) = {}^*P(f(V_0)-f(V_{0-})) - {}^*P\int_0^R f'(V_s)a(V_s)ds + {}^*P\int_0^R ds\int_{\mathbf{R}_+} N(V_s,dy)(f(y)-f(V_s)). \tag{3.6}$$

Since $V_{0-}=0$, $V_R=0$ $^*P$ a.e., the left-hand side of (3.6) is equal to 0. We now treat each term on the right-hand side of (3.6) separately.

The first term. Since the functional B_t and the kernel N take care of the jumps, and since we have assumed that $B_t=t$, then jumps out of zero may occur only if $\{t\colon V_t=0\}$ has positive Lebesgue measure. In this case there is a constant m so that the Lebesgue measure of $\{t\colon V_t=0\}$ is equal to m times L_t, where L_t is the local time at zero. Note that the constant m plays here the same role as that of 2.2. Hence for $H\in\mathcal{F}_R$

$$\begin{aligned} ^*P(H\cdot L_{\{V_0>0\}}) &= P_0\int_0^\infty e^{-t}1_{\{V_{t-}=0\}}\int K(0,dy)P_y(H)dt \\ &= \int K(0,dy)P_y(H)\cdot mP_0\int_0^\infty e^{-t}dL_t \\ &= m\int K(0,dy)P_y(H). \end{aligned}$$

In particular if f is any positive real function and $H=f(V_0)$ then

$$^*P(f(V_0)1_{\{V_0>0\}}) = m\int_{\mathbf{R}_+} K(0,dy)f(y).$$

Let $\gamma=m+{}^*P(R)$. Then $\pi\{0\}=m/\gamma$ and for A in $(0,\infty)$

$$\pi(A) = \frac{^*P\int_0^R 1_A(V_u)du}{\gamma}. \tag{3.7}$$

Hence

$$^*P(f(V_0)1_{\{V_0>0\}}) = \gamma\pi\{0\}\int_{\mathbf{R}_+} K(0,dy)f(y).$$

The first term on the right-hand side of (3.6) is therefore equal to

$$\gamma\pi\{0\}\int_{\mathbf{R}_+} K(0,dy)[f(y)-f(0)] = \gamma\pi\{0\}\int_{\mathbf{R}_+} f'(y)K(0,(y,\infty))dy.$$

The second term. By the definition of π in (3.7), the second term in (3.6) is equal to

$$\gamma \int_{(0,\infty)} \pi(dy) a(y) f'(y).$$

The third term. Recalling that $K(x,dy)=N(x,\ dy+x)$, and again the definition of π, this term is equal to

$$\gamma \int_{\mathbf{R}_+\setminus\{0\}} \pi(dy) \int_{\mathbf{R}_+} K(y,du)(f(y+u)-f(y)) + \gamma \int_{\mathbf{R}_+\setminus\{0\}} \pi(dy) \int_{[-y,0)} K(y,du)[f(y+u)-f(y)]$$

which after some algebra becomes

$$\gamma \int_{y=0}^{\infty} f'(y) \left[\int_{u\in(0,y)} \pi(du) K(u,(y-u,\infty)) - \int_{u\in(y,\infty)} \pi(du) K(u,(-\infty,y-u)) \right] dy.$$

Now substitute $f(y)=e^{-\lambda y}$ to obtain

$$-\int_{(0,\infty)} e^{-\lambda y} a(y)\pi(dy) = \int_{(0,\infty)} e^{-\lambda y} \left[\int_{u\in[0,y)} \pi(du) K(u,(y-u,\infty)) \right. \tag{3.8}$$

$$\left. - \int_{u\in[y,\infty)} \pi(du) K(u,(-\infty,\ y-u)) \right] dy$$

which shows that $a(y)\pi(dy)$ is absolutely continuous with respect to the Lebesgue measure and gives the desired form of the density. □

REMARKS

(i) Let $T_y=\inf\{t>0\colon\ V_t=y\}$ and suppose that points in $(0,\infty)$ are irregular for V_t (i.e. $P_y(T_y=0)=0$). Then $a(y)g(y)$ is the long run average of the difference between the number of discontinuous downcrossings and the number of discontinuous upcrossings of level y. This is the same as the long run average number of continuous crossings of that level. Hence, although a priori it is not clear that points in $(0,\infty)$ are actually hit (because the process may have jumps downwards), it turns out that when a is strictly negative (or positive) points are hit with probability 1. Let $N_y(t)$ denote the number of continuous crossings of level y in the interval $[0,t)$, then

$$\lim_{t\to\infty} \frac{N_y(t)}{t} \to a(y)g(y) \text{ a.s.} \tag{3.9}$$

where g is a derivative of $\pi(0,y]$.

(ii) The representation of π through (2.11) is not essential for the proof of Theorem 3.2. In fact, it can be shown, starting from (3.1) and performing computations of similar nature to the above, that Theorem 3.2 remains valid for any stationary distribution π, regardless of assumption (v). This assumption is invoked here to ensure the uniqueness of π and the ergodicity of the process. It is only under ergodicity that one may deduce (under the conditions of the previous remark) that (3.9) holds true. The computation used to prove Theorem 3.2 also shows that this is equal to

$$\frac{1}{m+{}^{*}P(R)}\,{}^{*}P[\text{number of continuous crossings of level } y \text{ during } (0,R)]. \tag{3.10}$$

In RUBINOVITCH and COHEN [21] (see also [3]), dam processes with input according to a process with stationary independent increments and continuous output, with rate function r, were considered. The lack of downwards jumps made it possible to use the discrete regeneration set $M=\{t\colon V_t=a\}$, for some $a>0$, and thus obtain

$$g(y)a(y) = \frac{1}{E_a(R)}E_a(\text{ number of continuous crossings of level } y \text{ during } [0,R)), \tag{3.11}$$

where $R=\inf\{t\colon V_t=a\}$. This equation was the key to the analysis of [21]. Note that in the special case of [21] the right-hand side of (3.11) is equal to (3.10).

ACKNOWLEDGEMENT

We are grateful to Ryszard Syski for his helpful comments on an earlier version of this paper.

REFERENCES

1. J.M. AZÉMA, M. KAPLAN-DUFLO, D. REVUZ (1967). Mesure invariante sur les classes recurrentes des processus de Markov. *Z. Wahrsch. Verw. Gebiete 8,* 157-181.
2. A. BENVENISTE, J. JACOD (1973). Systèmes de Lévy des processus de Markov. *Invent. Math. 21,* 183-198.
3. P.J. BROCKWELL, S.I. RESNICK, R.L. TWEEDIE (1982). Storage processes with general release and additive input. *Adv. in Appl. Probab. 14,* 392-433.
4. E. ÇINLAR (1975). *Introduction to Stochastic Processes,* Prentice Hall, Englewood Cliffs, N.J.
5. E. ÇINLAR (1975). A local time for a storage process. *Ann. Prob. 3,* 930-950.
6. E. ÇINLAR, J. JACOD, P. PROTTER, M.J. SHARPE (1980). Semimartingales and Markov processes. *Z. Wahrsch. Verw. Gebiete 54,* 161-219.
7. J.W. COHEN (1982). *The Single Server Queue,* revised edition, North-Holland, Amsterdam.
8. J.W. COHEN (1976). *On Regenerative Processes in Queueing Theory,* Springer-Verlag, Berlin.
9. J.W. COHEN, M. RUBINOVITCH (1977). On level crossings and cycles in dam processes. *Math. Oper. Res. 2,* 297-310.
10. J.W. COHEN (1977). On up- and downcrossings. *J. Appl. Probab. 14,* 405-410.

11. C. DELLACHERIE (1972). *Capacités et Processus Stochastiques,* Springer-Verlag, Berlin.
12. D. GEMAN, J. HOROWITZ (1973). Remarks on Palm measures. *Ann. Inst. Henri Poincaré Sect. B, IX,* 213-232.
13. R.K. GETOOR (1979). Excursion of a Markov process. *Ann. Prob. 7,* 244-266.
14. J.M. HARRISON, S.I. RESNICK (1976). The stationary distribution and the first exit probabilities of a storage process with general release rule. *Math. Oper. Res. 4,* 349-358.
15. H. KASPI, B. MAISONNEUVE. Regenerative systems on the real line. To appear in *Ann. Prob.*
16. H. KASPI (1984). On invariant measures and dual excursions of Markov processes. *Z. Wahrsch. Verw. Gebiete 66,* 185-206.
17. B. MAISONNEUVE (1975). Exit systems. *Ann. Prob. 3,* 399-411.
18. B. MAISONNEUVE (1974). *Systèmes Régénératifs,* Astérisque 15 S.M.F.
19. B. MAISONNEUVE (1983). Ensembles régénératifs de la droite. *Z. Wahrsch. Verw. Gebiete 63,* 501-510.
20. B. MAISONNEUVE (1987). *Ensembles Régénératifs Homogènes,* preprint.
21. M. RUBINOVITCH, J.W. COHEN (1980). Level crossings and stationary distributions for general dams. *J. Appl. Probab. 17,* 218-226.
22. M. SILVERSTEIN (1980). Classification of coharmonic and coinvariant functions for Lévy processes. *Ann. Prob. 8,* 539-575.

Estimation from an Infinite Server Queueing System with two Demands[1]

Michael J. Phelan[2] and N.U. Prabhu
School of Operations Research and Industrial Engineering
Cornell University, Upson Hall
Ithaca, NY 14853, USA

We consider an infinite server queueing system in which the arrival epochs of successive customers constitute an arbitrary point process. Each customer has two demands for service, the service times of customers being mutually independent. The system meets only the demand that needs the shorter service time. Situations described by this model arise in the context of clinical trials with staggered entries and competing risks. The objective is to estimate the probability distributions of the service times of the two demands. For this purpose the system is observed over a finite interval $[0, t]$, with t fixed. The data collected consist of the arrival epochs and service times of the customers who arrive during $[0, t]$, with some of the service times partially observed (censoring). The estimators are martingale estimators, for which we establish the consistency and weak convergence (as $t \to \infty$) of the normalized differences to a Gaussian process.

1. Introduction

We consider the following queueing system.

(1) Let $\tau_0, \tau_1, \tau_2, \ldots$ denote the arrival epochs of the successive customers. We assume that the point process $\tau = \{\tau_n, n \geqslant 0\}$ satisfies the following conditions:

(i) $\tau_0 = 0$,

(ii) $\tau_n < \infty$ $(n \geqslant 1)$, and

(iii) $\tau_{n+1} > \tau_n$, $\tau_n \uparrow \infty$.

Let $N = \{N_t, t \geqslant 0\}$ denote the counting process generated by the process τ. Then N_t gives the number of arrivals during a time-interval $(0, t]$. Here N has right-continuous sample paths with left limits, jumps of size 1 and $N_0 = 0$. Thus the input into the queueing system is described equivalently by N or τ.

(2) Each customer brings two demands for service. Let (X_n, Y_n) denote the

1. Partially supported by the US Army Research Office through the Mathematical Sciences Institute of Cornell University.

2. Current address: Department of Civil Engineering and Operations Research, Princeton University, Princeton, NJ 08544, USA.

service times of the two demands of the n-th customer ($n \geqslant 1$). We assume that $\{X_n, n \geqslant 1\}$ and $\{Y_n, n \geqslant 1\}$ are independent sequences of mutually independent random variables with common distributions F_1 and F_2, respectively, both concentrated on $(0, \infty)$. We also assume that these service times are independent of the input process.

(3) There are an infinite number of servers, so that there is no waiting line. However, each server meets only the demand that needs the shorter service time for the customer served.

Our objective is to estimate the distributions F_1 and F_2 of the service times of the two demands. For this purpose we observe the system over a time-interval $(0, t]$, with t fixed. The data consist of the arrival epochs and the service times actually received by the first N_t customers, but some of these service times may only be partially observed; namely, those customers with $\tau_n + \min(X_n, Y_n) > t$, for whom we only know that the service time $\min(X_n, Y_n) > t - \tau_n$.

In Section 2 we propose the estimators for F_1 and F_2, which turn out to be martingale estimators. Some preliminary results are derived in Section 3. The martingale property of the estimators leads to proofs of their consistency and the weak convergence (as $t \to \infty$) of the normalized differences to a Gaussian process. These results are established in Section 4. We make only a mild assumption concerning the rate at which N_t goes to infinity. Thus we avoid conditions such as $N_t/t \to$ constant >0 in some sense, as is often imposed in situations involving random sample sizes. This shows the advantage of our approach based on martingale properties.

In Section 5 we illustrate an application of our estimation procedure with mouse leukemia data found in KALBFLEISCH and PRENTICE [8].

The estimation problem discussed in this paper arises in the context of clinical trials with staggered entries, competing risks and fixed duration of study. Patients enter the study at random epochs and are immediately put on test (for example, the patient is treated by a drug therapy). Typically, the objective is to study a patient's time of death (survival time) from cause A, such as cancer or AIDS. Factors such as end of study or patient withdrawal make it impossible to observe patient's time of death (censoring). However, for some patients, death may actually be from a cause other than A, such as toxicity. Denoting this second cause as B, we say that A and B are competing risks. In such situations it may be important to study the hazard functions associated with both causes A and B, rather than A alone, as is usually done. Thus the observed survival time of each patient is the shorter of the survival times from A and B.

In the standard analysis of staggered entry clinical trials it is assumed that the epochs of entry $\{E_n\}$ of the patients are mutually independent random variables and the duration of study is fixed. In order to develop an asymptotic theory it is then assumed that the accrual rate of patient entry increases over this interval. In the terminology of this paper, $\{\tau_n\}$ are the order statistics generated by $\{E_n\}$, so that the input of patients constitutes a special type of a point process. Such a model is described by JENNISON and TURNBULL [7]. Reflection will show that for the asymptotic theory one looks for limit results as the 'amount of information' collected from the data increases to infinity in some manner. One way of doing it is to let the duration of study go to infinity. A second way is to collect data from a sequence of systems (described by our model), each with the same end of study period and let the accrual rate of this sequence go to infinity. In the interest of specificity of our model, we have chosen to follow the first approach.

On some occasions there is limited availability of the drug used in the therapy and consequently there is a delay between the patient's time of entry and the actual time he is put on test. Such a situation leads effectively to a finite-server queueing system. This is being investigated, and results will be presented elsewhere.

Finally, a problem related to the one considered here arises in the context of censored renewal processes, where it is desired to estimate the lifetime distribution from a renewal process partially observed over an expanding time horizon. For this problem SELKE [10] shows that an estimation procedure quite similar to ours provides a consistent and asymptotically Gaussian estimator of this distribution. His result is based on an asymptotic condition (see equation (12) of his paper), for which a sufficient condition in our context is that $N_t/t \to$ constant >0 in probability. This latter condition is stronger than our condition (3.11) and therefore would restrict the class of input processes allowed in this paper.

2. THE ESTIMATORS

The data described in Section 1 can be conveniently summarized as follows. For each $n \geq 1$ define

$$C_n^t = \max(0, t-\tau_n), \quad W_n^t = \min(X_n, Y_n, C_n^t)$$
$$\delta_n^t = 1(X_n = W_n^t), \qquad \eta_n^t = 1(Y_n = W_n^t).$$

In the terminology of survival analysis $\min(X_n, Y_n)$ is the survival time induced by two competing risks and C_n^t is called a random right-censoring time due to end of study in staggered entry clinical trials. The quantity $(1-\delta_n^t)(1-\eta_n^t)$ is the indicator of censoring and W_n^t is the observed randomly right-censored survival time. In the present context W_n^t is the observed service time, $(1-\delta_n^t)(1-\eta_n^t)=0$ iff the n-th customer has arrived and completed his service before time t, and $\delta_n^t=1$ $(\eta_n^t=1)$ iff the server meets this customer's first (second) demand, so that his service time is X_n (Y_n). Note that it is possible to have ties (that is, $\delta_n^t\eta_n^t=1$), in which case the server is indifferent as to which demand is met. Our observation scheme yields the data

$$\{(W_n^t, \delta_n^t, \eta_n^t), \quad n=1,2,\ldots, N_t\}, \tag{2.1}$$

from which we seek to estimate the distributions F_1 and F_2. The estimators are actually based on the statistics

$$\overline{N}(i, t) = \{\overline{N}_s(i, t), \ s \geq 0\} \ \ (i=1,2); \ \overline{Y}(t) = \{\overline{Y}_s(t), \ s \geq 0\}, \tag{2.2}$$

where

$$\overline{N}_s(1, t) = \sum_{n=1}^{N_t} 1(W_n^t \leq s, \ \delta_n^t = 1), \tag{2.2a}$$

$$\bar{N}_s(2, t) = \sum_{n=1}^{N_t} 1(W_n^t \leqslant s, \eta_n^t = 1), \tag{2.2b}$$

$$\bar{Y}_s(t) = \sum_{n=1}^{N_t} 1(W_n^t \geqslant s). \tag{2.3}$$

Here $\bar{N}_s(i, t)$ is the number of customers whose service times are less than s among those who arrive and complete their service before time t, and whose i-th demand is met $(i=1,2)$. Also, $\bar{Y}_s(t)$ is the number of arrivals in $(0,t]$ whose service times (complete or partial) exceed s. In the case of ties we have included the observation in both $\bar{N}(1)$ and $\bar{N}(2)$ to preclude biased estimation of F_1 or F_2.

Our estimation procedure yields estimators of F_1 and F_2, as well as their associated cumulative conditional rate functions b_1 and b_2 defined by

$$b_i(t) = \int_{(0,t]} [1-F_i(s-)]^{-1} dF_i(s), \quad i=1,2, \quad t \geqslant 0. \tag{2.4}$$

These estimators are provided by the processes $B^t(i)=\{B_s^t(i), s\geqslant 0\}$, and $\hat{F}_i(t)=\{\hat{F}_s^t(i), s\geqslant 0\}$ $(i=1,2)$ defined by

$$B_s^t(i) = \sum_{u\leqslant s} \bar{Y}_u^{-1}(t)\Delta\bar{N}_u(i,t) = \int_0^s \bar{Y}_u^{-1}(t)d\bar{N}_u(i,t), \quad i=1,2, \tag{2.5}$$

(where each term in the sum is interpreted as zero if both factors are zero), and

$$\hat{F}_s^t(i) = 1-\prod_{u\leqslant s}(1-\Delta B_u^t(i)), \quad i=1,2, \tag{2.6}$$

where

$$\Delta\bar{N}_u(i,t) = \bar{N}_u(i,t)-\bar{N}_{u-}(i,t),$$

$$\Delta B_u^t(i) = B_u^t(i)-B_{u-}^t(i), \quad i=1,2, \quad 0\leqslant u\leqslant s.$$

For $i=1,2$ in (2.5), each term in the sum is the proportion of completed service times of i-th demand equal to u (i.e. $\Delta\bar{N}_u(i,t)$) among those service times (complete or partial) which exceed u (i.e. $\bar{Y}_u(t)$) from among those customers who arrive during the time-interval $(0,t]$. Thus $B_s^t(i)$ estimates the cumulative conditional service rate for the i-th demand. The expression (2.6) is an estimator of the product integral of (2.4), which uniquely determines the distributions F_1 and F_2 from b_1 and b_2, respectively. The estimators $\hat{F}_i^t$ are called product-limit estimators, where by virtue of our observation scheme they are defined from a random number of partially observed service times (cf. GILL [5], who studies this type of estimator from a fixed number of censored survival times).

3. Preliminary results

For a formal analysis of the estimators (2.5) and (2.6) we denote by $(\Omega, \mathcal{F}, P)$ the probability space on which all the random variables in the queueing system of Section 1 are defined. In order to study the properties of the estimators (2.5) and (2.6) we need a history of the partially observed service times from the arrivals in $(0, t]$. For each $t>0$ we introduce the filtration $H^t=\{\mathcal{H}_s^t, s\geqslant 0\}$ defined by

$$\mathcal{H}_s^t = \mathcal{N}\vee\sigma(N_u,\ 0\leqslant u\leqslant t;\ W_k^t\leqslant u,\ u\leqslant s;\ \delta_k^t 1(W_k^t\leqslant s), \qquad (3.1)$$
$$\eta_k^t 1(W_k^t\leqslant s),\ k=1,\ldots,N_t),$$

where $\mathcal{N}$ contains the P-null sets of $\mathcal{F}$ and their subsets. Clearly $\mathcal{H}_s^t\subset\mathcal{F}$, and H^t is right-continuous, complete, and nondecreasing. The family H^t provides the needed history and the properties of the estimators are derived relative to the stochastic basis $(\Omega, \mathcal{F}, \mathcal{H}_s^t, s\geqslant 0, P)$.

We proceed by showing that the counting processes defined by (2.2a) and (2.2b) are Poisson type counting processes with auxiliary process given by (2.3) and cumulative conditional rate function given by (2.4). This is carried out in Proposition 3.1 below, which is used to derive the martingale property for the estimators (2.5) and (2.6). We have the following.

PROPOSITION 3.1. *Let $(\Omega, \mathcal{F}, \mathcal{H}_s^t, s\geqslant 0, P)$ be the stochastic basis defined above, and let $\overline{N}(i, t)$, $(i=1,2)$ and $\overline{Y}(t)$ be the statistics defined by equations (2.2)-(2.3). Define the process $A(i, t)=\{A_s(i, t), s\geqslant 0\}$ by*

$$A_s(i, t) = \int_{(0, s]} \overline{Y}_u(t)db_i(u), \quad i=1,2, \qquad (3.2)$$

where b_1 and b_2 are the cumulative rate functions defined by (2.4). Then $\overline{N}(i, t)$ is a counting process having compensator $A(i, t)$, $(i=1,2)$. Moreover, the local martingales $M(i, t)=\{M_s(i, t), s\geqslant 0\}$ defined by

$$M_s(i, t) = \overline{N}_s(i, t)-A_s(i, t), \quad i=1,2, \qquad (3.3)$$

are orthogonal square-integrable local martingales with quadratic characteristic given by

$$\langle M(i, t), M(j, t)\rangle_s = \begin{cases} \int_{(0, s]} \overline{Y}_u(t)(1-\Delta b_i(u))db_i(u) & \text{for } i=j, \\ 0 & \text{for } i\neq j, \end{cases} \qquad (3.4)$$

where $\Delta b_i(u)=b_i(u)-b_i(u-)$, $i=1,2$, $0\leqslant u\leqslant s$.

PROOF. Obviously $\overline{N}(1, t)$ and $\overline{N}(2, t)$ are H^t-adapted counting processes, but in the sense of BROWN [4] in that they have multiple points with positive probability on the atoms of F_1 and F_2, respectively. Also, $\overline{Y}(t)$ is H^t-predictable, since it is H^t-adapted and left-continuous. Moreover, $A(i, t)$ is a nondecreasing H^t-predictable process, since it is the integral of a predictable process relative to a fixed Borel measure $(i=1,2)$. Therefore it remains to prove that $M(1, t)$ and $M(2, t)$ are local martingales with quadratic characteristic given by (3.4).

We introduce the history $K^t = \{\mathcal{K}_s^t,\ s \geqslant 0\}$ defined by

$$\mathcal{K}_s^t = \sigma(N_u,\ 0 \leqslant u \leqslant t;\ X_k \leqslant u,\ Y_k \leqslant u,\ u \leqslant s,\ k = 1, \ldots, N_t), \qquad (3.5)$$

where we note that $\mathcal{H}_s^t \subset \mathcal{K}_s^t$ for all s. For each $k = 1, 2, \ldots$ we define the processes $x^k = \{x_s^k,\ s \geqslant 0\}$ and $y^k = \{y_s^k,\ s \geqslant 0\}$ given by

$$x_s^k = [1(X_k \wedge Y_k \leqslant s,\ X_k \leqslant Y_k) - \int_{(o,\, s]} 1(X_k \wedge Y_k \geqslant u)\, db_1(u)]\, 1(\tau_k \leqslant t) \qquad (3.6a)$$

and

$$y_s^k = [1(X_k \wedge Y_k \leqslant s,\ Y_k \leqslant X_k) - \int_{(o,\, s]} 1(X_k \wedge Y_k \geqslant u)\, db_2(u)]\, 1(\tau_k \leqslant t). \qquad (3.6b)$$

Here assumption (2) implies via direct calculation that $\{x^k,\ k \geqslant 1\}$ and $\{y^k,\ k \geqslant 1\}$ are orthogonal families of orthogonal square-integrable martingales relative to K^t. Now for each $k = 1, 2, \ldots$ define the processes $\tilde{x}^k = \{\tilde{x}_s^k,\ s \geqslant 0\}$ and $\tilde{y}^k = \{\tilde{y}_s^k,\ s \geqslant 0\}$ given by the Stieltjes integrals

$$\tilde{x}_s^k = \int_0^s 1(C_k^t \geqslant u)\, dx_u^k \qquad (3.7a)$$

and

$$\tilde{y}_s^k = \int_0^s 1(C_k^t \geqslant u)\, dy_u^k. \qquad (3.7b)$$

Since $\{1(C_k^t \geqslant s),\ s \geqslant 0\}$ $(k \geqslant 1)$ is a bounded K^t-predictable process it follows from the theory of stochastic integration that $\{\tilde{x}^k,\ k \geqslant 1\}$ and $\{\tilde{y}^k,\ k \geqslant 1\}$ are orthogonal families of orthogonal square-integrable martingales relative to K^t. By assumption (1) $P(N_t < \infty) = 1$, so that careful evaluation of the integrals in (3.7) will verify the fact that

$$M_s(1, t) = \sum_{k \geqslant 1} \tilde{x}_s^k \quad \text{and} \quad M_s(2, t) = \sum_{k \geqslant 1} \tilde{y}_s^k \ (s \geqslant 0).$$

Therefore, $M(1, t)$ and $M(2, t)$ are orthogonal square-integrable local martingales relative to K^t, and since the family K^t contains H^t, the martingale property holds relative to H^t as well. A particular localizing sequence may be given by $\sigma_n(i) = \inf\{s\colon \bar{N}_s(i, t) + A_s(i, t) \geqslant n\}$, $n \geqslant 1$ $(i = 1, 2)$. Finally, equation (3.4) follows from Lemma 18.12 of LIPTSER and SHIRYAEV [9] for $i = j$, and by orthogonality for $i \neq j$. □

REMARK 3.1. In the terminology of LIPTSER and SHIRYAEV [9], Proposition 3.1 shows that $\bar{N}(i, t)$ is a Poisson type counting process with compensator $A(i, t)$ $(i = 1, 2)$. From the compensator $A(i, t)$ it is convenient to distinguish the predictable auxiliary process $\bar{Y}(t)$ and the cumulative conditional rate function b_i $(i = 1, 2)$. In this case the estimator (2.5) is called the martingale estimator of b_1 and b_2. The martingale property is developed below, where it is also shown that (2.6) is a martingale estimator of F_i $(i = 1, 2)$. The martingale property is used to prove asymptotic (as $t \to \infty$) properties of these estimators.

REMARK 3.2. Suppose, in particular, that assumption (2) is valid and further that the likelihood of the input process is independent of the service distributions F_1 and F_2. Then Proposition 3.1 together with Lemma 19.5 of LIPTSER and SHIRYAEV [9], and the factorization criterion implies that the observations $(\overline{N}(1, t), \overline{N}(2, t), \overline{Y}(t))$ are sufficient statistics for estimating (F_1, F_2) from the queueing system observed over $(0, t]$.

We note that the local martingales $M(1, t)$ and $M(2, t)$ are of local bounded variation. In addition, derivations based on those in Chapters 3 and 4 of GILL [5] show that on the set $\{s : \overline{Y}_s(t) > 0\}$ the process $(\hat{F}_i^t - F_i)/(1 - F_i)$ may be represented as Stieltjes stochastic integrals of a certain bounded H^t-predictable process with respect to $M(i, t)$ $(i = 1, 2)$. According to the general theory of stochastic integration relative to bounded variation martingales (see SHIRYAEV [11]) these Stieltjes integrals are also martingales. It turns out that the same is true of the processes $B^t(i) - b_i$ $(i = 1, 2)$. We summarize these remarks in the following proposition.

PROPOSITION 3.2. *For $i = 1, 2$, the processes given by the Stieltjes integrals*

$$B_s^t(i) - b_i(s) = \int_0^s \overline{Y}_u^{-1}(t) dM_u(i, t), \tag{3.8}$$

and

$$\frac{\hat{F}_s^t(i) - F_i(s)}{1 - F_i(s)} = \int_0^s \frac{1 - \hat{F}_{u-}^t(i)}{(1 - F_i(u-))(1 - \Delta b_i(u))} \overline{Y}_u^{-1}(t) dM_u(i, t), \tag{3.9}$$

for $s \in \{s : \overline{Y}_s(t) > 0\}$ are local square-integrable martingales relative to H^t.

The process $\overline{Y}(t) = \{\overline{Y}_s(t), s \geq 0\}$ defined by (2.3) can be represented as

$$\overline{Y}_s(t) = \sum_{k=1}^{N_t} 1(X_k \wedge Y_k \wedge (t - \tau_k) \geq s). \tag{3.10}$$

Recall that according to assumption (1) $N_t \to \infty$ as $t \to \infty$ and the arrival epochs are finite a.s., and according to assumption (2) the service times are mutually independent. Thus, it is intuitively obvious that the ratio $\overline{Y}_s(t)/N_t$ converges to $[1 - F_1(s-)][1 - F_2(s-)]$ in probability as $t \to \infty$. This is proved in Proposition 3.3 below where it is also shown that convergence to the limit is uniform in s. The result is used in Section 4 to prove that the estimators (2.5) and (2.6) are consistent and asymptotically Gaussian.

PROPOSITION 3.3. *For $s > 0$ define $y(s) = [1 - F_1(s-)][1 - F_2(s-)]$ and $\theta = \sup\{s : y(s) > 0\}$. Suppose $y(\theta) = 0$ and that for each $s \in [0, \theta)$*

$$(N_t - N_{t-s})/N_t \underset{P}{\to} 0 \quad \text{as} \quad t \to \infty. \tag{3.11}$$

Then $\bar{Y}_s(t)/N_t$ converges to $y(s)$ in probability uniformly in s as $t\to\infty$. If $y(\theta)>0$ we replace $[0,\theta)$ with $[0,\theta]$ in the statement above.

PROOF. We consider the case $y(\theta)=0$; the proof for $y(\theta)>0$ is similar and is therefore omitted. For any s and t define

$$\xi_t(s) = \sum_{l=1}^{N_t} 1(X_l\wedge Y_l\geqslant s)/N_t. \tag{3.12}$$

Clearly, assumptions (1) and (2) imply that $\xi_t(s)$ converges to $y(s)$ in probability as $t\to\infty$. Note that $\xi_t(s)\geqslant \bar{Y}_s(t)/N_t$ and consider the inequality

$$\xi_t(s)-\bar{Y}_s(t)/N_t\leqslant \sum_{l=1}^{N_t} 1(X_l\wedge Y_l\geqslant s)1(t-s\leqslant\tau_l)/N_t. \tag{3.13}$$

For $s\geqslant\theta$ we bound the right side of (3.13) by $\xi_t(s)$ and note that as $t\to\infty$, $\xi_t(s)$ converges to $y(s)=0$ in probability. Alternatively, for $s<\theta$ we bound the right side of (3.13) by $(N_t-N_{t-s})/N_t$ which by (3.11) converges to zero in probability as $t\to\infty$. Therefore, $\bar{Y}_s(t)/N_t$ converges to $y(s)$ in probability as $t\to\infty$. Finally, a standard Glivenko-Cantelli argument shows that this convergence is uniform. □

REMARK 3.3. The function y of Proposition 3.3 appears in the variance function of the limit Gaussian process in Section 4 (see, for example, equation (4.3)). For Theorem 4.2 which establishes weak convergence it suffices to have the conclusion of Proposition 3.3 for some monotone function y. The choice $y=(1-F_1)(1-F_2)$ is guaranteed by the condition (3.11) on the input process, which is seen to be a condition affecting the asymptotic variance of the estimators (2.5) and (2.6).

4. Asymptotic properties of the estimators

We are now ready to prove the asymptotic properties of the martingale estimators (2.5) and (2.6). We begin with the problem of consistency.

THEOREM 4.1 (CONSISTENCY). *Let θ and the function y be defined as in Proposition 3.3, and suppose that (3.11) is valid. Then for $u\in[0,\theta)$ we have, as $t\to\infty$,*

$$\sup_{s\in[0,u]}|\hat{F}^t_s(i)-F_i(s)|\underset{p}{\to}0,\quad i=1,2, \tag{4.1}$$

and

$$\sup_{s\in[0,u]}|B^t_s(i)-b_i(s)|\underset{p}{\to}0,\quad i=1,2. \tag{4.2}$$

PROOF. For each $u\in[0,\theta)$, $y(u)>0$. Hence Proposition 3.3 and the fact that $N_t\to\infty$ as $t\to\infty$ imply that $\bar{Y}_u(t)$ converges to infinity in probability as $t\to\infty$. This verifies condition (4.1.12) of Theorem 4.1.1 of GILL [5]. Therefore, by virtue of Proposition 3.2, the monotonicity of $\bar{Y}_s(t)$ in s, and the proof of

Theorem 4.1.1 of GILL [5], the desired results follow. □

According to Theorem 4.1, for large t, $\hat{F}^t(i)$ and $B^t(i)$ are uniformly close estimates of F_i and b_i $(i=1,2)$, respectively, on subintervals in the intersection of the support of F_1 and F_2. We now show that, in addition, the normalized differences in (4.1) and (4.2) converge weakly to Gaussian processes. For this purpose let $D(\theta)$ denote the space of right-continuous functions defined on $[0,\theta)$ and having left-limits. Also, let $Z^i=\{Z_s^i,\ s\in[0,\theta)\}$ and $W^i=\{W_s^i,\ s\in[0,\theta)\}$ $(i=1,2)$ denote mean zero Gaussian processes of independent increments and covariance functions

$$<W^i, W^j>(s) = \begin{cases} \int_0^s (1-\Delta b_i(u))(y(u))^{-1}db_i(u),\ s\in[0,\theta) & \text{for } i=j, \\ 0 & \text{for } i\neq j, \end{cases} \tag{4.3}$$

and

$$<Z^i, Z^j>(s) = \begin{cases} \int_0^s (1-\Delta b_i(s))^{-2}d<W^i, W^j>(s),\ s\in[0,\theta), & \text{for } i=j, \\ 0 & \text{for } i\neq j. \end{cases} \tag{4.4}$$

Note that Z^1 is independent of Z^2 and that W^1 is independent of W^2. We have the following theorem.

THEOREM 4.2 (ASYMPTOTIC NORMALITY). *Suppose the conclusion of Proposition 3.3 is valid. Consider the normalized processes*

$$\begin{aligned} U_i^t &= N_t^{1/2}(\hat{F}^t(i)-F_i)/(1-F_i), \\ V_i^t &= N_t^{1/2}(B^t(i)-b_i), \quad i=1,2. \end{aligned} \tag{4.5}$$

Then as $t\to\infty$,

$$(U_1^t, U_2^t) \xrightarrow{\mathfrak{D}} (Z^1, Z^2) \tag{4.6}$$

and

$$(V_1^t, V_2^t) \xrightarrow{\mathfrak{D}} (W^1, W^2) \tag{4.7}$$

in $D(\theta)\times D(\theta)$ *endowed with the Skorohod topology.*

PROOF. First observe that Proposition 3.2, Proposition 3.3, and Theorem 4.2.2 of GILL [5] imply that each of U_i^t and V_i^t converge weakly to Z^i and W^i, respectively, in $D(\theta)$ as $t\to\infty$. Since $D(\theta)$ is separable and complete, Prohorov's theorem (see BILLINGSLEY [3]) implies that each family $\{U_i^t\}$, $\{V_i^t\}$

($i=1,2$) is tight. This implies, in particular, that $\{(U_1^t, U_2^t)\}$ and $\{(V_1^t, V_2^t)\}$ are tight. Thus it remains to show that the finite dimensional distributions converge. To do this it suffices to use the Cramér-Wold argument of Theorem 7.7 of BILLINGSLEY [3]. The essence of this argument is carried out in detail in Theorem A.2 of AALEN [1] where it is necessary to invoke Proposition 3.2 and the fact that $M(1,t)$ and $M(2,t)$ are orthogonal martingales. □

Theorem 4.2 provides the asymptotic distribution theory of estimators (2.5) and (2.6). The asymptotic functions (4.3) and (4.4) are estimated by

$$<W^i, W^i>_s^t = N_t \int_0^s \overline{Y}_u^{-1}(t)(1-\Delta B_u^t(i))dB_u^t(i) \tag{4.7}$$

and

$$\begin{aligned} <Z^i, Z^i>_s^t \; &= N_t \int_0^s (\overline{Y}_u(t) - \Delta\overline{N}_u(i,t))^{-1} 1(\Delta N_u(i,t) \\ &< \overline{Y}_u(t))dB_u^t(i), \end{aligned} \tag{4.8}$$

for $s \geqslant 0$, $i=1,2$. It turns out that (4.7) and (4.8) are consistent estimators of (4.3) and (4.4), respectively. This result is proved using Proposition 3.3 and Theorem 4.1.

Theorem 4.2 can also be proved by an appeal to the general functional central limit theorems for semimartingales in JACOD, KLOPOTOWSKI, and MEMIN [6], but with some added technical details. However, these theorems provide a unified theory for handling various extensions, such as dependent service times, and avoid special constructions implicit in Gill's proof of his Theorem 4.2.2 [5].

5. ILLUSTRATION

We illustrate our estimation procedures with data on mouse leukemia found in KALBFLEISCH and PRENTICE [8]. The discussion here is intended solely to demonstrate the implementation of these procedures, not to carry any further statistical or scientific inference. The data arose in the laboratory of Dr. Robert Nowinski of the Fred Hutchinson Cancer Research Center, Seattle, Washington, in a study conducted to examine genetic and viral determinants in the development of spontaneous leukemia in mice with specific genetic characteristics. The mice entered the study on their date of birth and were followed for a two-year period until death due to leukemia (thymic and nonthymic), or other natural causes, with those mice still alive after two years being sacrificed.

For our purposes the mice are customers who arrive at the time of their birth and bring with them two types of demands; death due to leukemia and nonleukemic death with no tumors. We fix the study period at 743 days beginning at December 11, 1975 and ending at December 23, 1977. This means that those mice born after December 11, 1975 are only followed until

December 23, 1977, rather than for an additional two year period as in the original study. Thus our arrivals are those mice born within this two year period and their observed service times are given by the shorter of their survival times until leukemic and nonleukemic death or, if they survive, it is given by the number of days between their date of arrival and December 23, 1977. The mice actually arrive in litters of various sizes which corresponds to a bulk arrival queueing system. This is an obvious extension of our model which only allows for single arrivals. Therefore we have sampled one mouse at random from each litter to include in our calculations.

The data for the 13 mice included in our sample appear in Table 1 and include: date of birth (month, day, year), observed service time in days, service received, and an indicator of end of study censoring. The censoring variable $(1-\delta_n^t)(1-\eta_n^t)$ indicates that as many as 8 mice were censored at the end of the two year period. As it turns out, 7 mice were diagnosed as leukemic and the remaining 6 were nonleukemic without tumors.

case (n)	τ_n	W_n^t	$(1-\delta_n^t)(1-\eta_n^t)$	d_n
1	121175	743	1	0
2	121275	742	1	0
3	121575	507	0	1
4	121675	328	0	0
5	121775	737	1	0
6	122275	732	1	1
7	010576	567	0	1
8	011276	711	1	1
9	011576	715	1	0
10	012676	696	1	1
11	012776	427	0	1
12	020276	499	0	1
13	020976	682	1	0

TABLE 1. *Data from Study of Mouse Leukemia*. Here τ_n = date of birth of mouse; W_n^t = observed time (in days) until death or censoring; $(1-\delta_n^t)(1-\eta_n^t)=1$ if censored because of end of study, $=0$ otherwise. Finally, $d_n=1$ if diagnosed with leukemia, $=0$ otherwise, which is additional information concerning the diagnosis of censored mice; this need not be known in general.*

* Source: Data Set V of KALBFLEISCH and PRENTICE [8].

From the data in Table 1 we calculate (2.6), the estimators of the service time distributions for the demands. The graphs of these estimators appear in Fig. 1. Both estimators are step functions which increase to 1 at 744 days which is 1 day more than the maximum observed service time. Notice that nonleukemic deaths appear to take longer so that censoring has left us with very little information about the service distribution for the second demand. To apply our asymptotic theory of Section 4 we would need a longer study period and more arrivals. Then one could, for example, construct confidence bands to estimate F_1 and F_2 which include an indication of variability. For a discussion of various procedures along these lines to which our asymptotic results apply see ANDERSEN and BORGAN [2]. To carry out their procedures it is often easier to work with the estimator (2.5), for which purpose one needs the estimator (4.8) which is easy to calculate.

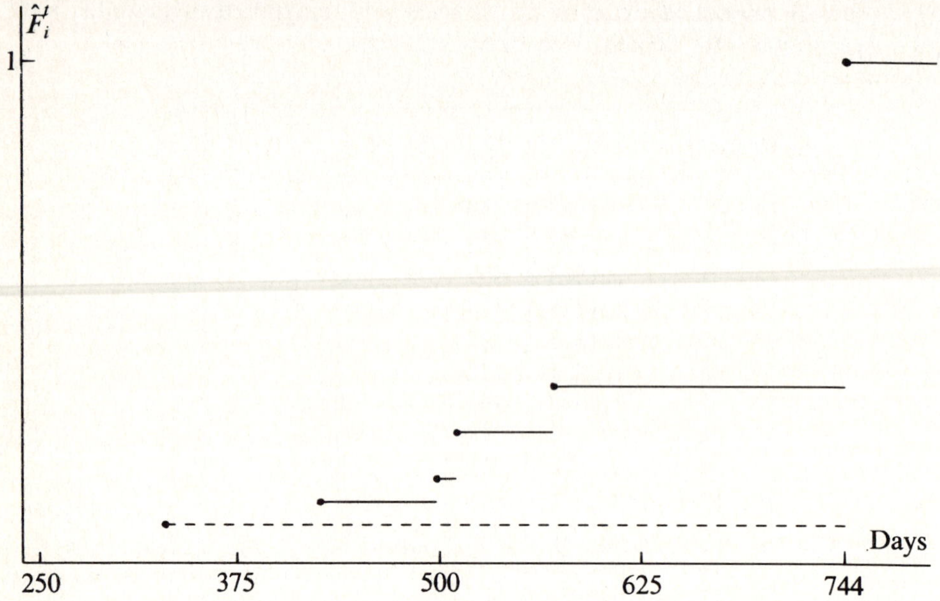

FIGURE 1. *Estimates of service distribution for the first (——) and the second (-----) demand*

REFERENCES

1. O. AALEN (1977). Weak convergence of stochastic integrals related to counting processes. *Z. Wahrsch. Verw. Gebiete 38*, 261-277.
2. P.K. ANDERSEN, O. BORGAN (1985). Counting process models for life history data: A review. *Scand. J. Statist. 12*, 97-158.
3. P. BILLINGSLEY (1968). *Convergence of Probability Measures*, J. Wiley, New York.
4. T. BROWN (1978). A martingale approach to the Poisson convergence of simple point processes. *Ann. Prob. 6*, 615-628.

5. R.D. GILL (1980). *Censoring and Stochastic Integrals,* Mathematical Centre Tracts 124, Mathematisch Centrum, Amsterdam.
6. J. JACOD, A. KLOPOTOWSKI, J. MEMIN (1982). Théorème de la limite centrale et convergence fonctionnelle vers un processus à accroissements indépendants: La méthode des martingales. *Ann. Inst. Henri Poincaré 18,* 1-45.
7. C. JENNISON, B.W. TURNBULL (1985). Repeated confidence intervals for the median survival time. *Biometrika 72,* 619-625.
8. J.D. KALBFLEISCH, R.L. PRENTICE (1980). *The Statistical Analysis of Failure Time Data,* J. Wiley, New York.
9. R.S. LIPTSER, A.N. SHIRYAEV (1978). *Statistics of Random Processes II,* Springer-Verlag, New York.
10. T. SELKE (1986). Weak convergence of the Aalen estimator for a censored renewal process. *Proc. Purdue Symposium on Decision Theory.*
11. A.N. SHIRYAEV (1981). Martingales: Recent developments, results and applications. *Int. Statist. Rev. 49,* 199-234.

EPILOGUE

Art and Science

The Greatness and Tragedy of Applied Science

The Environments of the International Teletraffic Congresses

Arne Jensen

Chairman of the International Advisory Council
of the International Teletraffic Congresses
Bldg. 321, The Technical University of Denmark
DK-2800 Lyngby, Denmark

Wim Cohen has managed to keep the mathematical fire burning in a most realistic down-to-earth way. Nothing is easier than to quit when you have reached a result. However, Wim Cohen's results, based on real problems, and reached through tenacious, tough work in order to find stringent solutions, have brought him forward to the frontline of applied mathematics. He is highly respected by both mathematicians and by technicians.

Teletraffic pioneers

The teletraffic pioneers at the turn of this century found themselves in the middle of real problems the solutions of which were of vast economic consequences. Solutions might mean neck or nothing for small telephone companies and their sub-contractors. Waiting times and blocking were realities, both in connection with material and staff. Changes took place every year with such speed that you always had to plan on the basis of unknown, large traffic loads. You had absolutely no experience on which you could draw. You did not know what the consequence was for this blocking, for this waiting time. Only at an advanced stage did the mathematical world realize the greatness of the work having been done by these pioneers. The Frenchman Vaulot learned to read Erlang's works in Danish by attending Swedish language courses at the Swedish Academy in Paris. This tells us a bit about the language barriers of the time. It also tells us how vital it was for the administrations to possess the modern tools for traffic theory.

Soon the fact that you had tools for everyday operation did not suffice. The administrations had to have tools which impartially were able to compare material of various companies, and which were able to compare different technical solutions to the same problems. The activity thus arising uncovered basic science results both within statistics and within probability theory. This was the forerunner of the mathematician's work within stochastic processes.

Today's young generation would say that the pioneers' work was characterized by the fact that the results were applied before the work finished. Even when the pioneers did finish their work, such as was the case with Felix Pollaczek, only very few read the works with an insight. But many applied the results, nevertheless.

Characteristic of the interdisciplinary environments in which traffic people lived at the time, is the fact that they lived their lives independent of the universities. The basis of our science was created by reality which again was inspired by the probability theory from Pascal to Maxwell's demon, but incidentally was not greatly influenced by the academic world of the time. Academic work carried out in harsh reality had a hard time being recognized by the narrow circle of the university world. Maybe the reason is that they apply a different yardstick from what is beautiful? For people in the ivory tower the simple solution is the beautiful goal. Outside, beautiful means the simple problem formulation which might only require well-known methodology for the solution. For some of those who step out of their ivory tower the sun is so dazzling that they are unable to see the problems. This makes it so very difficult to give a reply to the right problem instead of to the question actually asked. It makes it difficult to avoid involvement in a fight which is of no concern to you, and it is difficult to obtain recognition of your efforts.

It was certainly not among the universities' priesthood that the pioneers should find support and basis for free research. No, it had to get its charter from among a wide, understanding circle within the application area. It had to build on farsighted leaders' understanding attitude. No doubt, telephony would have survived without these traffic specialists, but not as well as with them.

The pioneers' youth was characterized by the mathematicians' lack of understanding of the efforts. When they grew older, and tried to finish their work, mathematically, they also lacked understanding of those who had applied the work of their youth.

Postwar developments, the emergence of ITC

This was the situation when Wim Cohen entered the stage as a young Philips engineer. A new generation of telephone managers who had obtained their ability to manage - to see things on the whole - from working with traffic, opened up doors for the applied mathematics and the interdisciplinary science, teletraffic. These managers supported the requirements of teletraffic, technical understanding, financial reality and strictly mathematical probability theory. To these you have got to add their respect for the subscribers' manifold wishes with respect to communication.

Cohen's work within queueing theory is very much characterized by his knowledge of what is relevant and what is irrelevant, by his deep insight into mathematics as a tool and by his ability to draw on the environments and surroundings created by the International Teletraffic Congresses, a circle created at the dinnertable of his countryman, Leendert Kosten - before dinner - once I passed through on a study trip as a Rockefeller Fellow to visit pioneers in the United Kingdom, France, and The Netherlands. Our initiative was strongly supported by the old personality in Delft, Professor Bähler. In this forum with a meeting every three years, Cohen soon found support for his efforts. He has participated in all the International Teletraffic Congresses. He has followed their developments. Wim Cohen is a part of these events. He has made the requirement to quality a MUST. This fruitful environment was composed of representatives from administrations, industries, and universities, of all ages. History was very much present, interacting with today's work, and with the problems of the future. The Dutch tradition in the area helped to establish the quiet environments for his work, so vital in order to obtain fundamental results.

The strength of these congresses was that they were close to prevailing problems. The users' view of the problems was part of the meeting. Friendship and cooperation was established not only within the small circle of specialists. Everybody, from engineers, economists, managers to mathematicians was soon a cooperating family with a common goal, good and cheap service to society. Managers picked up new knowledge before final results had been reached. The research workers became inspired by the problems in status nascendi.

The technical art in postwar-time required more and more from the basic sciences. Some even forgot the privilege of technique, that technique is a kind of art pointing at possible ways to reach your goal. They wanted to rub shoulders with the natural sciences, reveal the eternal truths instead of the art, to point out the many possible ways to the desired goal.

Such a development was greatly supported by society. But society became inclined to push choices among possibilities away from itself, and make them into technical problems - which they are not. However, with changed conditions the discipline of teletraffic and applied mathematics also gained respect from the technical universities, and recognition at the stronghold of the universities, mathematics.

Due to the development of accepting the interdisciplinary challenges at the universities, many of the teletraffic practitioners, including Cohen, took their teletraffic problems to the universities instead of, as had been the case in the past, let them be dealt with in the environments of industry and administrations. Under the changed conditions, they were given the possibility of finishing the science in their work while the art in their work had a harder time outside industry and administrations. Also, the extrovert, social involvement in the telephone customers' lives took up less space in their world.

THE FUTURE: ENTERING THE INFORMATION SOCIETY

We live in a transition period. We are on our way into the information society with its technical possibilities. This in a jump will change the technical environments created to satisfy the new customers' wishes. It also changes the basis for one of the cornerstones of our science, teletraffic.

Changes nowadays occur so rapidly that the interest revolves around the changes themselves instead of what has always been the real issue, i.e., meeting the subscribers' wish to communicate, one way or the other. To meet these requirements, it is necessary to possess a deep insight into the dynamic nature of traffic, and into those conditions on which traffic is built in our society, in our homes, and at our places of work. Once air and water were free benefit. This is no longer the case. Big movements in society have taught us that. Communication has never been a free benefit. Those who plan as if it is a free benefit are going to be confronted with movements in our society that will teach them that our task is different from and more than pure business. The issue is to realistically meet the individual's need for communication, in narrow and in wider circles, in a professional and cultural way, from the most banal to the most sophisticated way. The door to this future has already been opened. At the time, when Cohen believed that we had finished our work in queueing theory - there were no more problems in queueing theory - the new technique ran into unexpected problems. His door was knocked at. The technical art of choices in the information area is still governed by scarcity.

Often, many of the original pioneers felt that they had been forgotten even before the end of their working-days. For Cohen it must be a great satisfaction that he is as busy as ever, that so many are still knocking at his door. Above all, the circles around the International Teletraffic Congresses which throughout the years have drawn on his working power would like to express their gratitude. Wim Cohen has always been willing to share with us his deep understanding of the nature of traffic.

It has not become any easier for traffic people. Today, town and area planning distinctly split-up towns in industry and dwelling areas. A centralization of distribution systems is taking place. And the increased standard of living has changed the entire family pattern. We ask ourselves whether the town as such has come to an end.

One of the tasks of the traffic theory is to fit into a system for those subscribers who do not use the telephone very much. Thereby you create an inexpensive buffer for those subscribers who use the telephone a lot. Their overloading situation is also one of the tasks of the teletraffic.

The information society is certainly still going to be a challenge to the specialists within traffic theory, administration, and planning.

The first International Teletraffic Congress was closed with the wish that our dreams in the area of teletraffic and its application for the benefit of mankind should come true. Our dreams in the past did come true. I am sure that, when we in this transition period on our way into the information society repeat the wish as far as the young generation is concerned, 'Let their dreams come true', we are in accord with Wim Cohen as a scientist and as a human being.